Water Resources

The world faces huge challenges for water as population continues to grow, as emerging economies develop and as climate change alters the global and local water cycles. There are major questions to be answered about how we supply water in a sustainable and safe manner to fulfil our needs, while at the same time protecting vulnerable ecosystems from disaster.

Water Resources: An Integrated Approach provides students with a comprehensive overview of both natural and socio-economic processes associated with water. The book contains chapters written by 20 specialist contributors, providing expert depth of coverage to topics. The text guides the reader through the topic of water, starting with its unique properties and moving through environmental processes and human impacts upon them, including the changing water cycle, water movement in river basins, water quality, groundwater and aquatic ecosystems. The book then covers management strategies for water resources, water treatment and reuse, and the role of water in human health before covering water economics and water conflict. The text concludes with a chapter that examines new concepts such as virtual water that help us understand current and future water resource use and availability across interconnected local and global scales.

This book provides a novel interdisciplinary approach to water in a changing world, from an environmental change perspective and interrelated social, political and economic dimensions. It includes global examples from both the developing and developed world. Each chapter is supplemented with boxed case studies, questions, project ideas and further reading, as well as a glossary of terms. The text is richly illustrated throughout with over 150 full-colour diagrams and photos.

Joseph Holden holds the Chair of Physical Geography at the University of Leeds. He is Head of water@leeds, the largest interdisciplinary water research centre in the UK, and he is also Director of Research for the School of Geography.

Water Resources

An integrated approach

Edited by
Joseph Holden

LONDON AND NEW YORK

First published 2014
by Routledge
2 Park Square, Milton Park, Abingdon, Oxon OX14 4RN

and by Routledge
711 Third Avenue, New York, NY 10017

Routledge is an imprint of the Taylor & Francis Group, an informa business

© 2014 Editorial matter and selection: Joseph Holden; individual chapters: the contributors

The right of the editor to be identified as the author of the editorial material, and of the authors for their individual chapters, has been asserted in accordance with sections 77 and 78 of the Copyright, Designs and Patents Act 1988.

All rights reserved. No part of this book may be reprinted or reproduced or utilised in any form or by any electronic, mechanical, or other means, now known or hereafter invented, including photocopying and recording, or in any information storage or retrieval system, without permission in writing from the publishers.

Trademark notice: Product or corporate names may be trademarks or registered trademarks, and are used only for identification and explanation without intent to infringe.

British Library Cataloguing in Publication Data
A catalogue record for this book is available from the British Library

Library of Congress Cataloging in Publication Data
Water resources/edited by Joseph Holden.
   pages cm
   Includes bibliographical references and index.
   1. Water-supply. 2. Water resources development. 3. Water security.
   I. Holden, Joseph, 1975– editor of compilation.
   TD346.W28 2014
   333.91—dc23
   2013004283

ISBN: 978-0-415-60281-5 (hbk)
ISBN: 978-0-415-60282-2 (pbk)
ISBN: 978-0-203-48941-3 (ebk)

Typeset in Minion and Univers
by Florence Production Ltd, Stoodleigh, Devon, UK

Printed and bound in India by Replika Press Pvt. Ltd.

Contents

List of figures, tables and boxes	vii
List of contributors	xv
Preface	xvii
Acknowledgements	xix

1	**Water fundamentals** Joseph Holden	1
2	**The changing water cycle** Stephen Sitch and Frances Drake	19
3	**River basin hydrology** Joseph Holden	49
4	**Surface water quality** Pippa J. Chapman, Paul Kay, Gordon Mitchell and Colin S. Pitts	79
5	**Groundwater** L. Jared West and Noelle E. Odling	123
6	**Aquatic ecosystems** Lee E. Brown, Colin S. Pitts and Alison M. Dunn	161

CONTENTS

7 Water demand planning and management   203
Adrian T. McDonald and Gordon Mitchell

8 Water and health   223
Rebecca J. Slack

9 Potable water and wastewater treatment   265
Nigel J. Horan

10 Water economics   293
Sonja S. Teelucksingh, Nesha C. Beharry-Borg and Dabo Guan

11 Water conflict, law and governance   315
Kitriphar Tongper and Anamika Barua

12 The future of water: water footprints and virtual water   333
Martin R. Tillotson, Megan Beresford, Dabo Guan and Joseph Holden

Glossary   351
Index   365

Figures, tables and boxes

FIGURES

1.1	A schematic diagram showing water molecules and the covalent and weak hydrogen bonds	2
1.2	Pond skaters are invertebrates that move across the water surface	4
1.3	Water forms droplets on most solid materials it comes into contact with	4
1.4	Capillary action in two glass tubes	5
1.5	When dissolved in water, each of the sodium and chloride ions is hydrated	5
1.6	Categories of world water resources	10
1.7	Declining per capita water availability in India	13
1.8	Competing water uses for main income groups of countries	14
2.1	The global hydrological cycle, with estimated flows and storage	20
2.2	The Clausius–Clapeyron (CC) relation	22
2.3	Saturated pressure–temperature curves for water and ice	23
2.4	Land–atmosphere coupling strength in June, July and August	25
2.5	The static jet stream of summer 2010, which led to different extreme events across Europe and Asia	26
2.6	Summer 2007 rainfall amount as a percentage of the 1971–2000 average	27
2.7	Conditions in the Pacific during (a) normal periods and (b) El Niño	29
2.8	Winter weather anomalies during (a) El Niño and (b) La Niña over North America	29
2.9	Severe conditions during the 1930s Dust Bowl of North America	31
2.10	Temperature and precipitation anomalies for the Dust Bowl drought from the Climate Research Unit (CRU) climate model simulations	32
2.11	Precipitation difference from long-term average 1900–2011 in the Sahel region showing the prolonged dry period since the 1970s	34
2.12	Changes in glacial extent for selected glaciers around the world	36

2.13	Leaf stomata regulate the uptake of carbon dioxide for photosynthesis, and water loss through transpiration	38
2.14	Mechanisms connecting changes in leaf conductance to canopy evapotranspiration and soil moisture changes	38
2.15	Population exposed to severe water scarcity (red)	40
2.16	Schematic showing example feedbacks of enhanced water vapour in the atmosphere	42
2.17	Robust findings on regional climate change for mean and extreme precipitation, drought and snow	42
3.1	An example gentle rainfall event resulting in a reduction in the infiltration capacity of the soil over time and the subsequent production of infiltration-excess overland flow	51
3.2	Production of saturation-excess overland flow	51
3.3	Soil water energy status on a slope	53
3.4	Macropore flow is indicated by the staining white dye solution where fingers of dye extend downward from the main body of dye in a soil profile	54
3.5	A large soil pipe with some flowing water emerging from the pipe which is at the head of a gully	54
3.6	A rain gauge flush with the ground surface surrounded by a metal mesh to reduce errors from turbulence and splash	55
3.7	Storm hydrographs (a) where infiltration-excess overland flow dominates, and (b) where throughflow dominates	57
3.8	Two example annual river flow records showing regimes of each river	58
3.9	The endorheic river basins of the Earth	59
3.10	A worked velocity-area calculation for river discharge	60
3.11	Predicted river storm flow discharge using the unit hydrograph model	62
3.12	New housing development in Calgary with a sustainable urban drainage system with a storm water storage lake	64
3.13	Mean monthly total discharge change (%) in the Pinios River for a 24-year modelling period	65
3.14	A river basin showing what happens to the main channel flood hydrograph after forestry plantation within a tributary basin due to flood wave synchronisation	67
3.15	The main types of drainage network pattern	68
3.16	Examples of river channel forms: (a) braided river; (b) meandering river, and (c) a straight river	69
3.17	Helicoidal flow through a meandering channel	70
3.18	River meanders tend to become more exaggerated until eventually the meander is cut through by the river, leaving behind an oxbow lake	71
3.19	Temporal variation of water and sediment discharge in the main channel of the Indus River below Kotri	73
3.20	Re-meandering on the River Kissimmee	74
4.1	The different hydrological pathways that precipitation may take to reach surface waters and their effect on solute concentrations	80
4.2	Eh-pH diagram for the simple ions and hydroxides of iron at atmospheric pressure and 25°C	82

4.3	Relationship between mean annual runoff and (a) discharge weighted mean total dissolved solids concentrations, and (b) mean annual dissolved load for a sample of 496 world rivers	85
4.4	Cross-section of a thermally stratified eutrophic lake in summer	87
4.5	The influence of geology and climate on the total dissolved solids (TDS) and major chemical composition of surface waters	88
4.6	Dissolved oxygen concentrations and temperature recorded at 15 minute intervals at a stream in the North Pennines, UK	90
4.7	Changes in stream discharge and silicon and dissolved organic carbon concentrations during a storm event in a headwater stream in NE Scotland	90
4.8	Mean monthly concentration in stream water collected weekly from the headwaters of the River Tees, northern England (1993–2010)	91
4.9	Long-term trends in the concentration of (a) sulphate (SO_4), (b) hydrogen (H^+) ions and (c) dissolved organic carbon in weekly samples collected from 11 streams	92
4.10	Slurry being injected into a field by a specialist machine	96
4.11	Farm advice activities being undertaken as part of the England Catchment Sensitive Farming Delivery Initiative	98
4.12	Long-term NO_3 concentrations and fluxes in the River Stour	99
4.13	Acid mine discharge at Jackson Bridge on New Mill Dyke, Yorkshire, UK	106
4.14	(a) Semi-natural wetland and (b) artificial reed bed wetlands, River Pelenna, South Wales, UK	108
4.15	Critical load exceedance maps for the UK freshwaters in 2004	112
5.1	Role of groundwater in the hydrological cycle	124
5.2	Factors influencing rainfall recharge to groundwater	125
5.3	Interactions between streams and groundwater	126
5.4	Map of the Nubian Sandstone Aquifer System	128
5.5	Well drilling in Egypt	129
5.6	Confined and unconfined aquifers	131
5.7	Sedimentary rock aquifers in the United Kingdom: (a) Permo-Triassic Sandstone; (b) Cretaceous Chalk	132
5.8	Column experiment that can be set up in the laboratory illustrating Darcy's Law for flow in sand	133
5.9	Hydraulic gradients in unconfined aquifers	134
5.10	Flow in a fracture of aperture b and hydraulic gradient i is described by the Cubic Law	135
5.11	Two-dimensional computer simulation of flow through a fractured rock where the rock matrix is itself permeable and porous	136
5.12	Hydrogeological map showing the River Wylye flowing over unconfined Cretaceous Chalk aquifer in southern England	137
5.13	Impact of groundwater abstraction on flow at a gauging station in the River Worfe, West Midlands, UK	137
5.14	Hand dug wells	138
5.15	Cross-section illustrating the construction of a qanat	139
5.16	Modern groundwater abstraction wells	141
5.17	Well loss and well efficiency	142

FIGURES, TABLES AND BOXES

5.18	Buildup of iron bacteria on the intake of a submersible pump	143
5.19	Artificial recharge in the Llobregat Delta Aquifer, Barcelona, Spain	144
5.20	Schematic diagram of dominant chemistry in the Cretaceous Chalk aquifer of southern England	148
5.21	Acid mine drainage showing precipitation of ferric hydroxides	149
5.22	Contaminant source characteristics: (a) landfill site representing point source; (b) agrochemical application representing diffuse source	150
5.23	Trends in nitrate fertiliser application in the United Kingdom	151
5.24	Behaviour of Non Aqueous Phase Liquids in groundwater aquifers	153
5.25	Well-head or source protection zones as defined by Environment Agency, England	155
6.1	Hierarchical structure of river ecosystems	162
6.2	Visible changes occur to the river environment from headwaters to lowland: (a) an open canopy UK moorland stream; (b) a meltwater-fed river in the French Pyrenees; (c) a densely forested stream in southeast Alaska; (d) a forested river in northwest North America; (e) a mid-reach on the River Aire, UK; (f) the River Danube in the lowlands of Austria; and (g) the braided Matukituki River in New Zealand	163
6.3	Diversity of lake habitats in the Windermere catchment, northwest England	164
6.4	A general scheme of littoral and profundal habitats in a freshwater lake	165
6.5	Examples of aquatic producers	167
6.6	The life cycle of the trophically transmitted parasitic worm *Echinorhyncus truttae*	170
6.7	Schematic outline of (a) a simple aquatic food chain, and (b) a simple aquatic food web	171
6.8	Connectance food webs from (a) Felbrigg Hall Lake, UK, and (b) Broadstone Stream, UK	173
6.9	Food web structure of Muskingham Brook, New Jersey, USA, during fall 2003	174
6.10	Examples of the various methods used by freshwater ecologists to study primary production	176
6.11	A battery-powered peristaltic pump being used to deliver small quantities of dissolved NO_3^-, NH_4^+ and organic carbon from a tank into an Arctic tundra stream as part of an experiment to measure nutrient spiralling lengths and uptake velocity	179
6.12	Nutrient-diffusing pots with wooden (brown) and clay (orange) substrate	179
6.13	River channel fragmentation and flow regulation effects of dams on 292 of the world's major river systems	181
6.14	Regulated river channel below Digley Reservoir, West Yorkshire, UK	182
6.15	The Spöl River, Switzerland, (a) before and (b) during an artificial flood generated by water release from a reservoir	183
6.16	Channelised urban streams	184
6.17	Intense algal growth on a eutrophic water body next to a public footpath	187
6.18	Fish farm in a sheltered bay off the Isle of Skye, northern Scotland	192
7.1	(a) A dumb meter, which simply records accumulated usage, and (b) a smart meter, which provides continuous data on usage through time	206
7.2	Current and forecast global water demand and resource availability	210
7.3	Populations living in areas of water stress	211

7.4	Global water scarcity distribution map	211
7.5	Water demand in different countries – average water use per person per day	212
7.6	Household water use indexed against water prices across a range of countries	213
8.1	John Snow's map of the area surrounding the Broad Street pump, representing each cholera death by a bar	225
8.2	Faecal contamination of water supplies is one of the commonest routes for the spread of water-related disease	232
8.3	The life cycle of the *Schistosoma* spp. trematode worm	236
8.4	Map showing countries affected by malaria	237
8.5	The life cycle of the mosquito is dependent on the presence of a water body	238
8.6	Adult female mosquitoes become infected after feeding on human blood	238
8.7	Human sewage effluent can alter both the biological and chemical content of water	240
8.8	Cyanobacteria blooms can have considerable impacts on water quality	244
8.9	Too much water can have health consequences	250
8.10	Cross-section through the city of New Orleans, showing dependence on levees and floodwalls	251
8.11	Population tracking and the impact of Hurricane Katrina on New Orleans	252
8.12	The immediate aftermath of the Japanese tsunami, March 2011	253
8.13	Irrigation of crops may be needed to obtain food with optimal nutritional content	254
8.14	Collecting water can often involve a walk of several kilometres	254
8.15	Age weightings applied to calculation of DALYs	256
8.16	An appropriate hand-washing technique to minimise the spread of disease	257
9.1	Typical process flow train for producing water of potable quality that meets the WHO guidelines summarised in Table 9.1	268
9.2	Schematic diagram of (a) rapid and (b) slow gravity filters that are able to take out smaller particles from water	269
9.3	Underdrain design for a rapid gravity filter showing the many roles this has to perform	270
9.4	Arrangement of membranes used in a range of water treatment applications	272
9.5	The oxygen sag curve that develops when wastewater is discharged to a watercourse with a concentration of organic material in excess of the available oxygen dissolved in the water	275
9.6	Annual distribution of biochemical oxygen demand in the effluent from a wastewater treatment plant	277
9.7	(a) A combined sewer overflow serves to protect the sewer from flooding; and (b) innovation processes to remove coarse material from combined sewer overflows	279
9.8	Screens are the first stage of treatment, and remove solid material which is larger than the holes in the screen	280
9.9	Primary settlement tanks proved a relatively cheap and simple option for removing up to half the pollution in sewage	281
9.10	Attached growth processes maintain a viable population of microorganisms by the provision of an inert support such as plastic or mineral media	282

9.11	The activated sludge process with air introduced into the tank	283
9.12	A belt thickener allows the water contained in primary and waste activated sludge to drain away to produce a thickened sludge	286
9.13	A flow scheme for the anaerobic digestion of sludge, demonstrating the range of options for energy recovery	287
10.1	The three-tiered approach to economic value	294
10.2	The balance between water demand and water supply	296
10.3	A typology of economic value	302
10.4	Linkages between ecosystem services, human well-being and drivers of change	303
10.5	Part of Tram Chim National Park	304
10.6	The goods and services provided by wetland ecosystems	305
10.7	Total Economic Value versus ecosystem services of wetland ecosystems	305
10.8	Economic valuation techniques for water resources	306
10.9	Ditch through a peatland in Nidderdale, UK	309
11.1	The Farakka Barrage	319
12.1	Examples of indirect and direct water use for a home	335
12.2	Green, blue and grey water footprint contributions for sugar beet, sugar cane and high fructose maize syrup	338
12.3	The water footprint of nations averaged per person for drinking, washing, food consumption and consumption of other consumer goods	340
12.4	The total water footprint of nations	340
12.5	The proportion of national water footprints dependent on imports	341
12.6	Virtual water balances for major global regions for the water flows associated with agricultural trade	341
12.7	Map showing China separated into seven regions, shaded by their freshwater availability per person per year in m^3	343

TABLES

1.1	Typical water use to produce some common human foodstuffs	7
1.2	Water-rich and poor countries of the world	11
2.1	Average residence time in major water stores	21
4.1	Nitrate-N concentrations and discharge for a 24-hour period 2–3 January 2013 for Eagle Creek, at Zionsville, Indiana	86
4.2	The nature, sources, effects and control of some major types of pollutants	94
4.3	Some urban diffuse pollutants and their sources	100
4.4	Pollutant concentrations in wastewater and storm water across Europe and North America	100
4.5	Causes of low river quality in Scotland	101
4.6	Typical sustainable urban drainage system devices	103
4.7	Some chemical data from mine discharges	107
4.8	Changes in typical chemical composition of the Bullhouse mine discharge into the River Don, Yorkshire, UK, before and after remediation	109
6.1	Functional Feeding Groups (FFG) of invertebrates, their feeding mechanisms, food sources and typical size range of particles ingested	169

6.2	Food web summary statistics for rivers, estuaries and lakes	172
6.3	Critical thresholds of classification for trophic status	186
6.4	Examples of algal types associated with trophic status	186
6.5	Comparison of aquaculture and capture production (tons) between 1990 and 2009	192
7.1	Biases in domestic consumption monitors (DCM)	207
7.2	A very simplified domestic consumption monitor table	208
7.3	Potential water savings through change of habit proposed by Northumbrian Water and Essex & Suffolk Water, UK	217
8.1	Water-related diseases: pathogens, treatment and prevention strategies	227
8.2	World Health Organization guideline concentrations for chemicals found in drinking water	242
8.3	Limits and guidelines set for drinking water for selected pesticides and other organic substances	243
8.4	Global burden of disease for selected factors	255
9.1	Water quality guidelines for safe drinking water for a number of parameters as recommended by the World Health Organization	267
9.2	The performance of a range of wastewater treatment systems, showing improving performance as capital and operating cost increases	278
11.1	Causes of conflicts	316
12.1	Water footprint composition data for a selection of products	335
12.2	Water footprint (litres) of a hypothetical 0.5 litre PET bottle containing a sugar carbonated soft drink	337
12.3	Recent water accounting studies using environmental input–output models	339
12.4	Availability of water resources in China	344

BOXES

1.1	Japan's river culture	9
1.2	India's looming water crisis	13
1.3	Near-future drought and food security in Asia	15
2.1	Hot, dry weather in UK summer 2006, floods in UK summer 2007	27
2.2	Dust Bowl of 1930s North America	31
2.3	Sahel drought	34
2.4	Role of direct effects of CO_2 upon water usage by plants	37
3.1	Calculating the hydrological efficiency of a river basin	56
3.2	Measuring river flow from space	61
3.3	Modelling land use change impacts on river discharge	65
3.4	Impacts of river flow and sediment reduction on the Indus delta	73
3.5	River Kissimmee rehabilitation, Florida	74
4.1	pH and redox potential	83
4.2	Calculating solute loads for Eagle Creek, Indiana, USA	86
4.3	Long-term nitrate trends in rivers	99
4.4	De-icing agents and diffuse pollution	102
4.5	Emerging pollutants	103
4.6	Formation of acid mine drainage	106

4.7	Evaluating the acidification status of surface waters: the critical load approach	111
5.1	The Nubian Sandstone Aquifer System	128
5.2	Fractured aquifers	135
5.3	The qanats of Iran	139
6.1	Measuring aquatic primary productivity	176
6.2	Measuring aquatic secondary production	177
6.3	Warming of Lake Washington, USA	189
6.4	Biomonitoring and the European Water Framework Directive	193
7.1	The perfect storm	204
7.2	Intelligent metering	206
8.1	John Snow and cholera	225
8.2	Mosquitoes and malaria	238
8.3	Maintaining water quality	241
8.4	Hurricanes, earthquakes and tsunamis	251
8.5	Disease burden and socio-economic loss	255
8.6	Minimising disease	257
9.1	Comparing process efficiencies	269
9.2	Energy from warm sewers	278
9.3	Energy-neutral wastewater treatment	284
9.4	Recovering energy from waste by anaerobic digestion	288
10.1	Areas at risk: Small Island Developing States	298
10.2	Tram Chin National Park, Vietnam	304
10.3	Economic effects of regulating flow releases from Glen Canyon Dam, USA	308
10.4	Water-quality supply in the UK uplands	309
11.1	Conflict through consumptive use: Cauvery water dispute, India	317
11.2	Conflict through pollution: the Rhine	318
11.3	Relative distribution conflict: the Ganges basin	319
11.4	Absolute distribution conflict: the Jordan basin	321
11.5	Examples of international water agreements	328
12.1	Calculating the water footprint of a soft drink	337

Contributors

Dr Anamika Barua, Department of Humanities and Social Sciences, Indian Institute of Technology, Guwahati, 781039, India; abarua@iitg.ernet.in

Dr Nesha C. Beharry-Borg, water@leeds, School of Earth and Environment, University of Leeds, LS2 9JT, UK; n.c.beharry-borg@leeds.ac.uk

Megan Beresford, water@leeds, School of Earth and Environment, University of Leeds, LS2 9JT, UK; megan.beresford@britishsugar.com

Dr Lee E. Brown, water@leeds, School of Geography, University of Leeds, Leeds, LS2 9JT, UK; l.brown@leeds.ac.uk

Dr Pippa J. Chapman, water@leeds, School of Geography, University of Leeds, Leeds, LS2 9JT, UK; p.j.chapman@leeds.ac.uk

Dr Frances Drake, water@leeds, School of Geography, University of Leeds, Leeds, LS2 9JT, UK; f.drake@leeds.ac.uk

Dr Alison M. Dunn, water@leeds, Institute of Integrative and Comparative Biology, University of Leeds, Leeds, LS2 9JT, UK; a.dunn@leeds.ac.uk

Dr Dabo Guan, water@leeds, School of Earth and Environment, University of Leeds, Leeds, LS2 9JT; d.guan@leeds.ac.uk

Professor Joseph Holden, water@leeds, School of Geography, University of Leeds, Leeds, LS2 9JT, UK; j.holden@leeds.ac.uk

Dr Nigel J. Horan, water@leeds, School of Civil Engineering, University of Leeds, Leeds, LS2 9JT, UK; n.j.horan@leeds.ac.uk

Dr Paul Kay, water@leeds, School of Geography, University of Leeds, Leeds, LS2 9JT, UK; p.kay@leeds.ac.uk

Professor Adrian T. McDonald, water@leeds, School of Geography, University of Leeds, Leeds, LS2 9JT, UK; a.t.mcdonald@leeds.ac.uk

Dr Gordon Mitchell, water@leeds, School of Geography, University of Leeds, Leeds, LS2 9JT, UK; g.mitchell@leeds.ac.uk

Dr Noelle E. Odling, water@leeds, School of Earth and Environment, University of Leeds, Leeds, LS2 9JT, UK; n.e.odling@leeds.ac.uk

CONTRIBUTORS

Dr Colin S. Pitts, water@leeds, School of Earth and Environment, University of Leeds, Leeds, LS2 9JT, UK; c.pitts@leeds.ac.uk

Professor Stephen Sitch, Geography, College of Life and Environmental Sciences, University of Exeter, EX4 4RJ, UK; s.a.sitch@exeter.ac.uk

Dr Rebecca J. Slack, water@leeds, School of Geography, University of Leeds, Leeds, LS2 9JT, UK; r.j.slack@leeds.ac.uk

Dr Sonja S. Teelucksingh, Economics Department and Sir Arthur Lewis Institute for Social and Economic Studies, University of the West Indies, St. Augustine, Trinidad and Tobago; sonja.teelucksingh@sta.uwi.edu

Dr Martin R. Tillotson, water@leeds, School of Civil Engineering, University of Leeds, Leeds, LS2 9JT, UK; m.r.tillotson@leeds.ac.uk

Kitriphar Tongper, water@leeds, School of Earth and Environment, University of Leeds, Leeds, LS2 9JT, UK; ktongper@gmail.com

Dr L. Jared West, water@leeds, School of Earth and Environment, University of Leeds, Leeds, LS2 9JT, UK; l.j.west@leeds.ac.uk

Preface

Water is of fundamental importance to life on Earth. It also has huge economic and cultural significance. This book examines water and water resources from scientific, economic and social perspectives. It is aimed at university students of all levels and water practitioners and policy makers who want to obtain a good grounding in the subject of water across the disciplines. The world faces grand challenges for water as population continues to grow, as emerging economies develop and as climate change alters the global and local water cycles. There are major questions to be answered about how we supply water in a sustainable and safe manner to fulfil our needs, while at the same time protecting vulnerable ecosystems from disaster. These grand challenges require an interdisciplinary approach to address them because there are scientific and technological issues to be addressed, there are economic and political issues to be addressed and there are social and cultural issues to be addressed and these all interconnect. Solving a technological problem on water supply may be futile if there is no political or social will, nor the economic means to utilise that technological advance.

The team of twenty authors who have contributed their knowledge and understanding to this book have compiled their experiences from around the world. This book begins by outlining the nature of water, some of its unique properties, the challenges for water resources and the role of water in society. It then moves in Chapter 2 to examine the global water cycle and the importance of water in moving energy around the planet and how climate change is affecting the water cycle through a series of feedback mechanisms. Chapter 3 covers the water cycle at a more local scale, looking at processes within river basins, including water movements through and over soils and in rivers. The pathways for water through river basins and the way we manage the landscape affect both water quality in rivers and lakes and such surface water quality issues are covered in Chapter 4. The discussion is supplemented by a detailed treatment of groundwater processes, water supply and groundwater quality in Chapter 5. Water bodies form an important part of ecosystems and also host a diverse community of organisms. Aquatic ecosystems and their modifications through human action are given attention in Chapter 6.

The demand for water needs to be managed and the supply of water to people, industry and agriculture needs to be planned for. Thus, Chapter 7 provides an overview of issues around water resource management. The quality and quantity of water both for drinking water supply and within the local environment has a fundamental role to play in human health. Droughts and periods of water scarcity can lead to famine and death. Too much water through flooding can kill not only by drowning but through the spread of disease, food shortages, loss of shelter and livelihood disruption. Infectious diseases associated with water and chemicals carried within water can have huge impacts on human health and therefore these topics are covered in Chapter 8. Techniques for providing clean water for consumption and for treating wastewater (and utilising the resources that wastewater provides) are described in Chapter 9.

Water is an economic good and is crucial for agriculture, industry and many other things from which we derive benefit. It is therefore essential, as described by Chapter 10, to examine the principles of economics and different types of valuation techniques which might be applied to water in order to understand some of our water problems and potential water solutions. Because water is so fundamental to life, is in such demand by humans and has economic value, it is also a cause of conflict around the world. Determining who has the rights to access, extract and use water, a substance that is fluid and mobile, is a complex issue with different traditions operating in different parts of the world. Chapter 11 looks at different types of water conflict and water rights issues from around the world and how these conflicts can be managed. Finally, in Chapter 12, the book looks at new concepts that can help us to understand current and future water resource use and availability across interconnected local and global scales, including the water footprint and 'virtual' water flows.

Each chapter contains some boxed features, which are grouped into one of four themes: case studies, contemporary challenges, techniques or the future of water. These boxed features allow interested readers to study more detail on the selected topics, should they wish. As you read the book you will also notice some words that are typeset in bold within the text. These words are highlighted the first time they appear in each chapter and can be found in the glossary with an explanation of their meaning. Each chapter also contains reflective questions and some project ideas.

The book's interdisciplinary nature reflects my own personal role as head of water@leeds, which is the largest water research group in any UK university. water@leeds members work together as scientists, social scientists and humanities experts to tackle water challenges facing the world. Research is often in partnership with external bodies such as government bodies, NGOs, industry and practitioners to ensure that the research is applicable to societal needs. The water challenges we face are crucial to the survival of the human race, stability and peace around the world and to the sustainability of the Earth's ecosystems. It is therefore vital that around the world we improve understanding of water resource issues from an interdisciplinary perspective. This book is part of that mission and I hope that you will feel inspired to join us in trying to make a real difference to people's lives through sharing an understanding of water resource issues with others.

Acknowledgements

Alison Manson, University of Leeds, is thanked for her work in producing most of the drawings within the book. Additional figure and table source acknowledgements are provided where known. Kathryn Smith is thanked for helping to compile and check the references and glossary and obtaining figure permissions. Samantha Bowman and Rianne Dubois are thanked for permissions management and compiling the index. Thanks are also given to the editorial production team at Routledge for their support in producing this volume. Kitriphar Tongper and Dr Anamika Barua are thanked for contributing some materials for Chapter 1. Dr Tanya A. Warnaars, Centre for Ecology and Hydrology, Wallingford is thanked for support on Chapter 2 in providing the latest update on glaciers in South Asia and material on the EU WATCH project. Every effort has been made to contact copyright holders for their permission to reprint material in this book. The publishers would be grateful to hear from any copyright holder who is not here acknowledged and will undertake to rectify any errors or omissions in future editions of this book.

CHAPTER ONE

Water fundamentals

Joseph Holden

> **LEARNING OUTCOMES**
>
> After reading this chapter you should be able to:
>
> ■ describe the nature of water
> ■ describe some of the roles of water in historic and contemporary society
> ■ outline some of the major global water resource issues.

A INTRODUCTION

Water is a very common substance on Earth, covering 73% of the Earth's surface in oceans, rivers and lakes. However, water is also a special substance. It is one of only a handful of substances that expands when it freezes. Water is one of the few substances found in solid (ice), liquid and gas (water vapour) form within our natural environment. It is the best naturally occurring solvent, meaning that it acts to dissolve and carry more different types of material within it than anything else, being good for cleaning and helping to sustain life in **plants** by providing them with nutrients. Indeed, water in solid, liquid and gaseous forms has been very important in shaping the landscapes on Earth through its role in **weathering**, erosion and transport of materials. Water is also fundamental to life on Earth. Humans, for example, are around 70% water by mass and we quickly dehydrate if we do not drink; people will die within a few days without water. Water can also be harnessed for other purposes, such as being a cooling agent for industry and power generation, and it thus serves a wide range of economic functions. Water also takes on a cultural and religious significance for many societies in terms of rituals, blessings, concepts of nice vistas or leisure activities and was fundamental to the development of agriculture and, hence, civilisations. This chapter deals with the properties of water, issues around global water availability and introduces the role of water in society.

B WHAT IS WATER?

Water is not an element in itself. It is a compound made from two other elements: hydrogen and oxygen. Each water molecule contains two hydrogen atoms and one oxygen atom (H_2O), which are strongly bonded together. Each hydrogen atom shares a pair of electrons with the oxygen atom. Such pairing is called a **covalent bond**. All of the electrons position themselves as far apart from each other as they can, as they are all negatively charged and hence repel one another. However, the electrons that are not bonded to the hydrogen atom are closer to the oxygen atom than the ones being shared with the hydrogen atom and therefore the repulsion forces are slightly stronger for the unpaired electrons. This stronger force pushes the hydrogen atoms slightly closer together so that the two hydrogen atoms rest apart at 104.5° (**Figure 1.1**). While the overall charge of the water molecule is neutral, the electron arrangement shown in **Figure 1.1** indicates that more of the negative (electron) charge is concentrated towards the oxygen side of the molecule. This means that the slightly positive hydrogen side in one water molecule is attracted to the slightly negative concentration on the oxygen side of another water molecule, forming a 'hydrogen bond'. Hydrogen bonds are between 10 and 50 times weaker than covalent bonds.

Both the strong covalent and the weak hydrogen bonds are important properties of water, and the latter in particular make water unusual in its nature. The former means that it is very difficult to force apart the hydrogen and oxygen atoms within water molecules, which is why water is so prevalent on the planet and has not disappeared. The latter means that the water molecules themselves can move between one another relatively easily, meaning that it is highly mobile, enabling it to flow rapidly and allowing materials (e.g. fish, people, ships, plastic bags, excrement) to pass through it easily. The hydrogen bonds also enable water to act as an excellent solvent, as the molecules can attach to other compounds, enabling them to be taken up into solution (see below). Finally, the hydrogen bonds mean that water is present as a liquid over most parts of the planet's surface. If it were not for these bonds, water would boil at -80°C rather than 100°C.

1 Physical properties of water

1.1 Thermal properties

Water is at its most dense at 4°C. This temperature–density property is highly unusual. At this temperature water is seen to be at a standard volume of 1 g per cm^3 or 1 kg per litre. Water at 4°C will tend to sink towards the bottom of water bodies and there will be thermal stratification. Upon freezing, water expands by about 9% (~0.91 g cm^{-3}) and the hydrogen bonds form a strong structure. Thus, ice, which is less dense than water, floats upon the surface of water. This is important, because if ice were denser than water it would sink, thereby crushing the creatures living in water bodies or leaving them stranded at the top of the water body. Indeed, lakes would freeze solid quite quickly during winter if it were not for the special properties of water. The ice at the surface insulates the lower liquid water and also reflects more of the Sun's energy, thereby reducing the chance of the lake freezing solid.

Normally, as liquids are heated, the molecules gain energy and bounce around more, thereby

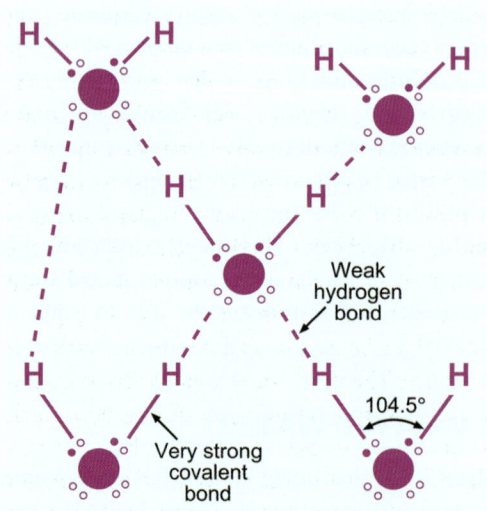

Figure 1.1 A schematic diagram showing water molecules and the covalent and weak hydrogen bonds.

taking up more space. However, this does not happen until > 4°C for water. As water warms between 0 and 4°C, the hydrogen bonds bend or break frequently, causing the water molecules to pack more tightly. The expansion of water at temperatures greater than 4°C is very small in comparison to that caused by freezing. For example, in warming from 4°C to 10°C, the expansion of water is less than 0.03%. Nevertheless, this thermal expansion is still important for sea level rise under climate change. It is estimated that half of the sea level rise of up to 0.44 m during the twenty-first century will be due to thermal expansion of water in the oceans (IPCC, 2007). The rest will come from the melting of ice on the land surface. Note that melting of sea ice does not directly affect global sea levels, since the ice is already part of the water body and sea levels are already displaced by the mass of ice sitting in the water. It should also be noted that seawater does not behave like freshwater, due to its high salt content, so that the density is not at a maximum at 4°C and instead it behaves more like a normal liquid (Berner and Berner, 1987).

To raise the temperature of 1 kg of water by 1°C requires 4200 joules of energy (the **specific heat** of water). This is the highest of any liquid except ammonia. Cooling water by 1°C would release the same amount of energy. This is why water takes a long time to heat up and cool down and why coastal areas have a moderated climate compared to inland areas (McClatchey, 2012). Coastal areas often have seawater temperatures warmer than air temperature during winter, as the sea cools slowly from the previous summer's heat. During summer, water temperatures can be a lot lower than air temperature, as the water warms only very slowly.

Energy is also used or released when water changes phase between solid, liquid and gas forms. It takes a lot of energy to break the hydrogen bonds in water to form water vapour. In fact it takes 2.3 million Joules to evaporate 1 kg of water without any change in temperature. This energy used to turn a liquid into a gas is known as **latent heat of vaporisation**, and for water it is one of the highest known because the molecule has to escape both the normal molecular attraction forces of the liquid and the hydrogen bonds. Water vapour may then travel large distances as a gas before it condenses. When it condenses (e.g. as droplets in a cloud) the same amount of energy is released (e.g. into the atmosphere). Thus, energy can be moved around the planet and away from the Tropics by water in the atmosphere (see Chapter 2). To put this into context, it takes six times more energy to evaporate water from liquid to gas (at any temperature, even when the liquid is already at 100°C) than it takes to raise the temperature of water from 0°C to 100°C.

While the boiling point of water at normal sea level atmospheric pressure is 100°C, water vapour is still present all around us. Evaporation of water occurs at the surface of water bodies where sufficient energy has been imparted to break the hydrogen bonds and allow the water vapour to escape. People often think of white clouds or white trails from kettles as being water vapour, but in fact water vapour is invisible. Clouds and kettle trails are liquid forms of water where condensation has occurred. Evaporation is a crucial part of the water cycle and invisible water vapour is a very powerful **greenhouse gas**. Short-wave radiation comes from the Sun and hits the Earth's surface. It then radiates back up into the atmosphere from the Earth as long-wave radiation which gets absorbed by several gases, thereby warming the air from below. Water vapour is good at absorbing radiation released from the Earth's surface. This absorption of radiation means that radiation is kept in the atmosphere rather than being allowed to escape back out into space. Therefore water vapour acts as a good 'greenhouse' gas, warming the atmosphere. Such greenhouse gases have made our planet hospitable, but changes in the concentrations of water vapour and other greenhouse gases in our atmosphere alter the balance and can lead to further warming. Chapter 2 provides more detail on this process and the impacts that water vapour feedbacks may have on climate change. Note that most water vapour in the atmosphere is produced from transfers from the oceans via evaporation of surface water.

1.2 Surface tension

Surface tension is common to liquids (i.e. it is not unique to water). Within the main volume of a liquid, molecules are attracted in all directions by their neighbours. However, at the surface of a liquid there are no molecules outside and so the molecules are only pulled inward. This has the effect of forcing the liquid to adopt the minimum surface area, which is a spherical shape. Without gravity, liquid would form a sphere. The force pulling the liquid in towards itself is greater than the gravitational force of some objects such as pond skaters which might rest on the water surface. These invertebrates can be seen skimming across the water surface without sinking (**Figure 1.2**). The surface tension of water is the greatest of all liquids that are not metals.

For liquids contacting solid matter, the behaviour will be a result of the balance between surface tension (cohesive forces) and any attractive forces between the molecules of the liquid and those of the solid matter (adhesive forces). For many solid materials water is more strongly attracted to

Figure 1.3 Water forms droplets on most solid materials it comes into contact with, as the internal attractive forces are greater than those between the water and the solid.
Source: Photo courtesy of Shutterstock.

itself and so water forms droplets on the surface (**Figure 1.3**). However, for other materials the water is attracted to the material more than to itself and here it will spread out and wet the surface. In other words, if the adhesive force between the liquid and solid exceeds that of the liquid's cohesive force, the liquid wets the surface. Adding a **detergent** to water makes the water reduce its surface tension and therefore allows a substance to be wetted by spreading water.

The attractive forces of liquids result in a phenomenon known as **capillary action** whereby liquids flow within narrow gaps between solid material, no matter which way gravity acts. In other words, capillary action means that water can flow upward. This is vital for plant growth. As plants lose water from their leaves, new water can be drawn up from the soil. Capillary action can be seen if you dip a tissue into some water – the water rises up the tissue. If the diameter of the pore (e.g. glass tube, soil or rock pore space, tissue pore space) is sufficiently small, then surface

Figure 1.2 Pond skaters are invertebrates that move across the water surface using the surface tension of water to maintain their position.

WATER FUNDAMENTALS 5

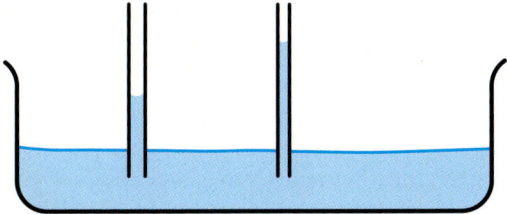

Figure 1.4 Capillary action in two glass tubes. The curved meniscus can be seen where water contacts the solid material it is wetting. Surface tension tries to counteract this attraction between the water and solid material by pulling upward in an attempt to move itself back towards a spherical shape where the surface area is minimised. This results in water moving up the narrow glass tube against the force of gravity.

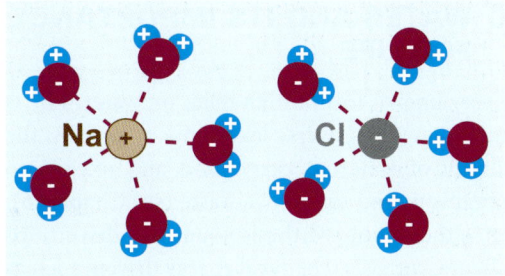

Figure 1.5 When dissolved in water, each of the sodium and chloride ions is hydrated as it is surrounded by water molecules.

tension combined with adhesive forces between the liquid and solid acts to force the water into the pore. The narrower the pore space (e.g. a narrow glass tube), the greater the rise of liquid. The surface of the liquid near the edge of solid matter will curve upward if it is wetting that surface (**Figure 1.4**). Surface tension then acts as a force to pull the liquid upward as it tries to counteract the curved edge which is in the opposite direction to the natural spherical shape which the cohesive forces naturally pull towards. Chapter 3 provides more information on this topic.

2 Chemical properties of water

Water is an excellent solvent. Pure water does not readily occur in nature because water so quickly dissolves other substances that it comes into contact with. A substance will dissolve in water if its molecules can interact with charged water molecules. The hydrogen side of each water molecule is positively charged, while the oxygen side is negatively charged. These charges are stronger than the charges that hold many other molecules together. For example, a salt such as sodium (Na) chloride (Cl) is torn apart by water in that the Na^+ **ion** (an ion is a charged atom or molecule, due to an unequal number of protons (positive charge) and electrons (negative charge)) is attracted to the negative charge and the Cl^- is attracted to the positive charge on the water molecule. This breaks down the solid crystals and the salt dissolves into the water. Water tends to reduce the electrical attraction between oppositely charged ions by 80-fold (Ghali, 2010). The fairly small size of water molecules allows many water molecules to surround one molecule of a solute (**Figure 1.5**), enabling the ion to be hydrated. Substances such as sugars and alcohols are water soluble not because they are ionic like salts, but because they have polar molecules (rather like water) with some weak electrostatic charge on parts of their molecules which attracts opposite charges from parts of water molecules. Substances that are electrically neutral (e.g. oil) dissolve poorly in water. Chapter 4 describes in more detail the chemical properties of water and how water quality is affected by flow paths and human intervention.

> **REFLECTIVE QUESTION**
>
> Why does water often move upward in soils in defiance of gravity?

C WATER AND ITS IMPORTANCE FOR LIFE

Every known form of life relies on water. Life is maintained in rivers, lakes and oceans by the ability of water to carry food and oxygen to organisms and to remove waste (see Chapter 6). In either photosynthesis (sunlight driven) or chemosynthesis (chemical energy driven, such as around deep ocean hydrothermal vents), water is combined with carbon dioxide (or methane for some chemosynthesis) to create carbohydrates and release oxygen:

$6CO_{2(gas)} + 12H_2O_{(liquid)} + photons \rightarrow C_6H_{12}O_{6(glucose)} + 6O_{2(gas)} + 6H_2O_{(liquid)}$

carbon dioxide + water + light energy $\rightarrow$ glucose + oxygen + water

Leaves use the Sun's energy to enable chemical reactions resulting in carbon dioxide being absorbed from the atmosphere, mixed with water and converted into simple sugars, such as glucose (see also **Box 2.4** in Chapter 2). Glucose is made into compounds for storage such as starch, and compounds for structure such as cellulose and lignin (which make up cell walls for leaves and stems). These carbohydrates can then be consumed by higher parts of the food chain as the basis of life on Earth.

Plants are able to obtain water from the soil through capillary action. As water is transpired through leaves, new water containing essential dissolved nutrients to allow the plant to grow is drawn up from below. For land plants, water is additionally important because they are mechanically supported by the water pressure inside them. Without water, plants wilt. Only a few plants can survive drying, by switching off the photosynthetic process when drought comes and then restarting it when water is abundant. Some plants cope with drought by shedding leaves or are adapted to drought by having thick, waxy coatings on their leaf surfaces, or small, thick leaves and special water storage areas.

All life is made up from cells. Some species have one cell (e.g. bacteria) while others have hundreds of billions (e.g. humans). Cells are the building blocks for life. They provide structure for the organism, take in nutrients from food, convert those nutrients into energy and carry out specialised processes. Cells also contain the blueprint for the living thing and information from its hereditary past and can make copies of themselves. The major biochemical processes that operate within cells rely on water, enabling cell structure, function and division.

For humans 70% of our body is water. The human brain is 95% water. Human function is all about the flow of water through our system. Water is important to transport chemical signals and nutrients to vital organs of the body and is necessary for our body to digest and absorb vitamins and nutrients and flush out wastes. We generally feel thirsty at about 2–3% water loss. When a human suffers 5–6% water loss they will feel unwell and may have headache, nausea or tingling in the limbs. At dehydration levels greater than this the skin will shrivel, vision dims, delirium may occur, muscle control will be lost and, normally, dehydration beyond 15% of normal body water is fatal.

Drinking water directly is not the only way to obtain water. Food contains water within it and typically one-fifth of human water intake comes from eating food. Typically a lettuce is 95% water and an apple 84%. Water is also essential in the production of food for humans. Some foodstuffs are low in their water requirements, while others have heavy water requirements (**Table 1.1**), although water use for a particular crop type will depend on soil type and climate. Meat production requires 6–20 times more water than cereal production. Cereals are the most important source of total food consumption worldwide, accounting for over half of the diet of most developing countries. However, a large proportion of cereals are used for animal feed. As countries develop they also tend to demand more food that uses water more intensively (e.g. meat products) and so, as the world's population both expands and develops, more water will be required for agriculture.

Table 1.1 Typical water use to produce some common human foodstuffs

Product	Water use, m³ per kg of food/drink
Fresh beef	15
Fresh lamb	10
Fresh poultry	6
Eggs	3.4
Rice	1.5
Cereals	1.5
Bread	1.3
Coffee	1.2
Citrus fruits	1.0
Pulses, roots and tubers	1.0
Milk	1.0
Wine	0.9
Beer	0.3
Potatoes	0.3

> **REFLECTIVE QUESTION**
>
> What do water and carbon dioxide combine to produce as part of photosynthesis?

D WATER'S ROLE IN SOCIETY

1 The development of civilisations

Gathering foods and hunting animals originally depended on the natural distribution of plants and animals. However, around 13,000 years ago agriculture developed. Agriculture involves cultivating plants and herding animals. The animals and plants in most cases underwent **domestication**, indicating that humans have taken control of the reproduction of the individual plants and animals concerned. Almost all domestication occurred over the last 13,000 years (Blockley and Pinhasi, 2011). Domestication of goats, cattle, sheep, horses and so on provided the ability to convert inedible plants (e.g. grass) into food which humans could eat, such as meat and milk. It also provided materials for clothing and the ability to travel more quickly, and provided power (e.g. horses). Most of the earliest sites of domestication were in the 'fertile crescent' of Mesopotamia. Here, around 12,000 years ago, hunter-gatherers foraged for the large seeds of certain wild grasses. They must have realised that planting the seed led to a more predictable crop the following year. They favoured the planting of seeds from those with the largest seeds, and with non-shattering seed heads as these were easier to gather. While most grasses drop their seed to the ground when ripe, a few mutants in any population will retain the seeds on the head. Consequently, over time, the average seed size of the wheat and barley that they grew increased and, as the seeds remained on the seed heads, these new strains could not reproduce without human help. The new varieties of cereals were joined by other crops such as peas, lentils and beans and by domesticated animals, notably goats in the first instance. Elsewhere in the world, people also began domesticating plants and animals.

As a result of agriculture, populations of farming communities grew rapidly because food supplies were more reliable and there was no longer a need to be nomadic. The over-production of food by farming communities meant that they needed storage containers, and hence this encouraged technological innovation. It also meant that such food could be traded with others, who could in turn specialise in other activities rather than have to gather and hunt for food. These specialised activities might be woodwork, pottery, metalwork and so on. Thus we have the start of trading and an economy, such that some people accumulate more goods than others. This encouraged stratified societies whereby some people became powerful. The powerful were then able to organise others to contribute to large-scale projects such as irrigation schemes and developing monuments (Hassan, 2012).

For some groups, locating settlements close to permanent springs and streams meant that they could enjoy more stability for their farming and be less reliant on rainfall variations from year to year. Thus, those farming communities who settled around water resources had an upper hand over others. As well as having a natural water supply, humans living around watercourses realised that they could divert the flow by building channels across the landscape to move water from the river into farmland. Some of the most famous early civilisations developed because they controlled that water supply. From around 10,000 years ago, groups of farmers came together on the floodplains of the Tigris and Euphrates rivers and diverted the courses of rivers and embanked and arranged fields so that water could be retained and made to flow from one field to another. The combination of a hot climate, fertile floodplain soils and a carefully controlled water supply made these farming techniques very productive, and empires were built on the surpluses that they produced. In the long run, however, problems could emerge. Disputes over water rights, and the political difficulty of building and maintaining such large shared resources, could lead to conflict (Hassan, 2012). On the whole, though, the transformation of Mesopotamia from semi-desert to a bread-basket was an outstanding example of the landscape engineering that allowed complex civilisations to flourish for many millennia (Lawson, 2012).

Civilisations such as the Roman Empire, Egyptian rule and the Omayyad Dynasty were all founded on their access to water. The rise of state societies with kings to organise, defend and rule also had cultural and religious significance. Leaders of civilisations were often seen as representatives of gods or God, or in some cases as gods in themselves. Social practices and rituals focused on water are common in history. During the Pharaonic period in Egypt, the River Nile was worshipped and floods were seen as being sent from the gods. The river levels were also used to determine the level of tax that should be levied on farmers, because the higher the flood, the more land was under agricultural production. The development of cities became possible because of agriculture and the use of water. Water was used for hygiene as well as drinking water and food supply. Water was also culturally important to early cities. Baths in ancient Rome were socially very important, and enormous volumes of water were channelled into the city in spectacular fashion, while excess water was used to flush out sewage. Water was used in architectural designs for streets and gardens. The Romans also drained wetland to make it agriculturally more productive or to reduce the threat from water-borne disease.

2 Contemporary society

As with the development of civilisations, control of water resources today provides an economic advantage that enables food production, industry, navigation and sustenance for the population. Careful management of water also provides health benefits, through disease control, for example (Chapter 8). Chapter 10 details how we value water economically, while Chapter 11 provides some information on water resource use around the world and the differentials between nations and regions. Chapter 12 outlines how water is used to produce products that are traded around the world and looks into concepts that help us to understand this water use and how they might change the culture of water use around the world.

In many religions, water is of great importance. Water is used in ritual washing in Hinduism, Islam, Shinto, Taoism and many other religions. In Shinto, waterfalls are sacred and an individual can become purified by standing under one. For Jews, water is used in ritual washing for several different purposes. Muslims wash with water before performing ritual prayers or handling the

Qur'an. Buddhists use water at funeral services. Water is also important to Christian baptism and for blessings or cleansing from evil. In Sikhism, water is used to make Amrit, ready for use in an initiation into the Khalsa through Amrit Sanskar. Some water bodies such as the Ganges River (Hinduism) and the Well of Zamzam (Islam) are sacred.

Water also has wider cultural significance (e.g. see **Box 1.1**). Water can represent different things to different people or groups of people. For example, for some North American natives, water represents mobility, providing a highway in ancestral tribal cultures where tribes moved between different camps following seasonal patterns of resource availability. Water features in art such as music, paintings, writing and poetry. It is used aesthetically in the layout of gardens (ponds, fountains, etc.) or landscaping. Many sports focus on water, such as swimming or sailing.

In the developed world, the control of water and its use for industrial, navigational or flood control purposes has been partly replaced by recognition that the wider ecosystem and environment is a user of water. Hence water is seen to be useful for environmental conservation. Rather than drain wetlands, there is a move to preserve or restore them. For example, large parts of the Florida Everglades were drained and cut through by canals in the late nineteenth and early twentieth centuries. Pollution from urban areas and agriculture also threatened the ecology. Since the 1970s there has been conservation protection status on part of the Everglades, but in 2000 a comprehensive package of restoration, to take place over 30 years and estimated to cost more than $12 billion, was agreed. This is thought to be the world's largest ecosystem restoration. Actions include rewetting large areas, infilling drains and canals and reuse of wastewater to safeguard water quality. In many other parts of the world, rivers

BOX 1.1 CASE STUDIES

Japan's river culture

For Japanese people there is a symbolic representation of their culture embedded within scenes of flowing water, especially in rivers and streams. There is a nostalgia associated with aesthetic landscapes of water which perhaps represent a more rural Japan of two or three generations ago, before rapid urbanisation (Shinmura, 1995). Now, however, many watercourses are channelised (concrete linings, straight or buried channels, etc.) in Japan to protect people from flooding and to enable navigation or water resource management. However, the nostalgic and symbolic importance of rivers to Japanese people has meant that there is a growing shift in the agenda among community groups, some government officials and some academics to reconnect people with rivers and enable interaction with watercourses through the provision of amenity areas (e.g. areas for children to play) and to soften the hard engineering by creating more natural-looking riverine landscapes (Waley, 2000). In Japan, as in many parts of the world, there is therefore a strong movement around river rehabilitation. Events such as firework displays located along waterfronts, lantern boats for parties, and community events including sports and educational events around rivers serve to reinforce the cultural significance of rivers and build a sense of community around water bodies. However, the tension between ecological engineering and river culture and issues of flooding and public health is still strong, but through continued engagement between a wide variety of agencies these tensions are slowly being weakened (Waley and Åberg, 2011).

that had previously been straightened and have become rather dull and monotonous, with low ecological and geomorphological diversity, are now being rehabilitated through the installation of meander bends, removal of concrete bank linings and addition of woody debris (Palmer and Bernhardt, 2006). The EU Water Framework Directive requires that water bodies must be in good ecological status and efforts must be undertaken by all EU member states to improve the status of water bodies. In other words, legislation is in place to protect water bodies for their own sake as well as for the sake of society, which enjoys and relies on water bodies for the wider benefits it receives from them. Further information about the EU Water Framework Directive is provided in Chapters 4, 6 and 10. However, there is still a tension between rehabilitating rivers or having 'natural-looking' water landscapes and societal expectations around issues such as flood management, where flood defences are still expected to be in place (see Chapter 3).

> **REFLECTIVE QUESTION**
>
> How did control of water resources enable civilisations to develop?

E Global water resources

1 The amount of water available

The Earth's hydrosphere contains a huge amount of water – about 1388 million cubic kilometres (see also Chapter 2). However, approximately 97% of this amount is saline water and only 3% is freshwater (**Figure 1.6**), although as Chapter 2 notes there are large uncertainties as to the exact values of the stores. The greater portion of this freshwater (between 48 and 69%) is in the form of ice and permanent snow cover in the Antarctic, Greenland, other parts of the Arctic and the mountainous regions. While 30% of freshwater exists as groundwater, <1% of the total amount of freshwater on the Earth is concentrated in lakes, reservoirs and river systems. Hence <1% of the total available water is most easily accessible for our economic needs, and at the same time it is vital for land-based aquatic ecosystems.

Water is the most widely distributed substance on our planet. Although it is available everywhere its availability varies across the world. The mean value of the renewable global water resource is estimated at 42,700 km^3 per year, and this is extremely variable in space and time (Shiklomanov, 1998). At the continental level, the Americas have the largest share of the world's total freshwater resources with 45%, followed by Asia with 28%,

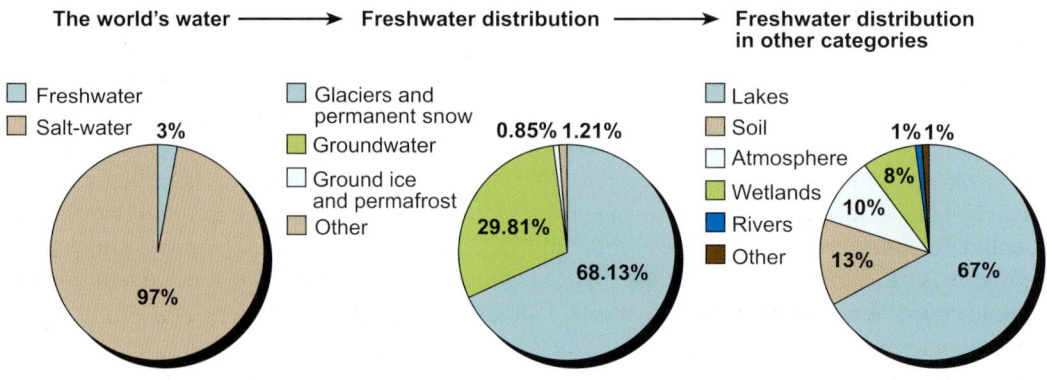

Figure 1.6 Categories of world water resources.

Table 1.2 Water-rich and poor countries of the world

Water-rich countries	Total resources (km³ yr⁻¹)	Water-poor countries	Total resources (km³ yr⁻¹)
Brazil	8233	Israel	1.67
Russia	4507	Jordan	0.88
Canada	2902	Libyan Arab Jamahiriya	0.60
Indonesia	2838	Mauritania	11.40
China (mainland)	2076–2830	Cape Verde	0.30
Columbia	2132	Djibouti	0.30
USA	2071	United Arab Emirates	0.15
Peru	1913	Qatar	0.05
India	1897	Malta	0.05
		Gaza Strip	0.06
		Bahrain	0.12
		Kuwait	0.02

Source: Data from FAO (2003).

Note: These are just total values and not based on per person water availability.

Europe with 16% and Africa with 9%. A country-level analysis shows that nine countries are world giants in terms of internal water resources (total surface and groundwater, not accounting for transboundary waters or border rivers), accounting for 60% of the world's natural freshwater (**Table 1.2**). As shown in **Table 1.2**, twelve countries are water-poor countries and they are mostly small (notably islands) or arid ones. However, the values shown in **Table 1.2** do not entirely reflect the amount of water available for each person, as countries differ so much in area and in population.

Look at **Figure 7.4** in Chapter 7, which maps the amount of freshwater available per person per year in different parts of the world. Falkenmark (1986) proposed a threshold of 1000 and 500 m³ of freshwater per inhabitant to correspond respectively to the *water stress* and *water scarcity* levels. Note that in **Figure 7.4** the OECD has used less severe thresholds to define scarcity (<1000 m³) and stress (<1700 m³) although it does define **severe water scarcity** as freshwater <500 m³ per person per year (see section D in Chapter 7). Therefore you need to be aware of different definitions for these terms. Falkenmark (1986) felt that in an average year, 1000 m³ of water per inhabitant can be considered as a minimum to sustain life and ensure agricultural production in countries with climates that require irrigation for agriculture. There are currently 28 countries which have less than 500 m³ of water internally available to them per person per year and 45 countries with less than 1000 m³ per person per year. Among these countries are Egypt (22), Saudi Arabia (85), Singapore (115), Niger (218), Pakistan (311), Hungary (602), the Netherlands (660) and South Africa (887) (values in brackets are m³ per person per year). Some countries are undergoing very rapid industrialisation and population growth, which puts their water resources under stress even if the total water resources available within the country are large. India, for example, currently has around 1165 m³ per person per year of internally available renewable freshwater (compare this to 9000 for the USA or 70,000 for New Zealand), but this has decreased from 3100 in the early 1960s and 2000 in the early 1980s (**Box 1.1**). See also **Figure 2.15** in Chapter 2, which provides a map showing the number of people across the world estimated to be subject to severe water scarcity.

There are 44 countries in the world which depend on other countries for over 50% of their renewable water resources. These countries are: Argentina, Azerbaijan, Bahrain, Bangladesh, Benin, Bolivia, Botswana, Cambodia, Chad, Congo, Djibouti, Egypt, Eritrea, Gambia, Iraq, Israel, Kuwait, Latvia, Mauritania, Mozambique, Namibia, Netherlands, Niger, Pakistan, Paraguay, Portugal, Republic of Moldova, Romania, Senegal, Somalia, Sudan, Syrian Arab Republic, Turkmenistan, Ukraine, Uruguay, Uzbekistan, Vietnam and the seven independent countries that made up the former country of Yugoslavia.

2 Increasing demand for water use and consumption

Water can be used both in-stream and off-stream. In-stream water use refers to direct use of water in rivers or lakes where there is no withdrawal of the water. For example, the uses of water for navigation, hydropower, tourism and fisheries are all in-stream water uses. Off-stream water use involves withdrawal of the water from the source. Where water is withdrawn and then not returned to the resource that provided it, this is known as **consumptive water use**. Examples might include manufacturing, where water is abstracted for cooling, which results in evaporation of the water, or water used in food preparation, where the water is not returned to the stream. Agricultural consumptive water use includes the water transpired by plants plus losses from evaporation from the soil surface and leaves in the crop area. Good farming practice attempts to reduce crop water consumption to its most efficient minimum.

The world's population is growing by about 80 million people a year, implying increased freshwater demand of about 64 billion cubic metres a year (UNESCO, 2009). An estimated 90% of the 2.5 billion people who are expected to be added to the population by 2050 will be in developing countries, many in regions where the current population does not have sustainable access to safe drinking water and adequate sanitation. Good sanitation is important for health (see Chapter 8), yet 2.5 billion people have to use hygienically unsafe toilets or have to defecate on open land. Poor sanitation causes a myriad of water-borne diseases, including severe diarrhoea which kills around 1.5 million children each year. As populations grow, demand for water increases, due to escalating demand from domestic, agricultural and industrial sectors. While the worldwide demand for water is increasing every year, water supply cannot remotely keep pace with the demand.

By 2025, 52 countries containing two-thirds of the global population are expected to be short of water. To many, this is seen as a water crisis. According the Asian Development Bank, China and India (**Box 1.2**) alone are forecast to have a combined supply shortfall of one trillion cubic metres in 2030. Within Asia, other countries at or near water stress conditions are Bangladesh, Cambodia, Nepal, Pakistan, the Philippines and Vietnam (Richardson, 2010). With the addition of another 1.5 billion people to feed in Asia by 2050 there will be unprecedented stress on water supply in the region.

In developing countries, industry is an essential engine of economic growth, and the industrial demand for water is likely to increase. Global annual water use by industry is expected to rise from an estimated 725 km^3 in 1995 to about 1170 km^3 by 2025, by which time industrial water usage will represent 24% of all water abstractions. See also **Figure 7.2** in Chapter 7 for a 2030 prediction from major industrial organisations. Much of the increase will be in developing countries now experiencing rapid industrial development (UNESCO, 2003). This will further intensify competition between the domestic, industrial and agricultural sectors for water, leading to sectoral water conflicts (see Chapter 11). **Figure 1.8** shows industrial water use per region, compared with other main uses determined by UNESCO (2003). Clearly, such figures depend on exactly how you

BOX 1.2 THE FUTURE OF WATER

India's looming water crisis

India's demand for water is growing at an alarming rate. India currently has the world's second largest population, which is expected to reach a staggering 1.6 billion by 2050, putting enormous strain on water resources. The burgeoning population, unplanned urbanisation, and growing sectoral demands put tremendous pressure on freshwater resources. Estimates of total usable water resources range from 668 billion cubic metres to 1086 billion cubic metres (NCIWRD, 1999; Narasimhan, 2008; Garg and Hassan, 2007). Whichever estimate one accepts, there is no escaping the fact that water availability in India is inexorably approaching the scarcity benchmark of 1000 cubic metres per capita (Figure 1.7). With unabated growth in irrigation and even more rapid growth in industrial and domestic water demand, water shortages seem inevitable.

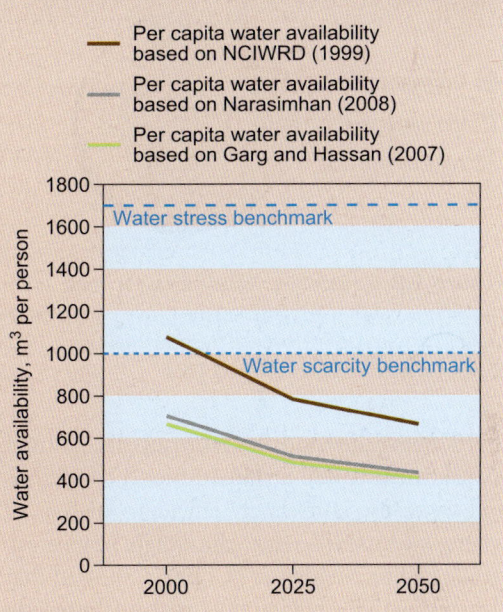

Figure 1.7 Declining per capita water availability in India against the benchmarks of water stress and water scarcity.

carve up the sectors and, as shown in Chapter 12, there are other ways of accounting global water use. Hoekstra and Chapagain (2008), for example, estimated that 86% of global water use is for agriculture and processing of agricultural products, 10% is for industry and the rest is for domestic consumption.

Agriculture is the largest consumer of water for human use. Most water for agriculture comes from rainfall stored in the soil profile and only 15% is provided through irrigation. However, of the 3800 km³ water abstracted from rivers and groundwater, 70% is used for agricultural irrigation (Molden et al., 2007). Additional water required for agriculture under forecasted population growth will strain ecosystems and increase competition for water resources. While efforts to reduce water consumption in Europe and North America for industry and domestic use are sensible, for the rest of the world the focus needs to be on using water more efficiently in agricultural processes.

There appear to be five key ways of increasing water efficiency for global agriculture:

- focus efforts in areas with currently low agricultural yields
- improve soil fertility in arid and semi-arid areas
- use trade to manage water efficiency
- reduce evaporation
- use biotechnology.

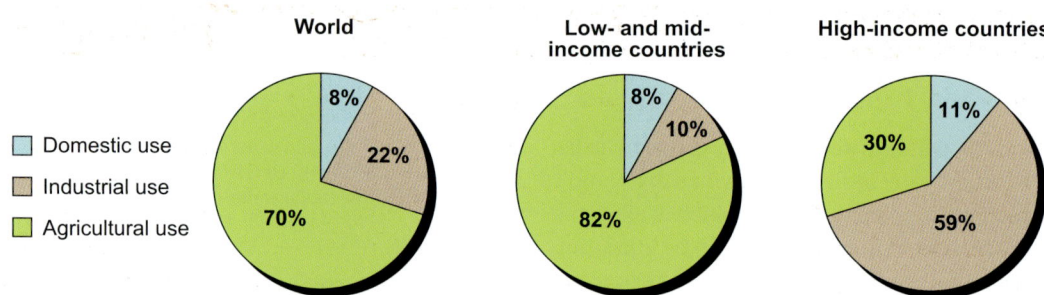

Figure 1.8 Competing water uses for main income groups of countries. Note that the global figure for agricultural water use, including the processing of agricultural products, is 86%. However, in this chart, agricultural processing has been covered by industrial use.
Source: Based on data from UNESCO, 2003.

In many developed countries agricultural management is well adapted to local conditions and water use is relatively efficient. Analysis of where water use efficiency gains might be greatest, conducted by Molden *et al.* (2007), showed that the areas with the highest potential are those with currently very low yields, such as sub-Saharan Africa, which are also the poorest areas of the world. Hence a focus on these areas could reduce global water use and reduce poverty at the same time. In many of the most productive areas of the world, such as the lower Yellow River Basin, Molden *et al.* (2007) suggested that large improvements have already been made and the remaining scope for water efficiency improvements is small. For arid and semi-arid regions, improvements in soil fertility would also help to make water use more efficient, as lack of nutrients may limit plant growth, thereby reducing water efficiency. Global improvements in water efficiency for agriculture can also come about by growing crops in places where climate and management practices enable high water efficiency and trading them to places with lower water efficiency. Reducing evaporation while increasing transpiration from productive crops can enhance water productivity. Drip and sprinkler irrigation systems do not necessarily result in less evaporation than good surface irrigation systems, but practices such as mulching, ploughing or breeding plants that quickly expand their leaves to shade the ground reduce evaporation and increase productive transpiration. In arid environments up to 90% of rainfall evaporates back into the atmosphere, leaving just 10% for productive transpiration. Water-harvesting techniques such as collecting water from roofs or plastic sheets on the ground and channelling it into storage tanks can capture more water for crops and livestock. Crop breeding of grains such as wheat, maize and rice allowed excellent advances in water efficiency during the 1960s to 1980s. However, further advances in these grains may make only very moderate gains. Nevertheless, breeding for resistance to disease, pests and **salinity** (which reduces the need for agrochemical inputs) and boosting the harvest index for crops such as millet and sorghum that have not received as much attention as other grains will help with water efficiency.

While population growth and industrialisation are two concerns for changing water resource availability, climate change poses an additional challenge. In some places this may not be because of changes to the overall water resource availability but because of changes to the risk of extreme conditions (e.g. drought). **Box 1.3** describes a recent analysis relating to food security impacted upon by potential drought risk change in Asia for the 2020s. Note that the analysis in **Box 1.3** represents a project which combined scientific

> **BOX 1.3 CONTEMPORARY CHALLENGES**
>
> ### Near-future drought and food security in Asia
>
> The latest climate projections from 14 of the world's leading climate modelling centres were used by Forster *et al.* (2012) to examine the potential risks to food production for wheat, maize and rice across Asia from climate change-driven drought in the 2020s. An analysis of adaptive capacity was also conducted, based on projections of seven key socio-economic drivers, to determine which crop-producing regions are most vulnerable to climate change-driven drought. Forster *et al.* (2012) showed that:
>
> 1. *There is an increased risk of more severe droughts in the 2020s.* Compared to the period 1990–2005, the 2020s will bring marked increases in drought severity across much of Asia, with larger deficits in soil moisture over longer periods of time. Immediate actions are needed so as to adapt. Water resource management can mitigate impacts of all but the most severe and prolonged droughts. This would include sustainable use and safeguarding of groundwater supplies as well as improved harvesting of rainfall.
>
> 2. *The increased risk of drought severity presents a global-scale challenge to food security.* The 2020s will bring significant increases in drought severity for northern parts of China and India, Afghanistan, Mongolia and Pakistan. This presents a global risk to food security, as China and India are the world's largest food producers. China is currently one of the best placed to adapt effectively, but other countries may suffer without effective adaptation strategies. India is forecast to get slightly wetter, but is predicted to have low adaptive capacity, putting its large harvests of wheat and maize under continued, but not increasing, threat.
>
> 3. *Projected adaptive capacity is driven by local and regional characteristics.* The ability to adapt to droughts in the very near future depends on local contexts. Regions with the greatest reductions in adaptive capacity from 1990–2005 to the 2020s are those with authoritarian regimes and/or arid ecosystems. Adaptive capacity is projected to be relatively strong in China and relatively weak in India, Afghanistan and Western Russia. Additional fertiliser use in tropical and arid countries may buffer wheat yields from drought, while high rural populations in temperate climates may help to buffer maize yields against drought through labour-intensive adaption strategies. Forster *et al.* (2012) found no adaptive capacity relationship for rice. They suggest that this is possibly because much of the world's rice is irrigated, uses improved varieties and benefits from fertiliser inputs, and is therefore not as affected by changes in soil moisture as are wheat or maize.

approaches (e.g. on climate change effects) and social science approaches (e.g. on adaptation capacity) to determine risks to food security. Such interdisciplinary approaches are increasingly necessary to tackle global and regional water resource issues.

> **REFLECTIVE QUESTION**
>
> Why do some people state that there is a global water crisis?

F SUMMARY

Water is a unique substance having important physical and chemical properties, brought about by its covalent bonds and its polar hydrogen bonds, which enable it to act as an excellent solvent, to weather and erode the landscape and to regulate climate. Water is necessary for photosynthesis and for the survival of all forms of life. Water has been pivotal to the development of civilisations through its use in developing societal power relations formed by the onset of agriculture and developed over the past 13,000 years. Control of water resources enabled agriculture, which allowed non-farmers to specialise in other activities, some of which could be traded for food. In modern society, as in ancient civilisations, water has economic, religious and symbolic significance. Today the main use of water by humans worldwide, by volume, is in agriculture and the processing of agricultural products. Hence a focus on agricultural water efficiency is crucial for the future as the world's population grows and places enormous pressure on global water resources. There are already 45 countries that have less than 1000 m^3 of renewable water resource available per person per year. This water is needed for agriculture, industry and domestic use. It is anticipated that many more countries will reach this water 'stress' threshold in the coming years as the water crisis hits hard. These include large, populous countries in Asia such as India and China. Around 44 countries rely on other countries to provide more than half of their water supply. Thus water security for individual nations is of great importance and typically requires a balance of technological, socio-economic and political solutions. Interdisciplinary water research is therefore necessary to enable us to reduce the impacts of the global water crisis.

FURTHER READING

Hassan, F.A. 2012. *Water management and early civilizations: from cooperation to conflict*. Report prepared for the UNESCO–Green Cross International project From Potential Conflict to Cooperation Potential (PCCP): Water for Peace [online]. Available from: http://tinyurl.com/6tebpg5.

An interesting essay on the development of civilisations and water cooperation and water conflict.

Molden, D. (ed.). 2007. *Water for food, water for life*. Earthscan; London and International Water Management Institute; Colombo.

An excellent resource full of useful facts, theory and necessary adaptation strategies for the future.

Shiklomanov, I. 1998. *World water resources*. UNESCO; Milton Keynes.

Important text demonstrating water resource inequalities.

Waley, P. and Åberg, E.U. 2011. Finding space for flowing water in Japan's densely populated landscapes. *Environment and Planning A* 43: 2321–2336.

This is a useful article that illustrates some interesting tensions between symbolic and cultural identity and pragmatic protection of people from water threats. It provides more information relevant to **Box 1.1**.

Classic papers

On the properties of water:

Bernal, J.D. and Fowler, R.H. 1933. A theory of water and ionic solution, with particular reference to hydrogen and hydroxyl ions. *Journal of Chemical Physics* 1: 515–549.

Pople, J.A. 1951. Molecular association in liquids. II. A theory of the structure of water. *Proceedings of the Royal Society of London A* 205: 163–178.

On global water resources:

Vörösmarty, C.J., Green, P., Salisbury, J. and Lammers, R.B. 2000. Global water resources: vulnerability from climate change and population growth, *Science* 289: 284–288.

PROJECT IDEAS

- Look for reports and data on the Internet that report past or predict future water use for different countries over time. Try to partition the relative roles of climate change, population growth and development on changing water use for some selected countries.

- Produce a list of agricultural measures to reduce water abstraction and consumption for different environmental conditions or crops and evaluate what might be the most effective measures in different regions.

REFERENCES

Bernal, J.D. and Fowler, R.H. 1933. A theory of water and ionic solution, with particular reference to hydrogen and hydroxyl ions. *Journal of Chemical Physics* 1: 515–549.

Berner, E.K. and Berner, R.A. 1987. *Global water cycle: Geochemistry and environment.* Prentice Hall; Englewood Cliffs, NJ.

Blockley, S.P.E. and Pinhasi, R. 2011. A revised chronology for the adoption of agriculture in the Southern Levant and the role of Lateglacial climatic change. *Quaternary Science Reviews* 30: 98–108.

Falkenmark, M. 1986. Fresh water – time for a modified approach. *Ambio* 15: 192–200.

FAO. 2003. *Review of world water resources by country.* Food and Agriculture Organisation of the United Nations; Rome.

Forster, P., Jackson, L., Lorenz, S., Simelton, E., Fraser, E. and Bahadur, K. 2012. *Near future drought and food security in Asia.* Centre for Low Carbon Futures; York.

Garg, N.K. and Hassan, Q. 2007. Alarming scarcity of water in India. *Current Science* 93: 932–937.

Ghali, E. 2010. *Corrosion resistance of aluminium and magnesium alloys: Understanding, performance and testing.* John Wiley and Sons; New York.

Hassan, F.A. 2012. *Water management and early civilizations: from cooperation to conflict.* Report prepared for the UNESCO–Green Cross International project From Potential Conflict to Co-operation Potential (PCCP): Water for Peace [online]. Available from: http://tinyurl.com/6tebpg5.

Hoekstra, A.Y. and Chapagain, A.K. 2008. *Globalization of water: sharing the planet's freshwater resources.* Blackwell Publishing; Oxford.

IPCC. 2007. *Climate Change 2007: Impacts, Adaptation and Vulnerability. Contribution of Working Group II to the Fourth Assessment Report of the Intergovernmental Panel on Climate Change.* Parry, M.L. et al. (eds), Cambridge University Press; Cambridge.

Lawson, I.T. 2012. The Holocene. *In:* Holden, J. (ed.), *An introduction to physical geography and the environment (3rd edition).* Pearson Education; Harlow, 670–699.

McClatchey, J. 2012. Regional and local climates. *In:* Holden, J. (ed.), *An introduction to physical geography and the environment (3rd edition).* Pearson Education; Harlow, 157–182.

Molden, D. et al. 2007. Pathways for increasing agricultural water productivity. *In:* Molden, D. (ed.), *Water for food, water for life: a comprehensive assessment of water management in agriculture.* Earthscan; London and International Water Management Institute; Colombo, 279–310.

Narasimhan, T. 2008. A note on India's water budget and evapotranspiration. *Journal of Earth System Science* 117: 237–240.

NCIWRD (National Commission for Integrated Water Resource Development). 1999. *Integrated Water Resource Development: A Plan for Action. Report of the National Commission for Integrated Water Resource Development, Volume I.* Ministry of Water Resources, Government of India; New Delhi.

Palmer, M.A. and Bernhardt, E.S. 2006. Hydroecology and river restoration: ripe for research and synthesis. *Water Resources Research* 42: doi 10.1029/2005 WR004354.

Pople, J.A. 1951. Molecular association in liquids. II. A theory of the structure of water. *Proceedings of the Royal Society of London A* 205: 163–178.

Richardson, M. 2010. *The coming water crisis in Asia.* Institute of Southeast Asian Studies; Singapore.

Shiklomanov, I. 1998. *World water resources.* UNESCO; Milton Keynes.

Shinmura, J. 1995. 'Machi-zukuri, hito-zukuri to kasen' (Community planning, people planning and rivers). *In*: Ouchi, T., Takahashi, Y. and Shinmura, J. (eds), *Ryu-iki no jidai: mori to kawa no fukken o mezashite (The era of river basins: aiming for a rehabilitation of forests and rivers).* Gyo-sei; Tokyo.

UNESCO. 2003. *Water for people, water for life.* UNESCO; Paris.

UNESCO. 2009. *World Water Development Report 3: Water in a changing world.* UNESCO; Paris.

Vörösmarty, C.J., Green, P., Salisbury, J. and Lammers, R.B. 2000. Global water resources: vulnerability from climate change and population growth. *Science* 289: 284–288.

Waley, P. 2000. Following the flow of Japan's river culture. *Japan Forum* 12: 199–217.

Waley, P. and Åberg, E.U. 2011. Finding space for flowing water in Japan's densely populated landscapes. *Environment and Planning A* 43: 2321–2336.

CHAPTER TWO

The changing water cycle

Stephen Sitch and Frances Drake

> **LEARNING OUTCOMES**
>
> After reading this chapter you should be able to:
>
> ■ understand the global water cycle and its main components
> ■ describe each component of the water cycle and how it has changed over recent decades, and how it is projected to change during this century
> ■ appreciate the importance of climate variability and extremes on the water cycle, and for the provision of freshwater
> ■ describe the main uncertainties in the present-day and future global water cycle.

A INTRODUCTION

This chapter introduces the main components of the global water cycle, describes the main stores and fluxes, and residence times, and presents the underlying physical processes. A discussion on uncertainties in the contemporary water cycle is included. The impacts of climate variability and extremes on the water cycle will cover different time and spatial scales, drawing on evidence from the past as indicative for possible future changes. Specific examples will be provided for North America, the UK, and West Africa. The impact of human modification of the water cycle will be discussed, including a direct role through land use, and indirectly through a modification of atmospheric composition and climate as a result of anthropogenic emissions of **greenhouse gases** and **aerosols**. Finally, this chapter outlines major challenges and unknowns in the global water cycle under the spectre of global environmental change.

B THE GLOBAL WATER CYCLE

The global water cycle is shown in **Figure 2.1**. Approximately 97% is saline water in sea and oceans (1,338,000 thousand km^3), leaving only 3% as freshwater. Approximately half of the freshwater is stored in glaciers and snow (excluding Antarctica, 24,064 thousand km^3), and a

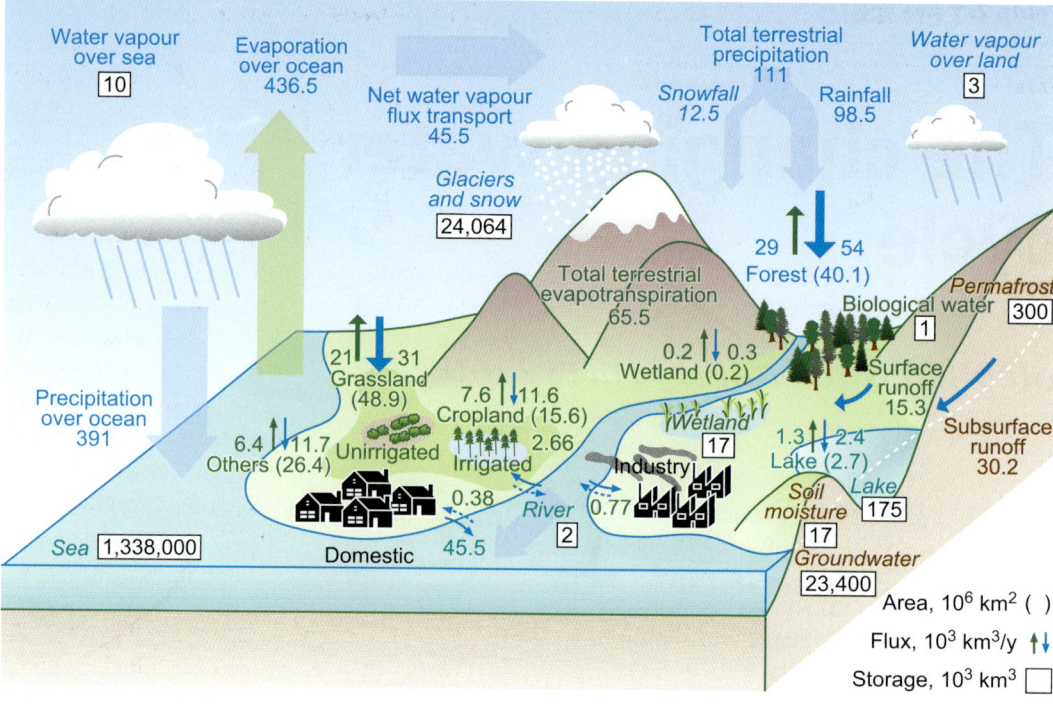

Figure 2.1 The global hydrological cycle, with estimated flows and storage. The terrestrial water balance does not include Antarctica. Big vertical arrows show total annual precipitation and evapotranspiration over land and ocean (thousand km³ yr⁻¹), which include annual precipitation and evapotranspiration in major landscapes (thousand km³ yr⁻¹) presented by small vertical arrows; brackets indicate area (million km²). The direct groundwater discharge, which is estimated to be about 10% of total river discharge globally, is included in river discharge.
Source: Oki, T. and Kanae, S. 2006. Global hydrological cycles and world water resources. *Science* 313: 1068–1072. Reprinted with permission from AAAS.

substantial (although highly uncertain) amount in groundwater (23,400 thousand km³).

It is important to consider not only the size of each storage zone (or reservoir) but the exchange or flux of water between reservoirs; after all, it is the exchange of water between reservoirs that forms the global water cycle. Both reservoir size and residence time need to be considered. Average residence times can be computed as the reservoir size divided by the flux (**Table 2.1**). For example, water vapour stays in the atmosphere on average for only nine days, and plays an active role in the water cycle. Note that only a small fraction of the total water is in fact 'available' to drive the water cycle.

1 Evaporation, condensation and transpiration

Water is transferred into the atmosphere through evaporation over the ocean (436.5 thousand km³ per year) and **evapotranspiration** over land (65.5 thousand km³ per year). Evaporation is the transformation of water from the liquid to gas phase directly above water bodies, from the soil surface, or from plant canopy **interception** (rainfall intercepted and subsequently evaporated off leaves). Imagine a half-filled jar of water where the water is initially separated from the air. If we then remove the separation device within the jar, evaporation will take place into the air within the

Table 2.1 Average residence time in major water stores

Reservoir	Average residence time
Oceans	1500 years
Glaciers	1000 years
Soil moisture	6 months
Deep groundwater	10,000 years
Shallow groundwater	200 years
Lakes	50 years
Rivers	14 days
Atmosphere	9 days

2 Linking water and energy

The water cycle is integrally linked to the global energy budget. Water vapour is a greenhouse gas and modulates the radiative cooling of the atmosphere (i.e. the absorption or emission of **infrared radiation**; Allan, 2011). In fact water vapour contributes 60% of the **natural greenhouse effect** (Kiehl and Trenberth, 1997). The tropics receive a surplus and high latitudes a deficit in solar energy over the annual cycle. Energy and its redistribution due to these latitudinal energy imbalances drive atmospheric and oceanic circulation, and thus climate and weather. In other words, warmth from the equator is redistributed polewards. This redistribution is not in a straight northerly or southerly direction as air and water are fluids and the Earth's rotation (via the **Coriolis effect**) produces circulation effects in them. Approximately 60% of the energy imbalance is redistributed through atmospheric circulation and 40% via ocean circulation (Stevens, 2011).

During evaporation of water, energy is absorbed in the phase change from liquid to gas. During the reverse process, condensation, energy is released in the phase change from gas to liquid. This is known as the **latent heat of vaporisation**, with 2260 kJ kg^{-1} of energy released or absorbed during a phase change without a change in temperature (see Chapter 1). In this respect, water vapour acts as a global 'air-conditioner', acting to cool the planet's surface (Trenberth *et al.*, 2003), because energy is consumed in evapotranspiration (i.e. changing the bonds between molecules) rather than in raising surface temperatures (i.e. the kinetic energy of molecules). Similarly, we sweat to cool down during hot summer days. As a parcel of air rises in altitude, it cools as it expands (**adiabatic cooling**). Assuming no change in the total amount of water vapour in the parcel as the air parcel rises and cools, it will eventually reach its **dew-point temperature** (which represents cloud base) when the air parcel becomes saturated with water vapour (i.e. 100% relative humidity;

jar, while **condensation** will also occur. Initially, because the amount of water vapour in the air within the jar is low, the evaporation rate will exceed the condensation rate, but eventually the two will balance. This equilibrium state is called saturation (Aguado and Burt, 1999). The total pressure exerted by the atmosphere is the sum of the pressures exerted by each component gas of the atmosphere. These pressures are called the **partial pressures** of the individual gases. At any given temperature there is a limit to the density of water in the air and thus a limit on the partial pressure, often referred to in meteorological textbooks as the vapour pressure (e) that water vapour can exert. This limit is known as the **saturation vapour pressure** (e_s). If you try to add more water vapour to an atmosphere that is already saturated at that temperature, then the condensation rate exceeds the evaporation rate. Hence more water droplets will appear in the atmosphere and be able to form cloud water droplets. On the large spatial scale (10+ km^2), evaporation over land is predominantly a radiation-driven process (Priestley and Taylor, 1972). Transpiration represents the flux of water from the soil to the atmosphere, passing through plants. Water is lost to the atmosphere as water vapour through small pores in the leaves called **stomata**.

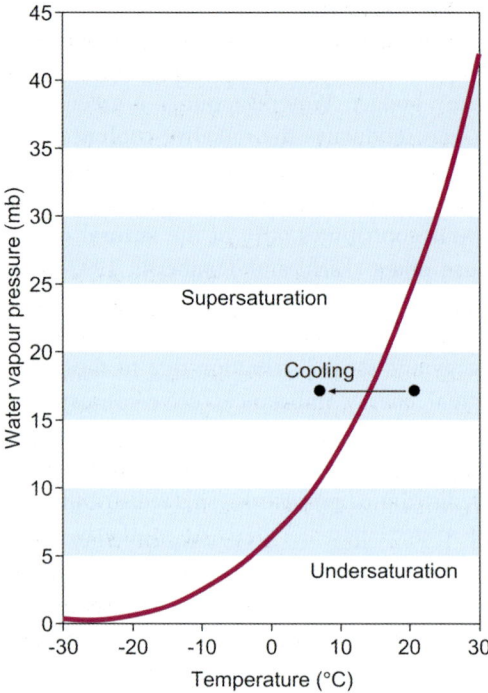

Figure 2.2 The Clausius–Clapeyron (CC) relation whereby saturated water vapour pressure (the pressure exerted by water vapour molecules when the air is saturated with water vapour) increases almost exponentially with temperature. Included is an example whereby a parcel of air at a given temperature and water vapour pressure can go from being undersaturated to supersaturated with cooling.

the Clausius–Clapeyron relationship describes the relationship between saturation vapour pressure of a gas and temperature, see **Figure 2.2**). At this point, water vapour condenses out, changing from the gaseous to the liquid phase, releasing energy to the atmosphere. The same process is responsible for surface dew on some cold winter mornings. Overnight the surface cools, especially strongly during cloudless nights, where energy can readily escape into space. As surface temperature drops the air may become saturated, and dew be formed. In the atmosphere, condensation leads to cloud water droplets, which themselves are not large enough to form **precipitation**. There are a number of physical processes which allow the cloud water (or ice) droplets to grow to precipitation size, as discussed below.

3 Precipitation

The majority of the water evaporated over oceans is precipitated back into the oceans, which is unsurprising given that 71% of the Earth's surface is covered by ocean. The distribution of precipitation also helps to partly explain **salinity** variations in the ocean. Transportation of water in the atmosphere is intimately linked to general atmospheric circulation. Approximately 10% (45.5 thousand km^3 per year) of the evaporated water over oceans is transported and precipitated over land. An equal amount is returned to the oceans via river runoff (an equal amount is required in a balanced system). This net water vapour transport from ocean to land represents approximately 40% of the total precipitation over land (111 thousand km^3 per year), with the remaining 60% originating from land surface evapotranspiration (resulting in convective precipitation).

Precipitation refers to all liquid and frozen forms of water: rain, snow, hail, dew, hoar-frost, fog-drip and rime, but only the first two make significant contributions to precipitation totals. Precipitation usually occurs in connection with three modes of uplift of air – **convective**, **cyclonic** (depression) and **orographic** (air forced over mountains). Condensation occurs easily in the air because of **hygroscopic aerosols**, such as dust, smoke, sulphur dioxide and salt. These hygroscopic aerosols are water soluble and this allow condensation to take place before the air is saturated. Small cloud droplets grow faster than large ones. As droplets get larger the growth by condensation lessens. Cloud droplets are typically less than 1 to 50μm in size and need to grow to 100μm (i.e. 100 times larger) if they are to fall as precipitation. The process of condensation alone is not sufficient to account for the growth of cloud droplets to precipitation size. The processes by which cloud droplets grow to precipitation size depend on the temperature of the cloud. For simplicity, we can distinguish three types of cloud as described below.

3.1 Warm clouds

In warm clouds, where the temperature is above 0°C (cloud top temperature less than −15°C), all the cloud droplets will be water. To reach a precipitation size the cloud droplet would have to continue growing by condensation for several hours, and most precipitating clouds have a lifetime far less than this. However, as the droplets grow, different droplet sizes evolve. The larger droplets will fall faster than the small droplets. As the large droplets (greater than 40μm) fall they will sweep the small droplets up and grow in size. This is the collision-coalescence process. The larger the droplet, the more efficient it is at collecting other droplets it encounters. Even with collision, growth occurs only if the two coalesce. This is more likely to happen if the two droplets vary greatly in size. As they get larger, droplet fall speed increases, sweeping out a larger volume of the cloud. In this case the growth rate increases with the size of droplet and precipitation-size droplets are quickly formed. The thicker the cloud (the longer the droplet has to collect other droplets), the larger the droplet grows.

3.2 Frozen clouds

If a cloud is below −40°C all the droplets will be frozen. Growth mainly occurs through **riming**, where supercooled water in the atmosphere is deposited onto the crystal as ice. Growth by this method is relatively rapid. In these 'glaciated clouds' crystal growth can occur by a process analogous to the collision-coalescence process, called aggregation, but growth by this method is slow. These glaciated clouds are so high in the atmosphere that the precipitation rarely reaches the ground.

3.3 Mixed clouds

In clouds between −40 and 0°C there will be a mixture of frozen and water cloud droplets. In these types of clouds the Bergeron–Findeisen process takes place. At temperatures below 0°C the saturation vapour pressure with respect to water is greater than that with respect to ice (**Figure 2.3**). Note in **Figure 2.3** the difference in saturated vapour pressure for supercooled liquid water and ice. A gas will **advect** from high pressure to low pressure (i.e. there will be a vapour flux along the pressure gradient from close to the supercooled water droplet towards the ice particle). Therefore, in a mixed cloud dominated by supercooled water droplets, the air, which is close to saturation with water, is supersaturated with respect to ice. Ice crystals therefore grow faster than water droplets. They may grow so much that the air becomes sub-saturated with respect to water and the remaining water droplets evaporate. The ice crystals then grow by **sublimation**, where water vapour is directly deposited onto the ice crystal as a solid, missing the liquid

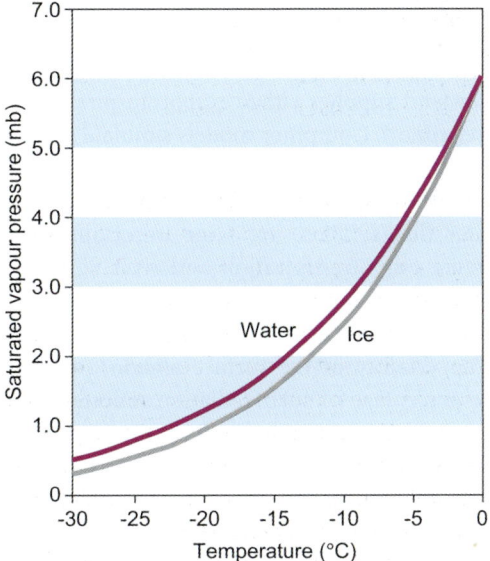

Figure 2.3 Saturated pressure–temperature curves for water and ice.

phase. This is the Bergeron–Findeisen process. To see this process in action, look at http://www.youtube.com/watch?v=-UXpJ3KRO_s. The ice crystals can grow by several means as they fall to the lower, warmer parts of the cloud by aggregation with other ice crystals and collision-coalescence with water droplets.

Clouds do not fall neatly into warm and cold types, and a mixture of both processes may occur within a cloud. The Bergeron–Findeisen process is believed to be the dominant process in mid-latitude precipitating clouds (Jonas, 1994). The type and size of precipitation that falls out of the cloud is determined by the cloud type. However, the precipitation type to reach the ground will be affected by the temperature structure of the air beneath the cloud.

4 Uncertainties in the global water cycle

Rather surprisingly, there are large uncertainties even in the main components of the global water cycle. For example, based on seven global datasets, mean annual land precipitation estimates vary between 96 and 118,000 km^3 per year^{-1} for the period 1979–1999 (Biemans *et al.*, 2009). Assuming no change in water stores, precipitation over land supplies either evapotranspiration or river runoff. Computer models simulate the key ecohydrological processes on land (see e.g. Gerten *et al.*, 2004) and are used to quantify land-based water fluxes. There are large uncertainties in annual evapotranspiration and river runoff of 17% and 20–25%, respectively (Haddeland *et al.*, 2011; Trenberth *et al.*, 2009). Furthermore, there is uncertainty in the surface **albedo**, which is the percentage of solar radiation reflected back into space. Snow and ice typically have high albedo (up to 90% for fresh snow, compared to only 17% for soil) and reflect a large portion of incoming solar radiation. As climate warms, spring snowmelt at high latitudes occurs earlier in springtime, and consequently less radiation is reflected back into space, thus further enhancing surface warming. This is known as the *snow-ice-albedo feedback* and is the main reason why warming is expected to be most pronounced at high latitudes, where snow is prevalent.

> **REFLECTIVE QUESTION**
>
> What are the different processes that can result in rainfall?

C CLIMATE VARIABILITY AND CHANGE

The earlier section described the average global water cycle. However, the water cycle varies both spatially and temporally, as we are all well aware when we witness floods and droughts. The Intergovernmental Panel for Climate Change defines climate variability as 'variations in the mean state and other statistics (such as standard deviations, the occurrence of extremes, etc.) of the *climate* on all spatial and temporal scales beyond that of individual weather events'. An extreme weather event is an event that is 'as rare as or rarer than the 10th or 90th percentile of the observed probability density function' (IPCC, 2007a). Climate extreme events occur when 'a pattern of extreme weather persists for some time, such as a season, especially if it yields an average or total that is itself extreme (e.g., drought or heavy rainfall over a season)' (IPCC, 2001). Here we will give examples of climate variability at multiple scales, relevant to the global water cycle, including extreme climate events and variability at multi-annual and multi-decadal scales.

1 Extremes

The summer heatwave of 2003 across western Europe was an extreme climate event. Areas experienced elevated temperatures of 3.5°C and below-average precipitation. Across Europe,

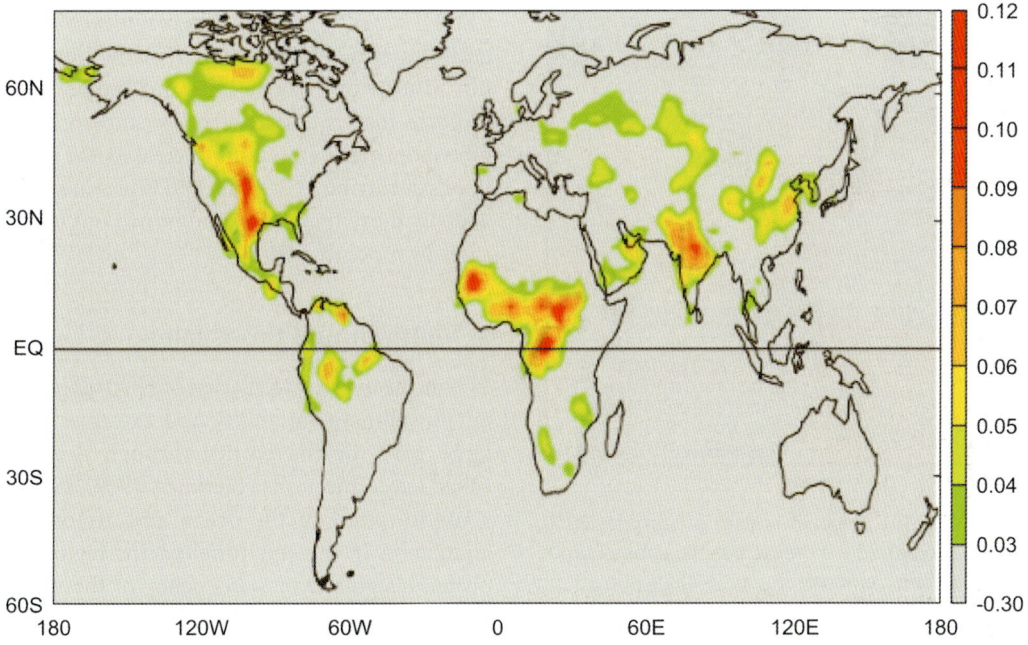

Figure 2.4 Land–atmosphere coupling strength in June, July and August. The units are dimensionless and just provide a relative picture of where coupling is greatest.
Source: Adapted from Koster, R.D. et al. 2004. Regions of strong coupling between soil moisture and precipitation. *Science* 305: 1138–1140. Reprinted with permission from AAAS.

30,000 to 50,000 deaths were attributed to this heatwave (Fedoroff *et al.*, 2010), **gross primary productivity** declined by 30% (Ciais *et al.*, 2005), and crop yields of grains and fruits were reduced by 20–36% (Fedoroff *et al.*, 2010). By studying the conditions surrounding the heatwave we can learn about feedbacks that make extreme events more extreme. Reduced evapotranspiration, due to the dry land surface conditions, prevented surface cooling, and thus even higher temperatures resulted; a positive feedback. Teuling *et al.* (2010) showed how initial heating was suppressed more over grasslands than forests because of enhanced evaporation; however, the more conservative use of soil moisture by forests mitigates extreme heat over the longer term. It is also important to consider the effect of soil moisture on precipitation. **Figure 2.4** shows the land–atmosphere coupling strength for the northern summer. Land–atmosphere coupling is particularly strong across the Great Plains, Sahel and South Asia (Koster *et al.*, 2004). Where such coupling occurs it is likely that when there is low soil moisture there is little supply to the atmosphere and dry conditions are enhanced. Such feedbacks may exacerbate (intensify and/or prolong) extreme conditions. Both soil moisture–temperature, and soil moisture–precipitation feedbacks are likely to be more important in Europe under future climate change (Seneviratne *et al.*, 2006).

Extreme events occurred in the summer of 2010 across central Europe, with drought, leading to fires, especially in peatlands, with smoke leading to a deterioration of air quality in places such as Moscow. Concurrently there were catastrophic floods across Pakistan. Both were associated with the same breakdown in the normal

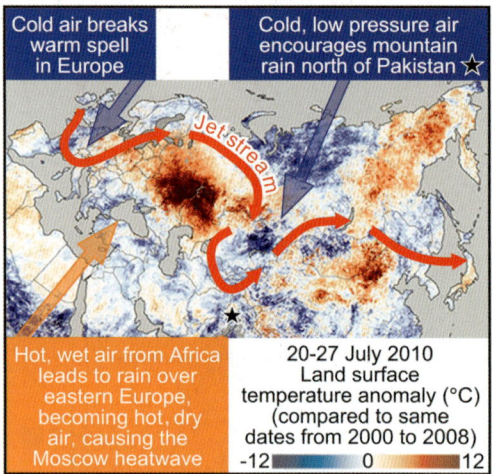

Figure 2.5 The static jet stream of summer 2010, which led to different extreme events across Europe and Asia. The temperature anomalies shown on the map are from 20–27 July 2010 compared to temperatures for the same dates from 2000 to 2008. The anomalies are based on land surface temperatures observed by the Moderate Resolution Imaging Spectroradiometer (MODIS) on NASA's Terra satellite. Areas with above-average temperatures appear in red and orange, and areas with below-average temperatures appear in shades of blue.

Source: From NASA/Earth Observatory; NASA image has been altered.

flow of the **jet stream** (high-velocity winds at the top of the lower atmosphere), as explained within **Figure 2.5**. Patterns in the upper **troposphere** can be important for determining extreme events and variability from year to year, as described in **Box 2.1**.

Modelling results provide evidence for a human contribution to more intense precipitation extremes over the northern hemisphere during the last 50 years (Min *et al.*, 2011), and **probabilistic techniques** are now used to attribute a contribution of anthropogenic climate change to increased flood risk, for example, during the UK floods in autumn 2000 (Pall *et al.*, 2011; Allan, 2011). For example, from results based on climate model simulations, Pall *et al.* (2011) stated: 'The precise magnitude of the anthropogenic contribution remains uncertain, but in nine out of ten cases our model results indicate that twentieth-century anthropogenic greenhouse gas emissions increased the risk of floods occurring in England and Wales in autumn 2000 by more than 20%, and in two out of three cases by more than 90%.'

2 Multi-annual timescales

El Niño Southern Oscillation (ENSO) is important for climate variability and extremes across large parts of the world, through a process called teleconnections (Bjerknes, 1969). ENSO is a modification of the ocean and atmospheric circulation in the equatorial pacific region, and occurs irregularly at a frequency of three to six years and with varying intensity.

In normal conditions (**Figure 2.7a**), warm water stacks up in the western Pacific, maintained by the easterly (from the east) surface trade winds. The warm water evaporates, producing the large convective rainfall systems over the western Pacific. The waters in the east are cool through evaporation and upwelling of cold, nutrient-rich water from below. During El Niño conditions the easterly winds slow and the body of warm water is shifted eastwards. Rainfall shifts eastwards, leaving the western Pacific region drier than normal, and heavy rainfall occurs over Peru. Upwelling of the cold, nutrient-rich water near the eastern Pacific coast weakens, and leads to drastic reductions in the anchovy populations (and thus harvest); the name, El Niño, is Spanish for 'the boy-child', and the term given by Peruvian fishermen, in reference to the Christ child, as this phenomenon starts around Christmas time. La Niña is essentially a shift in the ocean atmosphere circulation in the opposite direction, with stronger trade winds, a greater buildup of warm waters in the western Pacific, with heavier than normal rains in the western Pacific region.

> **BOX 2.1 CASE STUDIES**

Hot, dry weather in UK summer 2006, floods in UK summer 2007

Summer 2007 was one of the wettest UK summers on record (Mayes, 2008; Figure 2.6). Summer 2007 had average temperatures, while summer 2006 was one of the hottest on record for the UK, with below average rainfall (Prior and Beswick, 2007). Why such large interannual variability?

If we look at an average surface pressure chart for the mid-latitudes we see areas of low pressure. Travelling depression or low pressure systems dominate mid-latitude weather and are responsible for most frontal precipitation. Depression systems at mid-latitudes are associated with the convergence of air masses with different characteristics; warm and cold air. The two air masses do not easily mix and the boundary between them is called a front. The warmer, less dense air tends to rise over the cold dense air. As the warm air rises it adiabatically cools, leading to cloud formation, which can then result in precipitation associated with the front. In contrast, high-pressure systems result in dry, calm conditions. In winter this can lead to low cloud or fog but can also provide clear blue skies and cold, frosty weather. In the summer, high pressure leads to dry, bright and warm sunshine (Young, 1994a and b).

If we look at an average pressure chart for the top of the tropopause (top of the troposphere, which is the lower layer of the Earth's atmosphere) at mid-latitudes we do not see areas of low and high pressure; instead there are undulating waves. These are called Rossby waves and cause the troughs and ridges in the fast-flowing upper westerlies (jet streams are areas of maximum velocity) near the tropopause. The formation and direction of depression systems are controlled by these Rossby waves. The formation of depression systems is associated with troughs and high-pressure systems with the ridges of Rossby waves. Rossby waves are anchored by the Rocky Mountains. If the UK is stuck under a trough, as in summer 2007, then the UK can receive continuous depression systems. There is a complex relationship between the upper air flow, jet streams, and surface conditions such as temperature, which all affect the position and track of frontal depression systems and, consequently, the precipitation. Under a ridge, the UK receives high-pressure systems. Rossby waves go through a cycle over 20 to 60 days from a smooth zonal flow to undulating waves with pronounced troughs and ridges. The strength of the westerlies is measured by an index. A high index is associated with the strong zonal flow. Some of the warmest summers occur when the Rossby wave index is at its lowest and the waves are most pronounced. At this point cut-off can occur and a blocking anti-cyclone formed. This type of slow-moving high-pressure system leads to exceptionally warm and dry summers when they get stuck over the UK, as in summer 2003 or 2006 (Ogi et al., 2005).

The summer of 2012 saw very wet conditions in the UK and drought over large parts of the central USA. A useful explanation with diagrams of the jet stream position can be found at: http://tinyurl.com/8223wwz. An excellent video explaining jet streams can be found at: http://tinyurl.com/b4nxuvg.

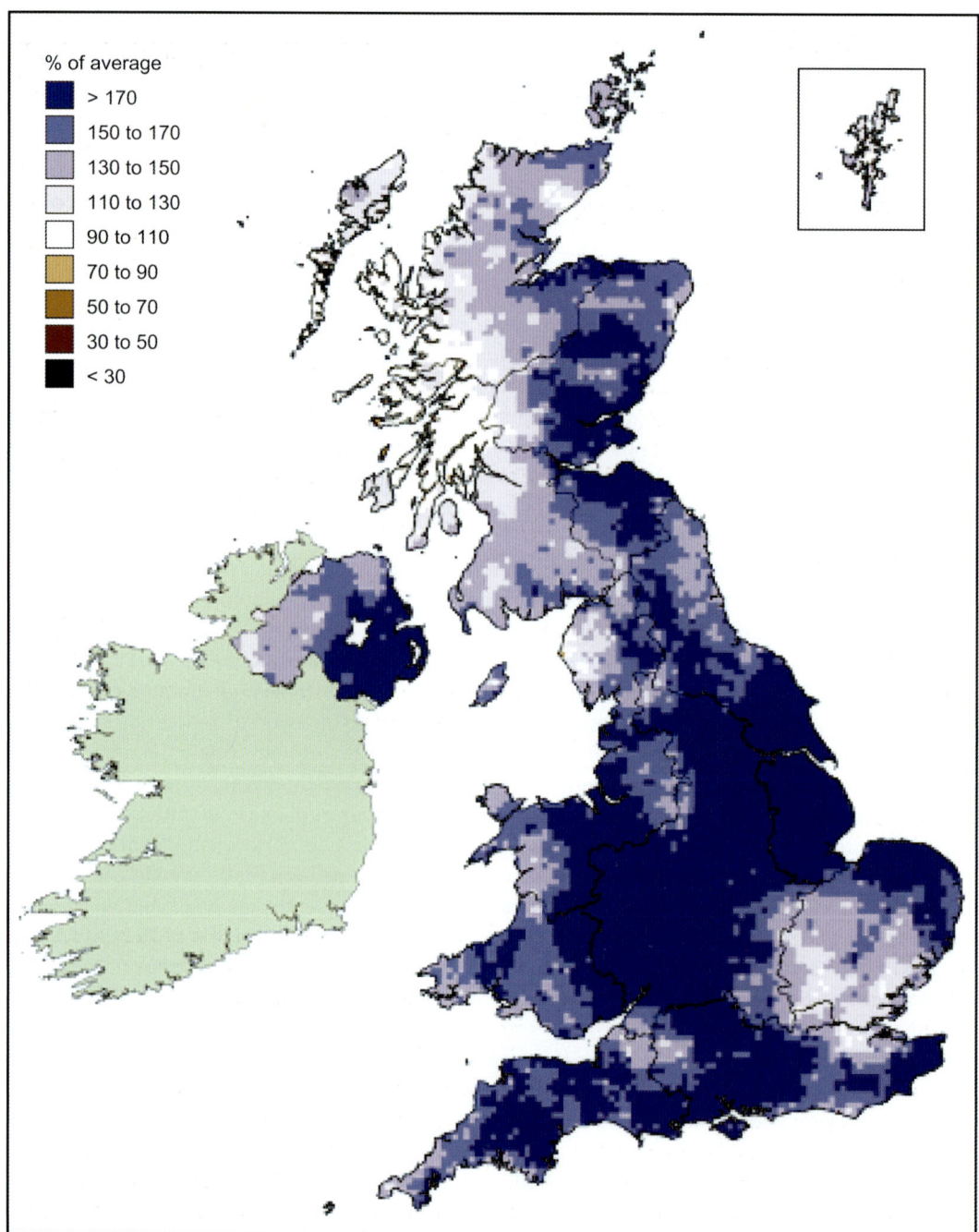

Figure 2.6 Summer 2007 rainfall amount as a percentage of the 1971–2000 average.
Source: From UK Met Office; the figure contains public sector information licensed under the Open Government Licence v1.0.

THE CHANGING WATER CYCLE

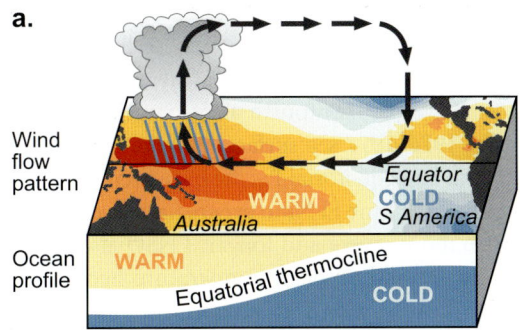

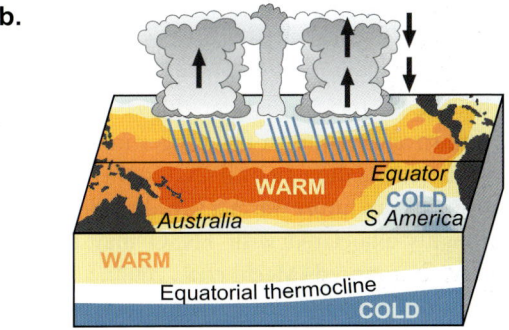

Figure 2.7 Conditions in the Pacific during (a) normal periods and (b) El Niño. During normal conditions strong trade winds keep warm water near New Guinea and Australia. During El Niño the trade winds relax along with the temperature gradient of the ocean waters and hence warm surface waters in the west can flow eastwards.

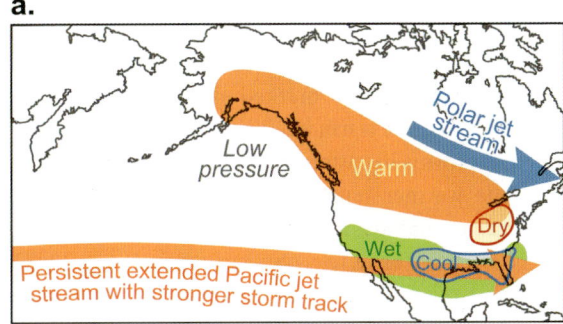

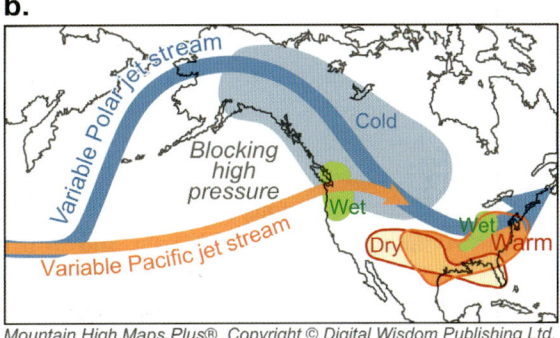

Figure 2.8 Winter weather anomalies during (a) El Niño and (b) La Niña over North America.

These periodic large-scale changes in ocean-atmosphere circulation have important ramifications for climate and the water cycle across the globe. Typically, global land precipitation is anomalously high and low during La Niña and El Niño periods, respectively, and ENSO is responsible for regional extremes, in terms of both flooding and drought. ENSO teleconnections have been well studied (Moron and Ward, 1998). ENSO has been shown to affect the sub-tropical Pacific basin, eastern and southern Africa. Indonesia, Australia, India and south-east Africa all experience drought (leading to potential crop failure and famine) during El Niño. During El Niño events the Amazon dries, forest fires are more prevalent in Indonesia, floods and landslides occur across Peru and southern California. Reverse weather conditions occur during La Niña events.

There are very clear changes in the North American seasonal weather pattern associated with ENSO due to displacement of the jet streams, leading to unusual weather outside of the tropics (Coghlan, 2002). The greatest change occurs during the winter months (**Figure 2.8**). During El Niño the USA experiences wet and windy weather along its west coast and southern states. California was particularly hit by flooding during the 1982/83 El Niño. However, El Niño events are linked to lowered hurricane activity in the

Gulf of Mexico (Pielke Jr and Landsea, 1999). The northern states experience mild winters and below normal precipitation during El Niño. This was particularly marked during the 1997/98 El Niño. The warm weather, paradoxically, caused ice storms in eastern Canada in January 1998, as the warmer temperatures led to freezing rain rather than the normal snowfall.

The clearest teleconnection is between ENSO and the Indian monsoon. Typically, during El Niño years the monsoon rains in the northern hemisphere summer are weakened and there is less precipitation, particularly in north-western India. As a consequence crops are adversely affected. There are various claims that El Niño affects the climate of northern Africa and Europe. Although climate anomalies can be shown to occur at these locations it is thought that the ENSO influence is small (Moron and Ward, 1998). The disturbance of upper air flow patterns may well lead to variations in stratospheric humidity and other chemical processes in the stratosphere.

Gergis and Fowler (2009) used proxy data to reproduce ENSO events over the centuries from 1525 and considered what the future might hold. Proxy data are measurements which can be used to reconstruct the climate, but in themselves do not measure climate variables such as temperature. A good example of proxy data are tree rings, which can be used in some circumstances to estimate past temperature and rainfall. There are many difficulties with using proxy data to form long-term data series, but used carefully they provide an invaluable record against which to compare present-day climate change. The ENSO reconstructed record revealed that although extreme ENSO events had occurred throughout the 478-year period, the twentieth century stood out for several reasons. The twentieth century contained around 43% of the most severe ENSOs and also 28% of the long-duration events (Gergis and Fowler, 2009). Furthermore, 30% of extreme ENSOs in the 478-year-long record have occurred since 1940 (Gergis and Fowler, 2009). The link between surface temperatures and the ENSO cycle means that climate change may well affect the occurrence of extreme and protracted ENSOs. The IPCC (2007a) fourth assessment noted that since 1976–1977 sea surface temperatures in the central and equatorial Pacific have increased and at the same time El Niño events have also tended to be more severe and longer. ENSO also influences the global mean temperature because of large heat exchanges between the ocean and atmosphere (IPCC, 2007a). Many of the warmest years on record have occurred in the last 25 years and several of these have had strong El Niño events, helping to increase the global mean temperature (Gergis and Fowler, 2009). If ENSO is more strongly coupled to changes in the global climate than is thought by IPCC (2007a), this may mean that in a future warmer world the floods and droughts associated with ENSO may become more pronounced (Meehl *et al.*, 2007).

Another example of variability on a multi-annual timescale is drought across western North America between 2000 and 2004, and the famous Great Plains 'Dust Bowl' of the 1930s (see **Box 2.2**). In the latter case, land management practices contributed to the severity of the drought.

3 Multi-decadal timescales and trends: long-term drought, CO_2 effects, river discharge, glacier melt

An example of multi-decadal variability is the drought occurring in the Sahel (see **Box 2.3**). Here too climate variations and human activities appear to play important roles in **desertification** of the region. Human activities include the impact of increasing population pressures on land and its management, through to the role of global dimming, as a result of increasing anthropogenic emissions of scattering aerosols, especially from the industrialized northern hemisphere countries. These aerosols scatter incoming solar radiation, sending a proportion back into space and reducing the incoming radiation reaching the Earth's

BOX 2.2 CASE STUDIES

Dust Bowl of 1930s North America

The Dust Bowl of the 1930s is an iconic period in US American history, forming the backdrop for the literary classic *The Grapes of Wrath* by John Steinbeck. A period of prolonged drought and dust storms devastated farming in the area known as the Great Plains, particularly in the states of Texas, Oklahoma, Kansas, Colorado and New Mexico (Figure 2.9).

By the 1930s the Great Plains had become subject to intensive agriculture, and the high price of wheat during the First World War had made the Southern Plains very prosperous. However, in 1931 the rains that had previously made farming possible failed, and by 1932 dust storms had begun to occur. In 1933 there were 38 huge dust storms. These led to hundreds of deaths from 'dust pneumonia', where dust infiltrates the lining of the lungs. It also caused millions of people to leave their homes, with many permanently emigrating to other states, particularly California. There were also severe economic consequences, with bank failures, large falls in land prices and an extension of the Great Depression.

Droughts are a common feature of the mid-latitudes, including those of North America. Observations of rainfall over the twentieth century show a great deal of variability, with long periods of both below and above average rainfall over the Great Plains (Garbrecht and Rossel, 2002). It has been suggested that it was anomalously high rainfall prior to the 1930s that encouraged settlers to believe that the climate was suitable for intensive agriculture. However, the observational record shows that during the first part of the twentieth century droughts tended to occur every 20 years. Longer records reconstructed from proxy data show evidence of severe droughts occurring in the Great Plains once or twice a century (Woodhouse and Overpeck, 1998). Particularly intense and prolonged

Figure 2.9 Severe conditions during the 1930s Dust Bowl of North America: (a) Dust piled up against a farm building, (b) a farmer digs in dust, Cimmaron County, Oklahoma.

Source: From Library of Congress, Prints & Photographs Division, FSA-OWI Collection, LC-DIG-fsa-8b38341 DLC (a) and LC-USZ62–131312 DLC (b).

continued

droughts occurred in the thirteenth and sixteenth centuries, with the former lasting for some 38 years (Woodhouse and Overpeck, 1998). It has been suggested that the epic droughts during the Medieval warm period (900–1300 AD) could provide a possible analogue for future droughts in the twenty-first century in western North America (Cook *et al.*, 2004; Woodhouse *et al.*, 2010).

What are the potential causes of these severe droughts, and the Dust Bowl in particular? Drier conditions tend to occur in the southwest and southeast USA during La Niña (Figure 2.9). This is when there are lower than average sea surface temperatures in the eastern tropical Pacific. La Niña occurs on an irregular basis, but typically every two to six years. The La Niña at the time of the Dust Bowl, however, was relatively small, given the warmth and intensity of the drought. Also the position of the low rainfall was farther north than might be expected. Typically, La Niña droughts occur in the southwest and Mexico. The drought was also much more widespread, extending as far north as Canada. Climate models have difficulty reproducing the features of the Dust Bowl by simply including La Niña and thus it is likely that there were other contributing factors (Cook *et al.*, 2008). Droughts in North America are also linked to anomalously warm sea surface temperatures (SSTs) in the north Atlantic and these occur on a decadal timescale. By including the SST anomalies for the Dust Bowl period in a climate model run, Schubert *et al.* (2004) were able to improve the rainfall results. They found that the tropical Pacific and tropical Atlantic SST anomalies were the most important contributing factors.

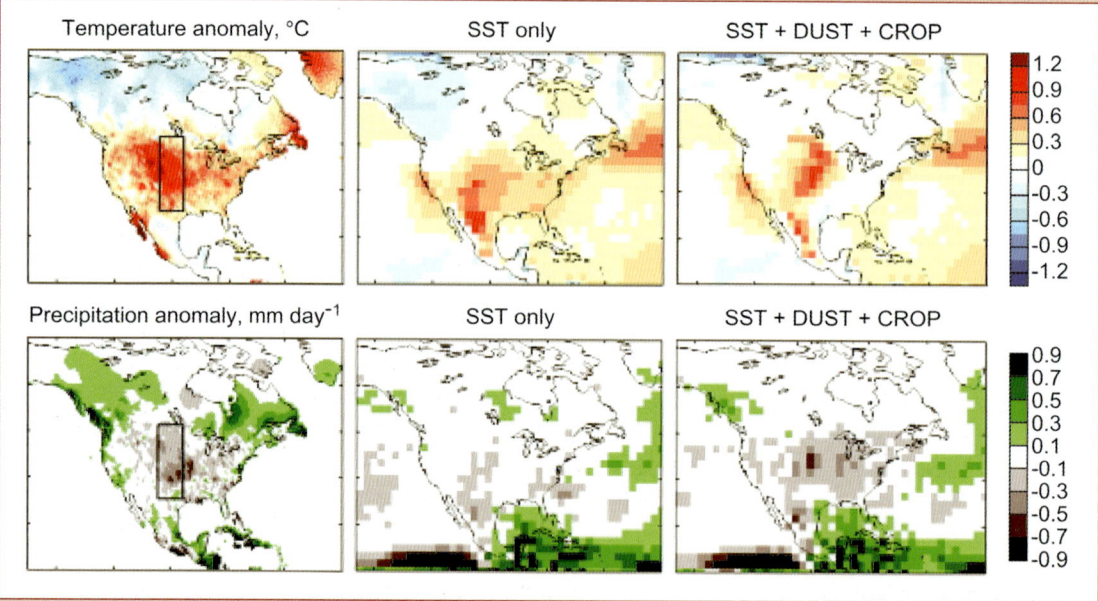

Figure 2.10 Temperature (°C) (top row) and precipitation (mm day^{-1}) (bottom row) anomalies for the Dust Bowl drought from the Climate Research Unit (CRU) and climate model simulations. The left panel shows actual data. The middle and right panels try to reproduce the actual data using model runs to investigate the causes of the Dust Bowl. The middle panel shows sea surface temperature effects only and the right panel shows sea surface temperature, dust aerosol plus crop removal effects combined.

Source: From Cook, B.I., Miller, R.L. and Seager R. 2009. Amplification of the North American 'Dust Bowl' drought through human-induced land degradation. *Proceedings of the National Academy of Sciences* 106: 4997–5001.

continued

> Schubert *et al.* (2004) also pointed out that seasonality is important. In an average year most precipitation falls on the Great Plains during spring and summer. Cold SSTs in the Pacific in the summer lead to changes on a global scale in the upper troposphere which suppress rainfall in the Great Plains. Warm SSTs in the Atlantic in summer and autumn produce anomalous cyclonic conditions at the surface which stop moisture from the Gulf of Mexico entering North America, further suppressing rainfall.
>
> A final factor which may help to explain why the Dust Bowl was so intense are the dust storms themselves. Prior to human settlement, the Great Plains had been dominated by grasses and other flora, which were suited to the area and able to withstand prolonged drought. Their roots helped to bind together the topsoil. The farmers planted wheat, which was not drought tolerant and was unable to hold the soil together. This change in plant cover, poor land use practice and the low soil moisture content following the lack of rainfall all helped to form dust storms (Cook *et al.*, 2008). Not only did the loss of vegetation lead to an increase in surface albedo, but the dust in the atmosphere blocked out the incoming solar radiation. This meant there was less energy available for evaporation into the atmosphere and therefore the potential for precipitation was reduced (i.e. soil moisture–precipitation feedbacks). By including the effect of dust together with the cold Pacific and warm Atlantic SST anomalies, Cook *et al.* (2008, 2009) were better able to reproduce the Dust Bowl with a climate model, suggesting that all these factors are important (**Figure 2.10**).

surface (reducing evapotranspiration). They also modify the temperature structure of the atmosphere, make brighter (more reflective) clouds, reduce the precipitation efficiency of clouds and increase their lifetime (Ramanathan *et al.*, 2001). Global dimming has been hypothesised to explain observed global trends in reduced pan-evaporation (evaporation from a pan filled with water – a widely used meteorological instrument) over the last 50 years, despite concurrent increases in global temperature. This further demonstrates the key role of solar radiation in driving evaporation (Roderick and Farquhar, 2002; Roderick *et al.*, 2007). Aerosols impact upon precipitation by forming tiny condensation nuclei which result in smaller cloud droplets, meaning that fewer droplets are large enough to form rain. The overall effect of aerosols is to weaken the hydrological cycle. Indeed large reductions in precipitation over land and river runoff during 1991 and 1992 have been attributed to the effect of aerosols emitted from the Mount Pinatubo eruption in June 1991 (IPCC, 2007a). However, in some areas brightening has occurred from the 1980s onwards, especially in former industrialised regions where implementation of air-quality measures has reduced air pollution and aerosol emissions. Global evapotranspiration increased between 1982 and 1997, with a small decline between 1998 and 2008 attributed to limited soil moisture supply, as observed from space satellites (Jung *et al.*, 2010).

Over the twentieth century, **river discharge** increased (Gedney *et al.*, 2006; Peterson *et al.*, 2002). This was mainly due to increasing global precipitation and physiological effects of rising atmospheric CO_2 content reducing plant transpiration (Gedney *et al.*, 2006; Gerten *et al.*, 2008). The role of direct effects of CO_2 on water usage by plants and feedbacks is discussed in detail in **Box 2.4**. Clearly, these effects would not be dominant in areas where vegetation cover was sparse (e.g. Sahel, **Box 2.3**). While increased irrigation has reduced river discharge in many areas (see Chapter 3), Gerten *et al.* (2008) found that land use change had increased river discharge overall.

Another important consideration for river runoff, aside from changes in precipitation and

BOX 2.3 CONTEMPORARY CHALLENGES

Sahel drought

The Sahel is a region in sub-Saharan West Africa – an area between the Sahara desert at 18°N and the equatorial forests at 15°N. The Sahel covers several African countries including Senegal, Mauritania, Mali, Burkina Faso, Niger, Nigeria, Chad, Sudan, Somalia, Ethiopia and Eritrea. It has a population of around 50 million inhabitants and a surface area over 5 million km^2. The vegetation is diverse and contains trees, shrubs and grasslands. In some ways the Sahel acts as a buffer zone from the arid Sahara north of the region. The climate of the area is controlled by the West African monsoon, itself controlled by the seasonal migration of an atmospheric phenomenon known as the Intertropical Convergence Zone (ITCZ). The location of the ITCZ broadly follows the seasonal variations in the receipt of the Sun's energy, and is located north of the equator in the northern summer and south in winter. As the ITCZ moves northwards it brings a short period of intense rain between late June and mid-September. Rapid vegetation growth results in a lowering of the surface albedo (as bare soil is light, it has a higher albedo than vegetation). Farming in the Sahel is almost entirely dependent on this rainfall. About 95% of the land in the Sahel is devoted to agriculture and 65% of people depend on agriculture for a livelihood (Druyan, 2011). Traditionally, people in the area maintained a subsistence living by moving to accommodate this climatic variability.

In the twentieth century colonial and post-colonial governments encouraged intensive modern food production in the area. This move accelerated in the 1950s and 1960s, which also coincided with a period of unusually high rainfall in the area (Figure 2.11). By the 1970s, however, the Sahel had entered a period of severe drought and there was a 40% decline in

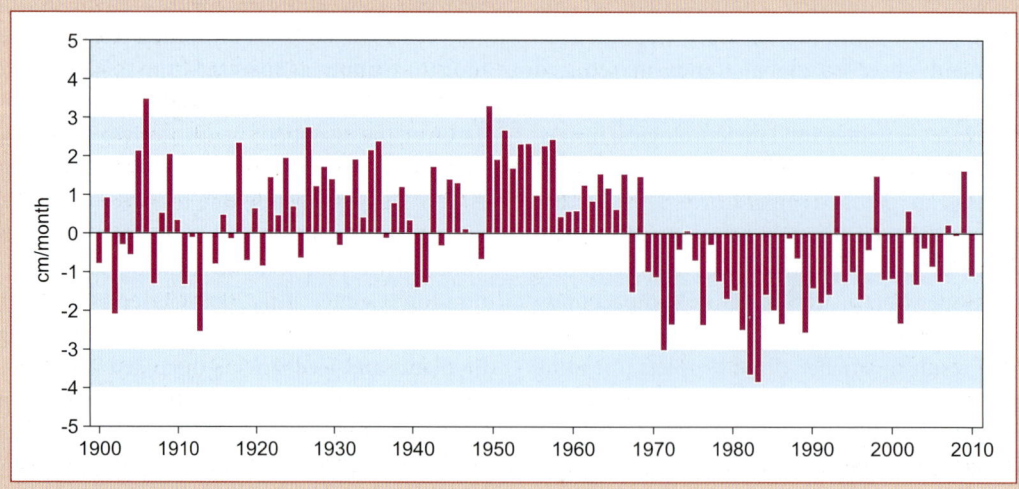

Figure 2.11 Precipitation difference from long-term average 1900–2011 in the Sahel region showing the prolonged dry period since the 1970s.

Source: Figure reproduced with kind permission of the Joint Institute for the Study of the Atmosphere and Ocean at the University of Washington.

continued

rainfall (Biasutti, 2011). Drought is associated with either a late onset of the rains, an early finish to the rains or weak rain systems (Druyan, 2011). The Sahel was affected by drought from the 1970s to the 1980s, and although rainfall totals have recovered they are still not back to 1950 levels (Figure 2.11). It was thought that such a large drop in rainfall could only be due to the result of human activity. Overgrazing, which led to an increase in albedo and therefore a cooler surface, would mean less evaporation and a decrease in precipitation (Charney, 1975). Note from Figure 2.4 the strong land–atmosphere coupling over the Sahel, which means there is potential for strong soil moisture–precipitation feedbacks. This represents a positive feedback reinforcing desertification. The blame was placed upon environmental degradation due to traditional grazing methods, suggesting that the people in the area ought to modernise further.

Whether overgrazing was the dominant cause of precipitation changes has always been open to debate. It is now recognised that the area has a very variable rainfall both spatially and temporally. Rainfall over the Sahel varies year to year and on a decadal timescale. These variations are linked to SST anomalies in the Pacific, Indian and Atlantic ocean basins. In the Pacific, the El Niño phase of ENSO leads to drying and La Niña to positive rainfall anomalies. Indian ocean warming and changes in the cross-equatorial temperature gradient in the Atlantic all affect rainfall in the Sahel (Biasutti, 2011). On a decadal timescale the temperature gradient across the northern and southern hemisphere oceans is important, with decreased rainfall when it is colder in the north than south (Biasutti, 2011). Land surface changes due to vegetation dieback are still thought to play a role (Caminade and Terray, 2010).

Recent work shows that precipitation changes over the Sahel may well have been enhanced through the release of sulphate aerosols from fossil fuel burning in North America and Europe (Ackerley et al., 2011). Sulphate aerosols have both a direct and an indirect effect on the radiation (and thus energy budget) of the Earth. They reflect more solar radiation away from the Earth, thus cooling the Earth's surface. They also cool the surface by changing the radiative properties of clouds by making them brighter and also increase cloud lifetimes (IPCC AR4). Such cooling will reduce evaporation, leading to less water in the atmosphere and so reducing precipitation. Ackerley et al. (2011) performed ensemble climate model runs; this is where a climate model is run many times but each time with small changes to the physics parameters within their known ranges, to see how parameter value assumptions affect the results. They found that a combination of increasing greenhouse gases and sulphate aerosols led to a net decrease in rainfall over the Sahel, mainly by affecting the SST gradient.

Now, it is generally acknowledged that the Sahel disaster of the 1970s and 1980s was about the implementation of development policies which paid little attention to the environment of the area. There was a failure in recognising that there is a complex interweaving of human socio-economic systems with environmental systems. It is important that we continue to more fully explore the impact of anthropogenic changes of the climate to such vulnerable areas. It is also important that we look to improve water and land management in the Sahel so that millions of people are not regularly pushed into starvation.

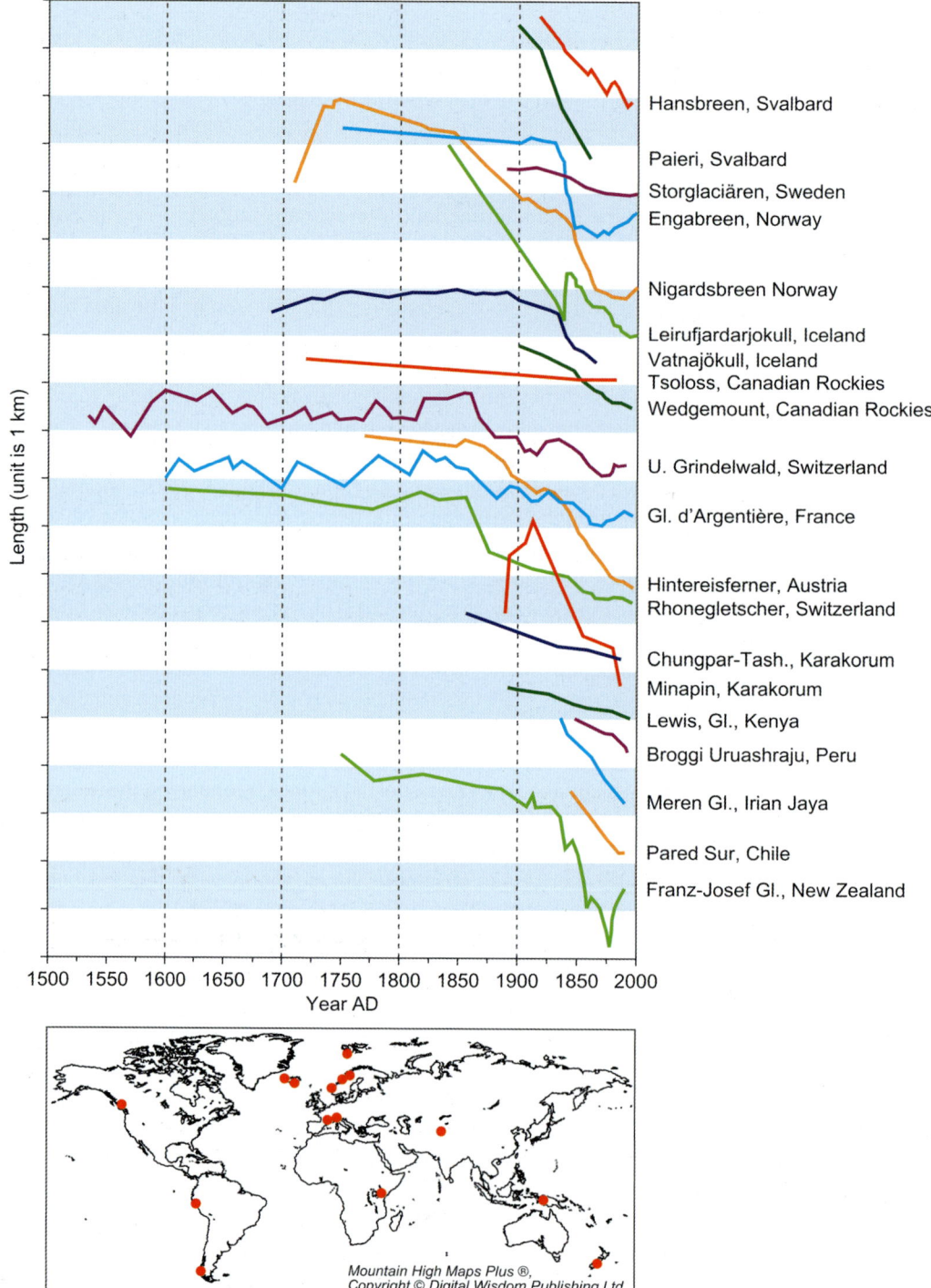

Figure 2.12 Changes in glacial extent for selected glaciers around the world.

Source: From *Climate Change 2001: The Scientific Basis. Contribution of Working Group I to the Third Assessment Report of the Intergovernmental Panel on Climate Change,* Figure 2.18. Cambridge University Press.

CO_2-induced changes in transpiration, relates to changes in snow cover, the timing of seasonal snow, and glacial melt. Over 50% of the world's **potable water** comes from rivers (Barnett *et al.*, 2005). Glaciers and snow packs act as reservoirs for water storage and are especially important for freshwater supply in snow-dominated regions, supplying water to more than one-sixth of the world's population (Barnett *et al.*, 2005). In South Asia more than 1.4 billion people rely on water from the Indus, Ganges, Brahmaputra, Yangtze and Yellow rivers, with the Himalayas at their source (Immerzeel *et al.*, 2010). Snow and glacial melt are particularly important for the Indus and Brahmaputra river basins (Immerzeel *et al.*, 2010). Satellite data have been used to estimate an earlier onset of snowmelt by ~0.5 days per year, an earlier finish of spring melt by ~1 day per year, and a shortened length of melting season by ~0.6 days per year over the Pan-Arctic over the last 30 years (Tedesco *et al.*, 2009). Over this period snow cover extent has decreased in May and June by 14% and 46%, respectively, in response to earlier snow melt (Brown *et al.*, 2010). For the western USA 60% of the climate-related trends of river flow over the second half of the twentieth century have been attributed to human-induced climate change (Barnett *et al.*, 2008).

Across the globe, glaciers have been in retreat over the last century (**Figure 2.12**). For example, there has been an estimated 80% reduction in the ice extent at the main summit of Mount Kilimanjaro between 1912 and 2003 (IPCC, 2007a), and glacial shrinkage over the Himalayas (Singh *et al.*, 2011), the Americas and Europe (IPCC, 2007a). In fact, information on changes in glacial length has been used to reconstruct temperature changes (Oerlemans, 2005). Important for **glacial mass balance** is the consideration of changes in both temperature and precipitation. For example, some of the losses at Kilimanjaro may be attributed to changes in precipitation rather than to direct warming (Kaser *et al.*, 2004). Excluding the major ice-caps (Greenland and Antarctica), between 1961 and 1990 glaciers lost 219 ± 112 kg m^{-2} yr^{-1}, equivalent to 0.33 ± 0.17 mm yr^{-1} of sea-level rise (Kaser *et al.*, 2006).

BOX 2.4 THE FUTURE OF WATER

Role of direct effects of CO_2 upon water usage by plants

Water provides a medium for most cell functions and is thus vital for plant growth. Small pores, called stomata, regulate the flux of CO_2 going into leaves for **photosynthesis** and water loss to the atmosphere, known as **transpiration** (**Figure 2.13**). There is thus a trade-off for the plants between the needs to take up CO_2 for photosynthesis and to avoid water loss. It has been postulated that leaves optimally regulate their degree of stomatal opening or closing (known as stomatal conductance), in order to maximise daily photosynthesis for a given water loss (Cowan and Farquhar, 1977; Barton *et al.*, 2011). This introduces the concept of water-use efficiency; the intrinsic water-use efficiency is defined as the ratio of the instantaneous rates of CO_2 assimilation and transpiration through the stomata (Condon *et al.*, 2002). Improvements in plant water-use efficiency would lead to increased crop yields in many dry environments (Condon *et al.*, 2002).

As it is a greenhouse gas, elevated CO_2 concentrations in the atmosphere will have indirect effects on transpiration and the global water cycle by changing the surface energy balance and temperature. Elevated CO_2 concentrations will also directly affect leaf physiology, for example, leaf photosynthesis. CO_2 and O_2 compete as substrates for the photosynthetic enzyme Rubisco (Prentice *et al.*, 2001). O_2

continued

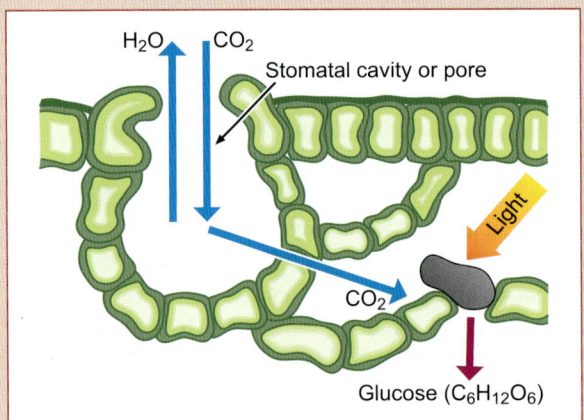

Figure 2.13 Leaf stomata regulate the uptake of carbon dioxide for photosynthesis, and water loss through transpiration.

reactions (oxygenation) lead to CO_2 release through photorespiration, and thus reduced efficiency of photosynthesis in C_3 plants (i.e. all tree and many herbaceous species). Elevated atmospheric CO_2 concentrations will increase the rate of carbon-fixing reactions (carboxylation), and decrease oxygenation, and thus, in addition, reduce photorespiration. In Free-Air-Carbon-Enrichment (FACE) experiments elevated CO_2 concentration stimulated photosynthesis by on average 31% (Ainsworth and Rodgers, 2007). There are also direct effects of elevated CO_2 on water usage by plants (Field *et al.*, 1995). Stomatal conductance decreases on average by just over 20% under elevated CO_2 across a wide range of vegetation types (Medlyn *et al.*, 2001; Ainsworth and Rodgers, 2007). However, these changes may not translate to similarly large changes in regional evapotranspiration or river runoff, due to the net effect of numerous feedbacks, as shown in Figure 2.14.

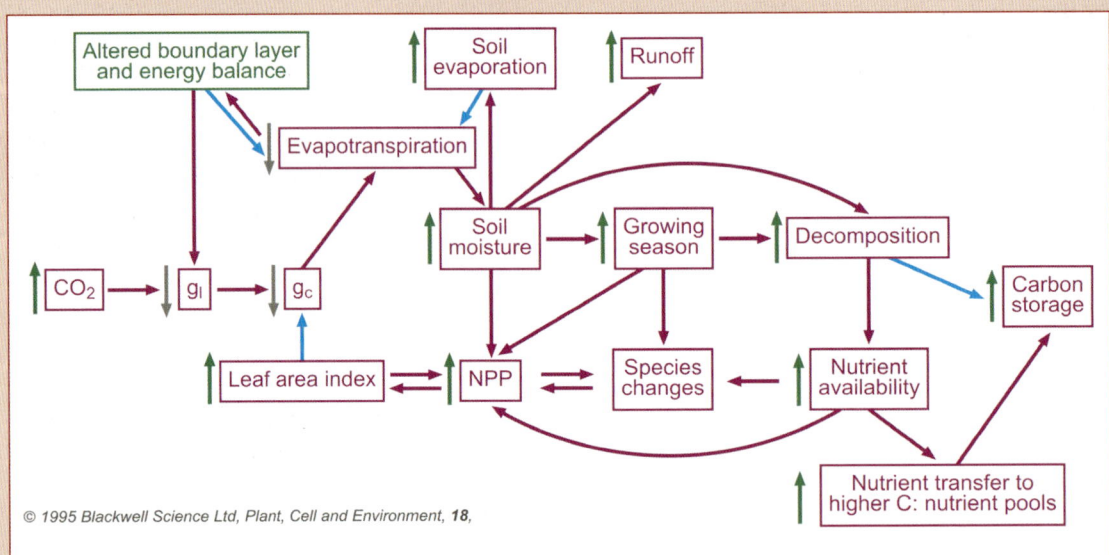

Figure 2.14 Mechanisms connecting changes in leaf conductance to canopy evapotranspiration and soil moisture changes. Positive effects are shown in purple arrows with negative effects in blue arrows. Upward arrows (green) indicate an increase and downward arrows (grey) indicate a decrease in that process (after Field *et al.*, 1995).

Source: Reproduced with permission of John Wiley and Sons based on Field, C.B., Jackson, R.B. and Mooney H.A. 1995. Stomatal responses to increased CO2: implications from the plant to the global scale. *Plant, Cell and Environment* 18: 1214–1225.

continued

Vapour pressure deficit (VPD) is defined as the difference between the pressure exerted by water vapour in saturated air and the actual water vapour pressure. It is a measure of the 'drying power of air'. The ratio of CO_2 uptake to transpiration rate at the canopy scale appears to be strongly dependent on VPD (Barton *et al.*, 2011). Two negative feedbacks act to reduce the sensitivity of transpiration to leaf conductance. Firstly, a reduction in leaf conductance leads to reductions in transpiration and thus to a drier **boundary layer**, which in turn 'increases the driving gradient for transpiration' (Field *et al.*, 1995), i.e. VPD. Also a reduction in transpiration means a reduction in the latent heat flux. A commensurate increase in **sensible heat** flux is needed to balance net radiation. An increase in sensible heat flux implies an increase in leaf temperature and thus an increase in the VPD between leaf and air, which in turn acts to increase transpiration. However, transpiration does not increase linearly with VPD, due to the physiological response of stomata to VPD (i.e. stomatal conductance itself decreases with increasing VPD, an example of a positive feedback) (Oren *et al.*, 1999). Accounting for these canopy-scale feedbacks, it has been estimated that for many cropland locations a 40% decrease in stomatal conductance would translate into only a 10% decrease in evapotranspiration (Schulze *et al.*, 1994; Field *et al.*, 1995).

Other feedbacks which relate to responses of ecosystem processes to reduced stomatal conductance under elevated CO_2 concentrations need to be considered (**Figure 2.14**). For example, a reduction in transpiration will lead to increased soil moisture, which may lead to increases in river runoff (Gedney *et al.*, 2006; Betts *et al.*, 2007). Therefore, the physiological effect of future elevated atmospheric CO_2 concentration would be to increase future freshwater availability. In water-limited regions increases in soil moisture can lead to reduced plant water stress and potentially to longer growing seasons. This will lead to increased annual **net primary production** (NPP), and increases in leaf area, and consequently to an increase in canopy conductance and transpiration. This represents a negative feedback, whereby the response of the system in growing more leaf area counteracts the initial leaf-level reductions in stomatal conductance and transpiration. Other feedbacks include soil moisture-induced changes in species composition, changes in decomposition rates and nutrient availability for plants, leading to changes in NPP and leaf area. A comprehensive account of feedbacks is given in Field *et al.* (1995) and references therein.

REFLECTIVE QUESTION

What examples of climate variability having an impact on the local water cycle can you think of that have occurred in the last 12 months? What categories of climate variability would you consider them to come under?

D IMPACT OF LAND MANAGEMENT ON THE WATER CYCLE

We have already seen how humans affect the global water cycle through changing atmospheric composition (CO_2 and aerosols) and climate. Direct human impacts on water resources include the construction of dams, water extraction, and modification of river channels (Harding *et al.*, 2011), with implications for seasonal water supply, water temperatures, and freshwater biodiversity (Lytle and Poff, 2004). In addition, changes in land use and human water use greatly affect the global water cycle and freshwater resources. Land use and land cover changes include regional desertification, deforestation (originally in mid-latitudes, Europe and North America, and more recently in tropical regions), mid-latitude reforestation and afforestation (e.g. during recent decades in China), and agricultural expansion throughout the twentieth century. Humans currently use less than 10% of renewable freshwater, and agriculture accounts for one-third of total land evapotranspiration (Oki and Kanae, 2006). Nevertheless, nearly 80% of the world's population live in areas exposed to either high levels of water insecurity or threat to biodiversity (Vörösmarty *et al.*, 2010).

Demand for water outweighs supply from river runoff in many regions (**Figure 2.15**, Vörösmarty *et al.*, 2000; Oki and Kanae, 2006; Harding *et al.*, 2011), which can lead to water conflict (see Chapter 11). To help sustain increases in agricultural production over the last 40 years the area of land under irrigation has doubled (Oki and Kanae, 2006); and even tripled across India between 1970 and 1999 (Rodell *et al.*, 2009; see also Chapter 11). In some regions this has led to potentially unsustainable use of groundwater reserves, for example, in parts of South Asia (Rodell *et al.*, 2009). Arid regions, such as Pakistan, are particularly reliant on groundwater for irrigation (Rodell *et al.*, 2009; http://www.youtube.com/watch?v=o1QsCa7RmmU). Thus a focus on water-efficiency measures for agriculture is urgently required (see Chapter 1). Future changes in the global water cycle and freshwater resources will depend on population growth. Economic development will impact upon land use and water demand (see also Chapters 1, 11 and 12). Economic development will also indirectly impact upon climate change and atmospheric composition.

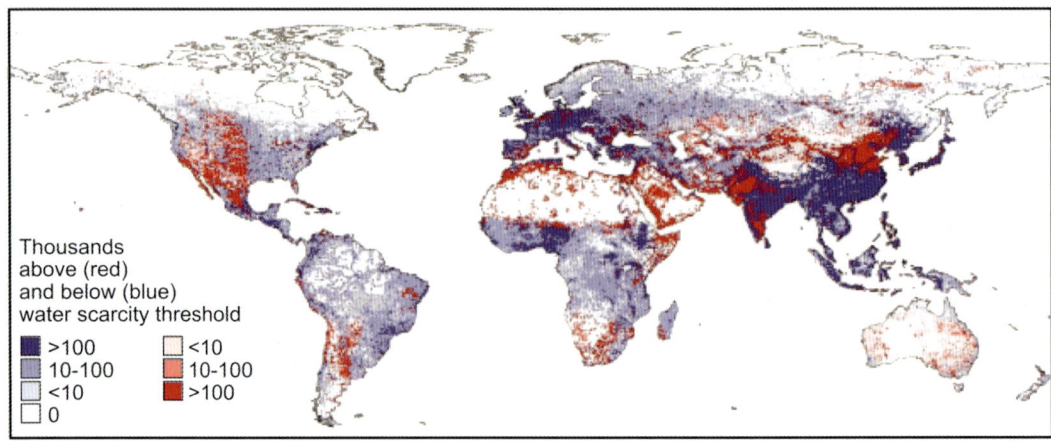

Figure 2.15 Population exposed to severe water scarcity (red).

Source: From Vörösmarty, C.J., Green, P., Salisbury, J. and Lammers, R.B. 2000. Global water resources: vulnerability from climate change and population growth. *Science* 289, doi:10.1126/science.289.5477.284. Reprinted with permission from AAAS.

> **REFLECTIVE QUESTION**
>
> Humans do not directly drink much water each year. So why does population growth pose such a major threat to water resources and the water cycle?

E IMPACTS OF FUTURE ENVIRONMENTAL CHANGE

1 Climate change: physical basis

Climate change is often discussed in terms of changes in temperature. Arguably, changes in precipitation will be as, or more, important for many essential **ecosystem services**, for example, freshwater availability and food security, in the coming century. Precipitation must be considered in terms of its season, intensity, and frequency rather than annual total alone. For example, rainfall intensity governs the amount of water that percolates into soils. Heavy rainfall in wintertime, rather than percolating into soils, is lost as river runoff, and is therefore not available for farmers during the growing season.

With global warming, surface evaporation will increase (absorbing energy in the phase change from liquid to gas), supplying more water vapour into the atmosphere. In addition, a warmer atmosphere can carry more water vapour. This is based on the Clausius–Clapeyron relation, mentioned earlier, whereby the saturated water vapour pressure increases near exponentially with temperature (**Figure 2.2**). The atmosphere can carry 7% more water vapour per degree temperature increase. Increases in evaporation, together with the ability of the atmosphere to carry more water vapour, lead to an overall intensification of the hydrological cycle. Intensity of heavy rainfall is expected to increase and, while heavy rainfall is local in character, moisture will be gathered from far afield, on average drawing moisture from distances around ~4 times the radius of the precipitating area (Trenberth et al., 2003). An extratropical cyclone with a radius of typically 800 km will draw moisture from 3200 km away, or 30° latitude (Trenberth et al., 2003). This implies less moisture to drive moderate rainfall elsewhere (Allan, 2011). Therefore, an increase in rainfall intensity in one region (e.g. equatorial tropics) implies reduction in duration and/or frequency elsewhere (e.g. the sub-tropics) (Allan, 2011). Climate change will amplify existing regional differences in precipitation–evaporation, leading to an intensification of the global hydrological cycle (Allen and Ingram, 2002; Held and Soden, 2006).

However, as mentioned earlier, water fluxes and energy are linked, through latent heat exchange during phase change (evaporation at the surface, condensation in the troposphere), and therefore changes in global precipitation are tied to the surface and tropospheric energy budgets (Lambert and Allen, 2009). Water vapour is a powerful greenhouse gas, and clouds affect planetary albedo. Consequently water vapour, other greenhouse gases, and aerosols, control precipitation through their effect on the surface and tropospheric energy budget, and the radiative cooling of the atmosphere. In fact, energy provides a stronger constraint to increases in global precipitation than does moisture availability (Lambert and Allen, 2009; Trenberth, 2011). As a consequence an overall more moderate 1–3% increase (rather than 7% implied by the Clausius–Clapeyron relation) in global precipitation per degree warming is expected (Held and Soden, 2006; Lambert and Allen, 2009; Lambert et al., 2008).

An increase in atmospheric moisture content further enhances global warming. This is known as the *water vapour feedback*, and represents the most important feedback in the climate system. In the absence of feedbacks, the climate would warm by approximately 1 degree for a doubling of atmospheric CO_2 concentration. The water vapour feedback acting alone approximately doubles the warming due to CO_2 alone (IPCC,

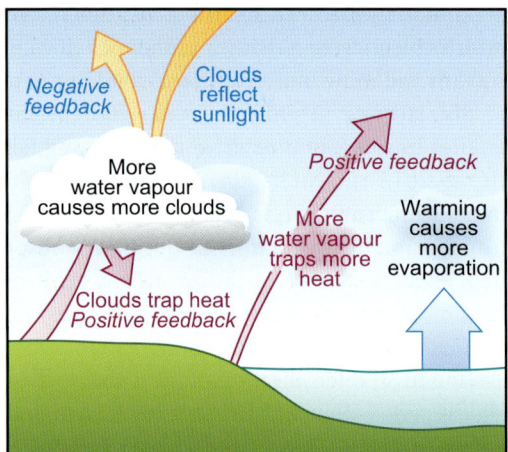

Figure 2.16 Schematic showing example feedbacks of enhanced water vapour in the atmosphere.

2007a). One of the main uncertainties in climate science relates to the uncertainties in the various feedback mechanisms and their strength (see **Figure 2.16**).

2 Climate change impacts

With an increase in the water vapour content, rainfall intensity will increase. This will lead to more floods and droughts (Burke *et al.*, 2006). Extreme rainfall events will become more frequent than they are now. However, there may be longer intervals between (non-extreme) individual rainfall events. Climate model projections suggest future increases in precipitation, more intense rainfall, more droughts, and in general wet regions getting wetter, and dry regions becoming drier (IPCC 2007a, **Figure 2.17**). This leads to an increase in flood risk, which will pressure physical infrastructure and impact upon water quality. IPCC (2007b) estimates that 20% of the world population will live in areas with increasing river flood risk by the 2080s.

Regionally, the tropics and high latitudes are expected to get wetter, the sub-tropics drier (i.e. enhanced precipitation–evapotranspiration deficit), with a poleward extension of the sub-tropics in both hemispheres (impacting upon the Mediterranean, South Australia, southwest USA). In mid-latitudes, changes in the seasonality of precipitation are expected, with drier summers (likely to impact upon agriculture and wildfire frequency and severity), and wetter winters (implications for flooding). Models project enhancements in the East African monsoon and summer monsoons in South and South East Asia.

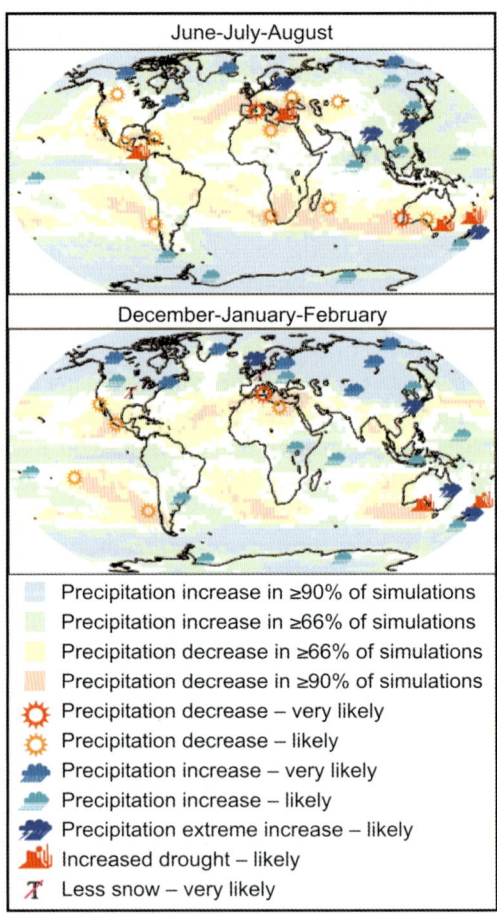

Figure 2.17 Robust findings on regional climate change for mean and extreme precipitation, drought and snow.

Source: From *Climate Change 2007: The Physical Science Basis. Working Group I Contribution to the Fourth Assessment Report of the Intergovernmental Panel on Climate Change*, **Box 11.1**, Figure 2. Cambridge University Press.

The main uncertainty in precipitation predictions among climate models is at the boundary of these climate zones.

Another important consideration is whether the precipitation falls as rain or snow. With warming, the balance shifts to relatively more rain (consequently more river runoff) and earlier snowpack melt, and a general contraction of snow coverage. Snow effectively acts as freshwater storage, reducing seasonality in water input into soils, and reducing wintertime runoff losses. This is likely to have significant implications for soil moisture content and freshwater availability in spring and summertime, for example, increased requirements for crop irrigation. With global warming, water supplies from glaciers and snow are projected to decrease, reducing water availability in regions dependent on glacier and snow melt. For South Asia this translates into a future decrease in mean water supply of −8.4%, −17.6%, and −19.6% for the upper Indus, Ganges, and Brahmaputra, respectively, despite consideration of the partial offsets from projected increases in monsoon rains (Immerzeel *et al.*, 2010). Accounting for irrigation requirements for crops and population, the Brahmaputra and Indus basins are most susceptible to future decreased flows, putting an estimated 60 million people at risk of food insecurity (Immerzeel *et al.*, 2010). Warming is likely to decrease **permafrost** (see Chapter 1) extent, challenging infrastructure in affected regions (ACIA, 2004). IPCC AR4 projects permafrost area to decrease by 20–35% by the mid-twenty-first century, and seasonal thaw depth to decrease by 30–50% by 2080.

The combined effects of future increases in precipitation with the competing effects of CO_2 on plant physiology and warming acting to decrease and increase transpiration, respectively, lead to projections of increased river runoff (Betts *et al.*, 2007). The IPCC Assessment Report 4, Working Group II, Chapter 3, 'Freshwater resources and their management', includes the following statements in the executive summary (IPCC, 2007b):

- Future river discharge will increase by 10 to 40% by mid-century at higher latitudes and in some wet tropical areas; decrease by 10 to 30% over some dry regions at mid-latitudes and dry tropics, due to decreases in rainfall and higher rates of evapotranspiration.
- Many semi-arid areas (e.g. the Mediterranean Basin, western United States, southern Africa and north-eastern Brazil) will suffer a decrease in water resources due to climate change.
- The negative impacts of climate change on freshwater systems outweigh its benefits. Increased annual runoff in some areas but increased precipitation variability and seasonal runoff shifts on water supply, water quality and flood risk.
- Increased temperatures will further affect the physical, chemical and biological properties of freshwater lakes and rivers, with predominantly adverse impacts on many individual freshwater species, community composition and water quality.

Uncertainties in the magnitude of future climate change relate to uncertainties in future population growth and socio-economic projections, and differences in climate model sensitivity to fossil fuel emissions. The latter relates to uncertainties in the strength of feedbacks in the climate system. Also, although all climate models predict a future increase in global precipitation with warming, there is less consensus on the specific location of precipitation changes, not only the magnitude but even the sign of change differing among models (see **Figure 2.17**). Projections of future land use and land cover changes vary markedly (Sitch *et al.*, 2005) and, together with uncertainty in changes in water demand and climate change, these provide major challenges for future water cycle and freshwater security assessments.

> **REFLECTIVE QUESTION**
>
> What are the major changes to the water cycle that you would expect to see in the next 50–100 years as a result of climate change?

F SUMMARY

The global water cycle is intimately linked to energy balance and transfers across the Earth for a number of reasons, including the latent heat transfer processes associated with evaporation and condensation and the fact that water vapour is an important greenhouse gas. There are large uncertainties in the estimates of the amounts of water being stored and transferred in parts of the global water cycle. There are also large variations between years, between inter-annual periods (e.g. due to ENSO), and between decadal periods (e.g. due to aerosols and land use change). Future changes in climate, land use, and water demand are likely to profoundly affect the global water cycle. There are multiple positive and negative feedbacks operating between factors such as increased CO_2 and plant physiology, aerosols, temperature, precipitation and evapotranspiration, changes in sea surface temperatures and atmospheric circulation patterns. The IPCC (2007b) report projects an intensification of the global water cycle, with future increases in precipitation and evaporation, and more intense rainfall, yet more droughts. Wet regions will get wetter and dry regions will get drier. The impacts of climate change include increases in annual river runoff, with changes in the seasonality of runoff in snow- and glacier-fed river basins, with implications for freshwater supply and food security for human populations and biodiversity.

FURTHER READING

Look at the latest IPCC reports available online for the latest scientific reports on climate change impacts on the water cycle: http://www.ipcc.ch/publications_and_data/publications_and_data_reports.shtml.

Classic papers

Field, C.B., Jackson, R.B. and Mooney, H.A. 1995. Stomatal responses to increased CO_2: implications from the plant to the global scale. *Plant, Cell and Environment* 18: 1214–1225.

A comprehensive account of feedbacks on the water cycle of changes in plant responses to CO_2 concentrations.

Held, I.M. and Soden, B.J. 2006. Robust responses of the hydrological cycle to global warming. *Journal of Climate* 19: 5686–5699.

The authors compare results from many models to see where the models agree that there are likely to be strong changes in the water cycle in the future.

Roderick, M.L. and Farquhar, G.D. 2002. The cause of decreased pan evaporation over the past 50 years. *Science* 298: 1410–1411.

Another example of a feedback effect – here increased cloud cover and aerosols are shown to be linked to reduced measured evaporation.

> **PROJECT IDEAS**
>
> ■ Using data in **Figure 2.1** and using a bucket of water to represent all of the water on Earth, design a poster that shows how much water from the bucket would be found in different stores (riverflow, lakes, atmosphere, oceans, etc.).

continued

- For your local river basin produce a list of all possible impacts and feedback effects of climate change on the water cycle. Use IPCC reports or local climate models for your region to inform your work. What mitigation measures might be adopted in the river basin?

- Produce a list of extreme events that have impacted upon the water cycle in your region over the past 20 years. Investigate the evidence as to the type of climate variability that might have been associated with them.

REFERENCES

ACIA. 2004. *Impacts of a Warming Arctic: Arctic Climate Impact Assessment.* Cambridge University Press; Cambridge.

Ackerley, D., Booth, B.B.B., Knight, S.H.E., Highwood, E.J., Frame, D.J., Allen, M.R. and Rowell, D.P. 2011. Sensitivity of twentieth-century Sahel rainfall to sulfate aerosol and CO_2 forcing. *Journal of Climate* 24: 4999–5014.

Aguado, E. and Burt, J.E. 1999. *Understanding Weather and Climate.* Prentice-Hall Inc., New Jersey.

Ainsworth, E.A. and Rodgers, A. 2007. The response of photosynthesis and stomatal conductance to rising [CO_2]: mechanisms and environmental interactions. *Plant, Cell and Environment* 30: 258–270.

Allan, R.P. 2011. Human influence on rainfall. *Nature* 470: 344–345.

Allen, M. and Ingram, W. 2002. Constraints on future changes in climate and the hydrologic cycle. *Nature* 419: 224–232.

Barnett, T.P., Adam, J.C. and Lettenmaier, D.P. 2005. Potential impacts of a warming climate on water availability in snow-dominated regions. *Nature* 438: 303–309.

Barnett, T.P. *et al.* 2008. Human-induced changes in the hydrology of the western United States. *Science* 319: 1080, doi:10.1126/science.1152538.

Barton, C.V.M. *et al.* 2011. Effects of elevated atmospheric [CO_2] on instantaneous transpiration efficiency at leaf and canopy scales in *Eucalyptus saligna*. *Global Change Biology*: doi:10.1111/j.1365–2486. 2011.02526.x.

Betts, R.A., Boucher, O., Collins, M., Cox, P.M., Falloon, P.D., Gedney, N., Hemming, D.L., Huntingford, C., Jones, C.D., Sexton, D.M.H. and Webb, M.J. 2007. Projected increase in continental runoff due to plant responses to increasing carbon dioxide. *Nature* 448: doi:10.1038/nature06045.

Biasutti, M. 2011. Atmospheric science: a man-made drought. *Nature Climate Change* 1: 197–198.

Biemans, H., Hutjes, R.W.A., Kabat, P., Strengers, B.J., Gerten, D. and Rost, S. 2009. Effects of precipitation uncertainty on discharge calculations for main river basins. *Journal of Hydrometeorology* 10: 1011–1025.

Bjerknes, J. 1969. Atmospheric teleconnection from the equatorial pacific. *Monthly Weather Review* 97: 163–172.

Brown, R., Derksen, C. and Wang, L. 2010. A multi-data set analysis of variability and change in Arctic spring snow cover extent, 1967–2008. *Journal of Geophysical Research* 115: D16111 doi:10.1029/2010JD013975.

Burke, E.J., Brown, S.J. and Christidis, N. 2006. Modeling the recent evolution of global drought and projections for the twenty-first century with the Hadley Centre climate model. *Journal of Hydrometeorology* 7: 1113–1125.

Caminade, C. and Terray, L. 2010. Twentieth century Sahel rainfall variability as simulated by the ARPEGE AGCM, and future changes. *Climate Dynamics* 35 (Special Issue): 75–94.

Charney, J.G. 1975. Dynamics of deserts and drought in Sahel. *Quarterly Journal of the Royal Meteorological Society* 101: 193–202.

Ciais, P. *et al.* 2005. Europe-wide reduction in primary productivity caused by the heat and drought in 2003. *Nature* 437: doi:10.1038/nature0972.

Coghlan, C. 2002. El Niño – causes, consequences and solutions. *Weather* 57: 209–215.

Condon, A.G., Richards, R.A., Rebetzke, G.J. and Farquhar, G.D. 2002. Improving intrinsic water-use efficiency and crop yield. *Crop Science* 42: 122–131.

Cook, B.I., Miller, R.L. and Seager, R. 2008. Dust and sea surface temperature forcing of the 1930s 'Dust Bowl' drought. *Geophysical Research Letters* 35: L08710, doi10.1029/2008GL033486.

Cook, B.I., Miller, R.L. and Seager, R. 2009. Amplification of the North American 'Dust Bowl' drought through human-induced land degradation. *Proceedings of the National Academy of Sciences* 106: 4997–5001.

Cook, E.R., Woodhouse, C.A., Eakin, C.M., Meko, D.M. and Stahle, D.W. 2004. Long-term aridity changes in the western United States. *Science* 306: 1015–1018.

Cowan, I. and Farquhar, G.D. 1977. Stomatal function in relation to leaf metabolism and environment. *Symposia of the Society for Experimental Biology* 31: 471–505.

Druyan, L.M. 2011. Studies of 21st-century precipitation trends over West Africa. *International Journal of Climatology* 31: 1415–1424.

Fedoroff, N.V. *et al.* 2010. Radically rethinking agriculture for the twenty-first century. *Science* 327: 833–834.

Field, C.B., Jackson, R.B. and Mooney, H.A. 1995. Stomatal responses to increased CO_2: implications from the plant to the global scale. *Plant, Cell and Environment* 18: 1214–1225.

Garbrecht, J.D. and Rossel, F.E. 2002. Decade-scale precipitation increase in Great Plains at end of 20th century. *Journal of Hydrologic Engineering* 7: 64–75.

Gedney, N., Cox, P.M., Betts, R.A., Boucher, O., Huntingford, C. and Stott, P.A. 2006. Detection of a direct carbon dioxide effect in continental river runoff records. *Nature* 439: 835–838.

Gergis, J.L. and Fowler, A.M. 2009. A history of ENSO events since AD 1525: implications for future climate change. *Climatic Change* 92: 343–387.

Gerten, D., Schaphoff, S., Haberlandt, U., Lucht, W. and Sitch, S. 2004. Terrestrial vegetation and water balance – hydrological evaluation of a dynamic global vegetation model. *Journal of Hydrology* 286: 249–270.

Gerten, D., Rost, S., von Bloh, W. and Lucht, W. 2008. Causes of change in twentieth century global river discharge *Geophysical Research Letters* 35: L20405, doi:10.1029/2008GL035258.

Haddeland, I. *et al.* 2011. Multi-model estimate of the global water balance: setup and first results. *Journal of Hydrometeorology* 12: 869–884.

Harding, R.J. *et al.* 2011. Preface to the Water and Global Change (WATCH) special collection: current knowledge of the terrestrial global water cycle. *Journal of Hydrometeorology* 12: 1149–1156.

Held, I.M. and Soden, B.J. 2006. Robust responses of the hydrological cycle to global warming. *Journal of Climate* 19: 5686–5699.

Immerzeel, W.W., van Beek, L.P.H. and Bierkens, M.F.P. 2010. Climate change will affect the Asian water towers. *Science* 328: 1382–1385.

IPCC. 2001. *Climate Change 2001: The scientific basis. Contribution of Working Group I to the Third Assessment Report of the Intergovernmental Panel on Climate Change* (Houghton, J.T. *et al.*). Cambridge University Press; Cambridge.

IPCC. 2007a. *Climate Change 2007: The Physical Science Basis. Contribution of Working Group I to the Fourth Assessment Report of the Intergovernmental Panel on Climate Change* (Solomon, S. *et al.* (eds)). Cambridge University Press; Cambridge.

IPCC. 2007b. *Climate Change 2007: Impacts, Adaptation and Vulnerability. Contribution of Working Group II to the Fourth Assessment Report of the Intergovernmental Panel on Climate Change* (Parry, M.L. *et al.* (eds)). Cambridge University Press; Cambridge.

Jonas, P.R. 1994. Why does it rain? *Weather* 49: 258–260.

Jung, M. *et al.* 2010. Recent decline in the global land evapotranspiration trend due to limited moisture supply. *Nature* 467: 951–954.

Kaser, G., Hardy, D.R., Mölg, T., Bradley, R.S. and Hyera, T.M. 2004. Modern glacier retreat on Kilimanjaro as evidence of climate change: observations and facts. *International Journal of Climatology* 24: 329–339.

Kaser, G., Cogley, J.G., Dyurgerov, M.B., Meier, M.F. and Ohmura, A. 2006. Mass balance of glaciers and ice caps: consensus estimates for 1961–2004. *Geophysical Research Letters* 33: L19501 doi:10.1029/2006GL027511.

Kiehl, J.T. and Trenberth, K.E. 1997. Earth's annual global mean energy budget. *Bulletin of the American Meteorological Society* 78: 197–208.

Koster, R.D. *et al.* 2004. Regions of strong coupling between soil moisture and precipitation. *Science* 305: 1138–1140.

Lambert, F.H. and Allen, M.R. 2009. Are changes in global precipitation constrained by the tropospheric energy budget? *Journal of Climate* 22: 499–517.

Lambert, F.H., Stine, A.R., Krakauer, N.Y. and Chiang, J.C.H. 2008. How much will precipitation increase with global warming? *EOS Transactions American Geophysical Union* 89: 193.

Lytle, D.A. and Poff, N.L. 2004. Adaptation to natural flow regimes. *Trends in Ecology and Evolution* 19: doi:10.1016/j.tree.2003.10.002.

Mayes, J. 2008. Editorial: special issue – the wet summer of 2007 in the UK. *Weather* 63: 251–252.

Medlyn, B.E. *et al.* 2001. Stomatal conductance of forest species after long-term exposure to elevated CO_2 concentration: a synthesis. *New Phytologist* 149: 247–264.

Meehl, G.A. *et al.* 2007. Global climate projections. *In*: S. Solomon, D. Qin, M. Manning, Z. Chen, M. Marquis, K.B. Averyt, M. Tignor and H.L. Miller (eds) *Climate Change 2007: The Physical Science Basis. Contribution of Working Group I to the Fourth Assessment Report of the Intergovernmental Panel on Climate Change.* Cambridge University Press; Cambridge.

Min, S-K., Zwiers, F.W. and Hegerl, G.C. 2011. Human contribution to more-intense precipitation extremes. *Nature* 470: 378–381.

Moron, V. and Ward, M.N. 1998. ENSO teleconnections with climate variability in the European and African sectors. *Weather* 53: 287–295.

Oerlemans, J. 2005. Extracting a climate signal from 169 glacier records. *Science* 308: 657–677.

Ogi, M., Yamazaki, K. and Tachibana, Y. 2005. The summer northern annular mode and abnormal summer weather in 2003. *Geophysical Research Letters* 32: L04706 doi:10.1029/2004GL021528.

Oki, T. and Kanae, S. 2006. Global hydrological cycles and world water resources. *Science* 313: 1068–1072.

Oren, R., Sperry, J.S., Katul, G.G., Pataki, D.E., Ewers, B.E., Philipps, N. and Schäfer, K.V.R. 1999. Survey and synthesis of intra- and interspecific variation in stomatal sensitivity to vapour pressure deficit. *Plant, Cell and Environment* 22: 1515–1526.

Pall, P. *et al.* 2011. Anthropogenic greenhouse gas contribution to flood risk in England and Wales in autumn 2000. *Nature* 470: 382–386.

Peterson, B.J. *et al.* 2002. Increasing river discharge to the Arctic Ocean. *Science* 298: 2171–2173.

Pielke Jr, R.A. and Landsea, C.N. 1999. La Niña, El Niño, and Atlantic hurricane damages in the United States. *Bulletin of the American Meteorological Society* 80: 2027–2033.

Prentice, I.C. *et al.* 2001. The carbon cycle and atmospheric carbon dioxide. *In*: J.T. Houghton *et al.* (eds) *Climate Change 2001: The Scientific Basis.* Cambridge University Press; Cambridge.

Priestley, C.H.B. and Taylor, R.J. 1972. On the assessment of surface heat flux and evaporation using large-scale parameters. *Monthly Weather Review* 100: 81–92.

Prior, J. and Beswick, M. 2007. The record-breaking heat and sunshine of July 2006. *Weather* 62: 174–182.

Ramanathan, V., Crutzen, P.J., Kiehl, J.T. and Rosenfeld, D. 2001. Aerosols, climate, and the hydrological cycle. *Science* 294: 2119–2124.

Rodell, M., Velicogna, I. and Famiglietti, J.S. 2009. Satellite-based estimates of groundwater depletion in India. *Nature* 460: 999–1002.

Roderick, M.L., Rotstayne, L.D., Farquahar, G.D. and Hobbins, M.T. 2007. On the attribution of changing pan evaporation. *Geophysical Research Letters* 34: L17403 doi:10.1029/2007GL031166.

Schubert, S.D., Suarez, M.J., Pegion, P.J., Koster, R.D. and Bacmeister, J.T. 2004. On the cause of the 1930s Dust Bowl. *Science* 303: 1855–1859.

Schulze, E.-D., Kelliher, F.M., Körner, C., Lloyd, J. and Leuning, R. 1994. Relationship among maximum stomatal conductance, ecosystem surface conductance, carbon assimilation rate, and plant nitrogen nutrition: a global ecology scaling exercise. *Annual Review of Ecology and Systematics* 25: 629–660.

Seneviratne, S.I., Lüthi, D., Litschi, M. and Schär, C. 2006. Land–atmosphere coupling and climate change in Europe. *Nature* 443: 205–209.

Singh, S.P., Bassignana-Khadka, I., Karky, B.S. and Sharma, E. 2011. Climate change in the Hindu Kush-Himalayas: the state of current knowledge. ICIMOD; Kathmandu.

Sitch, S., Brovkin, V., von Bloh, W., Van Vuuren, D., Eickhout, B. and Ganopolski, A. 2005. Impacts of future land cover on atmospheric CO2 and climate. *Global Biogeochemical Cycles* 19: GB2013 doi:10.1029/2004GB002311.

Stevens, A. 2011. Introduction to the basic drivers of climate. *Nature Education Knowledge* 2: 6.

Tedesco, M., Brodzik, M., Armstrong, R., Savoie, M. and Ramage, J. 2009. Pan arctic terrestrial snowmelt trends (1979–2008) from spaceborne passive microwave data and correlation with the Arctic Oscillation. *Geophysical Research Letters* 36: L21402 doi:10.1029/ 2009GL039672.

Teuling, A.J. *et al.* 2010. Contrasting response of European forest and grassland energy exchange to heatwaves. *Nature Geosciences* 3: 722–727.

Trenberth, K.E. 2011. Changes in precipitation with climate change. *Climate Research* 47: 123–138.

Trenberth, K.E., Dai, A., Rasmussen, R.M. and Parsons, D.B. 2003. The changing character of precipitation. *Bulletin of the American Meteorological Society* 84: 1205–1217.

Trenberth, K.E., Fasullo, J.T. and Kiehl, J., 2009. Earth's global energy budget. *Bulletin of the American Meteorological Society* 90: 311–323.

Vörösmarty, C.J., Green, P., Salisbury, J. and Lammers, R.B. 2000. Global water resources: vulnerability from climate change and population growth. *Science* 289: doi:10.1126/science.289.5477.284.

Vörösmarty, C.J. *et al.* 2010. Global threats to human water security and river biodiversity. *Nature* 467: 555–562.

Woodhouse, C.A. and Overpeck, J.T. 1998. 2000 years of drought variability in the central United States. *Bulletin of the American Meteorological Society* 79: 2693–2714.

Woodhouse, C.A., Meko, D.M., MacDonald, G.M., Stahle, D.W. and Cook, E.R. 2010. A 1,200-year perspective of twenty-first century drought in southwestern North America. *Proceedings of the National Academy of Sciences* 107: 21283–21288.

Young, M.V. 1994a. Depressions and anticyclones: Part 1. *Weather* 49: 306–311.

Young, M.V. 1994b. Depressions and anticyclones: Part 2. Life cycles and weather characteristics. *Weather* 49: 362–370.

CHAPTER THREE

River basin hydrology

Joseph Holden

> **LEARNING OUTCOMES**
>
> After reading this chapter you should be able to:
>
> - outline the main hydrological pathways to rivers and lakes
> - explain the form of river hydrographs and how these vary under different environmental settings and management
> - understand how river flow is measured and how to produce a river basin water budget
> - show a knowledge of different types of flood event and issues around managing floods
> - describe the controls on river channel size and shape, and processes of water and sediment transport in rivers.

A INTRODUCTION

The way water moves across and through a landscape is very important for determining river flow, water quality, and even the evolution of landscapes themselves. It is common to try to understand water movements across landscapes within topographically confined units. We call these units **river basins** (also known as **watersheds** or **catchments**), which are defined by the upslope area draining into a given point on a river. These can vary in scale from hillslopes to the size of the Amazon basin at 6 160 000 km^2. While the Amazon's seemingly vast river basin area amounts to only 2% of the Earth's land mass, the Amazon carries around one-fifth of the world's annual river discharge because the majority of its area rests in a tropical region with plentiful rainfall.

It is possible to create a water budget for a river basin, whereby we try to measure the inputs, outputs and stores of water for the system and understand how efficient the system is at turning rainfall into river flow. Depending on the environmental conditions, almost 100% of **precipitation** may reach the river, or as a little as zero. The rate at which water is delivered to the ground surface and the pathways and speed at which water travels across and through a landscape may also be impacted upon by climate change and land management activity, and these factors may affect flood risk. It is therefore important to understand the flow pathways for water across river basins and how environmental change may impact upon these processes. There are several types of flooding, of which water pouring out over river banks is only one, and so management strategies for coping with floods need to be well thought through. In addition, river channels themselves tend to be naturally dynamic, changing course, size and shape through time and migrating across landscapes. Management of river channels has traditionally tried to tie down the channels so that their movements are restricted in order to protect infrastructure or landholdings. However, such modifications impact upon the natural processes within that area and upstream and downstream, resulting in knock-on effects on biodiversity, sediment dynamics and water flows. Hence many river restoration projects are now trying to reverse such negative management impacts and require a more sensitive understanding of channel sediment and water dynamics. This chapter deals with the above topics in order to provide an understanding of river basin processes to improve the basis for management decisions. While the topic of groundwater is an important component of river basin hydrology it is dealt with in its own chapter (see Chapter 5) and therefore we will not deal with groundwater in great detail within this chapter, but will instead refer the reader to Chapter 5 at relevant points.

B HYDROLOGICAL PATHWAYS

1 Infiltration-excess overland flow

Precipitation may land on the ground surface or it may be intercepted by vegetation. Some of the intercepted water may evaporate before it reaches the ground. The water that reaches the land surface may **infiltrate** (see **infiltration**) into the soil or some of it may pond up and run over the land surface as overland flow. Infiltration is affected by the topography, vegetation cover, soil texture, soil structure, the amount and connectivity of soil **pore spaces** and surface compaction. The **infiltration rate**, which is a measure of the volume of water passing into the soil per unit area per unit time, is important because it determines whether all, part or none of the water reaching the ground surface will enter the soil or will move over the land surface as overland flow. Water moving as overland flow generally tends to enter rivers more quickly than water moving below the ground surface. Therefore, if a river basin has a large proportion of land with a low **infiltration capacity** (maximum rate at which water will enter the soil under plentiful water supply), then it is likely that the **river discharge** will increase rapidly in response to rainfall.

The infiltration capacity of a soil tends not to be steady (Holden, 2005a). Soils are made up of solid particles between which there are spaces. These pore spaces contain some air and, even in the driest deserts, also contain a small amount of water bound to edges of soil particles, known as **hygroscopic water**. As water infiltrates and then **percolates** through a soil, the pore spaces can fill with water. Once the pore spaces are full of water, then the soil is saturated. As the pore spaces fill up, then the infiltration capacity decreases and when the infiltration capacity drops below the rate of rainfall supply, then some of the water will start to pond up on the surface and flow (**Figure 3.1**). This process is called **infiltration-excess overland flow** (also known as Hortonian overland flow). This type of flow is uncommon in many temperate areas except in urban locations, along roads and paths or perhaps along compacted soils in arable fields, created by tractor wheels or overgrazing by animals. Infiltration-excess overland flow may become more common with climate change if rainfall intensities increase due to warming or if surface crusts form after long, dry spells (see Chapter 2). Infiltration-excess overland flow is more common in semi-arid regions where soil surface crusts have developed and rainfall rates can be rapid. It is also more likely in areas where the ground surface is often frozen. Soils with a deep litter layer, such as those within tropical rainforests, tend to have large infiltration capacities and so, despite high rainfall intensities, infiltration-excess overland flow is restricted. Only some parts of a river basin produce infiltration-excess overland flow at any given time and this is known as the **partial contributing area concept** (Betson, 1964). The areas producing infiltration-excess overland flow tend to be the same from storm to storm.

2 Saturation-excess overland flow

When the soil is saturated, the **water table** (highest point below which the soil or rock is saturated) is at the surface. Water can therefore leave the soil and run out over the surface, producing saturation-excess overland flow. Saturation-excess overland flow is common in shallow soils or at the bottom of hillslopes, where water running through the soil will collect (**Figure 3.2**). In these zones the soil is kept saturated for longer, due to the delivery of water from a large upslope area and low gradient. Water can return to the surface at this point after travelling within the soil, which is why it is sometimes referred to as 'return flow'. The chemistry of saturation-excess overland flow is more affected by the soil properties than is the chemistry of infiltration-excess overland flow, as the former will have had more contact with the surrounding soil (see Chapter 4). Saturation-excess overland flow can occur long after it has stopped raining. If it is raining, then the additional rainwater will find it difficult to enter the soil if it is saturated and so saturation-excess overland flow can be a mix of fresh rainwater and water that has been within the soil for some time.

The parts of a hillslope or river basin that produce saturation-excess overland flow will vary during an individual rainfall event. If the system starts off quite dry, then during any following rainfall perhaps only a tiny part of the basin will generate saturation-excess overland flow. However, if rainfall continues for some time, then, as more of the soil becomes saturated, especially in the valley bottoms, a larger area of the landscape will produce saturation-excess overland flow. The source areas for saturation-excess overland flow can also vary markedly with season, with larger areas tending to produce saturation-excess overland flow during cooler, wetter months.

This variability in the parts of the landscape that act as source areas for saturation-excess overland flow is known as the '**variable source area concept**' (Hewlett and Hibbert, 1967).

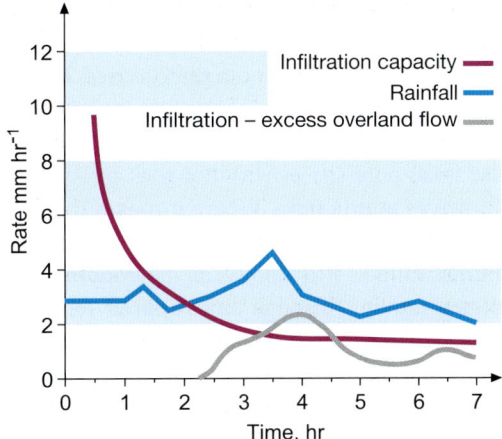

Figure 3.1 An example gentle rainfall event resulting in a reduction in the infiltration capacity of the soil over time and the subsequent production of infiltration-excess overland flow. Note that overland flow is not produced at the immediate point at which rainfall rates exceed infiltration capacity but is very slightly lagged as it takes time for water to pond in small depressions and overflow.

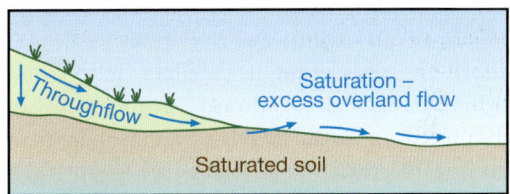

Figure 3.2 Production of saturation-excess overland flow.

3 Throughflow

Water moving through soils or rocks is called **throughflow** or **interflow**. This flow of water through the ground can maintain river flow during dry periods (the **baseflow** of the river). The pathway that water takes through the ground affects how quickly water can get to a river channel and, therefore, the typical nature of the response of a river to rainfall. Where water moves through the small pore spaces within a rock or soil it is known as **matrix flow**. Where it moves through larger connect pore spaces such as cracks or animal tunnels it is known as **macropore flow**. Water moving through **macropores** tends to travel at faster speeds than matrix flow.

Most soils do not remain saturated for long, as gravity drains some water out of the pore spaces. However, once gravity has played its part there is still water left within the soil. This is because of the combined attraction of the water molecules to each other and the water to the soil particles. Water that is held in soil against the force of gravity is known as **capillary water** and it will move within the soil from wet to dry areas. This means that water can move upslope and capillary action is how plants can obtain water from soil. Imagine you hold some tissue paper above a dish of water. When you dip the end of the tissue in the water you can see the water rise up the tissue well above the dish. That is capillary action and demonstrates that the upward 'suction' forces of the pore spaces within the tissue are stronger than downward gravitational forces (see Chapter 1). The forces holding water within soils are much greater in small pore spaces than larger ones, which is why sandy soils are generally drier than clay soils. When water is added to a soil it will tend to fill small pore spaces first, where the 'suction' or '**soil water tension**' is greater.

When all of the soil pore spaces are full of water the soil is saturated and there are no suction forces. Instead there will be a positive **pore water pressure** caused by the pressure of water from above and there will be a gravitational effect. When the soil is unsaturated there will be forces associated with gravity and negative forces associated with the suction effect. In fact it is possible to assess how all these forces balance to determine the direction in which water is likely to flow within a hillslope (Hillel, 1982). These effects may be expressed by assigning *potentials* to the soil water. Soil water potential is the energy per unit quantity of water in the soil. The energy state of soil water can be expressed as energy per mass of water (Joules per kg), as energy per water volume (Joules per m^3; which is also equivalent to pressure measured in Pascales) or as energy per unit weight of water (Joules per 9810 kgms^{-2}) (Duner and Or, 2005). If we use the latter, the results simplify down to units of length (m) since one Joule is equivalent to 1 kg m^2 s^{-2}. Therefore many hydrologists like to express water potential as a pressure head with units of water column length (m). An example is shown in **Figure 3.3**.

While there are many forces, when it comes to the dominant processes in the field they are matric potential (the soil suction effect we described above, which is inversely proportional to the pore space size), submergence potential (positive pore water pressure) and elevational potential (the relative effect of gravity). At any point in a soil profile either the soil is saturated or it is not. Therefore, that point in the soil can only either have a matric potential or a submergence potential and not both at the same time. Thus, to work out the **total hydraulic potential** of the soil water all that is needed is a measure of the relative elevation of that point (to determine elevational potential), and either the soil suction (if unsaturated) or the positive pore water pressure (if saturated) and to sum these values. Soil suction will always be negative, and elevation and submergence potentials will always be positive.

The pore water pressure is measured in situ using piezometers. Piezometers are tubes vertically installed into the ground with a thin perforated section located at the depth of interest in the soil or rock to allow water through at that point. The rest of the piezometer tube is not permeable. When there is positive pore water pressure (i.e. the point is saturated), then water will rise up the piezometer tube. The pore water pressure is equivalent to the elevation of water in the piezometer tube relative to the perforated part of the tube and so merely requires measurement of the water level. Alternatively, the pressure of the water in the tube can be measured by a pressure sensor housed within the piezometer. Soil suction is measured in situ using tensiometers. Tensiometers typically consist of a sealed tube filled with water, a porous ceramic tip and a vacuum gauge. The tensiometer is inserted into the soil so that the ceramic tip is at the depth of soil that is to be monitored. In drier conditions water is drawn out of the tensiometer through the ceramic tip, which creates a partial vacuum in the tube which can be measured by a gauge (either a visible dial or an electronic sensor). In wetter conditions, water is drawn back into the tube and the vacuum is reduced. Combined instruments that automatically measure both pore water pressure and soil suction for the same point (simply a tensiometer and piezometer connected together) using electrical recording sensors are now more

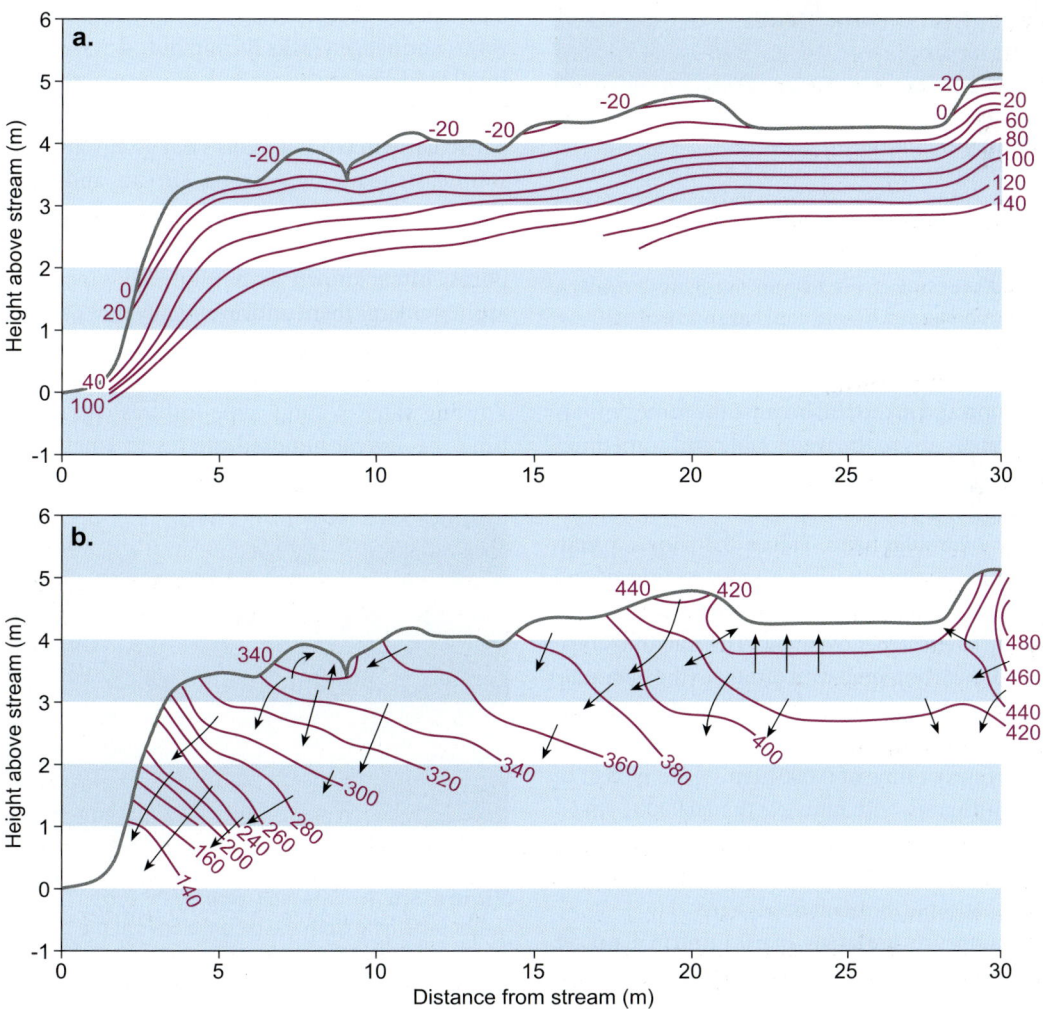

Figure 3.3 Soil water energy status on a slope: (a) contours of equal matric potential; (b) total potential where relative height has been added to matric potential.

Figure 3.4 Macropore flow is indicated by the staining white dye solution where there are fingers of dye extending downwards from the main body of dye in a soil profile. The dye was added evenly to the soil surface but more quickly percolated down macropore channels to produce this pattern and shows how flow in macropores can bypass the main mass of soil.

common and are useful because the same point is sometimes above the water table and sometimes below the water table.

The total hydraulic potential is also known as the **hydraulic head**. **Figure 3.3** shows a map of hydraulic potential for a cross-section of soil across a hillslope and this provides a way of predicting soil water movement. The rate of change of hydraulic potential with distance (the **hydraulic gradient**) determines the rate of flow. Typical rates of flow in saturated sandy soils are 0.01 cm s^{-1}, while rates of flow of 0.0001 cm s^{-1} are typical of silts and loams, with rates less than 0.00001 cm s^{-1} in clays. Deeper groundwater flow through rocks is described in Chapter 5, and so we do not deal with that topic in much detail here.

Macropores are larger than 0.1 mm in diameter and are also connected conduits for flow. The existence of macropores alone is insufficient for there to be a significant hydrological effect (Beven and Germann, 1982). A macropore has to be connected to a water supply in order for there to be flow and water must travel some distance along the macropore so that it bypasses part of the soil mass (**Figure 3.4**). In some environments, such as crusted sandy soils or peatlands, measurements have suggested that half of the subsurface water movement could be through macropores (Baird, 1997; Holden, 2009; Leonard *et al.*, 2001). Fertiliser applied to arable fields with lots of macropores may get washed through the soil and will not be available for crop use and may lead to river pollution. Ploughing can help break up macropores. Macropores can be formed by plant roots, shrinkage cracks during hot, dry weather, small landslips or soil animals. Some macropores can enlarge to be several metres in diameter, particularly in erodible soils such as in semi-arid southeast Spain or Arizona (Bryan and Yair, 1982), or in karst limestone cave systems. Soil water turbulent flow within these large macropores, often known as natural soil pipes, can rapidly enlarge them until eventually they collapse to form gullies (**Figure 3.5**). Macropores can also be enlarged by the dissolving of material into the flowing water. Natural pipe features can sometimes be several hundred metres in length and

Figure 3.5 A large soil pipe with some flowing water emerging from the pipe which is at the head of a gully. The pipe's collapse, followed by erosion at the pipe outlet, has led to gully formation. The photographer was standing in the gully to take the photograph. Note also the additional pipe on the left which flows only during extreme storm events.

have been reported to contribute large proportions (occasionally 10 to 50%) of the flow to rivers in some headwater systems (Smart *et al.*, 2012; Zhu, 1997).

4 Water balance

It is possible to calculate a water balance for river basin systems. The main input is precipitation (unless there is a large adjacent **aquifer** flowing into the basin, bypassing the surface topographical river basin boundary). The main outputs are river discharge or evaporation and **transpiration**. The remainder of the water budget consists of any changes in stores in the soil, rock, vegetation, rivers or lakes. However, measuring each of these components can be challenging, especially since there can be a large amount of spatial variability across a river basin. Therefore, estimates of river basin water budgets are subject to very large errors.

Rainfall is commonly measured using rain gauges and is normally presented in the form of a depth of water (e.g. mm, inches). Rain gauges have to be carefully sited to reduce impacts of turbulence around the gauge orifice affecting the measurements and so they tend either to be buried so that their tops are flush with the ground surface, or to have some aerodynamic shape to them or their surrounds (**Figure 3.6**). However, as their name suggests, rain gauges are not so good at measuring other forms of precipitation (snow, mist, hail, dew). In eastern Newfoundland, Price (1992) suggested that around half of the water being delivered to the ground surface was in the form of mist caught on vegetation which then trickled down to the surface, and this was not properly measured by rain gauges. Heated snow gauges are used in some locations to melt the snow as it arrives so that it can trickle into the measuring device to be recorded at the same time as it falls. Disdrometers are used in some places to measure precipitation drop sizes and consist of either a measurement plate that measures the impact of raindrops or a laser transmitter and a receiver placed at a known horizontal distance apart, which detects droplets passing between the two.

Rainfall can be highly variable across short horizontal distances and with altitude, and so if a river basin of, say, 100 km² has one rain gauge that gauge may not be representative of the true inputs to the basin. Thus, networks of gauges are normally installed during monitoring operations, but this still provides only a few values of point rainfall data which have to be interpolated spatially across the basin. Rainfall radar can be used in some locations to record the spatial pattern of rainfall being delivered to a catchment, but the actual amounts of rainfall are usually calibrated using ground rain gauges.

Flowing water which makes it out of a river basin via a river can be measured within the channel. Techniques for measuring river flow are described in section C. Water outputs in the form of evaporation and transpiration tend to be estimated rather than directly measured, based on data on temperature, humidity, wind and plant types. However, it is possible to set up monitoring of blocks of soil (with growing vegetation left

Figure 3.6 A rain gauge flush with the ground surface surrounded by a metal mesh to reduce errors from turbulence and splash.

on top) known as **lysimeters**, in which the soil is weighed in situ (i.e. a weighing device is inserted below the block of soil so that the soil is kept in the ground in its natural setting) and where changes in the weight are ascribed to changes in the soil water content. Water draining down through the soil is caught and weighed and so the remaining weight change of the soil is assumed to be due to evapotranspiration. Evaporation rates from open water bodies can be estimated by using evaporation pans, which are containers of water with open tops in which the water level is measured through time. Water inputs to the pan are known and the only output is by evaporation, and so changes in water level can easily be ascribed to evaporation rates.

Changes in water storage within soils and rock can be determined by measuring moisture content (there are dozens of methods – see Chapter 6 in Shaw *et al.* (2011) for a description of methods) or by the groundwater techniques described in Chapter 5. However, these processes are spatially highly variable and most soil water storage changes are simply assumed, based on the difference between measured inputs and outputs from the river basin.

Using a simple form of the water balance, it is possible to examine the water efficiency of a river basin. Such a measure is also known as the runoff efficiency or the rainfall to runoff ratio and tells us what proportion of inputs from precipitation is produced as outputs in the form of river discharge. Over annual cycles or longer, the change in storage is sometimes thought to be zero (unless heavily modified by humans, such as through water abstraction, or unusual weather conditions, such as a drought), and so the simplest form of the water budget can be used to provide the evapotranspiration from the basin as: precipitation – river discharge = evapotranspriation. **Box 3.1** provides a short example of a runoff efficiency calculation.

> **REFLECTIVE QUESTION**
>
> What are the main processes of water movement through and across the landscape to reach rivers and lakes?

BOX 3.1 TECHNIQUES

Calculating the hydrological efficiency of a river basin

Here we take a simple worked example to illustrate how runoff efficiency can be calculated. Assume a river basin area of 80 km², precipitation of 1600 mm in a year and that there was a river gauging station which told us that there had been 72 million cubic metres of water discharge in total during the year. What we need to do is put everything into the same units. One way of doing this is to put everything into units of length (depth of water, just like rainfall). So imagine that you spread all of the 72 million cubic metres of water across the river basin evenly; then you would have a depth of water equivalent to: 72 000 000 m³ divided by the area of the basin, which is 80 000 000 m², which gives us 0.9 m or 900 mm. If we compare this depth of water produced as river discharge from the basin to the precipitation that fell on the basin (1600 mm), then we can see that this equates to an efficiency of 56.25%. In other words, 0.5625 of the precipitation that fell on the basin was released from the river basin outlet as river discharge. The remaining 43.75% of water was either lost from the catchment as evapotranspiration or resulted in a change of storage in lakes or groundwater. Similar calculations can be performed for individual storm events.

C RIVER FLOW

1 Hydrographs and river regimes

Rivers are supplied by throughflow (including groundwater) and overland flow. The relative proportions of the different types of flow, along with the soil types, topography, size and type of drainage network, can determine how quickly river flow varies during rainfall events or seasonally. River flows can change during individual rainfall or snowmelt events, or remain fairly stable, depending on the nature of the river basin. There is a delay (lag) between the precipitation occurring and the peak discharge of a river.

Where infiltration-excess overland flow dominates the **runoff** response, then the hydrograph (graph showing discharge through time) is likely to have a short lag time and high peak flow (**Figure 3.7a**). The storm hydrograph showing river discharge through time in response to a rainfall or melt event will therefore be quite steep in shape, rising quickly from low flow to the peak flow. Urbanisation increases flood risk, as it reduces the infiltration capacity of the surface through construction, leading to rapid water flow to the river and resulting in steeply rising hydrographs with sharp peaks. If matrix throughflow dominates flow pathways within the river basin, then the river may rise and fall very slowly in response to precipitation and the peak may be small. However, since throughflow contributes to saturation-excess overland flow, then throughflow can in many circumstances still lead to rapid and large flood peaks. Even without saturation-excess overland flow, a throughflow-dominated catchment may still produce storm discharge if water can get through the soils quickly (e.g. if the soils are very permeable). In some soils, such as peat, only a small amount of infiltration may be needed to cause the water table to rise to the surface (Evans *et al.*, 1999). In other soils there may even be two river discharge peaks caused by one rainfall event (**Figure 3.7b**). This might occur where the first peak is saturation-excess overland flow

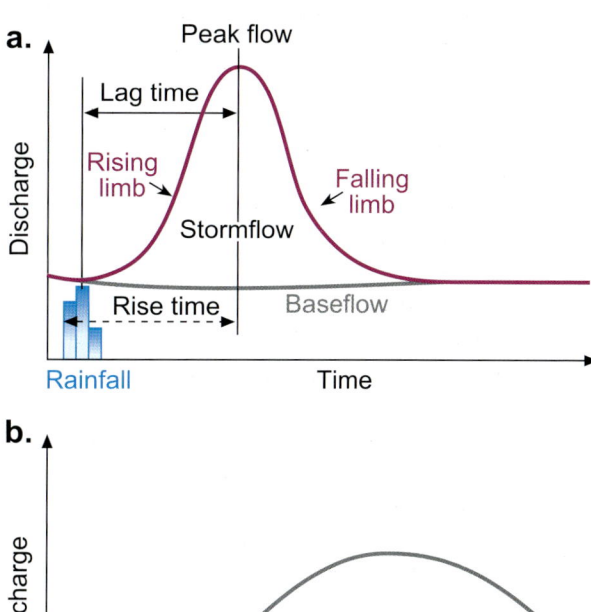

Figure 3.7 Storm hydrographs (a) where infiltration-excess overland flow dominates and (b) where throughflow dominates. Note that (a) also indicates typical hydrograph terminology.

dominated, or perhaps related to the direct delivery of rainfall into the river channel, and the second peak a little later may be much longer and larger and caused by subsurface throughflow accumulating at the bottom of hillslopes and in valley bottoms before entering the stream channel (Anderson and Burt, 1978). Throughflow may also contribute directly to storm hydrographs by a mechanism called piston or **displacement flow**. This is where soil water at the bottom of a slope is rapidly pushed out of the soil by new, freshly infiltrating water entering at the top of a slope.

Figure 3.8 shows flows in the form of hydrographs over one year for two nearby rivers where the climate is the same. Despite being in the same area, the flows are very different between the

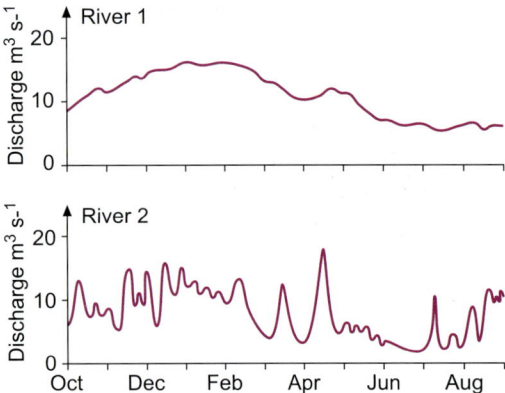

Figure 3.8 Two example annual river flow records showing regimes of each river (after Holden, 2011).

rivers. The flows in River 1 appear to be dominated by baseflow and there are no individual storm peaks, unlike for River 2, which appears to be more dominated by overland flow or rapid throughflow (e.g. via macropores). The River 1 basin overlies permeable bedrock, is gently sloping, and has soils that enable good infiltration and little chance of saturation to the surface. Therefore, infiltration-excess or saturation-excess overland flows are a rare occurrence here. For River 2, however, the soils are thin and sit above impermeable bedrock and so there is frequent saturation of the soils and generation of saturation-excess overland flow.

Seasonal variability in river flow is known as the **river regime**, for which there are four major global types:

1. Arid zones, especially in subtropical drylands, tend to experience very occasional but intense rainfall events. Intense rainfall along with little vegetation cover produces infiltration-excess overland flow, rapid runoff and high flood peaks. However, many dryland soils are coarse and sandy, with high infiltration capacities, resulting in little chance of overland flow. Therefore, there is a wide variation of response even if rainfall intensities are very high. In most drylands river flow will stop within a few days of the rainstorm and water often seeps into the river beds or is evaporated. River flow is therefore highly intermittent in these systems.

2. Where snow and ice melt dominate, then there can be a major peak of river flow during the late spring or early summer (e.g. Danube in Hungary, or Mackenzie in North America). River discharge can be extremely low during the winter months on some rivers, even though precipitation may be continuing, as this precipitation is stored on glaciers or snowpacks in winter. There can also be a strong daily change in river discharge due to daily melt cycles of the snow and ice. Night-time discharge immediately downstream from glaciers tends to be much lower than that of the mid-afternoon.

3. In temperate, oceanic areas precipitation occurs all year, perhaps with seasonal maximums. The river flow regime in these areas can change in response to either seasonal changes in groundwater storage and release, or higher evaporation and transpiration rates during the summer months (e.g. Seine, France).

4. Rivers in equatorial areas tend to have a fairly regular regime, while tropical river systems outside of the equatorial areas receive high precipitation during the summer but experience a marked dry season during the winter. Evaporation and transpiration are high at all times so that the streamflow mirrors the seasonal pattern of rainfall. Rivers such as the Brahmaputra or Mekong in south Asia have summer peaks associated with monsoon rains.

2 Closed basins

While most rivers around the world drain into the oceans (**exorheic basins**), there are many that do not, and these are often termed 'closed basins' or **endorheic basins**. Almost one-fifth of the Earth's land surface does not drain to an ocean but instead drains to an inland 'sink' (**Figure 3.9**). These tend to be in the interior of continents, in dryland regions, or where surrounding topo-

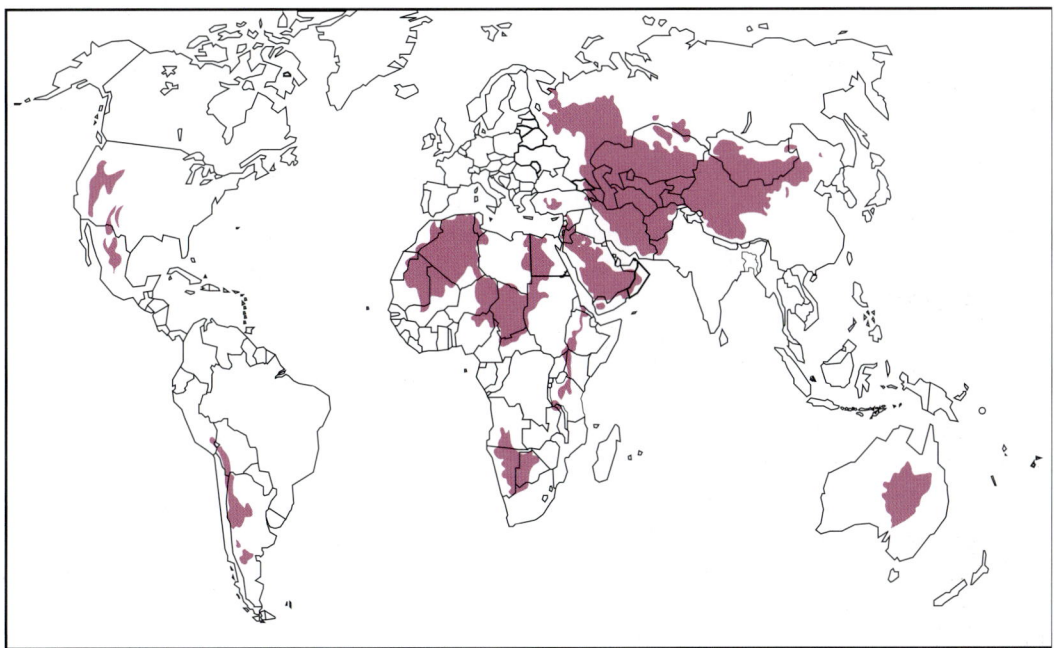

Figure 3.9 The endorheic river basins of the Earth.

graphy such as a mountain range prevents water from reaching the ocean. The largest of these endorheic basins covers much of continental Asia and river water here can end up in the Caspian Sea and Aral Sea. The Great Basin (which is actually several adjacent closed basins) in the USA and much of central Australia (e.g. Lake Eyre basin) are other examples. The water in lakes into which endorheic rivers drain tends to be very saline because evaporation is a major water loss, leaving behind concentrated salts. For example, the Dead Sea, at the end of the River Jordan, which is located at the lowest point on the surface of the Earth, is so saline that people can lie in the water without sinking. In the centre of many endorheic basins there is no permanent standing water body, but there can often be a white 'salt pan' where once a lake existed, or where the lake occurs only seasonally.

3 Measuring river discharge

Good measurement of river discharge is important for water resource management, flood warnings and monitoring for impacts of environmental change. Rivers act as a point of concentrated water movement where the confinement of water to a channel can assist in measuring the discharge. Typically, gauging stations consist of a method to measure the water levels (such as a pressure sensor) and a method to convert water levels into discharge. To help with this process many gauging sites use weirs or flumes, although these are not always practical (e.g. on large rivers). The weirs and flumes provide more control over the relationship between water level and discharge. This relationship is known as a **rating equation** or a stage-discharge equation. Sometimes a design of weir has a standard engineering rating equation, but even if this is the case, testing in the field is recommended. To derive a stage-discharge equation, so that from any water level at the point in the river being studied we can calculate the discharge, requires us to measure the discharge of the river at a range of water levels.

Various techniques can be used to measure river discharge, including dilution gauging, whereby a known volume and concentration of a tracer, such as salt solution, is added to the river. The more water in the river, the more diluted the salt solution will become. Therefore, measuring the salt's concentration slightly downstream enables us to work out the dilution of the salt and hence what the river discharge is. For a more detailed explanation of salt dilution gauging see the short video at http://tinyurl.com/aopjqxb. Alternatively, the cross-sectional area of the river can be measured and then the river cross-section is split into different segments and the velocity of the water passing each segment is measured. The water velocity may be measured using an impeller meter (typically at 0.6 of the depth as an assumed average of the water velocity in the entire water column; water tends to be slower near the bed, due to friction, and faster near the river surface) or other device (often done by someone standing in the water holding the device). Multiplying the velocity (m s^{-1}) by the area of the segment (m^2) produces the discharge (m^3 s^{-1}) for that segment of the cross-section (**Figure 3.10**). Summing across all segments in the cross-section then yields the total river discharge. This sort of technique can be impractical at very high river flows or where the river is highly vegetated, has lots of boulders or has several shallow channels.

Ultrasonic discharge gauges record the time for acoustic pulses to cross a river at different depths. Devices are placed on opposite banks of the river and the travel time for the pulse is used to calculate the streamflow velocity. A water level sensor is also needed and the cross-sectional area has to be surveyed in advance for different water levels. Doppler probes record the change in frequency when acoustic waves are reflected from natural suspended sediment or gas bubbles in river water. The change in frequency is proportional to the velocity of the particle or gas bubble. The Doppler method therefore works well in murky rivers, unlike ultrasonic discharge gauges described above. Sometimes Doppler profilers are deployed from boats or from cables across rivers to provide velocity measurements across the channel. Measuring river discharge can be very costly and is impractical in many very remote places or for large rivers. In fact the

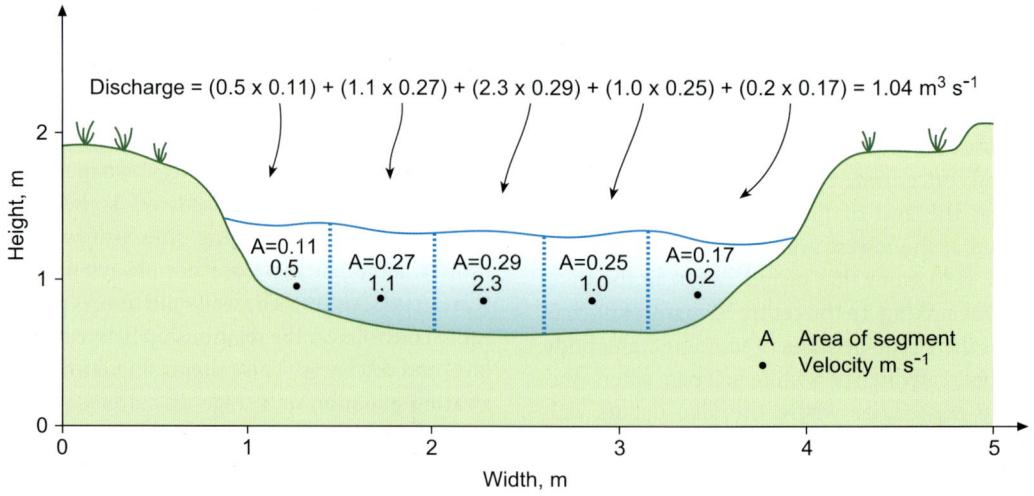

Figure 3.10 A worked velocity-area calculation for river discharge.

number of gauging stations is reducing worldwide due to cost (Vörösmarty *et al.*, 2000). However, work is ongoing to refine the use of satellite data to determine river discharge, as described in **Box 3.2**.

4 Models of river discharge

Models that can predict how rainfall (either real or simulated) produces river discharge can be very useful for water resource planning, flood forecasting, designing flood defence systems and in estimating river flow in basins that have no river gauges. They can also help us understand how river basin systems work, and we can perform landscape-scale experiments using models that we might not be able to do in the real world. Predicting river discharge from rainfall is not as simple as it might at first sound, particularly if the basin is large, because that will most likely mean that the basin has many different types of landscape within it. The river discharge that results from rainfall will vary with antecedent conditions because these affect infiltration rates and saturation conditions, and will vary with the distribution of rainfall across the basin through time. Many hundreds of models have been developed that are used to predict river discharge, and these vary in complexity. However, the quality of data input into models is an important consideration that will affect how good the prediction of discharge is. All models, no matter how complex, are just approximations of reality, and so the choice of model to be used depends on the questions being asked and the type of river basin being studied.

One of the simplest discharge models is known as the unit hydrograph, and was developed by Sherman (1932, 1942). The unit hydrograph is the stormflow that would be produced from an **effective rainfall** (that amount of rainfall which is released by the river as stormflow) over a unit of time (such as one hour or one day). The model assumes that the rainfall has occurred uniformly through time and across the basin. In essence, the unit hydrograph model requires users to obtain a storm hydrograph, work out the total volume of stormflow (in units of millimetres; i.e. average depth of water if spread across the catchment), and call that the effective rainfall. Then users should divide the hydrograph size (i.e. the y axis) by the effective rainfall so that what we are left with is the size and shape of hydrograph produced by 1 mm of effective rainfall. The hydrograph will now have units of $m^3 s^{-1}$ per mm of rain. This is now the unit hydrograph for the river basin. In order to predict the river discharge for new rainfall events (either real ones or ones that we have made up to test, for example, how big a flood might be) we need to establish how much of the rainfall is effective rainfall. This can be done by using the first storm we derived the unit hydrograph from by subtracting effective rainfall from actual rainfall

---

BOX 3.2 THE FUTURE OF WATER

Measuring river flow from space

The Surface Water Ocean Topography (SWOT) mission, supported by NASA and Centre National d'Etudes Spatiales, seeks to directly measure water heights from space with an accuracy of a few centimetres. The mission plans to measure water levels every 100 m across the entire globe every 10 days. Such a dataset would have the potential to hugely improve our understanding of the global water cycle and to monitor river and lake levels. Additionally, by measuring the topography of the river at low- and high-water levels from space, cross-sectional areas can be derived for different water levels. Combining water level data with cross-sectional information and the slope of the river, all derived from space, could enable river discharge to be estimated for rivers across the globe at 10-day intervals. For further information please see http://swot.jpl.nasa.gov.

and averaging the remainder over time to give us an average 'infiltration rate' in mm hr^{-1}. This infiltration rate is really just the average amount of rainfall that does not produce storm discharge and is not a real infiltration rate for the soils in the river basin. It is often called a phi-index. If we use that same phi-index for future rainfall events, we can then determine how much of the rainfall is effective rainfall. If we are then left with, say, 1 mm, 3 mm, 2 mm of effective rainfall over three consecutive hours, then we can establish the predicted hydrograph by multiplying the unit hydrograph by one in hour one and then making a new hydrograph starting one hour later which is three times the size of the unit hydrograph and finally, in the third hour, we have a hydrograph which is twice the unit hydrograph (**Figure 3.11**). The next step is simply to add the three storm hydrographs together to get overall predicted storm discharge from the rainfall event. If we wanted to predict overall river discharge, then the final step is to add baseflow.

Clearly, the unit hydrograph model is simple to use, but it also entails a lot of assumptions about uniformity in the river basin and does not deal with spatial processes such as the variable source area concept and issues around antecedent conditions. Models such as the unit hydrograph suffer when tested for different conditions because the same amount of rainfall, even if uniformly distributed across the river basin, can produce a very different storm discharge depending on antecedent conditions (e.g. wet soils or dry soils). There has therefore been development of a suite of models, such as Topmodel and SHE, that try to capture elements of spatial flow processes and saturation processes. Basic overviews of those and other models are provided in Shaw *et al.* (2011), with more detail in Beven (2001), but in essence such models use the topography of the river basin and some information about soil types and saturation conditions to move water through and across the landscape to the river system.

River flow models are typically tested by comparing a predicted river discharge with the observed river discharge. After some fine tuning to optimise the model to work best (it will never be perfect), then it can be applied to situations where we do not yet have river discharge data (e.g. future events, simulated storm rainfall, a change in land management (e.g. see **Box 3.3**), ungauged river basins). However, users of such models must always be aware that the predictions are just approximate estimates and it is useful for any predictions to be presented along with an estimate of how far we think the prediction might be wrong.

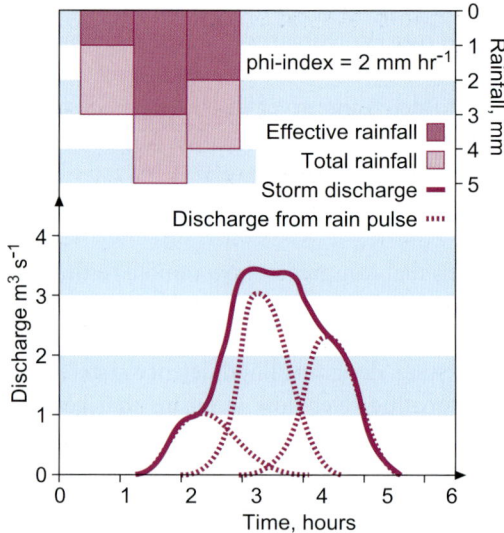

Figure 3.11 Predicted river storm flow discharge using the unit hydrograph model. Each hour of rainfall is associated with its own hydrograph. The first hydrograph produced from the first pulse of rainfall results from an effective rainfall of 1 mm and is therefore the basic unit hydrograph. The second hydrograph results from the second pulse of effective rainfall, which is 3 mm, and the hydrograph is therefore three times the size of the first one. All of the smaller hydrographs resulting from each pulse of effective rainfall are added together to produce the overall storm hydrograph for the catchment predicted to result from the rainfall event shown. Baseflow is not shown but if you wanted to predict total river discharge you would simply add baseflow to total stormflow.

5 Impacts of land management on river flow

Land management change can dramatically alter the flows in river systems. Building dams and changing or diverting stream courses alters river regimes. In fact the Colorado River, which is the main river in the southwest of the USA, has been so heavily modified through major dams and abstraction schemes that discharge along its lower course is massively reduced so that it now rarely flows to its natural exit in the Gulf of California.

Major land management change may alter the water budgets of river basins. For example, deforestation may result in a large reduction in transpiration rates. In 1934 a hydrological monitoring station was set up at Coweeta in the southern Appalachian Mountains to look at how forest management affected river flows. The research showed that the amount of water flowing from a forested basin increases after timber harvest, mainly because of reduced transpiration from trees. The volume of water from storms (efficiency of the river basin) and peak flow increases after deforestation. As the vegetation grows back and trees begin to recover, the peak flows and the basin efficiency start to decline again.

Any management activity that alters the flowpaths for water and the proportion of water moving via that flowpath – infiltration/saturation-excess overland flow, matrix or macropore flow – has the potential to alter the river hydrograph responses to storm events or the regime of the river. Agricultural activity can have a major impact on soil infiltration. Ploughing can often increase the infiltration capacity of the soil, which may then decline during the cropping season until the soil is ploughed once more (e.g. Imeson and Kwaad, 1990). It has been shown that livestock grazing has profoundly affected soils throughout the world, from arid and semi-arid rangelands to temperate moorlands. A review by Greenwood and McKenzie (2001) suggested that grazing animals exert pressure on the ground comparable to that of agricultural machinery, which leads to soil compaction. Research has generally shown that as grazing pressure increases, vegetation cover declines, soil bulk density increases, water infiltration rates decrease and surface runoff levels increase (Abdel-Magid *et al.*, 1987; Mwendera and Saleem, 1997). Evidence from a UK study showed that alterations of vegetation and soil properties due to grazing greatly increased active source areas for overland flow and discharge during storm events (Meyles *et al.*, 2006).

Field under-drainage and the use of open ditches have been common land management techniques in agricultural areas. These systems would tend to be installed where the land is deemed to be too wet for optimum agricultural performance and their aim is to lower the water table. Such drainage often contributes to river regime by increasing baseflows and potentially increasing the flood peak. Drains and ditches act as preferential flowpaths for the fast channelling of water to watercourses. In peatland systems the use of open ditches has lowered the water table and caused organic soils to dry and crack, which has resulted in enhanced flow of water through macropores and the expansion of natural pipe networks (Holden, 2005b). Flow through these pipes can erode the system from below, deliver a lot of carbon to the stream (Holden *et al.*, 2012) and lead to gully development when pipes collapse.

Urban development often reduces the local infiltration capacity to zero, which can increase the propensity for downstream flooding (Perry and Nawaz, 2008). The loss of vegetation cover also reduces transpiration and hence increases the annual and storm efficiency of the river basin. Urban areas tend to have a subsurface drainage system which takes water away from the surface using sinks which pipe it to water courses. Traditionally these underground drainage systems have served as fast pathways for water movement across the basin and have increased flood peaks. More modern developments use **sustainable urban drainage systems**, which try to minimise the downstream impact on river flows of the urban development (see Chapter 4, section F2.2). They often do this by encouraging the use of more permeable surfaces and also by having pipes and conduits that discharge into a series of ponds or storm water-collecting channels which then drain more slowly into the river network at a later stage. In essence, sustainable urban drainage systems have made space for storm water in their design. However, during intense rainfall events these systems can still be overwhelmed; clearly, the costs and space required for large storage are greater and so a judgement has to be made as to how much space to allocate for sustainable urban drainage, without overly compromising the aesthetics of the landscape or the commercial value of the land. Many sustainable urban drainage systems also seek to clean the water before returning it to rivers (e.g. by using lake or wetland systems to settle out sediment). These systems can be designed to look aesthetically pleasing and perhaps also serve other purposes such as providing recreational and leisure services (**Figure 3.12**).

Predicting the effects of land management change on river flow is not straightforward because river basins are not uniform and have varying soils, topography and drainage networks. Therefore river basin models are often deployed to investigate the potential impacts on river flow of management interventions (e.g. **Box 3.3**). In fact the same land management can have a very different impact on river discharge (e.g. the flood peak), depending on where in the river basin that management takes place. Section D provides further explanation of this issue. These spatial effects, related to how water is routed through the drainage basin and the timing of delivery of water

Figure 3.12 New housing development in Calgary with a sustainable urban drainage system with a storm water storage lake that creates an aesthetic scene for the houses.
Source: Photo courtesy of Shutterstock/Jeff Whyte.

BOX 3.3 CONTEMPORARY CHALLENGES

Modelling land use change impacts on river discharge: a case study from the Pinios River, Greece

Pikounis *et al.* (2003) used the Soil and Water Assessment Tool (SWAT), which is a hydrological model that requires inputs of land use on a digital map, topography, soils and climate data among other datasets. They wanted to examine the effect of potential land management change on monthly total river flow in the Ali Efenti basin, which is a tributary of the Pinios River in Thessaly, Greece. Three land management scenarios were tested: (a) agricultural expansion by 21% with a decrease in forest cover by 15%; (b) deforestation of a whole sub-basin, accounting for 9% of the Ali Efenti basin area; and (c) an increase in urban area from 3% to 7% in a sub-basin whose area covers only 9% of the whole basin. The latter was associated with a small decrease in forest and agricultural land. These land use changes were applied to areas of the river basin where it was felt most suitable for the change to take place (i.e. agricultural expansion on a steep mountain slope would not be sensible). Figure 3.13 presents the model predictions run over a 24-year period during which there was real streamflow data produced under the present land use. The figure shows that the SWAT model predicts for all three management scenarios an increase in discharge during wet months (October–April) and a decrease during drier periods. The deforestation scenario seems to have resulted in the largest changes, with an increased flow of around 23% during wet months and a decreased flow of 38% in the summer (May–September). These sorts of modelling results can be used by government agencies to support planning decisions, to understand potential future river flows and to allocate resources to investigate ways of minimising the impacts of land management change.

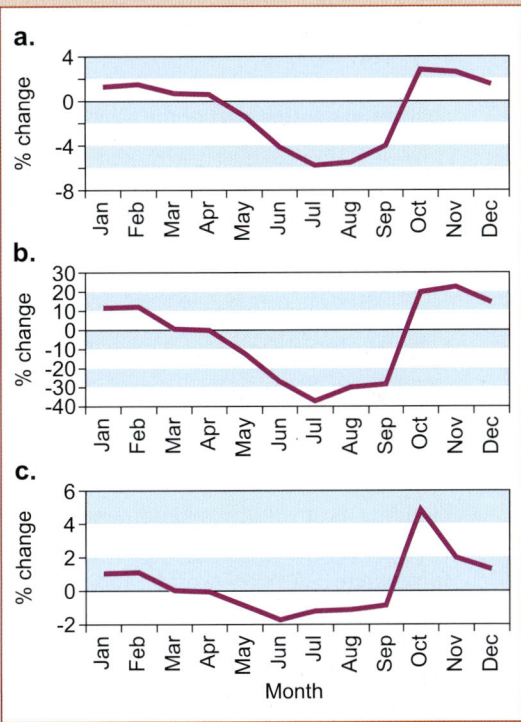

Figure 3.13 Mean monthly total discharge change (%) in the Pinios River for a 24-year modelling period compared to the observed data based on current land use for (a) agricultural expansion, (b) deforestation of the Trikala sub-basin and (c) urban expansion. Note the change in scale on the y-axes for each plot.

Source: Figure reproduced with kind permission of Global Nest from Pikounis *et al.* (2003). Application of the SWAT model in the Pinios River basin under different land-use scenarios. *Global Nest: The International Journal* 5: 71–79.

from different parts of the basin, are another reason why using models to help understand the impacts of land management change on river flows is important. Models can be used to tell us which parts of the river basin might be more suitable for certain types of management change, as the changes in those locations might have a much lower impact on river flow than the same changes in another part of the river basin.

> **REFLECTIVE QUESTION**
>
> What methods could be used to estimate river discharge for an accessible site and for a completely inaccessible site?

D FLOODING

Flooding is a natural phenomenon and is to be expected. There is always a flood occurring somewhere in the world and http://floodobservatory.colorado.edu/ provides maps of current floods and a database of flood events. Flooding can bring benefits to agricultural zones through supplying nutrients to the soil, but can also bring devastation when it affects residential areas. Flooding on the Limpopo River (Mozambique) in February 2012, which did not make any serious international media coverage, killed at least 37 people and displaced 100 000. In July 2010, 2.4 million people were displaced by flooding and over 600 killed in southern China. Details about the health impacts of flooding are provided in section E1 of Chapter 8.

When rivers overtop their banks and spill onto surrounding land it is known as fluvial flooding. Pluvial flooding occurs when there is heavy rainfall which leads to concentrated overland flow inundating an area. This is particularly common in urban areas, due to impermeable surfaces quickly generating overland flow. Groundwater flooding can also occur where saturation-excess overland flow concentrates. Coastal flooding caused by very high tides, storm surges or **tsunamis** is also important.

Often the fluvial flood record is analysed for the number of times the water level in a river peaked above a given (often critical) level. Sometimes this water level value is not equivalent to a discharge value because the lower parts of many catchments are affected by tides or because the channel has changed dimension (e.g. become silted up). If a storm flow upstream coincides with a spring high tide, then the flood risk may be enhanced. If long-term river-level records are available, then simple return period or **recurrence interval** calculations are possible. For example, if a water level on a river greater than 10 m occurred five times in five years we would say that the return frequency of the 10 m flood at that given point is on average once per year. Of course we could get a 10 m flood occurring three times in one year, and the return frequency is just an average value. Similarly, it is possible to estimate the size of the one in 100-year flood and to map its likely extent. These sorts of maps might be used by the insurance industry and management agencies to manage flood risk. Traditionally, flood defences were designed to deal with a flood of a particular magnitude or recurrence interval. However, inferring what might be expected in the future from flood events that have happened in the past may not be reliable, given that land management change and climate change might impact hydrological processes operating within a given catchment. Hence, flood risk mapping often requires some future climate modelling scenarios. Additionally, there has to be some acceptance of the risk of flooding and that engineering design does not rule out all possible flood events.

Humans have decided to live in low-lying areas subject to flooding. This is because these areas tend to be where there are fertile soils (often made more fertile by regular flooding) suitable for crops and where navigation of rivers by boats allows transport of goods and people. In many places the reaction to flooding has been the construction of flood defences and major river engineering projects. Many flood solutions have involved building **levees** or embankments next to rivers or straightening the river channel and clearing out the sediment and vegetation to allow faster, more uniform river flows. However, in many cases these techniques have led to worse flooding downstream because sending the water more quickly through one part of the river system simply reduces the lag time downstream and increases the overall flood peak. Floodplains, when functioning naturally, act as a temporary store of water. Therefore, if we reduce the space for flood waters upstream we can make flooding worse downstream. However, the demand for building or farming on the floodplains of the world means that fewer and fewer of them are freely available for a river to store its water during wet periods. Flooding in lowland areas is therefore a bigger problem, as the extra water is brought downstream more quickly and in greater quantities than ever before. Solutions to the flood problem need to be mixed, and include land management solutions, flood defence solutions and also getting people to accept the risk of flooding and designing coping strategies. Understanding risk means undertaking to identify the consequences (e.g. cost, potential lives lost), the probability of the consequences and the significance of the risk. In the EU, the Floods Directive requires EU member states to complete flood risk assessments (topographical maps, descriptions of previous floods, land use, etc.), produce flood risk maps (for extreme events with low probability, medium events with a return period of at least 100 years and high probability events) and to produce flood risk management plans. The latter are meant to set out how local agencies will reduce impacts of flooding on the economy, human health, the environment and cultural heritage. The plans are not meant to focus on improving hard-engineered protection from flooding, but to focus on all aspects, including prevention, preparedness and protection. This work should have been completed by 2015. Dealing with the flood problem means that we need to build resilience into the built environment in flood-prone areas. An example of resilience might be designing buildings that have very flexible ground floor use, such as parking areas for cars. When there is a flood warning, then the parking area can be evacuated of cars and flooding can be allowed to occur without any damage to residential dwellings on higher levels.

Changes in flood risk caused by land management change are often complex to understand and measure. It is not possible to say that if you cover 10% of a catchment with trees to take up more water you will reduce the flood risk by 10%. In fact, activities that reduce flooding in a small area of a catchment may lead to the opposite effect downstream and vice versa, depending on the configuration of the river basin. **Figure 3.14** shows a small tributary catchment within a larger basin.

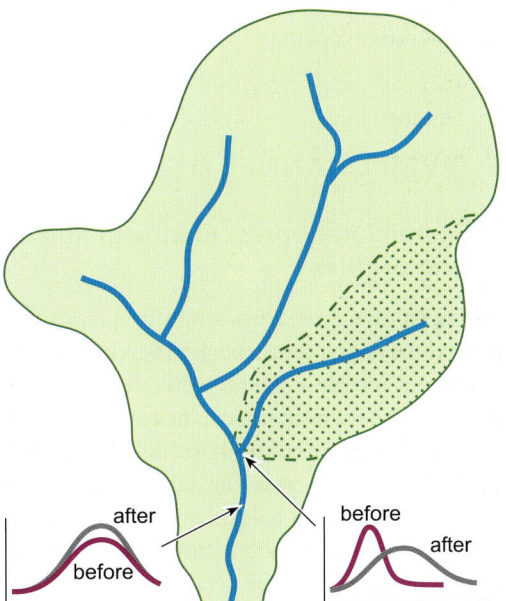

Figure 3.14 A river basin showing what happens to the main channel flood hydrograph after forestry plantation within a tributary basin due to flood wave synchronisation. The growth of trees both lowers and slows the flood peak from the tributary basin. However, because the flood peak timing from the tributary as it enters the main channel now coincides with the main channel flood peak this increases the flood peak in the main channel.

If land management were to introduce a change in the tributary catchment (e.g. a forestry plantation) that resulted in a lower flood peak and delayed lag times from the tributary, this might not necessarily be beneficial. If the peak from the small tributary has been delayed, it may peak at the same time as the discharge peak in the main river channel. Therefore, the tributary will contribute most discharge at the same time as the main river. It is therefore important to understand how all parts of the drainage basin interact and how the whole drainage network operates in order to fully understand the impacts of management change on flood risk.

> **REFLECTIVE QUESTION**
>
> What coinciding conditions would make a flood more severe?

E RIVER CHANNEL DYNAMICS

1 Channel networks, planform and long profiles

Rivers are dynamic. They often move position around the landscape through time, change their shape and move sediment, water and dissolved materials. They are an important agent in removing weathered material from the land surface and redistributing it across the landscape or into oceans. This process balances out the mountain building caused by plate tectonics so that, over long time periods, plate tectonics might build mountains but weathering, erosion and removal by river systems or ice masses (see later in this chapter) smooths out the landscape.

River channels are rarely singular in form and tend instead to be part of branching networks. **Stream order** is a way of categorising the network so that headwater streams are denoted as first-order streams and when two first-order streams meet that produces a second-order stream. Where two second-order streams meet that produces a third-order stream, and so on. A river basin that has more river channels in it than another basin of exactly the same area is likely to be the more efficiently drained of the two. The **drainage density** is defined as the length of stream channel

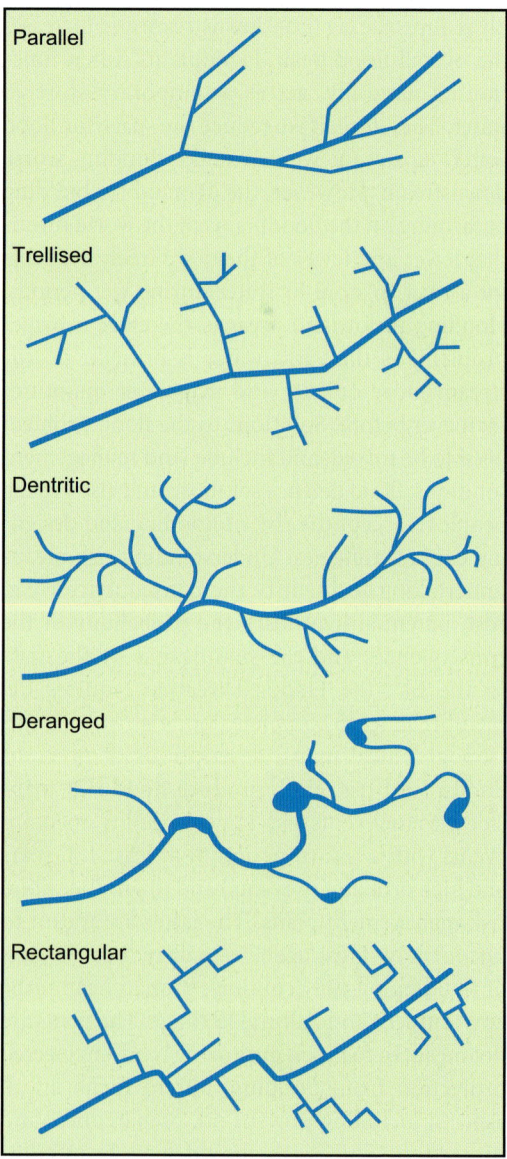

Figure 3.15 The main types of drainage network pattern.

divided by the drainage area. If you look down at the landscape from above, say from an airplane, or using a website that provides aerial images of the Earth, it is possible to see a wide range of patterns and sizes of rivers. From a great distance you can see the whole drainage network, and these are often different in form depending on topography and geology (**Figure 3.15**). Dendritic drainage networks are the most common. Rectangular networks develop where rocks have two directions of jointing at right angles. Joints tend to be eroded more easily, and so streams can develop along the lower-lying eroded joints which meet every now and then at approximately right angles.

Parallel drainage networks tend to occur where the slopes are steep and flow in the same direction. The steepness of the slopes means there are few branching tributaries and the channels tend to be straight. Trellised drainage systems are characteristic of **folded mountains** (such as the Appalachians), where there is a strong structural control on the direction of streams, due to geology. Deranged systems occur in areas where the topography is very irregular and there are lots of low points in which lakes can form.

Individual channels also have distinctive forms, the main ones being braided, meandering and straight (**Figure 3.16**) but one river may have each of these patterns in different sections along its course. Braided channels consist of lots of individual smaller channels which separate and come together, with the islands in between known as bars. The channels can rapidly move locations, with bars eroding on one side and growing on the other. Braided rivers are common where there is a lot of mobile sediment, such as downstream of a glacier. A drop of water running down the surface of a perfectly smooth piece of glass will take a wavy, looping course. It is therefore unsurprising that most rivers are meandering. The meandering pattern is measured by sinuosity, where sinuosity is the ratio of the length of river between two points compared to the length of the straight-line valley between these two points. Straight channels are defined as having a sinuosity of less than 1.5. A meandering channel refers to a single channel with a number of bends which

Figure 3.16 Examples of river channel forms: (a) braided river – Glacier Bay, Alaska; (b) meandering river, and (c) a straight river – the River Tara, Montenegro, occurring in a natural steep-sided valley.
Sources: (a) photo courtesy of Lee Brown; (b) photo courtesy of Shutterstock/Graham Prentice; (c) photo courtesy of Shutterstock/Kurt.

result in a channel sinuosity in excess of 1.5. Straight rivers are more controlled by human action and natural straight rivers are often unstable and become meandering, unless there is a very strong geological control such as in parallel drainage networks.

Channel slope and discharge exert a control on the planform of river channels. For a certain discharge there is often a critical slope above which channels will meander and then a further threshold above which they will braid. These slope thresholds decrease with increasing discharge. Thus, braided sections are usually found on large rivers or on small rivers with steep slopes. Braided rivers will occur where there is coarse sediment, erodible banks and where the main sediment transport mechanism is **bedload transport** rather than suspended transport.

The slopes of river channels (**long profile**s) from their source to mouth are mainly concave, with progressively lower gradients downstream. The long profile varies with geology and water flow. Other profile shapes also occur when there is interruption by lakes or very resistant rocks (which often result in waterfalls), or through large changes in sea level. If sea levels fall, then the whole river may start to erode its bed downward in response. If sea levels rise, the river may deposit more sediment and bed levels will be raised along its length.

2 River cross-sections

The shape and size of channel cross-sections adjusts to the discharge and sediment load conditions of the river. Channels are typically wider and deeper if the discharge is greater. Further downstream, river channel cross-sections tend to get larger, although there is not a perfect relationship between discharge and channel cross-sectional area because larger channels are more efficient at carrying water (less friction around the channel edges per volume of water). Channels with a high percentage of silt or clay in their banks (which are often more characteristic of lowland sections of river), and rivers transporting much of the sediment load in suspension, tend to be narrower and deeper than sand and gravel-bed rivers, where most of the sediment transport occurs via movements of sediment close to the bed and banks. Vegetation can also be important in controlling cross-section shapes by influencing bank resistance through root systems that bind the sediment. Removing bankside vegetation can lead to rapid bank erosion.

Within channel cross-sections there can be large variations in the velocity of water flow. Close to the bed or banks of the river the velocities tend to be slower. On bends of rivers there can be forces exerted that increase the pressure on the outer bank as the water flows by with less pressure on the inner bank of the bend. In addition, a circulatory flow called **helicoidal flow** (corkscrew-type circulation) is superimposed on the downstream flow of water (**Figure 3.17**). On the meander bend there will be an outward flow close to the water surface, a downward flow close to the outer bank of the bend and an inward flow towards the inner bank of the bend. Bank erosion is more likely on the outer bank, as faster water has

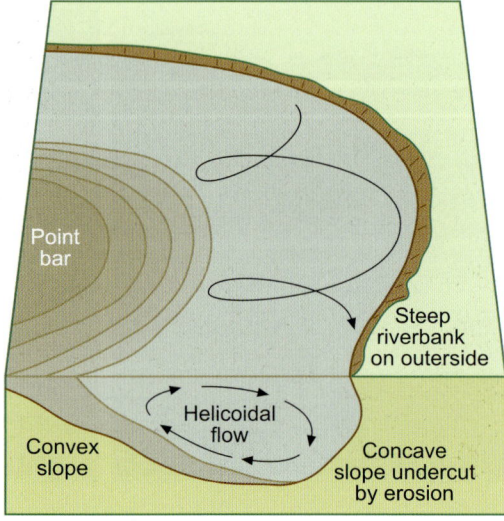

Figure 3.17 Helicoidal flow through a meandering channel.

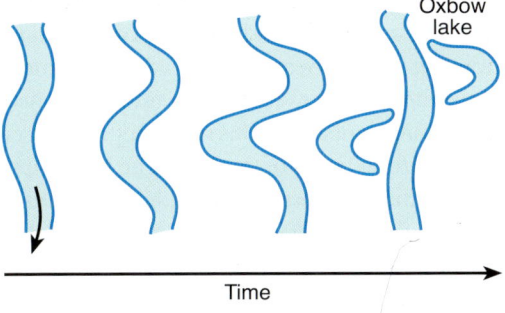

Figure 3.18 River meanders tend to become more exaggerated until eventually the meander is cut through by the river, leaving behind an oxbow lake. Meandering rivers can, over time, follow a course that slowly wanders across the entire flood plain.

more chance of picking up sediment. On the inner bank, sediment is more likely to be deposited. This causes further development of the meander (**Figure 3.18**) and means that the river is continuously eroding and depositing sediment as a natural process, and so the exact position of the river bank will change through time. Within meandering channels, because water velocity is greatest close to the outer bank, sediment size tends to be greatest here and slowly decreases towards the inner bank. When meander bends become too exaggerated the river will cut through the meander to a direct course downstream for that short section, leaving behind a crescent-shaped section of channel that is cut off from the main river, known as an **oxbow lake**.

3 Sediment transport

If some sediment is to be picked up (entrained) from the river bed or bank by flowing water a critical threshold has to be passed above which the water velocity or **shear stress** is sufficiently large to overcome frictional forces that resist erosion. The transport of materials close to the bed is known as bedload transport. Particles move by sliding, rolling or **saltating** (hopping) along or close to the bed. If the flow velocity does not change, a particle will come to rest only if it becomes lodged against an obstruction or falls into a sheltered area of a larger particle. With further increases in the strength of flow, the smaller particles may be carried upward into the main body of water and transported in suspension. Finer particles are preferentially moved downstream. The characteristics of river beds tend to change downstream. Often there are bedrock channels in the upper section of a river network or large boulders and cobbles. There is a sharp decline of bed material size downstream. This pattern is because smaller particles are transported downstream more easily and because abrasion of larger sediments within the river by colliding and grinding causes material to get smaller and more rounded downstream. However, these patterns are not seen everywhere and may be disturbed by local sediment inputs to the river network.

Where erosion exceeds deposition within a particular section of a river, there will be lowering of the river bed or widening of the banks. If erosion and deposition occur at the same rate, then the river will stay at the same level. Eroding channels may undermine structures such as bridges, while depositing channels may submerge structures such as roads. Stable channels, especially those whose beds are lined with bedrock, are less likely to be a problem to engineering structures, but fluctuations in river channel dimensions and locations caused by flooding or sediment pulses moving down the river can be problematic.

Within rivers there are several erosional and depositional landforms. In bedrock channels potholes can be found, formed by: mechanical wearing and grinding of small particles, enlarging an existing small depression or weakness in the rock; pressure changes, due to bubble collapse in turbulent flow; and chemical weathering. Within rivers that have gravels on their bed the most common landforms on the bed are sequences of pools and **riffles**. Pools are deep sections with relatively slow-flowing water with fine bed material. Riffles are formed by accumulation of coarse sediment with shallow, fast-flowing water.

The spacing of pools and riffles is often five to seven times the channel width, but this does vary. The beds of sandy sections of rivers can have small ripples in the sand which are less than four centimetres in height, and then also larger dune features. The size and shape of these dune features change with discharge during rainfall events. The dunes and ripples tend to migrate downstream as sand is carried up the upstream-facing side of the ridge of the feature and then falls down the downstream-facing side. At very high flow velocities, a flat river bed can be formed or dunes can even migrate upstream, since erosion from the downstream side of the dune allows suspension of material in the water, which occurs faster than it can be replenished from upstream.

4 Modification of river channels

River channels change their slope, cross-section, planform and bed forms in response to environmental change. Humans have modified river channels over the past few thousand years, although changes during the last two centuries have been greatest. Activities such as dam construction, urbanisation, mining, land drainage and deforestation can all impact upon river channels. Faster, more peaky flow from urbanisation or deforestation can accelerate erosion. Dams have a major impact on river flows and sediment dynamics. The Nile now transports only 8% of its natural load of silt below the Aswan Dam, thereby reducing the fertility of the downstream floodplains and accelerating river bed and coastal erosion, as the lack of sediment entering the sea no longer replenishes the sediment being eroded by wave action in the **delta**. There are concerns for channel change in many major rivers in the world, including in their deltas. **Box 3.4** provides the example of the Indus.

Humans have directly modified river channels for purposes of flood control, navigation and protection of property and infrastructure from bank erosion. Many river channels have been straightened, dredged, widened, culverted or lined along their bed or banks, had vegetation or other obstacles removed and had levee construction. However, rivers are naturally dynamic and move across the landscape. There are many places where river channel modifications were simply reversed by the natural forces of the river. The Mississippi has regained much of its meandering sinuosity since channel straightening took place at the start of the twentieth century. In other places it is clear that many river channel modifications have worked against the natural processes in rivers and have resulted in major problems. For example, where channels are straightened it shortens their length and therefore steepens the slope of the river channel. This speeds up flow and hence gives the river a better ability to pick up sediment from the banks and bed, and so there can be enhanced local erosion. If the bed or banks erode due to faster water velocities in the straightened section, there is a possibility that this will have upstream effects, as the lowered bed elevation in the straightened reach may incise upstream, lowering the elevation of the river bed upslope. This may also result in enhanced sediment deposition downstream of the straightened section, causing channel instability there. When river channels get choked with sediment they are more liable to suddenly jump to a new course, in a process known as channel **avulsion**.

In addition to the geomorphological impacts of channel modification, there have also been biodiversity impacts. The creation of uniform channels lacking in diversity of flow velocities, sediments, shading or backwaters means that the habitats for many aquatic species have been lost on modified rivers. If the diversity of flow velocities and diversity of physical features within river channels are reduced this inevitably also reduces the diversity of species present.

The realisation that rivers could be valued for their biodiversity and aesthetic qualities and that many channel modifications in the past have not dealt with river channel processes properly has encouraged the practice of river rehabilitation. Large sums are being spent on river

RIVER BASIN HYDROLOGY 73

BOX 3.4 CASE STUDIES

Impacts of river flow and sediment reduction on the Indus delta

The Indus is around 3000 km long; it has its source in Tibet and flows to the Arabian Sea. Over the past 10 000 years the Indus formed a huge delta of 6200 km^2. However, 60% of the river water is now used for irrigation of agriculture. There has been a dramatic reduction of water and sediment delivery to the lower Indus (**Figure 3.19**). The Indus flowed continuously into its delta until the 1960s and now it may flow to the delta for only two months of the year, during the monsoon season. Delta systems normally have river freshwater flow which forces itself up against seawater. The lack of water now flowing into the delta means that more salty seawater can move inland up the delta. There has therefore been incursion of seawater into groundwater supplies and damage to farmland by salt deposits, displacing several hundred thousand people (Inam *et al.*, 2007). Mangrove forests in the delta have also been affected by reductions of water flow. This is of vital importance because, of the 1.2 million people living around the forests, 130 000 rely on the mangroves for their livelihoods (Shah, 1999). Prosperous ports such as Keti Bandar have become small villages. Currently the Indus contributes hardly any sediment to the delta and the area of active delta has reduced to one-fifth of its former extent. Coastal erosion via wave action and lack of sediment and flow from inland is resulting in shoreline retreat of around 50 m per year.

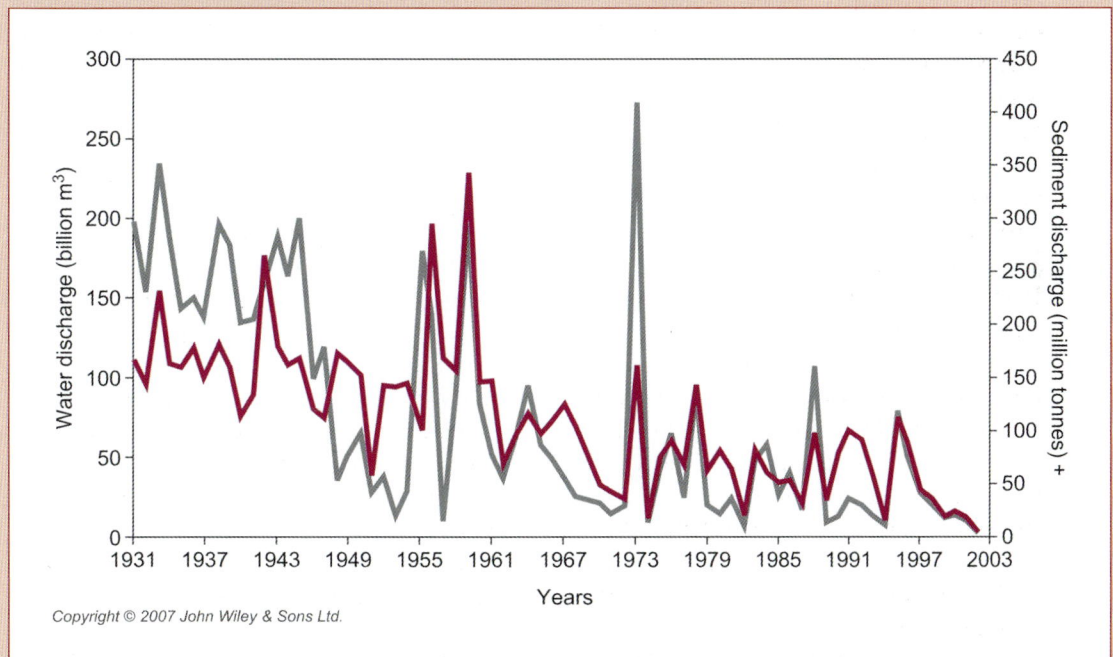

Figure 3.19 Temporal variation of water and sediment discharge in the main channel of the Indus River below Kotri.
Source: After Inam, A., Clift, P.D., Giosan, L., Tabrez, A.R., Tahir, M., Rabbani, M.M. and Danish, M. 2007. The geographic, geological and oceanographic setting of the Indus River. In Gupta, A. (ed.) *Large rivers: geomorphology and management.* John Wiley and Sons, Chichester, Figure 16.6, p. 341.

BOX 3.5 CASE STUDIES

River Kissimmee rehabilitation, Florida

The Kissimmee was once a meandering river which traversed 100 miles of central Florida. It was rich in wildlife. Major flooding in the 1940s led to government backing for engineering changes to deepen, straighten and widen the waterway. In the 1960s, the River Kissimmee was channellised into the C-38 canal. This destroyed the ecosystem. In excess of 90% of the waterfowl disappeared and the fish community changed dramatically.

A $500 million restoration project began in 1999 to rehabilitate the ecological integrity of the river system by reconstructing the river channel (Figure 3.19) and re-establishing hydrological processes. The C-38 canal is being filled in and the river is being allowed to flow in meanders once more. The project is due to be completed in 2015 and will return flow to 50 km of the historic channel. Part of the work has required the purchase of land to enable the restoration to take place and allow the floodplain wetlands to recover. This land purchase is typical of many landscape restoration projects around the world.

Figure 3.20 Re-meandering on the River Kissimmee. The straight canal which can be seen in the top left has been infilled.
Source: Photo courtesy of the Jacksonville District, U.S. Army Corps of Engineers/Mark Bias, http://www.saj.usace.army.mil.

rehabilitation, which often involves actions such as reinstating meander bends, *riparian* planting, removal of hard banks or beds, bank re-profiling, and the addition of pool and riffle sequences. For an example of a river rehabilitation project see **Box 3.5**. The scale of rehabilitation projects varies from tens of metres to tens of kilometres. One of the major problems that river rehabilitation practitioners face is how to compromise between different stakeholder wishes. Conflicts easily arise between water companies, fisheries, landowners, local residents and environmental organisations. Hence, rehabilitating a degraded river requires the coming together of disciplines such as ecology, hydrology, geomorphology and sociology as well as parties from all sides.

> **REFLECTIVE QUESTION**
>
> How might the complete removal of a dam which has become full of sediment over the past 100 years impact upon downstream river channel dynamics?

F SUMMARY

Water reaches rivers and lakes by direct deposition into the channel, or movement across and through the landscape. Overland flow can occur when surface water supply is greater than the infiltration capacity of the ground. Overland flow also occurs when the ground surface is saturated and water returns to the surface as saturation-excess overland flow. Saturation-excess overland flow can occur long after rainfall has stopped, especially at the foot of long, gentle-gradient slopes. Throughflow occurs through small pores as matrix flow and through large pores as macropore flow. Throughflow can contribute to river stormflow response as well as the baseflow of the river.

There is a wide spectrum of river regimes and responses to rainfall events controlled by climate, topography, drainage networks, geology, soils, vegetation and human management. Fast-responding systems with a rapid rise and fall of the hydrograph tend to be those with steep slopes, impermeable surfaces or quickly saturating soils. It is possible to take measurements to calculate the annual, seasonal or storm efficiency of river basins in delivering river discharge resulting from precipitation. Land management can alter the hydrological efficiency of the system by utilising water for agriculture, industry or domestic use or by changing the land cover and soil characteristics which affect the flow paths for water across the river basin. Modelling approaches can be useful for understanding how land management or climate change might influence river regimes in the future.

River flooding is a natural process and should be expected. There are other types of flooding too, including pluvial flooding, groundwater flooding and coastal flooding. Managing flood risk requires more than hard-engineering solutions, particularly as allowing some areas to flood may make the impacts of flooding elsewhere much less severe. Resilience needs to be built into flood management solutions, which include improving perceptions of risk, developing flood warning systems and building landscapes that allow flooding to occur with minimal loss to the economy. Numerical modelling is often required to understand how changes in land management or channel configuration will affect flood risk, because river basins are not simple systems and a change in management in one location will not have the same impact on flooding as the same management change in another location of the river basin.

River channels can take on a number of planforms with different densities of drainage and different channel patterns, including braided and meandering forms. Modification of river channels, such as channel straightening or building flood protection embankments, can alter the nature of the river regime or storm hydrographs by altering the amount and velocity of water flows through the channel. These changes can also impact upon channel sediment dynamics, with knock-on impacts for river channel dynamics upstream and downstream of the managed river reach, as well as negative impacts on biodiversity. Thus river rehabilitation schemes are being developed which try to re-engage rivers with their natural processes. One of the major problems for many of the world's largest rivers is that of reduced water and sediment discharge, due to human abstraction from dams or diversion channels for agriculture. In many cases, such as the Colorado or the Indus, this means that those large rivers now rarely flow to their original mouth at the sea. Starving the river mouth of freshwater and sediment can have huge knock-on effects for coastal erosion and intrusions of saline water inland.

FURTHER READING

Arnell, N. 2002. *Hydrology and global environmental change*. Pearson Education; Harlow.

An excellent book full of useful examples with very clear explanation of processes.

Gupta, A. (ed.) 2007. *Large rivers: geomorphology and management*. John Wiley and Sons; Chichester.

An excellent resource with descriptions and research results from many of the world's large rivers. The book includes discussion of potential future changes due to human modification and climate change.

O'Connell, E., Ewen, J., O'Donnell, G. and Quinn, P. 2007. Is there a link between agricultural land-use management and flooding? *Hydrology and Earth Systems Sciences* 11: 96–107.

A paper reviewing links between farming and flood risk.

Shaw, E.M., Beven, K.J., Chappell, N.A. and Lamb, R. 2011. *Hydrology in practice (4th edition)*. Spon Press; London.

A very detailed book dealing with hydrological measurements and modelling.

Thorne, C.R., Hey, R.D. and Newson, M.D. 1997. *Applied fluvial geomorphology for river engineering and management*. John Wiley and Sons; Chichester.

An overview of fluvial geomorphology as a basis for effective river management and engineering.

Ward, R.C. and Robinson, M. 2000. *Principles of hydrology*. McGraw-Hill; London.

One of the best textbooks on river basin hydrology.

Classic papers

Anderson, M.G. and Burt, T.P. 1978. Role of topography in controlling throughflow generation. *Earth Surface Processes and Landforms* 3: 331–344.

One of the early papers on throughflow mapping.

Brookes, A. 1985. River channelisation, traditional engineering methods, physical consequences and alternative practices. *Progress in Physical Geography* 9: 44–73.

A review article that shows how traditional engineering approaches to channelisation lead to geomorphic problems and environmental degradation.

Leopold, L. and Maddock, T. 1953. The hydraulic geometry of stream channels and some physiographic implications. *Geological Survey Professional Paper 252*. US Department of Interior; Washington.

The paper which identified simple power law relationships between channel width, depth, velocity and discharge.

PROJECT IDEAS

- Measure infiltration rates (by maintaining a shallow pond of water in a ring on the surface and determining the rate of water loss into the soil) for a given field site. Run the experiments until the infiltration rate is steady (this may take 20–30 minutes). Test whether the steady state infiltration rate varies through seasons or with land use. Compare infiltration rates to rainfall intensities in the area and establish how often you think infiltration-excess overland flow is likely to occur.

■ Find out about a river restoration programme near your location and determine the people and organisations involved, why the restoration is planned or has taken place and what the anticipated benefits might be. Find out if any monitoring is taking place to establish benefits for both the natural environment and local communities.

■ Measure the water efficiency for a small headwater river basin. Measure it also for a slightly larger area (which contains the smaller one) and again at another slightly larger scale. Are there differences in efficiency between scales? Why might this be?

REFERENCES

Abdel-Magid, A.H., Schuman, G.E. and Hart, R.H. 1987. Soil bulk density and water infiltration rates as affected by three grazing systems. *Journal of Range Management* 40: 307–310.

Anderson, M.G. and Burt, T.P. 1978. Role of topography in controlling throughflow generation. *Earth Surface Processes and Landforms* 3: 331–344.

Arnell, N. 2002. *Hydrology and global environmental change.* Pearson Education; Harlow.

Baird, A.J. 1997. Field estimation of macropore functioning and surface hydraulic conductivity in a fen peat. *Hydrological Processes* 11: 287–295.

Betson, R.P. 1964. What is watershed runoff? *Journal of Geophysical Research* 69: 1541–1552.

Beven, K.J. 2001. *Rainfall-runoff modelling – the primer.* Wiley; Chichester.

Beven, K. and Germann, P. 1982. Macropores and water-flow in soils. *Water Resources Research* 18: 1311–1325.

Brookes, A. 1985. River channelisation, traditional engineering methods, physical consequences and alternative practices. *Progress in Physical Geography* 9: 44–73.

Bryan, R.B. and Yair, A. 1982. Perspectives on studies of badland geomorphology. *In*: Bryan, R.B. and Yair, A. *Badland Geomorphology and Piping.* GeoBooks; Norwich, 1–12.

Durner, W. and Or. D. 2005. Soil water potential measurement. *In*: Anderson, M.G. and McDonnell, J.J. (eds) *Encyclopedia of hydrological sciences.* John Wiley and Sons; New York, 1089–1102.

Evans, M.G., Burt, T.P., Holden, J. and Adamson, J. 1999. Runoff generation and water table fluctuations in blanket peat: evidence from UK data spanning the dry summer of 1995. *Journal of Hydrology* 221: 141–160.

Greenwood, K.L. and McKenzie, B.M. 2001. Grazing effects on soil physical properties and the consequences for pastures: a review. *Australian Journal of Experimental Agriculture* 41: 1231–1250.

Gupta, A. (ed.) 2007. *Large rivers: geomorphology and management.* John Wiley and Sons; Chichester.

Hewlett, J.D. and Hibbert, A.R. 1967. Factors affecting the response of small watersheds to precipitation in humid areas. *In*: Sopper, W.E. and Lull, H.W. (eds) *Forest hydrology.* Pergamon Press; New York, 275–290.

Hillel, D. 1982. *Introduction to soil physics.* Academic Press; New York.

Holden, J. 2005a. Infiltration, infiltration rate and infiltration capacity. *In*: Lehr, J.H. and Keeley, J. (eds) *Water Encyclopedia, Volume 5.* John Wiley and Sons; New York, 212–214.

Holden, J. 2005b. Controls of soil pipe density in blanket peat uplands. *Journal of Geophysical Research* 110: F010002 doi: 10.1029/2004JF000143.

Holden, J. 2009. Flow through macropores of different size classes in blanket peat. *Journal of Hydrology* 364: 342–348.

Holden, J. 2011. *Physical geography: the basics.* Routledge; London.

Holden, J., Smart, R.P., Dinsmore, K., Baird, A., Billett, M.F. and Chapman, P.J. 2012. Natural pipes in blanket peatlands: major point sources for the release of carbon to the aquatic system. *Global Change Biology* 18: 3568–3580. doi: 10.1111/gcb.12004.

Imeson, A.C. and Kwaad, F.J.P.M. 1990. The response of tilled soils to wetting by rainfall and the dynamic character of soil erodibility. *In*: Boardman, J., Foster, I.D.L. and Dearing, J.A. (eds) *Soil erosion on agricultural land*. John Wiley and Sons; Chichester, 3–14.

Inam, A., Clift, P.D., Giosan, L., Tabrez, A.R., Tahir, M., Rabbani, M.M. and Danish, M. 2007. The geographic, geological and oceanographic setting of the Indus River. *In*: Gupta, A. (ed.) *Large rivers: geomorphology and management*. John Wiley and Sons; Chichester, 333–346.

Leonard, J., Perrier, E. and de Marsily, G. 2001. A model for simulating the influence of a spatial distribution of large circular macropores on surface runoff. *Water Resources Research* 37: 3217–3225.

Leopold, L. and Maddock, T. 1953. The hydraulic geometry of stream channels and some physiographic implications. *Geological Survey Professional Paper 252*. US Department of Interior; Washington.

Meyles, E.W., Williams, A.G., Ternan, J.L., Anderson, J.M. and Dowd, J.F. 2006. The influences of grazing on vegetation, soil properties and stream discharge in a small Dartmoor catchment, southwest England, UK. *Earth Surface Processes and Landforms* 31: 6221–6631.

Mwendera, E.J. and Saleem, M.A. 1997. Infiltration rates, surface runoff, and soil loss as influenced by grazing pressure in the Ethiopian highlands. *Soil Use and Management* 13: 29–35.

O'Connell, E., Ewen, J., O'Donnell, G. and Quinn, P. 2007. Is there a link between agricultural land-use management and flooding? *Hydrology and Earth Systems Sciences* 11: 96–107.

Perry, T. and Nawaz, N.R. 2008. An investigation into the extent and impacts of hard-surfacing of domestic gardens in an area of Leeds, United Kingdom. *Landscape and Urban Planning* 86: 1–13.

Pikounis, M., Varanou, E., Baltas, E., Dassaklis, A. and Mimikou, M. 2003. Application of the SWAT model in the Pinios river basin under different land-use scenarios. *Global Nest: The International Journal* 5: 71–79.

Price, J.S. 1992. Blanket bog in Newfoundland: Part 1. The occurrence and accumulation of fog-water deposits. *Journal of Hydrology* 135: 87–101.

Shah, G.R. 1999. Sociological conditions of Indus Delta Rehabilitation and Replanting of Mangrove Project (IDRRMP) community and the use of mangrove ecosystem. *In*: *Proceedings of the National Seminar on Mangrove Ecosystem Dynamics of the Indus Delta*. Sindh Forest and Wildlife Department; Karachi, 112–123.

Shaw, E.M., Beven, K.J., Chappell, N.A. and Lamb, R. 2011. *Hydrology in practice (4th edition)*. Spon Press; London.

Sherman, L.K. 1932. Streamflow from rainfall by the unit graph method. *Engineering News Record* 108: 501–505.

Sherman, L.K. 1942. The unit hydrograph method. *In*: Meinzer, O.E. (ed.) *Hydrology*. Dover Publications; Englewood, CO.

Smart, R.P., Holden, J., Dinsmore, K., Baird A.J., Billett, M.F., Chapman, P.J. and Grayson, R. 2012. The dynamics of natural pipe hydrological behaviour in blanket peat. *Hydrological Processes* doi: 10.1002/hyp.9242.

Thorne, C.R., Hey, R.D. and Newson, M.D. 1997. *Applied fluvial geomorphology for river engineering and management*. John Wiley and Sons; Chichester.

Vörösmarty, C.J., Green, P., Salisbury, J. and Lammers, R.B. 2000. Global water resources: vulnerability from climate change and population growth. *Science* 289: 284.

Ward, R.C. and Robinson, M. 2000. *Principles of hydrology*. McGraw-Hill; London.

Zhu, T.X. 1997. Deep-seated, complex tunnel systems – a hydrological study in a semi-arid catchment, Loess plateau, China. *Geomorphology* 20: 255–267.

CHAPTER FOUR

Surface water quality

Pippa J. Chapman, Paul Kay, Gordon Mitchell and Colin S. Pitts

LEARNING OUTCOMES

After reading this chapter you should be able to:

- understand the natural factors that control surface water chemistry
- understand how spatial and temporal patterns in surface water chemistry occur
- explain how water use by humans leads to a deterioration in water quality
- define the different sources of pollution of surface waters
- explain how changes in agricultural practices and urbanisation affect water quality
- discuss the causes of acid rain and acid mine drainage, and their impact on water quality
- understand the role of national and international policies and legislation in reducing water pollution.

A INTRODUCTION

Surface waters refer to rivers, streams, lakes, ponds and reservoirs. When rain falls on the land it either seeps into the ground to recharge groundwater aquifers (see Chapter 5) or becomes **runoff** which flows downhill over and through the soil into streams, rivers, ponds and lakes (see Chapter 3). However, surface water bodies do not just receive water from runoff; many receive inputs from groundwater (see **Figure 4.1**), the contribution of which generally increases during periods of low flow. Streams and rivers form where surface water accumulates and flows from land of higher altitude to lower altitude on its journey towards the oceans. Lakes or ponds form where surface runoff accumulates in a flat area, relative to the surrounding land, and the water entering the lake or pond comes in faster than it can escape, whether via outflow in a river, seepage to groundwater, or by evaporation. This means that lakes and ponds are standing or very slow-moving bodies of water, while rivers and streams are distinguished by a fast-moving current. While most lakes contain freshwater, some, especially those where water cannot escape via a river, are salty. In fact, some lakes are saltier than the oceans (see also Chapter 3, section C2 on closed basins). The terms 'lakes' and 'ponds', and 'rivers' and 'streams' are often used interchangeably, because in reality there is no obvious distinction between them, although the latter term is typically used when describing a 'smaller' standing body of water or running watercourse, respectively. Reservoirs, also called impoundments, are human-made lakes. However, they can display characteristics of both rivers and lakes because they are created by building a dam across a river and flooding the valley. This damming and flooding creates an artificial lake, filled by the river inflow. Thus the upstream section of the reservoir has predominantly river-like qualities, meaning there is often still some current, while the area closest to the dam is more lake-like.

Stream and river networks drain more than 75% of the Earth's land surface. The precipitation that falls on the land percolates over and through vegetation and soil, picking up **solutes** (dissolved matter) and sediment along its route to surface waters. This runoff delivers different amounts of solutes and sediment to rivers depending on the **hydrological pathway** it has taken through the catchment and the characteristics of the surrounding landscape (see **Figure 4.1**). Surface water chemistry is therefore controlled by processes occurring in the river's basin. Thus any changes that occur in the catchment also lead to a change in surface water chemistry. River systems also provide a vital linkage between the terrestrial and aquatic ecosystems. Thus streams and rivers have been referred to as 'the environment's circulatory system' (Wetzel, 2001).

This chapter will first consider the natural processes and factors that control the spatial and temporal patterns of surface water chemistry before considering how water use by humans impacts upon water quality. In particular, it will explain how changes in agricultural practices and urbanisation affect water quality, and the mitigation factors that are being used to try to reduce water pollution from these sources. This chapter also describes the causes of acid mine drainage, its impact on water quality and mitigation options that exist. The causes of acid rain, its impact on water quality and how legislation in Europe and North America has led to the reduction of acid rain are discussed. Finally, the chapter will look at the role of national and inter-

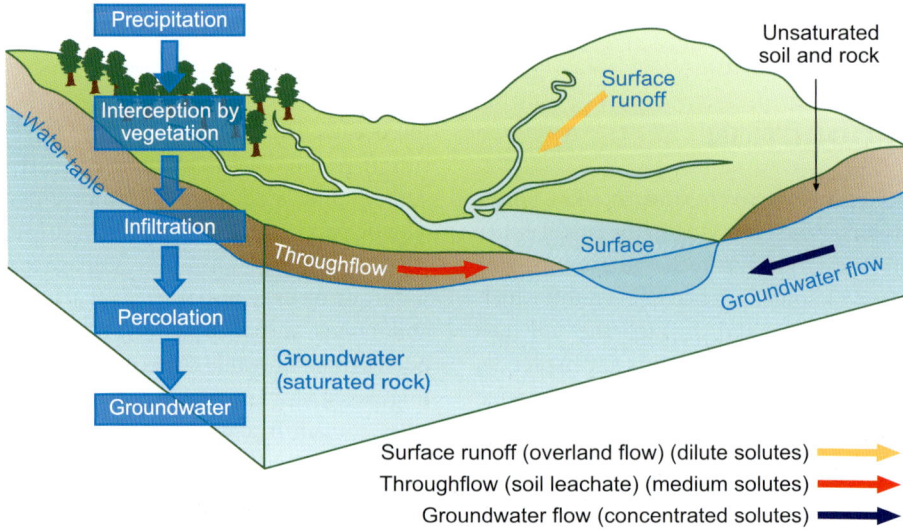

Figure 4.1 The different hydrological pathways that precipitation may take to reach surface waters and their effect on solute concentrations.

national policies and legislation in reducing water pollution and how climate change and population growth may affect surface water quality in the future.

B CHARACTERISTICS OF SURFACE WATERS

It is important to understand the natural processes that control surface water chemistry before exploring the effect of human activities on water quality. The concentration and form of chemical elements and compounds are constantly changing as a result of hydrological, physical, chemical and biological processes. This section briefly discusses the roles of these processes in controlling surface water chemistry.

Precipitation falling on the land takes a variety of different routes, known as hydrological pathways, through the catchment to reach surface waters, as shown in **Figure 4.1**. Precipitation, which has low solute concentrations, may flow downhill over the soil surface, as **infiltration-excess overland flow** (see Chapter 3), or rapidly through the soil via **macropores** to reach rivers and lakes. As the residence time of this water within the catchment is short, the solute concentration of this runoff is usually very similar to that of precipitation. Alternatively, precipitation may flow through the soil horizons, where the residence time is longer, before reaching surface waters, and solute concentrations increase due to inputs from weathering reactions and microbial activity. Sometimes, when the soil is saturated, then **saturation-excess overland flow** occurs (see Chapter 3) and this water can be a mix of fresh precipitation and water which has more solutes from within the soil. Lastly, precipitation may percolate through the soil into the bedrock below, if it is permeable. At a certain depth below the land surface, called the **water table**, the ground becomes saturated with water, whereas the ground above the water table is unsaturated. If a river cuts into this saturated layer, as shown in **Figure 4.1**, then water may flow out of the groundwater into the river. This is why even during dry periods there is usually some water flowing in streams and rivers. In addition, due to the longer residence time of this pathway, groundwater tends to be enriched in solutes derived from weathering reactions.

1 Chemistry

All surface waters contain dissolved (solute) and suspended (particulate) inorganic and organic substances. The distinction between dissolved and suspended substances is based on filtration, usually through a 0.45 μm membrane filter, although other sized filters are also used. The concentration of suspended particles (usually referred to as **suspended sediment**) can have a large impact on water quality. Suspended sediments include soil particles (clay, silt and sand), algae, plankton, microbes and other substances, and are typically in the size range of 0.004 mm (clay) to 1.0 mm (sand). In rivers, the main sources of suspended sediment include soil erosion, waste discharge, urban runoff and eroding stream banks, whereas in lakes decaying algae, plants and animals and bottom-feeding fish play a major role in controlling suspended sediment concentrations. High concentrations of suspended sediment can block sunlight from reaching submerged vegetation, which slows down **photosynthesis**. This results in less dissolved oxygen being released into the water by the plants, and in extreme cases can result in the death of the plants. In addition, suspended sediment causes the water temperature to increase, as it absorbs heat from sunlight. This can result in the dissolved oxygen content reducing further because warm water holds less dissolved oxygen than cold water. If the dissolved oxygen content of a water body becomes very low, fish may die. As suspended sediments settle they can cover the river/lake bottom, especially in slower waters, and smother fish eggs and **benthic** macroinvertebrates. The greater the amount of suspended sediment in the water, the murkier it looks and the higher the measure of **turbidity**. Turbidity refers to the clarity of the water and how much the particles suspended in water decrease the passage of light through the water. Turbidity is not a measurement of the amount of suspended sediment present, since it measures only the amount of light that is scattered by suspended particles. Measurement of total solids is a more direct measure of the amount of material suspended and dissolved in water. A turbidity meter consists of a light source that illuminates a water sample and a photoelectric cell that measures the intensity of light scattered at a 90° angle by the particles in the sample. It measures turbidity in nephelometric turbidity units or NTUs. In rivers, turbidity readings are usually less than 10 NTU during periods of low flow (baseflow) but these can increase markedly during storms, when particles from the surrounding land are washed into the river. In addition, during storms, water velocities are faster, which can re-suspend material from the stream bed. Chapter 6 describes in more detail the impact of changes of turbidity on aquatic ecosystems.

When some compounds (mainly inorganic) dissolve in water they break down (dissociate) to form **ions**, for example, NaCl (sodium chloride) added to water produces $Na^+ + Cl^-$ ions (see Chapter 1). Ions are charged atoms or molecules caused by having an unequal number of protons (positively charged) and electrons (negatively charged); positively and negatively charged ions are normally attracted to each other. Cations are positively charged, as they have more protons than electrons. Anions are negatively charged. Some elements and ions are very soluble, while others have a strong affinity to stay as particulate matter or become attached to suspended matter. The major dissolved ions in surface waters, which occur at concentrations exceeding 1 mg L^{-1}, are bicarbonate (HCO_3^-), sulphate (SO_4^{2-}), chloride (Cl^-), calcium (Ca^{2+}), magnesium (Mg^{2+}), sodium (Na) and potassium (K^+). In fact these seven ions along with silica (which occurs as $Si(OH)_4$ at the pH of most natural waters) constitute ~95% of the total dissolved inorganic solutes in surface waters. This reflects their relative abundance in the Earth's crust and the fact that they are moderately to very soluble. In contrast, the metals and **metalloids**, which occur at much lower concentrations, are generally found in or bound to the particulate matter in surface waters.

Some elements exist in a number of different forms while others occur in more than one oxidation state. For example, the most common ionic forms of dissolved inorganic nitrogen in aquatic ecosystems are ammonium (NH_4^+), nitrite (NO_2^-) and nitrate (NO_3^-), while iron can be present as Fe^{2+} (ferrous iron) or Fe^{3+} (ferric iron). The form and oxidation state of an element is mainly controlled by environmental factors, particularly pH and redox potential (see **Box 4.1** for definitions of these terms), and has an important control on the solubility and **toxicity** of an

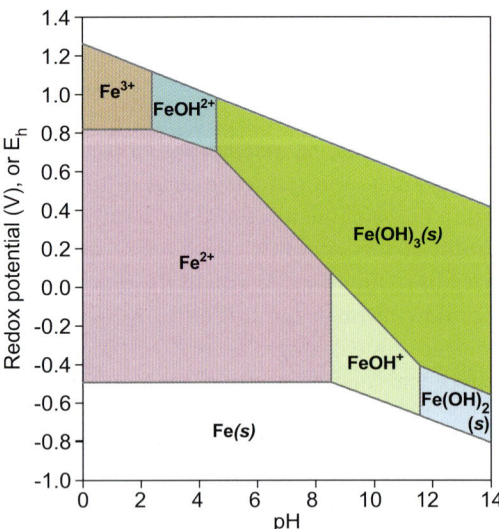

Figure 4.2 Eh-pH diagram for the simple ions and hydroxides of iron at atmospheric pressure and 25°C. If the water pH and redox potential are known the figure shows the form of iron present.

Source: After Morgan and Stumm (1965). Reprinted with permission from Elsevier from *Advances in Water Pollution Research, Proceedings of the Second International Conference on Water Pollution Research*, Morgan, J.J. and Stumm, W. The role of multivalent metal oxides in limnological transformations, as exemplified by iron and manganese. Pergamon Press, Oxford, 1, 103–118. Copyright 1965.

SURFACE WATER QUALITY

> **BOX 4.1 TECHNIQUES**
>
> ## pH and redox potential
>
> Whether a water body is acidic, neutral or alkaline is determined by measuring the hydrogen ion concentration in solution. In pure water at 24°C, water ionises (forms ions) to give equal concentrations of hydrogen (H^+) and hydroxide (OH^-) ions: $H_2O \leftrightarrow H^+ + OH^-$. The concentration of both H and OH^- ions is 1×10^{-7} (0.000 000 1) moles per litre. To overcome using these very small concentrations of H^+ ions, a simpler method of using the negative logarithmic of the hydrogen ion concentration was developed, known as pH: $pH = -\log_{10}[H^+]$. Although the pH scale ranges from 1 to 14, with low values the most acidic and high values the most alkaline, most surface waters have a pH range of 4 to 9. As the pH scale is logarithmic it should be remembered that a change of one unit represents a 10-fold change in H^+ ion concentration.
>
> A redox reaction is one where the oxygen state of the substance is changed. For example, a redox reaction occurs when carbon reacts with oxygen to become carbon dioxide. Oxidation involves the increase of oxidation state of a substance (such as in carbon to carbon dioxide) and is associated in a loss of electrons. Reduction is the reverse case. Redox potential (also known as oxidising or reducing potential or Eh) is a measure (in volts) of the tendency of a substance to acquire electrons compared with a standard. The standard used is hydrogen and its redox potential is set to zero:
>
> $$E^+ + e^- \leftrightarrow \frac{1}{2}H_{2(g)}$$
>
> As the redox potential decreases, the solution is more reduced (i.e. has more electrons to give), and as the redox potential increases, the solution is more oxidised (i.e. will accept more electrons).

element. The influence of pH and redox potential on the oxidation state, and hence form, of solute in freshwaters is illustrated for iron in **Figure 4.2**. It can be seen that soluble Fe^{2+} occurs in highly acidic but well-oxidised waters, such as acid mine drainage, and also waters of neutral pH and reducing conditions. However, at neutral pH and oxidising conditions Fe^{2+} is converted to the insoluble Fe^{3+}, in the form of iron hydroxide ($Fe(OH)_3$). Therefore the iron will precipitate out of solution to become solid matter within the water. This is what occurs when acid mine drainage mixes with surface waters of a higher pH and the iron hydroxide can be observed as a red precipitate on the bed of the river (see **Figure 4.13**). Other major elements found in surface waters that are strongly influenced by redox reactions are carbon, nitrogen, sulphur and manganese.

The sum of all the dissolved solutes plus silica (SiO_2) present in a water body is known as the **total dissolved solids (TDS)**. The TDS can be determined gravimetrically by evaporating a known volume of water and measuring the mass of the residue left. Alternatively it can be determined by the measurement of electrical conductivity, as the dissolved ions present in the water create the ability for water to conduct an electric current. When correlated with gravimetric measurements of TDS, conductivity provides an approximation of the TDS concentration of a water sample. However, in very low electrical conductivity waters where the TDS is small, such as in some peatland stream waters, a normal conductivity meter will not work because an electrical circuit will not be completed. In these circumstances special conductivity probes are required where the electrodes are placed very close together to measure low TDS waters (including rainwater). Total dissolved solids in streams and rivers can vary between 50 and 1000 mg L^{-1}, compared with a typical value of less than 20 mg L^{-1} for rainwater (Meybeck et al., 1996). In lakes with outlets, TDS are similar to those of streams and rivers, whereas for lakes without outlets, TDS can range from 1000 to 100 000 mg L^{-1} (Meybeck et al., 1996). Most bottles of mineral water will indicate the value of TDS on their label. In the United States, bottled water must contain greater than 250 mg L^{-1} TDS to be labelled as 'mineral water'. If the TDS exceed 1500 mg L^{-1} then a 'high mineral content' label is used. You should find that the higher the TDS, the more distinct the taste of the water.

The **biochemical oxygen demand** (**BOD**) is a commonly measured feature of water quality in surface waters. It equates to the amount of dissolved oxygen (in milligrams) needed by biological organisms in a water body so that they can break down organic matter in a sample of water (normally 1 litre) over a set time period (normally 5 days) for a given temperature (normally 20°C). Typically, if there are a lot of organic compounds present in the water sample the BOD will be very high. Most unpolluted rivers are thought to have a BOD of less than 1 mg L^{-1} while untreated sewage might have a BOD of 50–500 mg L^{-1} (see section C1 in Chapter 9). If there is a discharge of organic pollutants (e.g. failing septic tank, urban storm water runoff, effluent from pulp and paper mill) into a river the dissolved oxygen concentration will quickly drop, which might result in the deaths of many aquatic animals that need to extract oxygen from the water in order to survive. This drop in dissolved oxygen concentration after a pollution incident is often known as a sag curve and is shown in Chapter 9 (**Figure 9.5**).

Useful methods for measuring BOD can be found at http://water.epa.gov/type/rsl/monitoring/vms 52.cfm.

Another variation of the BOD is the sediment oxygen demand (SOD), which is the usage of dissolved oxygen in the overlying water by sediment chemicals and organisms that live in the sediment on the bed of the water body. Such organisms include burrowing fauna, worms, insect larvae, bacteria, protozoa and fungi. To measure SOD, sediment cores are normally extracted and then oxygen use is measured in the laboratory over a certain period of time at a controlled temperature.

The **chemical oxygen demand** (**COD**) is another measure of pollution by organic compounds, but one which tests the chemical demand for oxygen within the water body without including the biological processing. The COD is a measure of the amount of dissolved oxygen needed to oxidise the organic matter within a litre of water using a standard chemical oxidising agent. COD is measured in milligrams (mg) of oxygen per litre (L) of water and is more commonly used in testing wastewater than surface waters in rivers and lakes (see Chapter 9).

2 Hydrological processes

Surface water chemistry is strongly influenced by the hydrology of a water body. The residence time of water in streams and rivers usually ranges between two and six months, while that of lakes can vary from months, for shallow lakes, to 100 years, for deep lakes. River and stream flow is unidirectional, usually with good lateral and vertical mixing, but may vary greatly depending on climatic conditions and season. In general, the higher the annual runoff of a river, the lower the concentration of solutes, and hence TDS, in a river (**Figure 4.3a**). This inverse relationship can be explained simply by the dilution of the available solutes as surface runoff volumes increase. However, as the volume of runoff increases so does the total amount of solutes released from the

SURFACE WATER QUALITY

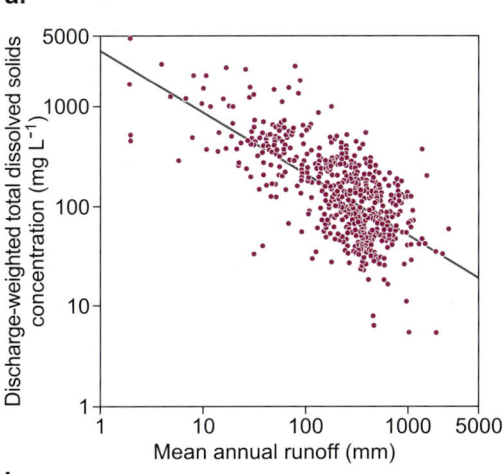

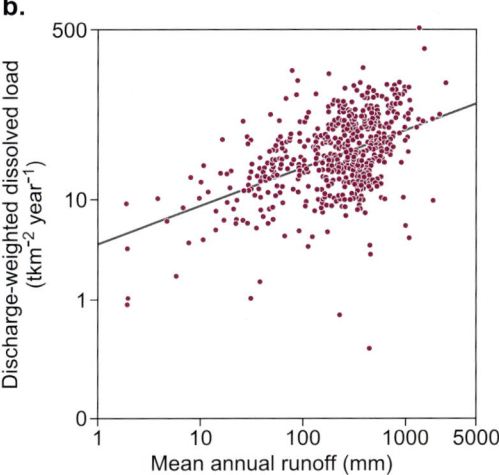

Figure 4.3 Relationship between mean annual runoff and (a) discharge weighted mean total dissolved solids concentrations, and (b) mean annual dissolved load for a sample of 496 world rivers.

Source: After Walling and Webb (1986); reproduced with permission from Walling, D.E. and Webb, B.W. 1986. Solutes in river systems. *In*: Trudgill, S.T. (ed.) *Solute Processes*, John Wiley and Sons Ltd; Chichester, 251–327.

catchment and hence available for transport (**Figure 4.3b**). This is known as the **solute load** or flux and is determined by multiplying the solute concentration by the discharge at a specific location on a river at a certain time. To calculate the annual load of a specific solute, continuous measurements of discharge and concentration are required. However, while discharge can be measured regularly, solute concentration is usually measured less frequently, due to the expense of collecting and analysing water samples. This means that solute concentration must be estimated between sampling periods. Several studies have compared different methods for estimating solute and sediment loads (e.g. Littlewood, 1992; Webb *et al.*, 1997) and have found that regular monitoring programmes tend to over-emphasise base flow components, which is compounded the longer the time interval between samples. Webb *et al.* (1997) investigated the effect of different methods to estimate chemical fluxes for the River Derwent in Yorkshire, UK, and found that for 20 out of the 36 determinands investigated, the difference between the minimum and maximum load estimate, expressed as a percentage of the minimum value, exceeded 50%, and for five determinands, the difference was greater than 100%.

In catchment studies, the annual dissolved load serves as an integrated measure of all the processes that occur within the river basin that affect stream water chemistry. While measurement of solute concentration informs us how much of a particular solute is present at a specific point in time and is useful for comparing with water quality standards, solute loads can be used to quantify the amount of solute entering a specific water body, such as a lake or ocean, from terrestrial sources, or the impact of environmental changes on the terrestrial cycle of nutrients. In order to compare solute loads from catchments of different size the load is divided by the area of the catchment and is generally expressed in tonnes per km^2 per year (which is equivalent to g m^{-2} yr^{-1}) (see **Box 4.2**).

Lakes and reservoirs are characterised by alternating periods of **stratification** and vertical mixing, which has an effect on water chemistry. Stratification in lakes or reservoirs occurs when the water acts as two different bodies with different densities, one floating on the other. It is most commonly caused by temperature differences within the water column, particularly in deep

BOX 4.2 TECHNIQUES

Calculating solute loads for Eagle Creek, Indiana, USA

Table 4.1 shows some example data for stream discharge and nitrate-N (NO_3-N) concentrations at hourly intervals for Eagle Creek at Zionsville in Indiana which were obtained from the United States Geological Survey (USGS) database, which is a public domain source. Data are for one day in January 2013. The drainage area is 274.5 km^2. We would like to determine the total flux of nitrate-N (nitrate-N refers to the amount of nitrogen present in the nitrate ion, so if we want to calculate the total nitrate (NO_3) flux we need to multiply the nitrate-N flux by 4.43, as 10 mg L^{-1} of NO_3-N is equivalent to 44.3 mg L^{-1} of NO_3) from the catchment and, so that we can fairly compare it to data from other sites, to calculate the solute load for the day per unit catchment area.

Multiplying the discharge value for each hour by the number of seconds in each hour gives us an estimate of total discharge in that hour in m^3. In order to put everything into the same units we can divide the concentration of nitrate by 1000 to express it as mg per m^3 of water (since there are 1000 litres in a m^3). If we then assume that the concentration of nitrate-N applies for the whole hour, then, multiplying that concentration by the total discharge for the hour, we end up with a value of mg of nitrate that has moved past the gauging point on the river. Taking the first row of data from Table 4.1 as an example gives us 7849.4 m^3 of water for that hour (multiply 2.18 m^3 s^{-1} by 3600 s). If we multiply 0.0032 mg m^{-3} of nitrate-N by 7849.4 m^3 of water we obtain a total flux for the hour of 25.3 mg of nitrate-N. Repeating this for each hour and summing the total provides us with an estimate of 747 mg of nitrate-N for the 24-hour period shown in the table. As the catchment area is 274.5 km^2 we can suggest that for one 24-hour period the nitrate flux from Eagle Creek was equivalent to 2.7 mg km^{-2}. Obviously we could repeat this for the whole year quite easily in a spreadsheet to provide a more realistic annual flux estimate.

In many cases we do not have data at such high frequency as is presented in Table 4.1. While we might have good discharge data collected at high frequency (e.g. every 15 minutes), often water quality samples for different solutes are taken once per week or sometimes less frequently, occasionally supplemented by sampling during some storm events. In these instances it is not suitable simply to multiply discharge at the time of sampling by concentration of the solute and then sum the values. That is because we do not have values of concentration for all of the times in between (when the discharge and concentrations might be quite different). Therefore we need to interpolate values between sample points to produce a value of annual solute load. The methods most commonly used vary depending on what data are available and are described in Walling and Webb (1985).

Table 4.1 Nitrate-N concentrations and discharge for a 24-hour period 2–3 January 2013 for Eagle Creek, at Zionsville, Indiana

Time	Discharge m^3 s^{-1}	Nitrate-N, mg L^{-1}
18:00	2.18	3.22
19:00	2.15	3.31
20:00	2.10	3.34
21:00	2.18	3.39
22:00	2.35	3.36
23:00	2.29	3.39
00:00	2.29	3.39
01:00	2.35	3.40
02:00	2.38	3.36
03:00	2.55	3.37
04:00	2.55	3.40
05:00	2.61	3.42
06:00	2.61	3.45
07:00	2.69	3.45
08:00	2.80	3.48
09:00	2.97	3.40
10:00	3.14	3.45
11:00	3.37	3.47
12:00	3.60	3.40
13:00	3.09	3.36
14:00	2.61	3.37
15:00	2.10	3.45
16:00	1.95	3.50
17:00	2.01	3.60

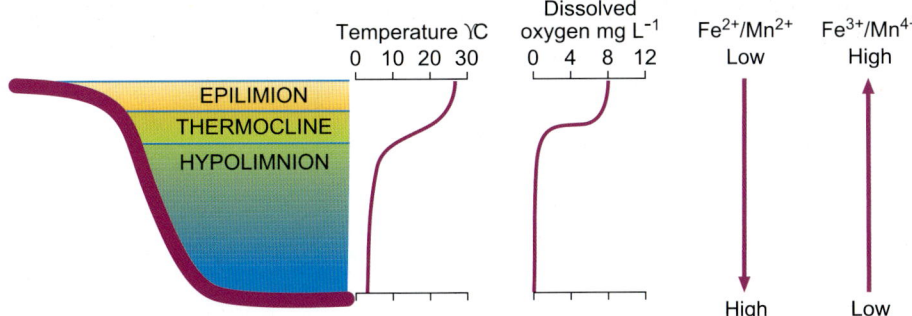

Figure 4.4 Cross-section of a thermally stratified eutrophic lake in summer showing location of the epilimnion, thermocline and hypolimnion and associated changes in temperature, dissolved oxygen and iron and manganese concentrations with depth.

(> 10 m) temperate lakes and reservoirs, because of seasonal changes in temperature. It is less common in tropical lakes, where temperature varies little over the year. Stratification results in different processes dominating in the different layers of the lake. In temperate regions during the summer, the surface layer, known as the **epilimnion**, receives more sunlight, is warmer and has an oxidising environment because it is exposed to the atmosphere. In contrast, the deeper layer, known as the **hypolimnion**, is separated from the epilimnion by a **thermocline** (see Chapter 6). Due to the isolation of the hypolimnion from the atmosphere, it is cooler and receives less oxygen, leading to anoxic conditions developing in the lake-bottom sediments. This gradient of dissolved oxygen means that elements that have more than one oxidation state occur in the oxidised state in the epilimnion and the reduced form in the hypolimnion (**Figure 4.4**). Thus stratification of reservoirs can lead to concentrations of iron and manganese exceeding drinking water quality standards. This is because the reduced forms of these metals are much more soluble than the oxidised forms. Therefore the metals within the sediments on the reservoir bottom will leach out and dissolve in the water, and if water is abstracted from the hypolimnion in the reservoir, then high concentrations of dissolved lead and manganese will be found in the abstracted water.

> **REFLECTIVE QUESTION**
>
> What is a redox reaction?

C SPATIAL VARIATION IN SURFACE WATER CHEMISTRY

In natural conditions, the chemistry of surface waters is controlled by three main processes (Gibbs, 1970; Meybeck et al., 1996; Walling and Webb, 1986): (i) atmospheric inputs, (ii) weathering of bedrock, and (iii) climate (evaporation–crystallisation processes). The main source of atmospheric inputs is from oceanic aerosols (Na and Cl); however, other sources include volcanic emissions (such as SO_2) and the dissolution of soil-derived particulates (such as Ca, HCO_3 and SO_4). The distance from the ocean will determine the extent to which sea-spray rich in Na, Cl, Mg and SO_4 influences surface water chemistry. In coastal areas, where oceanic inputs may dominate over all other sources, Na and Cl are the most abundant solutes in surface waters. These solutes may also dominate in areas where rocks are extremely resistant to weathering, such as in central Amazon.

In nearly all major rivers of the world, calcium is the dominate cation and bicarbonate the major anion. In fact, 80% of the dissolved load of the major rivers of the world is made up of just four solutes: calcium, bicarbonate, sulphate and silica (Walling and Webb, 1986). The major source of these solutes is the weathering of bedrock. The susceptibility of rocks to weathering varies from low for **plutonic** and **metamorphic** crystalline rocks to high for **sedimentary rocks** such as limestone and chalk. Thus TDS in rivers and streams draining catchments underlain by sedimentary rocks are five times greater than in rivers draining catchments dominated by crystalline rocks and about two and a half times greater than in catchments underlain by volcanic rocks (Meybeck, 1981). In some areas of the world, where rock type is very homogeneous, such as the Canadian and African shields and the Amazon and Congo basins, stream water chemistry may vary very little over large areas. In contrast, stream water chemistry can be highly variable in areas where geology varies on the finer scale. Overall, approximately 60% of all the solutes present in surface waters are derived from the weathering of rocks (Walling and Webb, 1986).

The influence of climate on surface water chemistry is very important at the global scale. As well as controlling the rate of weathering, through its influence on temperature and water availability, climate controls the water balance. In areas where evaporation exceeds precipitation, solutes derived from both weathering and atmospheric inputs are gradually concentrated. If evaporation is high enough, some minerals will start to precipitate, such as calcite ($CaCO_3$) and dolomite ($MgCO_3$). The precipitation of minerals results in the surface waters containing lower concentrations of Ca and HCO_3 and higher concentrations of Na, Cl, Mg and SO_4. In contrast, in areas where annual rainfall is high, the concentration of solutes in surface waters is diluted.

It has been proposed that because the above three processes influence the concentration of the major solutes in different ways, the relative abundance of the dominant cation and anion

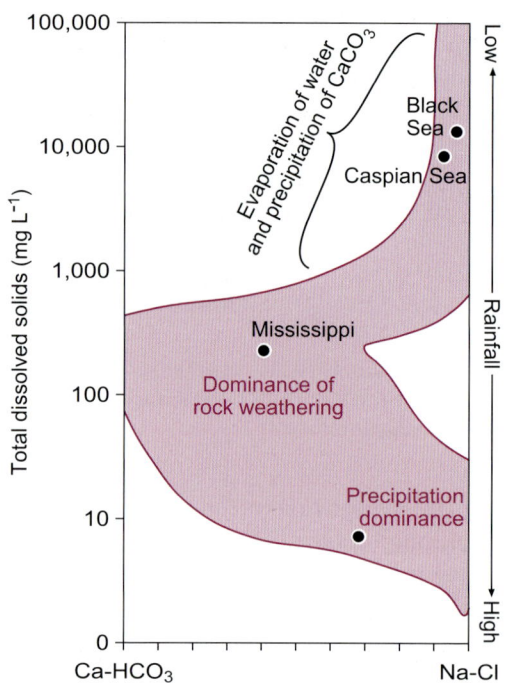

Figure 4.5 The influence of geology and climate on the total dissolved solids (TDS) and major chemical composition of surface waters.
Source: Adapted from Gibbs (1970) Mechanisms controlling world water chemistry. *Science* 170: 1088–1090. Reprinted with permission of AAAS.

can be used to evaluate which of the processes is most significant in controlling surface water chemistry (**Figure 4.5**). High concentrations of Na and Cl, with low TDS, are observed in wet coastal regions where atmospheric inputs are the dominant source of solutes, and inputs derived from weathering are low. Sodium and Cl can also dominate in semi-arid areas where evaporation–crystallisation processes dominate, resulting in very high TDS. However, Gibbs (1970) found that most surface waters were dominated by Ca and HCO_3 derived from the weathering of bedrock, and plot in the central region of **Figure 4.5**. This was supported by the findings of Meybeck (1981) in his survey of global rivers (representing about 60% of the water discharge to the oceans), where he observed that 97% of rivers were dominated by

Ca and bicarbonate and that Na was the dominant cation in only 1.7% of rivers and Cl and SO_4 were dominant in just over 1% of rivers. Thus the classification by Gibbs (1970), although attractive in its simplicity, is of limited value. In addition, other surface waters deviate from the plot in **Figure 4.5**, due to extreme geological or climatic conditions. For example, very acidic waters are often found in active volcanic regions or in sedimentary rocks rich in metal sulphides.

> **REFLECTIVE QUESTION**
>
> Are TDS in surface waters likely to be greater or smaller in warmer, drier climates than elsewhere?

D TEMPORAL VARIATIONS IN SURFACE WATER CHEMISTRY

Surface water chemistry varies over time for two main reasons. Firstly, the processes controlling the supply of solutes available for leaching vary over time, and secondly, the hydrological processes that generate runoff within a catchment are dynamic and vary with climate and season. The major temporal patterns in surface waters can be considered at the following timescales:

1. during storms (hours) as a result of changes in hydrological flowpaths delivering different sources and amounts of solutes
2. diurnal (24-hour) variations in, for example, pH and dissolved oxygen, resulting from biological processes and changes in daylight
3. seasonal variations associated with biological and hydrological cycles
4. year-to-year trends, usually as a result of anthropogenic activities in the catchment.

Fluctuations in dissolved oxygen concentration and pH are primarily related to photosynthesis and aerobic respiration of aquatic plants. Photosynthesis, which is driven by sunlight, removes carbon dioxide (CO_2) from the water and produces free oxygen, which causes an increase in surface water pH and dissolved oxygen. However, stream water pH and dissolved oxygen concentrations are also affected by changes in air pressure and temperature. The solubility of gases increases as air pressure increases; this can influence the pH of water, as more CO_2 will dissolve in the water as air pressure rises. In contrast, increases in temperature reduce the solubility of most gases; cold water can hold more dissolved oxygen than warm water. Therefore the dissolved oxygen content of surface water has both a seasonal and a diurnal cycle, as shown in **Figure 4.6**. This diurnal variation in dissolved oxygen is not generally observed from collecting regular water samples, as this typically occurs in the daytime. However, the use of continuous monitoring probes has enabled these small diurnal variations to be clearly observed (**Figure 4.6**).

During storms, changes in dominant hydrological flowpaths through the catchment (Chapter 3) can lead to large changes in solute concentrations, over relatively short periods of time. For some solutes, an increase in river discharge can lead to an increase in concentration as solutes are moved rapidly into the river by overland flow and shallow throughflow. For example, dissolved organic carbon (DOC) concentrations can increase with stream discharge during storms in some catchments (Chapman *et al.*, 1993; Hinton *et al.*, 1997; **Figure 4.7**). This reflects the change in hydrological flowpaths, with more flow through the upper organic surface soil horizon in the catchment where DOC is produced. In contrast, other solutes that are mainly derived from the weathering of bedrock, such as Ca, Mg and Si, and transported to surface waters by deep soil throughflow and groundwater display a negative relationship with discharge during storms (**Figure 4.7**), as their concentrations are diluted by surface runoff.

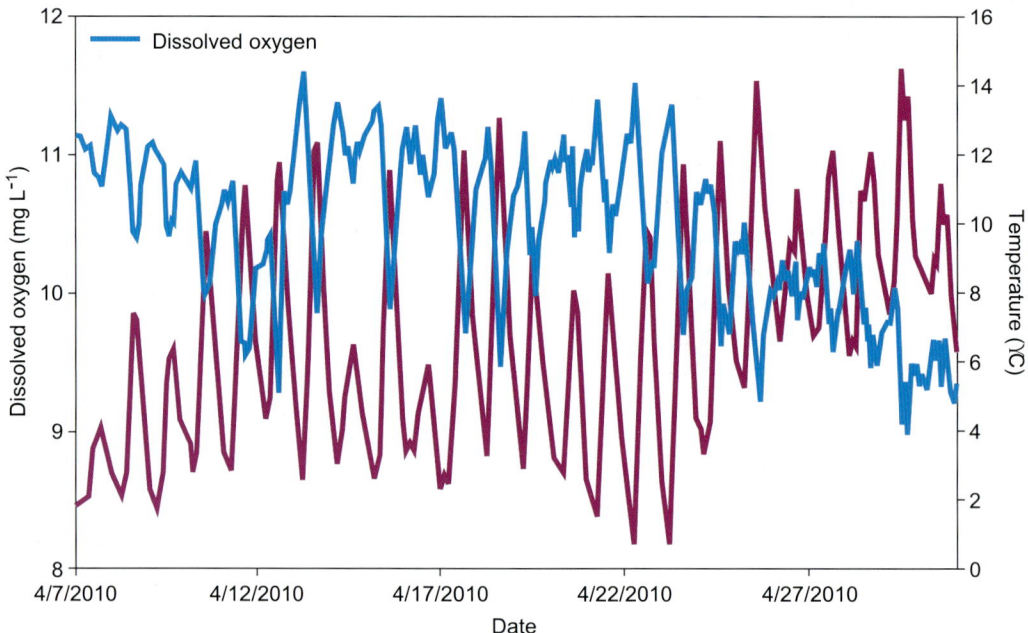

Figure 4.6 Dissolved oxygen concentrations and temperature recorded at 15-minute intervals at a stream in the North Pennines, UK, shows clear diurnal variations.
Source: Data kindly provided by Katie Aspray, University of Leeds.

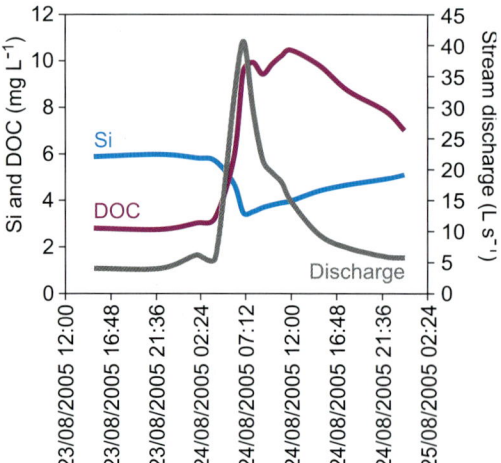

Figure 4.7 Changes in stream discharge and silicon and dissolved organic carbon (DOC) concentrations during a storm event in a headwater stream in NE Scotland.
Source: Adapted from Stutter *et al.* (2012b).

Many solutes display strong seasonal patterns (**Figure 4.8**). These patterns can be divided into those that are controlled by seasonal changes in hydrology and those that are mediated by biological processes. In streams and rivers, pH and concentrations of Ca and Mg often display a maximum in summer and minimum in winter, which reflects seasonal changes in the dominant source of water (**Figure 4.8a**). In summer, groundwater enriched in base cations often dominates stream flow, whereas in winter soil water and surface runoff, which is more acidic and contains lower concentrations of base cations, dominates. In natural waters, nitrate (NO_3) and potassium (K) also display strong seasonal patterns, with a summer minima and winter maxima (**Figure 4.8b**). This reflects seasonal changes in the availability of these nutrients for leaching, as plant uptake during the summer reduces the concentration of these nutrients in soil solution. DOC is also strongly mediated by biological processes, being produced via the decomposition of organic matter during summer and washed from the soil during the autumn. Hence DOC displays a maximum in late summer and autumn and a minimum in early spring (**Figure 4.8c**).

Long-term patterns in surface water chemistry are usually driven by climatic variability and anthropogenic activities, such as changes in land management or atmospheric pollution. However, it is often difficult to identify the main driver of change, as many of these factors are changing at the same time or the record of solute concentration (often referred to as a time-series) is of insufficient length. A number of studies have highlighted that caution is needed in interpretation of 'long-term' river and lake water chemistry and that misleading conclusions may be drawn from datasets of 10 years or less (e.g. Burt et al., 1988; Driscoll et al., 1989). For example, analysis of stream water NO_3 concentrations from the Hubbard Brook Experimental catchment, northeast United States, for the period 1964–1974 suggested a strong increasing trend (Likens et al., 1977). However, no long-term trend was observed in the dataset for the period 1964–1987 (Driscoll et al., 1989). This illustrates the importance of maintaining long-term sites and networks of surface water chemistry.

Long-term records of surface water NO_3 concentrations for semi-natural catchments generally show no evidence of time trends, despite an increase in atmospheric nitrogen deposition at many of these sites caused by air pollution. However, peak winter concentrations do vary markedly between years (e.g. Reynolds et al., 1992; Mitchell et al., 1996). This variability has been

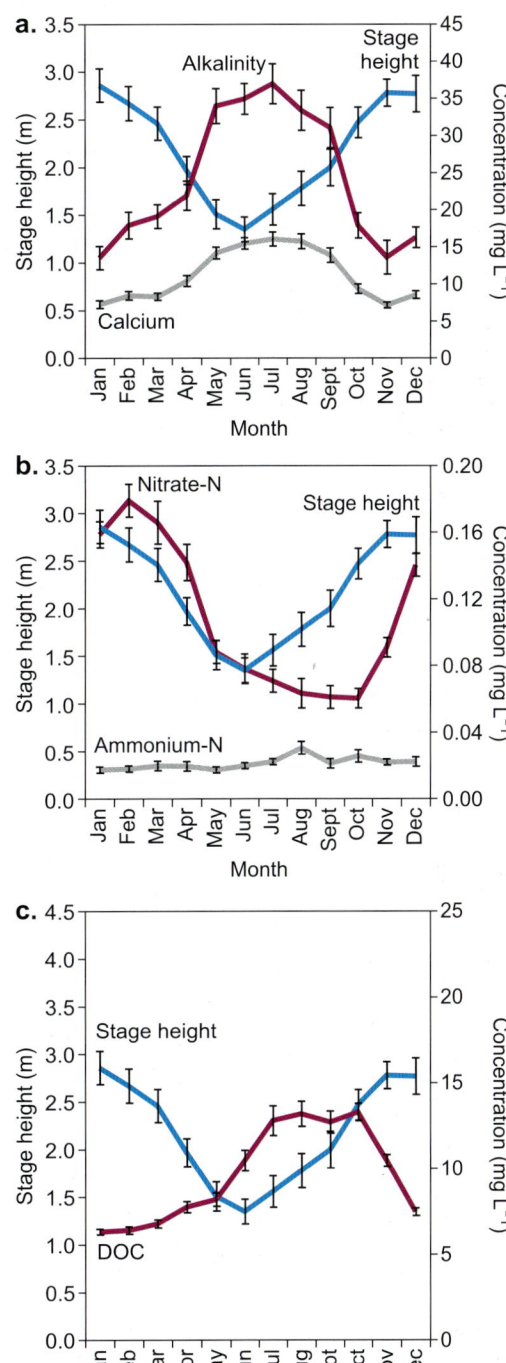

Figure 4.8 Mean monthly concentration in stream water collected weekly from the headwaters of the River Tees, northern England (1993–2010): (a) calcium and alkalinity, (b) nitrate and ammonium, and (c) DOC. Error bars represent the standard error of the means (n = between 72 and 79 per month).
Source: Data kindly provided by the Environmental Change Network, 2012, citation code: ECN:PC7/12.

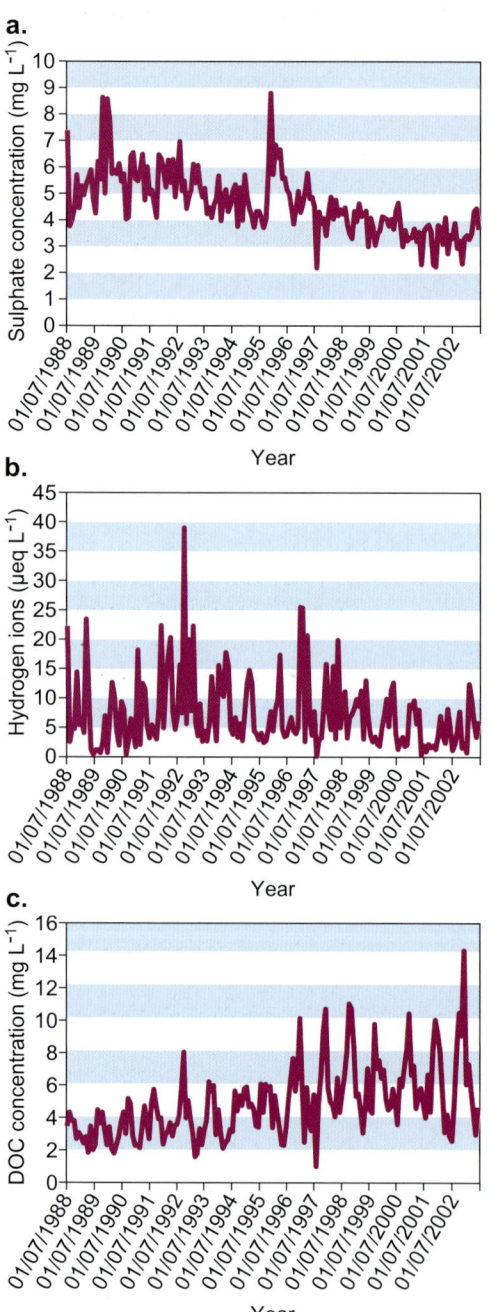

explained by inter-annual variation in climatic factors, including drought, summer temperature and winter freeze-thaw. For upland lakes and streams across the UK, Monteith *et al.* (2000) observed that the mean December–March **North Atlantic Oscillation (NAO) index** (a measure of large-scale climate conditions a bit like the El Niño Southern Oscillation) was a good indicator of peak NO_3 concentration. Highest peak NO_3 concentrations were observed when the NAO values were lowest, which indicates drier, colder weather in northwest Europe. Therefore, Monteith *et al.* (2000) suggested that freezing of the soil leads to the death of some of the soil microbial biomass and damage to plant roots, and subsequent thawing leads to the release of this material available for mineralisation and leaching to surface waters.

In the 1980s and 1990s, many countries set up atmospheric and stream or lake monitoring networks to determine the impact of emission controls implemented at coal-fired power stations on improving air and water quality. In response to rapid declines in atmospheric SO_4 deposition across North America and Europe since the late 1980s (e.g. Jeffries *et al.*, 2003; Fowler *et al.*, 2005), there have been widespread reductions in stream water sulphate (SO_4) and inorganic aluminium concentrations (Stoddard *et al.*, 1999; Davies *et al.*, 2005; Skjelkvåle *et al.*, 2005; Burns *et al.*, 2006; **Figure 4.9a**), and in some regions an increase in water pH. However, stream water recovery has been only partial and many waters remain in an acidified state. This lag in surface water acidification recovery (**Figure 4.9b**) has been attributed to a range of processes, including deposition of N remaining high, depletion of base cations and increase in DOC concentrations (**Figure 4.9c**). This demonstrates the long-term influence on surface water chemistry of biogeochemical processes occurring within a catchment.

Figure 4.9 Long-term trends in the concentration of (a) sulphate (SO_4), (b) hydrogen (H^+) ions and (c) DOC in weekly samples collected from 11 streams that form part of the UK Acid Water Monitoring Network (AWMN) (July 1988 to March 2002).
Source: Data kindly provided by CEH Lancaster as part of the AWMN.

> **REFLECTIVE QUESTION**
>
> Ignoring human influence, what are the key reasons that solute concentrations naturally vary over time in surface waters?

E WATER USE AND WATER QUALITY DETERIORATION

Although surface freshwater which is not locked up in ice sheets, glaciers or permafrost accounts for less than 1% of all the non-saline water on Earth, it is essential for the survival of terrestrial life, and without it humans could not survive. In addition to drinking water, humans use water for a wide variety of purposes that both consume and contaminate the water. Since ancient times, surface waters have been used as a convenient disposal route for our waste. However, this use (or abuse) of surface waters for waste disposal conflicts with almost all other uses of water and, most seriously, with the use of surface waters for drinking water (see Chapter 9).

The industrialisation of a country leads to a shift in the pattern of water use from rural and agricultural to urban and industrial, and the growth of large cities has been accompanied by increasing stress on the aquatic environment, leading to a reduction in the quality of the water (McDonald *et al.*, 2011 and see section F2 on urbanisation). Poorer water quality means water pollution.

Water pollution is usually defined as 'where one or more substances have built up in a water body to such an extent that it can be harmful to organisms and plants which live in the water body and to animals and humans that may drink the water'. Although water pollution usually refers to a deterioration in water quality caused by human activities, the same effect may occur naturally. For example, large amounts of sediment can be washed into streams and rivers during rainstorms, and toxic metals find their way into surface waters where concentrations of metal ores occur. These are natural processes and the environment will be able to neutralise the effects on water quality over time. Pollution caused by human activities, on the other hand, is usually on a larger scale, can often happen rapidly and can take a much wider variety of forms, as outlined in **Table 4.2**.

Pollution sources that affect surface water are usually separated into two categories:

1. point sources
2. non-point sources or **diffuse pollution**.

If pollution comes from a single location, such as a discharge pipe attached to a factory, it is known as **point-source pollution**. Other point sources include sewage treatment plants, industrial discharges or any other type of discharge from a specific location (commonly a pipe) into a stream. By contrast, **non-point sources** are **diffuse sources** of pollutants that are not as easily identified or measured as point sources. Common examples include:

- excess fertilisers and pesticides from agricultural and residential lands
- sediment from agricultural and forest lands, construction sites, and eroding stream and river banks
- bacteria and nutrients from livestock waste and leaking septic tank systems
- salt, oil, grease and heavy metals from urban runoff
- acid drainage from abandoned mines
- atmospheric deposition.

When point-source pollution enters the aquatic environment, the place most affected is usually the area immediately around the source. However, it is also dependent on the capacity of the receiving water body to dilute the input, and the ecological sensitivity of the receiving water. In contrast, the location most affected by diffuse pollution may be far downstream from the location where the

Table 4.2 The nature, sources, effects and control of some major types of pollutants

Pollutant	Nature	Source	Effects	Control
Bacteria	Disease-causing organisms	Domestic sewage and animal waste	Illness in humans and curtailed recreational use	Treatment of waste (e.g. at sewage treatment works) and chlorination of drinking water
Nutrients	Mainly refers to nitrogen and phosphorous species	Domestic sewage, industrial waste and agricultural land (mainly inorganic fertiliser)	Excessive growth of aquatic plants, leading to eutrophication; high nitrate concentrations in drinking water, potentially toxic	Phosphorus removed at sewage treatment works. Reduce nitrate leaching by controlling fertiliser inputs and good land management
Trace metals		Mining, industrial processes, runoff from roads	Toxic effects on aquatic ecosystems	Difficult
Organic chemicals	Pesticides, detergents, industrial by-products and pharmaceutics	Domestic sewage, industrial waste and farms	Toxic threat to fish and other aquatic fauna; possible impacts on human health	Difficult to control, as not removed by conventional sewage or water treatment works
Natural organic matter	Biodegradable organic material decomposed by aerobic bacteria (requires dissolved oxygen)	Domestic sewage, animal manure, food processing	Causes depletion in dissolved oxygen content, which has an impact on aquatic fauna and flora	Treatment of sewage, containment of sewage, animal slurry and manure and silage effluent
Sediment	Mainly soil material (organic and inorganic); industrial by-products	Land erosion by water and wind, construction sites, some quarrying and mining processes	Removal of fertile soil, filling of reservoirs and lakes, sedimentation of gravel beds; reduced aquatic life	Controlled by soil conservation and best management practices; use of settling ponds; flood control
Heat	Heated water returned to rivers and lakes	Power stations, steel mills and refineries and other industrial cooling units	Reduces dissolved oxygen content of water, with potential harm to aquatic life	Minimised by recirculation and reuse of industrial cooling waters

pollutant entered the watercourse, as it is often the cumulative effect of small amounts of contaminants gathered from a large area that results in an impact. Where pollution enters the environment in one place but has an effect hundreds or even thousands of miles away, this is known as transboundary pollution. An example of this is the emission of pollutants, such as sulphur dioxide (SO_2), into the atmosphere in one country which are then deposited in another country, having an impact on water quality (see section F4 on acid rain).

Different sources of pollution respond to storms in contrasting ways. Usually, the concentration of pollutants from non-point sources will increase as flow increases during storm runoff. In contrast, concentrations of pollutants from point sources generally decrease through dilution during storm runoff. Thus water monitoring programmes that collect water samples at regular time intervals (usually weekly or monthly) are much more likely to identify point sources of pollution which can be treated. In contrast, pollution from diffuse sources can be controlled by only prevention and remain the major challenge to improving surface water quality globally.

There are two main ways of measuring the quality of water. One is to take water samples and measure the concentrations of different solutes and sediments that they contain. If the concentrations exceed a certain limit, we can regard the water as polluted. Measurements like this are known as chemical indicators of water quality. An alternative method is to examine the fish, insects and other invertebrates that live in the aquatic environment. A widely used system focuses on macroinvertebrate benthic fauna and utilises the Biological Monitoring Working Party (BMWP) scoring system (Armitage *et al.*, 1983) to assess water quality. Measurements like this are called biological indicators of water quality. Often both types of measurements are made in order to classify the quality of a water body. For example, the European Water Framework Directive (WFD) requires member states to use chemical and biological indicators, along with hydrogeomorphological parameters, to rank the ecological status of rivers (see **Box 6.4** in Chapter 6).

> **REFLECTIVE QUESTION**
>
> Why might the people responsible for diffuse pollution often be difficult to trace?

F SURFACE WATER POLLUTION: CAUSES, CONSEQUENCES AND MANAGEMENT

1 Agriculture

Farm types can be split generally into arable and livestock. Common arable crops include wheat, barley, rice, oats and oil-seed rape, as well as vegetables, such as potatoes and other root crops. Livestock farms tend to focus on one particular type of production, which might be dairy, beef, sheep or pigs. Some of these farms will operate intensive indoor systems, while others will use more extensive outdoor production techniques.

1.1 Impacts of agriculture on water quality

Agriculture can have a huge impact on water quality, due to the vast array of chemicals that are used to increase production. Sediment is one of the most obvious agricultural pollutants, as fields can be left bare once a crop has been harvested, and even when the next crop has been sown it can take months before there is a vegetative cover, particularly when a winter crop is planted. Furthermore, overgrazing and trampling of soil by livestock can increase overland flow and erosion. Concentrations of suspended solids in agricultural streams typically reach several hundred mg L^{-1} during some storm events, which can lead to smothering of the **benthos** and impacts on fish spawning habitats.

It is thought that the aquatic life in the Gulf of Mexico is severely degraded by agricultural pollution brought down the Mississippi river from the heavily agricultural Midwest of the USA. The United States Geological Survey (USGS) estimates that 70% of the nitrogen and phosphorus that enters the Gulf of Mexico comes from agricultural sources in the USA. This compares to 9–12% from urban sources. Nutrients are one of the most well-studied agricultural pollutants and comprise largely nitrogen and phosphorus, which are applied to land in manure and slurry (**Figure 4.10**) as well as in inorganic fertiliser. Moreover, dung is deposited on fields by grazing livestock. Nitrate is the major form of nitrogen found in surface waters and is very mobile, unlike some other forms, such as ammonium, which is more strongly retained by soils (Hatch *et al.*, 2002). Nitrate leaching is a problem in most agricultural areas and peak concentrations in rivers usually exceed 10 mg N L^{-1}, although the other nitrogen species will be found at much lower concentrations (Neal *et al.*, 2006). Although it is not

Figure 4.10 Slurry being injected into a field by a specialist machine (a) with the deposited slurry shown in (b).

particularly toxic in the environment, nitrate must be removed from drinking water, which is costly to water utilities and consumers (Pretty *et al.*, 2000). In contrast, phosphate tends to form **complexes** with soil particles, and concentrations in rivers are thus often associated with levels of suspended solids. Nevertheless, orthophosphate is soluble and is therefore **bioavailable** to aquatic organisms, which can lead to **eutrophication** (an unwelcome bloom of plants in water, which has a detrimental effect on other organisms – see Chapter 6 for more details), as plants can quickly use the phosphate to encourage rapid growth (Jarvie *et al.*, 2007).

Pesticides, including herbicides, insecticides and fungicides, are applied to a wide variety of crops in order to improve growth by eliminating competitors and parasites. Residues of these chemicals are frequently transported from crops and soils to waters in runoff (Garrod *et al.*, 2007). Depending on the characteristics of a particular soil and chemical, certain amounts of a substance will be present in the solid and liquid phases. While some chemicals bond very strongly with soils, others tend to be found in the soil water and are much more mobile. The extent to which these substances are present in water will depend largely on their physicochemical properties, which determine their degradation and **sorption** (how one substance becomes attached to another) characteristics (Kladivko *et al.*, 1991). The degradation rate of a pesticide is usually expressed as its half-life (DT_{50}) (i.e. the time taken for the substance to degrade to 50% of its original concentration). Sorption characteristics are expressed as values representing the amount (per kg) of a substance that will attach to soil (abbreviated to K_d). As for phosphorus, sorption of a pesticide to soil does not mean that it will be immobile, as substances can be transported in the particulate phase in overland flow and subsurface drainflow (Øygarden *et al.*, 1997). Concentrations of pesticides detected in water bodies range from nanograms per litre, ng L^{-1} (billionths of a gram) up to tens of micrograms per litre, μg L^{-1} (millionths of a gram), although they are commonly measured in streams at peak concentrations below 10 μg L^{-1} (Du Preez *et al.*, 2005).

Veterinary medicines have similar physicochemical properties to pesticides, but reach the land following administration to livestock to prevent and treat disease. Residues of these compounds can be found in manure and slurry applied to land and in dung deposited by grazing livestock. Moreover, wash-off can occur following topical application. Antibiotics are the most commonly used group of veterinary medicines and have been measured in rivers at concentrations of up to a few µg L^{-1}. Toxic impacts are unusual in the aquatic environment, although concern exists about the spread of antibiotic resistance (Kay *et al.*, 2004).

Other important water quality issues resulting from agricultural production include sediment loss (Edwards and Withers, 2008) and organic pollution from manures and silage effluent, leading to significant reductions in the dissolved oxygen content of water bodies (Hooda *et al.*, 2000). See section D of Chapter 8 for a discussion of the health effects of nutrients, pesticides and veterinary medicines.

1.2 Agricultural stewardship

Due to the known impacts that agriculture can have on the aquatic environment, the implementation of stewardship measures is now commonplace and many of the subsidies available to farmers today are offered on this basis rather than being related to production. Specific measures have long been proposed to reduce the impacts of agriculture on water quality. Stewardship measures can be split into three general categories: those that reduce inputs of pollutants; others that limit transport to water bodies; and edge-of-field measures (Kay *et al.*, 2009). Hundreds of measures are available that can be implemented by farmers, but some of the most common include **buffer zones**, wetlands and specific soil management practices. Buffer zones are permanently vegetated areas of land perhaps 5–100 m in width, usually, but not exclusively, located adjacent to a watercourse, where farming practices are restricted with a view to protecting water from agricultural diffuse pollution. Restrictions can include one or more of the following: no fertiliser applied, no pesticides and herbicides applied, no cultivation, no livestock grazing allowed, no farming at all allowed, particular plants or types of plant must be grown/allowed to grow. Such a buffer may provide a biochemical and physical barrier against pollution inputs, such as sediment, nutrients and pesticides, and therefore reduces the connection between the potential pollution source and the receiving water body. A wealth of research is available from around the world that has quantified the impacts of buffer zones on water quality. This research has been reviewed by a number of people (e.g. Kay *et al.*, 2009; Arora *et al.*, 2010; Gumiere *et al.*, 2011) and shows that while the impact can be variable, sediment, nutrient and pesticide pollution is often reduced significantly (by up to 100% in some cases). Effects have been found to be site specific, though, and depend on the nature of the pollutant, soil characteristics and hydrology. Evidence suggests that more established vegetation, particularly the presence of trees, will lead to greater pollution reduction. However, the majority of this research remains biased towards consideration of single pollutants (see Gumiere *et al.*, 2011; Arora *et al.*, 2010). Recently we have started to value the landscape in terms of the provision of a range of interacting **ecosystem services** and goods. For buffer zones these services could include, among others, natural flood management, biodiversity, diffuse pollution mitigation and biofuel production. Therefore we should be revising our methodologies of designing, understanding and evaluating buffer zones to maximise these possible benefits (Stutter *et al.*, 2012a).

There are many examples of agricultural stewardship schemes around the world, including work undertaken by the Environmental Protection Agency in the USA and the Landcare scheme in Australia. Furthermore, many stewardship measures have been implemented in the UK in recent years through the England Catchment Sensitive Farming Delivery Initiative (ECSFDI) (**Figure 4.11**). This is a scheme set up to encourage farmers to adopt environmentally friendly practices through programmes of advice and the availability of subsidies paid through Cross Compliance and the Entry and Higher Level Schemes.

Figure 4.11 Farm advice activities being undertaken as part of the England Catchment Sensitive Farming Delivery Initiative (ECSFDI).

1.3 The future

There are a number of contemporary issues that are particularly pertinent as to future impacts of agriculture on water quality. While stewardship measures have been implemented for a number of years, and with renewed vigour since reform of the European Common Agricultural Policy in 2003, there is a limited amount of data to support their use to improve water quality (**Box 4.3**). While great impacts have been made in some cases these are likely to be site and pollutant specific, and for many measures no data are available. Crucially, limited data exist to describe the impact of stewardship at the catchment scale (Kay *et al.*, 2009).

Furthermore, subsidies for agricultural stewardship may change in the future, with little certainty that current regimes will remain. It has been postulated that human population growth will require agricultural production to be increased, and therefore subsidies for environmental protection could be replaced once more by production-related payments.

At present, agriculture remains one of the key sources of surface water pollution and there is currently little evidence to suggest that this situation is improving, despite concerted efforts to employ more environmentally friendly practices.

2 Urbanisation

2.1 Urban pollution and non-point sources

Many towns and cities develop in close proximity to lakes and rivers. The construction of housing, commercial buildings, roads and other infrastructure reduces the permeability of the surrounding land, and so alters the hydrological characteristics of storm water flowing to those lakes and rivers. Runoff volumes increase and, most obviously, discharges become 'flashy', with a shorter time to a greater peak discharge, resulting in increased bed and bank scour and increased risk of flooding. Loss of in-stream and **riparian** biodiversity also occurs, exacerbated by the low or zero flows that can then occur during dry weather. Urbanisation also has major impacts on water quality. Historically, the impacts of most concern were those associated with sewage disposal and end-of-pipe (e.g. factory) discharges. Investment in sewers and wastewater treatment processes, driven by public health imperatives (see Chapter 9), and greater regulation of commercial point-source discharges have led to dramatic improvements in the quality of urban surface waters in many countries over the last 150 years. However, urban waters often remain unacceptably polluted, due to emissions from non-point sources which originate over an area. These emissions comprise a diverse mix of pollutants that accumulate on built surfaces until they are washed off during storm events (see **Box 4.4** for an example). Impacts may then arise due to sedimentation of rivers (e.g. light loss, gill rot in fish), dissolved oxygen depletion, eutrophication, transmission of infectious pathogens and a wide range of toxic

SURFACE WATER QUALITY

BOX 4.3 CONTEMPORARY CHALLENGES

Long-term nitrate trends in rivers

The use of fertilisers in agriculture has helped to ensure the production of plentiful, cheap food supplies in the developed world, although one of the consequences of this has been nitrate pollution of rivers. Policies have been developed that have attempted to reduce this contamination, including the EU Nitrates Directive (EC, 1991) and the Clean Water Act in the US (US Senate, 2002), which have set regulatory limits for nitrate concentrations in rivers of 11.3 and 10 mg N L^{-1}, respectively. Recent studies to assess how effective these policies have been have determined that nitrate concentrations in polluted rivers have generally changed little in recent decades (since the 1970s) (Worrall *et al.*, 2009; Burt *et al.*, 2011; Kay *et al.*, 2012; Figure 4.12). In the Mississippi basin, for example, despite efforts to reduce pollution, nitrate losses to the Gulf of Mexico were actually 10% higher in 2008 than they were in 1980 (Sprague *et al.*, 2011). This has been attributed to increased concentrations at upstream sites, which, when normalised for changes in flow, have risen by around 75%. A particular problem is that nitrate can take decades to reach groundwater and then be transported to streams, and so the impacts of land management actions may not be evident for many years. Moreover, some work has highlighted that it may not be the inherent ability of some agricultural stewardship measures (e.g. reduced fertiliser applications) to reduce nitrate pollution that is the problem but, rather, the need for robust implementation of these by regulators and farmers (Kay *et al.*, 2012). Taking these factors into account, it may not be possible to achieve some policy goals, such as compliance with the EU Water Framework Directive, by 2015. If progress is to be made, though, continued efforts will be needed on farms to control nitrogen inputs and transport to waters, as will long-term (covering several decades) water quality datasets to assess their impacts.

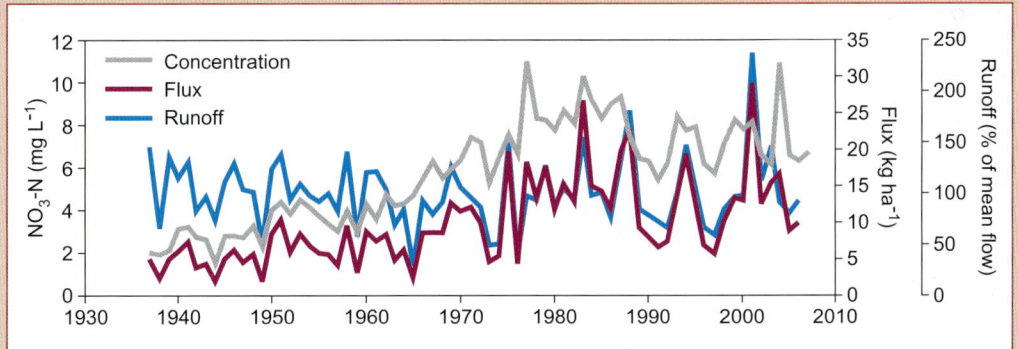

Figure 4.12 Long-term NO$_3$-N concentrations and fluxes in the River Stour.

Source: Reprinted (adapted) with permission from Burt, T.P., Howden, N.J.K., Worrall, F., Whelan, M.J. and Bieroza, M. 2011. Nitrate in United Kingdom rivers: policy and its outcomes since 1970. *Environmental, Science and Technology* 45: 175–181. Copyright 2010, American Chemical Society.

Table 4.3 Some urban diffuse pollutants and their sources

Source	Oil and hydrocarbons	PAH	Pesticides	Solvents/paints and dyes	Sediment	Bacteria and pathogens	Metals	Organic pollution	Nitrogen	Phosphorus
Residential runoff	2	2	2	1	3	2	2	2	2	2
Industrial and commercial runoff	2/3	2	1	2	2	1	2	1	1	1
Highway runoff	3	3	2	0	3	1	3	1	1	1
Construction industry	0	0	0	0	3	0	1	0	2	1
Rail track runoff	2	2	2	0	2	0	1	1	1	1
Garage/petrol/service stations	3	2	0	0	0	0	1	0	0	0
Road/rail weed control	0	0	3	0	0	0	0	0	1	0
Gardening	0	0	3	0	0	0	0	0	1	1
Pets/birds	0	0	0	0	0	2	0	3	3	3
Misconnections	2	1	1	1	1	1	2	2	2	2
Sewer leaks	0	0	0	0	1	2	1	2	2	2
Septic tanks	0	0	0	0	0	1	1	2	2	2
Litter/waste disposal	1	1	1	1	1	1	1	2	2/3	2
Car/vehicle emissions	2	2	0	0	0	0	2	0	2	2

Note: 0, unlikely source; 1, possible source; 2, documented source; 3, significant source

Source: From Ellis and Mitchell, 2006. Reproduced with kind permission of John Wiley & Sons from Ellis, J.B. and Mitchell, G. 2006. Urban diffuse pollution: key management issues for the Water Framework Directive. *Journal of the Chartered Institution of Water and Environmental Management* 20: 19–26.

Table 4.4 Pollutant concentrations in wastewater and storm water across Europe and North America

	Untreated sewage	CSO	Urban storm water
BOD mg L^{-1}	160	60–200	10–250
SS mg L^{-1}	225	100–1100	3000–11000
Total N mg L^{-1}	35	3–24	3–10
Total P mg L^{-1}	10	1–11	0.2–1.7
Coliforms MPN/100mL	10^7–10^9	10^5–10^7	10^3–10^8
Cadmium µg L^{-1}	0.15–0.38	1.4–9.6	0.2–63
Copper µg L^{-1}	110–240	80–170	2–1270
Zinc µg L^{-1}	85–180	100–1070	17–3550

Sources: From Mitchell, 2001; Novotny, 2002; Rule *et al.*, 2006b.

effects from metals, biocides and other persistent organic pollutants.

Table 4.3 indicates the range of urban pollutants involved and their principal sources. While most pesticide pollution is associated with agriculture, some pesticides are often used in gardens in urban areas, which can be problematic (e.g. metaldehyde products are used to kill slugs). Table 4.4 reveals that the concentration of many of these pollutants in urban storm water exceeds that of untreated sewage or from combined sewage overflows (CSOs), which comprise a mix of untreated sewage and storm water discharged to rivers when storm discharges would otherwise exceed sewer capacities (see **Figure 9.7a** in Chapter 9 for a photo of a CSO). Data limitations mean that it is difficult to understand precisely the role of urban diffuse sources in river degradation, but evidence (see, for example, **Table 4.5**) suggests that in developed countries urban diffuse pollution is now one of the main reasons why rivers fail to reach their water quality objectives. Note that when CSOs are also considered, urban storm water pollution accounts for more kilometres of poor water quality than any other cause.

It should also be noted that many possible urban water pollutants are not yet being monitored, as the detection techniques are only just becoming available. These pollutants include those derived from pharmaceuticals and personal care products such as steroids, antibiotics, fire retardants, fragrances and flavours, insect repellents, disinfectants, cosmetics, solvents, hair dyes and so on. In many cases it is not known what effects many of these substances might have on aquatic organisms or human health and this is an emerging area of study for water science (**Box 4.5**). However, it is known that many synthetic chemicals act as endocrine disrupters in the aquatic environment. This means that they can cause the sex of some creatures to be altered at the foetal stage. Section D3 of Chapter 8 provides more detail on the possible health impacts of these emerging contaminants.

Table 4.5 Causes of low river quality in Scotland

Cause	% of river length
Abandoned mines	22
Sewage from treatment works	21
Urban diffuse sources	20
Combined sewer overflows (CSOs)	12
Industry	12
Agricultural point sources	4
Agricultural diffuse sources	1
Other	8

Source: Data from SEPA, 1996.

2.2 Mitigation of urban pollution

The areal nature of diffuse sources means that they are much harder to control than point sources, which are amenable to end-of-pipe treatment. Urban diffuse sources are particularly challenging, as cities are highly heterogeneous in terms of land cover, land use and intensity of activity (e.g. housing density, traffic flow on roads), making effective targeting of mitigation measures more difficult. Such measures are of two fundamental types: structural and non-structural. In the USA both are covered by the term Best Management Practices (BMP), while in the UK the structural measures are known as **sustainable urban drainage systems** (SUDS). The term 'source control' is also commonly used, and recognises that BMPs and SUDS are most effective when applied close to the source of storm water.

The philosophy behind SUDS is that they act to reintroduce a more natural hydrological regime to the urban environment through detention and retention of storm discharges, through a variety of physical devices (**Table 4.6**; see also section C5 in Chapter 3). The term 'SUDS triangle' is often used to denote the primary benefits that SUDS offer. First, attenuation of storm discharge leads to reduced flood risk, and enhanced groundwater replenishment and dry-weather flow. Second, this discharge attenuation gives more time for pollutant degradation through natural processes of **volatilisation**, sedimentation and filtration, precipitation, adsorption, **photolysis**, microbial degradation and plant uptake. Finally, SUDS that include natural elements in their structure are able to enhance the biodiversity and amenity value of buildings and urban areas. The carbon sequestration function of vegetated SUDS is also now recognised as a valuable additional benefit, given the importance of climate change mitigation. The features and relative advantages of the many different types of sustainable urban drainage device, as well

---

BOX 4.4 CONTEMPORARY CHALLENGES

De-icing agents and diffuse pollution

In cold regions, highway authorities often undertake preventive measures to avoid road icing in winter. Salt is the most common de-icing agent and typically comprises about 90% sodium chloride, with the remainder made up of insoluble residues such as clay marl and trace amounts of metals. Colwill *et al.* (1984) estimated that de-icing contributed 25% of suspended loadings and 90% of soluble loadings in winter road runoff from a rural motorway in England, although clearly this will vary greatly with the frequency and extent of salt application. The chloride contained in road salt is of concern, due to its impact on ground and surface waters used for drinking supply and aquatic habitat. The metals contained in de-icer salt, including nickel, lead, zinc and chromium, are toxic in runoff, while the salt itself increases the solubility and mobility of metals previously deposited on road surfaces by vehicles. De-icer salt also contains sodium ferrocyanide, added as an anti-caking agent, which in strong sunlight and acidic conditions can generate toxic cyanide forms, although in surface waters this usually evaporates as hydrogen cyanide.

Monitoring in northeastern USA from 1991–2004 indicated that chloride concentrations exceeded environmental standards in around 2% of the 1300 groundwater wells analysed, while 15 of 100 catchments had winter chloride concentration in excess of aquatic habitat standards (Mullaney *et al.*, 2009). Concentrations were greatest in urban areas, where available data indicated an increasing historical trend in chloride loading. For the urban areas, chloride:bromide ratio analysis indicated that much of the chloride was due to de-icer application, although septic systems, wastewater discharge and saline groundwater plumes from landfills and salt storage areas were also significant sources. The increasing load in chloride, consistent with literature reports of long-term increases in chloride concentrations, was attributed to the expansion of road networks and impervious areas that require de-icing.

Reductions in de-icer application have been achieved through highway design, such as snow fences, from improved weather forecasting and ice prediction that allows more targeted applications in space and time, and by measurement of residual salt loadings, so that less salt is used when topping up. Because salt is corrosive, not only to vehicles but also to critical structures such as bridges, alternative de-icers, most commonly urea and ethylene glycol, may be used on parts of a highway. These de-icing chemicals are also used by the aviation industry on runways and aircraft. However, these de-icers also pose a risk to water quality. For example, urea can mobilise metals, such as copper, which occur in high concentrations on heavily trafficked roads. Urea also hydrolyses to ammonia, which exerts an oxygen demand on water and, in its un-ionised form, is also toxic. In the aviation industry, the use of urea as a de-icer is being replaced by calcium and magnesium acetate-based products.

> **BOX 4.5 THE FUTURE OF WATER**
>
> ### Emerging pollutants
>
> While it is the case that much progress has been made in cleaning up point-source discharges, new research is showing that there are many chemicals in sewage effluent that have, until recently, not been studied or regulated. Pharmaceuticals are the most prevalent group of these substances and antibiotics, painkillers, cardiovascular drugs and antidepressants have been measured in rivers at concentrations ranging between ng and mg L^{-1} (e.g. Kasprzyk-Hordern *et al.*, 2008). This results from their incomplete breakdown in the human body and excretion to the sewer system, as well as disposal of unused drugs. The effects on aquatic ecosystems are uncertain, although there is the potential for chronic impacts and the development of antibiotic resistance (Bundschuh *et al.*, 2009). One of the most striking of these has been the observed feminisation of male fish in the vicinity of sewage discharges in some UK rivers, thought to be due to exposure to endocrine disrupting chemicals (Tyler *et al.*, 2007) (see also Chapter 8). Further research will increase knowledge of the occurrence, fate and effects of emerging pollutants in aquatic systems, but regulators are already responding to the new information that is available. For instance, three pharmaceuticals (diclofenac, 17alpha-ethinylestradiol and 17beta-estradiol) are being considered for addition to the EU's List of Priority Substances (EC, 2011).

as a varied range of application case studies, can be found on the UK SUDS practitioners website: www.susdrain.org. There are plenty of photographs of different SUDS devices, and one example is also shown in **Figure 3.12** in Chapter 3.

SUDS differ in performance with respect to the main benefits they offer. While most offer multiple benefits, some perform particularly well with respect to flow attenuation, and others with respect to water quality remediation or ecological performance. This gives rise to a concept known as 'the treatment train', in which SUDS are developed in series, so as to maximise the benefits on offer from source to final discharge (whether watercourse or sewer). Thus a development might have green roofs at the building level, infiltration trenches and swales at the site level and ponds or wetland at the regional (neighbourhood) level.

Table 4.6 Typical SUDS devices

Position in control hierarchy	Typical SUDS devices
Source controls Control runoff at or very near the source	• Green roofs (can be extensive or intensive) • Downpipe storage (water butts) • Infiltration devices (soakaways, filter strips, infiltration trenches, pervious pavements)
Site controls Used to manage water in a localised area Often convey water from source to regional controls, or promote local infiltration	• Swales (broad, grassed channels) • Infiltration basins (dry depression storage) • Balancing pond (wet depression storage)
Regional controls Manage water from one or more sites	• Wet pond (primarily for flow attenuation) • Constructed wetland (primarily for water quality)

This approach builds robustness into the treatment of storm water, as it increases the range of pollutant remediation processes at work, and should one SUDS device fail, others are there to treat the runoff.

SUDS do pose particular problems, but these are now better understood and addressed. For example, safety fears with respect to urban ponds are tackled by careful design and edge planting; concerns over maintenance costs have been allayed, as experience shows that whole life-cycle costs are comparable to or less than those of conventional sewers; and a growing body of evidence on the performance of SUDS (both quality and quantity) means that SUDS designers are better able to develop urban drainage for new sites that meets specified performance criteria.

Although SUDS have received much attention, the importance of non-structural BMPS in urban storm water management must be recognised. These are measures designed to prevent discharge of polluted storm water, and should sit above source control in the BMP hierarchy. These include a very wide range of 'good housekeeping' measures, such as street sweeping, rainwater harvesting and reuse, using bunded tanks for storing hazardous materials, covering material in goods yards that might otherwise be washed to drains during rainfall, and having dedicated loading areas where a spill will not drain to a watercourse. Such measures can be incorporated into a site's environmental management plan, which may be a legal requirement of a consent to develop.

BMPs and SUDS are increasingly required by planning authorities as a condition of development consent. In the UK, for example, local planning policy is driven by national water management policies designed to address surface water quality objectives and flood risk. However, while BMPs are rightly considered essential, it is unlikely that they will lead to the restoration of urban river quality just yet. There are several reasons for this. Firstly, the rate of SUDS implementation remains low. For example, a 2009 survey indicated that in Greater Manchester, England, an area of 1100 km^2 with a population of 2.5 million people, there were only 36 SUDS sites in operation (White and Alacon, 2009). Secondly, where SUDS are installed the primary motivation is usually mitigation of flood risk, which poses the greater threat to human well-being. SUDS installed specifically for water quality reasons tend to be installed only to protect the most sensitive receiving waters. Thirdly, SUDS are being implemented principally in new urban development schemes which does nothing to combat the legacy effects of the existing built environment. Building stock renewal in developing countries is typically less than 1% a year, so most homes we will be inhabiting in 2050 are already built, hence there is a strong need to retrofit SUDS in the existing built environment. This is more challenging than for new build, but on-going research and case studies indicate that it is feasible (SNIFFER, 2006). Even high-density neighbourhoods offer opportunities for SUDS retrofit, through green roofs and devices such as infiltration trenches that occupy little surface space. Of course, funds for retrofit are limited, so knowing where to install SUDS for the greatest benefit is a key requirement.

SUDS are part of a wider goal of developing sustainable urban drainage, which has the main objectives of maintaining an effective public health barrier, avoiding flooding and pollution and minimising the use of resources, all in a socially acceptable manner. Strategies for achieving this goal include not mixing storm water and wastewater (e.g. via SUDS), not mixing wastewater with clean water (e.g. via water conservation, grey water recycling and use of no- or zero-flush toilets), and not mixing industrial and domestic wastes (e.g. via on-site commercial wastewater treatment), which makes effective treatment very difficult. Strategies such as these can add robustness to the urban system. For example, following these strategies should make it easier to deal with novel pollutants in domestic wastewater that arise from the use of pharmaceuticals and personal care products, which via CSOs find their way untreated into urban rivers. However, the transition to a more sustainable system will present problems to be overcome, as evidenced by the difficulties experienced in adequately flushing sewer deposits, as SUDS reduce sewer flows.

Clearly, SUDS are just one approach to delivering a more sustainable urban drainage system. Other technical measures exist that improve the operation of the conventional sewer system. These include partial treatment of wastewater in sewers, through the use of aeration, biofilm plates and activated sludge sites (see Chapter 9); real-time control of sewers, so that free capacity can be used to store water in the sewer and release it back to the treatment works after the storm event; and real-time control of discharge to river, to make best use of natural assimilative capacities.

Arguably, the main challenges facing the quality of urban rivers relate to the need for more integrated and radical solutions. The SUDS triangle neatly illustrates scope for addressing multiple objectives within the area of source control, but integration must be broader than this. Urban areas cannot be considered in isolation from the wider catchment, as too often happens. If programmes of measures to improve water quality (e.g. as required by the EU Water Framework Directive) are to be cost-effective, we must know the source area loading throughout the catchment, the risk that these spatially defined loadings pose to water quality and ecological objectives, and the cost and efficacy of different interventions, whether technical, regulatory (e.g. enforced rules on urban design, commercial activities, etc.), economic (e.g. raising tariffs for storm water disposal) or educational.

Integration can mean operating within a single sector, finding the best mix of interventions to deliver urban water objectives, but it must also be seen as a cross-sector approach too, applied at both project and strategy levels. This could mean, for example, incorporating water quality issues into traffic management (as is already done for air quality), working with car manufacturers to promote the use of more benign materials and reduce road-sourced pollution, or working with planners and urban designers so that city compaction to reduce travel and carbon emissions does not constrain delivery of SUDS benefits and exacerbate flooding. Thus, while SUDS have a role to play in restoring urban river water quality, arguably a greater one than they are currently given, they are only part of the solution, which requires management of multiple sources using multiple tools, both structural and non-structural in nature. Much good work is underway, particularly in terms of research and demonstration sites, that gives cause for optimism that more sustainable urban drainage can be delivered.

3 Acid mine drainage

3.1 Origins and processes generating pollution from mines

The pollution of surface waters from mines is a problem wherever mineral ores have been exploited. It is a problem throughout the world, including: China (He *et al.*, 1997); North America (Wren and Stephenson, 1991; DeNicola and Stapleton, 2002); South America (Van Damme *et al.*, 2008), and Australia (Battaglia *et al.*, 2005). In England and Wales there are over 300 abandoned mines, with discharges which impact on more than 700 km of streams and rivers. Mines which have been abandoned more commonly pose a greater problem than those that are still in operation. This is because working mines are dewatered by pumps (Younger, 2000). When mines are abandoned, dewatering ceases and a process of water table rebound occurs which floods the exposed mine workings. When the water table reaches the surface it emerges as springs, **adits** or surface seepage (**Figure 4.13**).

The activities of mining coal and minerals and mine dewatering expose underground pyrite- (FeS_2) bearing rock strata to oxygen. In the presence of oxygen and water, pyrite becomes oxidised into ferrous sulphate and sulphuric acid. Ferrous sulphate is further oxidised to ferric

Figure 4.13 Acid mine discharge at Jackson Bridge on New Mill Dyke, Yorkshire, UK.

BOX 4.6 CONTEMPORARY CHALLENGES

Formation of acid mine drainage

The following chemical reactions summarise the formation of acid mine drainage (Kelly, 1988):

$$FeS_2 + 7/2\ O_2 + H_2O \rightarrow Fe^{2+} + 2SO_4^{2-} \quad [1]$$

$$Fe^{2+} + 1/4\ O_2 + H^+ \rightarrow Fe^{3+} + 1/2\ H_2O \quad [2]$$

$$Fe^{3+} + 3H_2O \rightarrow Fe(OH) + 3H^+ \quad [3]$$

$$FeS_2 + 14Fe^{3+} + H_2O \rightarrow 15Fe^{2+} + 2SO_4^{2-} + 16H^+ \quad [4]$$

Reaction [1] shows the dissociation of iron pyrite (in the presence of oxygen and water). Pyrite breaks down to form sulphuric acid (SO_4^{2+}) and ferrous iron (Fe^{2+}). When surface waters combine with mine waters the ferrous iron (Fe^{2+}) forms the ferric variety (Fe^{3+}) [2]; a cycle is then established between these two reactions. The Fe^{3+} then further reacts with the available pyrite to form additional acidity, ferrous and sulphate ions [4]. The oxygenation of Fe^{2+} is the rate-determining step because reaction 2 is significantly slower than reaction 1.

---

sulphate; this then reacts with water to produce more acidity and ferric hydroxide (**Box 4.6**). The oxidation of the reduced forms of iron and sulphur is very slow. However, in the presence of acidophilic chemosynthetic bacteria the reactions are accelerated up to a million times. Bacteria which flourish at low pH, such as *Thiobaccilus ferroxidans* (Singer and Stumm, 1970) and *Thiobacillus thiooxidans* and *Metallogenium* (Kleinmann *et al.*, 1981), oxidise iron and sulphate respectively.

3.2 Environmental consequences of mine pollution

3.2.1 Physicochemical consequences

The waters that emerge from mines into receiving surface waters are very variable in chemical composition (**Table 4.7**) and may depend on the type of mine and the nature of the surrounding geology. They can be acidic and laden with suspended solids and dissolved metals (Gray, 1997). Indeed, they are frequently described as discharges of acid mine drainage. However, although many discharges are acidic, it is not a universal characteristic and in some cases they may be neutral or basic if the surrounding geology is high in carbonates and able to afford some neutralisation (Kelly, 1988; Schmiermund and Drozd, 1997). The contaminants lead to an increase in the acidity, turbidity, sedimentation and heavy metal concentrations of the receiving water body. They precipitate out as a coloured flocculate, known as ochre, which stains the substrate of receiving waters (e.g. **Figure 4.13**; see also **Figure 5.21**). Ochre can vary in colour from yellow to orange-brown.

It is difficult to predict the duration and chemical composition of mine water contamination. Younger (1997) separated the impacts into vestigial and juvenile components. Vestigial acidity is that acquired during the water table rebound phase and in theory should decline exponentially with time, with **asymptote** levels (where the curve of the line flattens out and almost

Table 4.7 Some chemical data from mine discharges

Chemical parameter	A	B	C	D
pH	2.7	5.3	3.18	7.75
Conductivity ($\mu S\ cm^{-1}$)	2590	295	2605	5571
Sulphate*	680	324	1850	3110
Iron*	493	29.04	165	43.7
Aluminium*	59.2	0.25	76	0.03
Zinc*	1.22	0.05	33	0.26
Lead*	—	0.04	—	—
Copper*	0.06	0.04	1.8	—
Cadmium ($\mu g\ L^{-1}$)	<0.005	—	49	—
Suspended solids	3158	—	—	—

Notes: * Readings in mg L^{-1}; A = Caulfield mine discharge, Pennsylvania, USA (Letterman and Mitsch, 1978); B = Stoney Heap coalmine water discharge, Durham, UK (Jarvis and Younger, 1997); C = Ballymurtagh mine water discharge (Gray, 1998); D = Whittle Colliery mine water discharge (Batty et al., 2005).

reaches zero) being achieved 40 years after the initial discharge. Juvenile acidity arises from the fluctuation of the water table and may persist for centuries.

3.2.2 Biological and ecological consequences

The impact of mine drainage contamination on aquatic ecosystems has been studied extensively. Processes of decomposition can be affected by mine drainage. Bacteria, fungi and decomposing invertebrates can be inhibited from colonising leaf litter, consequently reducing the rates of breakdown (Gray and Ward, 1983). Ferreira da Silva et al. (2009) reported that diatoms were rare at a location stressed by mine drainage with high levels of cadmium, lead and zinc. However, increased diversity and species richness was observed 6 km downstream of the impact. In Lynx Creek, Arizona, algal communities impacted upon by mine pollution exhibited reduced species richness as compared to locations upstream and those in areas of substantial recovery downstream (Lampkin and Sommerfield, 1982). Koryak and Reilly (1984) observed reduced growth of the aquatic macrophyte *Justica america* as a consequence of coalmine pollution in Ohio. Tremaine and Mills (1991) recorded that protozoan abundance and grazing rates on bacteria were substantially reduced in a lake affected by acid mine drainage. The impact on benthic invertebrates is widely reported, with reduced abundances, species richness and diversity being common factors. Communities in riffle habitats appear to be affected more than those in pool habitats (Van Damme et al., 2008). Furthermore, species belonging to the orders Ephemeroptera and Plecoptera also appear to be those most affected (Clements, 1994). Letterman and Mitsch (1978) noted that the standing crop of fish communities declined downstream of mine drainage on Ben's Creek, Pennsylvania from 228.2 kg ha^{-1} to 11.2 kg ha^{-1}. The sculpin, *Cottus bairdi*, showed the greatest decline, from 151.2 kg ha^{-1} to 0.3 kg ha^{-1}.

3.2.3 Social and economic consequences

The human impacts of mine drainage pollution are many and varied. Possible consequences include: reduced water availability for abstraction for public supply from both groundwater and surface water sources; surface waters unsuitable for irrigation and livestock watering; contamination of navigational rivers and canals, for both commercial and recreational use, impacts upon the maintenance of the navigational area and the waterborne craft; ecological damage can reduce the recreational potential for commercial fisheries and angling; and the general recreational and amenity value of watercourses is affected by the visual and aesthetic impact of highly coloured ferruginous mine waters.

3.3 Mitigation of mine pollution

The reduction of the impact of mine water pollution falls into three types of system: physical, chemical and biological. Physical systems involve water oxidation through impoundment, and cascades to facilitate the settlement of the contaminants into sludge (Diamond *et al.*, 1993). The costs of these systems can be high, as they require both high capital investment in terms of construction costs and there are further revenue costs from the disposal of the resultant sludge. Chemical systems use the addition of chemicals to actively promote the decontamination of the water. Examples typically include the addition of lime to produce flocculation and settlement of contaminants, followed by pH neutralisation (Chadwick *et al.*, 1986). Although capital costs are much reduced, they are expensive to run because of the costs of the treatment chemicals and sludge disposal. Biological systems are dominated by the construction of semi-natural or artificial habitats (**Figure 4.14**) which facilitate bacterial activity to reduce contaminants through reed beds and wetlands (Webb *et al.*, 1998). These systems are relatively cheap to construct and run. Furthermore, the ecological principles behind them also suggest that they may be a sustainable solution to mine drainage pollution (Kalin, 2004). However, such a solution requires the availability of suitable land areas close to the source of pollution and this is not always possible.

In reality, combinations and overlaps of treatment processes are frequently observed. For example, the mine remediation treatment at Bullhouse on the River Don in Yorkshire involves a physical settlement lagoon and a wetland area (Laine and Dudeney, 2000). Such systems can result in significant reductions in loadings of metals. The Bullhouse scheme illustrates this, with iron being the most effectively removed (**Table 4.8**). Wiseman *et al.* (2003) also noted similar reductions in iron loading (82% to 95%) on the River Pelenna in South Wales. Furthermore, ecological recovery has also been observed in the Pelenna study with the return of sensitive invertebrate species (e.g. the mayfly, *Empherella ignita*), brown trout (*Salmo trutta*) and birds such as dippers (*Cinclus cinclus*).

Figure 4.14 (a) Semi-natural wetland and (b) artificial reed bed wetlands, River Pelenna, South Wales, UK.

Table 4.8 Changes in typical chemical composition of the Bullhouse mine discharge into the River Don, Yorkshire, UK, (A) before and (B) after remediation

Chemical parameter	A	B
pH	5.9	5.9
Conductivity ($\mu S\ cm^{-1}$)	442	281
Sulphate*	788	475
Iron*	54.6	2.7
Aluminium*	2.31	0.32
Zinc*	91.5	65.3
Lead*	0.147	0.049
Copper*	0.0202	0.0042
Cadmium ($\mu g\ L^{-1}$)	—	0.28
Suspended solids	16	11

Source: Data from Laine and Dudeney, 2000.

Notes: * Readings in mg L^{-1}.

4 Atmospheric pollution

4.1 Source of acid rain

Over the last 100 years there have been significant changes to global atmospheric chemistry. The major atmospheric pollutants include acidifiers such as sulphur dioxide (SO_2) and nitrogen oxides (NO_x); toxic substances such as ozone, volatile organic compounds (VOCs) and heavy metals; and fertilising substances such as ammonia (NH_3) and nitrous oxides NO_x. The amount of acidifying pollutants present in the atmosphere has been affected by an increase in the combustion of fossil fuels, changes in agricultural practices and, more recently, by the introduction of new legislation to reduce emissions to the atmosphere (e.g. the 1999 Gothenburg Protocol, which set emission ceilings for 2010 for sulphur, NO_x, VOCs and ammonia).

Rainfall is naturally acidic, with a pH of about 5.6, due to its equilibrium with carbon dioxide in the atmosphere:

$$CO_2 + H_2O \leftrightarrow H_2CO_3 \leftrightarrow H^+ + HCO_3^-$$

However, the pH of rain decreases as the amounts of SO_2 and NO_x in the atmosphere increase. Acid rain is generated from the oxidation of SO_2 and NO_x to H_2SO_4 and HNO_3, respectively. Acid rain can be carried great distances in the atmosphere, not just between countries but also from continent to continent. Hence the impact of acid rain on freshwater ecosystems can be far from its source.

4.2 Impact of acid rain on surface water

In the 1960s and 1970s, observations of an increase in surface water acidity leading to a decline in the salmon population across areas of Southern Scandinavia were widely reported in the media. However, much research was required to establish the cause–effect relationship between acid rain and surface water acidification. **Palaeolimnological** techniques that reconstructed pH based on diatoms in lake sediment cores allowed the chemical histories of a large number of lakes across Britain, Scandinavia, northeast USA and Canada to be evaluated (Battarbee *et al.*, 1990). This showed that lakes in areas with base-poor geology that had received high rates of acidic deposition had experienced significant acidification.

The effect of surface water acidification on the aquatic flora and fauna is pronounced (e.g. Schindler, 1988). The effects on aquatic biota can be direct, such as toxicity associated with changes in water chemistry, or indirect, such as changes in habitat or food availability. Acidification also increases the solubility of heavy metals, such as aluminium, manganese, lead, cadmium and zinc, many of which are toxic to aquatic organisms. For example, acidity and high concentrations of aluminium (Al^{3+}) can lead to deterioration of aquatic life, with losses in the diversity and size of invertebrate and fish populations. At pH less than 5, aluminium is soluble and leached from the soil to surface waters as Al^{3+}. The toxicity of Al^{3+} has been mainly studied with reference to fish; at concentrations < 100 $\mu g\ L^{-1}$, it affects the **osmoregulation** of fish (i.e. their ability to regulate the amount of salt and water within their body) and at concentrations > 100 μg L^{-1} a gelatinous precipitate of aluminium hydroxide (Al(OH)$_3$) forms on their gills, leading to suffocation (Jackson and Jackson, 2000). Changes in the population of the dipper, *Cinclus cinclus*, a riparian bird, in areas that experienced stream water acidification, such as mid-Wales (Tyler and Ormerod, 1992), are an example of the indirect effect of acidification on aquatic biota.

The susceptibility of a particular surface water body to pollution from acid rain depends on its capacity to act as a buffer and neutralise the effects of acid rain. This capacity is largely determined by the chemical properties of surrounding geology and soil type. Rocks and soils that contain large amounts of base cations (**bases** are substances that can accept hydrogen ions (protons) and are the opposite of acids), such as calcium and magnesium, have a high **buffering capacity**. In addition, surface water bodies in chalk or limestone regions are able to buffer the increase in hydrogen ions (H$^+$), due to the presence of bicarbonate anions (HCO$_3^-$):

$$H^+ + HCO_3^- \leftrightarrow H_2CO_3 \leftrightarrow H_2O + CO_2$$

In contrast, water bodies underlain by base-poor **igneous** (e.g. granite) or ancient metamorphic rocks have a lower buffering capacity. Soil development on such lithologies generally results in thin, base-poor soils with a small base cation pool and slow release of base cations from the weathering of the bedrock. Base cation stores in such soils are rapidly depleted in response to incoming acidic rain, resulting in the acidification of soils and surface waters. Hence, areas dominated by igneous and metamorphic rocks and receiving high annual rainfalls that are acidic are particularly vulnerable, such as the Canadian shield, the Appalachians of America and the mountains of Scandinavia and western Britain. This differential ability of surface waters to cope with acid rain has been examined through use of a '**critical load**' approach, as discussed in **Box 4.7**.

4.3 Recovery from surface water acidification

Since the 1980s considerable national and international effort has been made to decrease emissions of acidifying pollutants. This has been challenging, since those countries suffering most from acid rain are not always the main polluters. In 1979, 34 European and North American countries adopted the Convention on Long-Range Transboundary Air Pollution, which bound them to reducing emissions. The convention agreed firm targets in 1983, when 21 European countries agreed to reduce their SO$_2$ emissions by 30% from the 1980 levels by 1993. Twelve countries reached this target by 1988. The UK did not ratify this agreement, but did eventually declare an intention to reduce SO$_2$ emissions by 30% in the late 1990s. In the United States, an 'Acid Rain Program' commenced in 1995 that aimed to achieve environmental and public health benefits through reductions in emissions of SO$_2$ and oxides of nitrogen (NO$_x$). Since the signing of these agreements there has been a substantial reduction in SO$_2$ emissions in Western Europe, although emissions of NO$_x$ continued to increase until the late 1980s, before starting to decline.

In the UK, sulphur deposition declined by 60% between 1986 and 2005 (Fowler *et al.*, 2005). The response of surface waters to this decline in sulphur deposition has been monitored via the UK Acid Water Monitoring Network (AWMN). The decline in sulphur deposition has resulted in a significant decline in the sulphate concentration of upland waters (**Figure 4.9a**; Davies *et al.*, 2005), and the majority of AWMN sites also display an increasing trend in pH, alkalinity and acid neutralising capacity (Monteith and Evans, 2005). Data from the International UN monitoring program ICP Waters (International Cooperative Programme on Assessment and Monitoring of Rivers and Lakes) for 12 regions of Europe and North America also display similar declines in sulphate concentrations and increases in alkalinity and pH (Skjelkvåle *et al.*, 2005) in response to decline in sulphur deposition.

BOX 4.7 TECHNIQUES

Evaluating the acidification status of surface waters: the critical load approach

The concept of the critical load approach is based on a 'dose–response' relationship and has been widely applied to addressing pollution problems (Fenn *et al.*, 2011). A critical load can be defined as a 'quantitative estimate of an exposure to one or more pollutants below which significant harmful effects on specified sensitive elements of the environment do not occur according to present knowledge' (Nilsson and Grennfelt, 1988). The critical load of acidity for a freshwater resource is exceeded when the total amount of acid deposition is greater than the ability of the system to neutralise the deposited acidity. Two models have been widely used in critical loads for surface waters: the steady-state water chemistry (SSWC) model (Henriksen *et al.*, 1992) and the first-order acidity balance (FAB) model (Posch *et al.*, 1997). Both models provide an estimate of the level of acid deposition below which the acid neutralising capacity (ANC) will remain above a pre-determined level. ANC, measured in microequivalents per litre (μeq L^{-1}), is a commonly used chemical indicator of sensitivity to acidification, as it characterises the ability of water to neutralise strong acids, including those introduced by atmospheric deposition, and it can be linked to biotic response thresholds such as the health and condition of aquatic biota. Current UK and European critical load applications use the FAB model, and in the UK an ANC limit of 20 μeq L^{-1} is applied to all sites, except those which are believed to be naturally acidic, where a value of ANC = 0 μeq L^{-1} has been used.

The acidity critical loads for UK freshwaters are based on data from a national survey of lakes or headwater streams, where a single site, judged to be the most sensitive (in terms of acidification) was sampled in each 10 km grid square of the country. In less sensitive regions (e.g. southeast England) the sampling generally consisted of one site in each 20 km grid square. In 2004 this 'mapping dataset' was updated to include sites from other surveys and networks, where appropriate data were available. To date, the FAB model has been applied to 1595 sites in Great Britain and 127 in Northern Ireland. Hence the freshwater critical load maps do not represent all waters in the UK and the results are mapped by site location (Figure 4.15). The critical load for freshwaters is expressed in terms of three terms: CLmaxS, CLmaxN and CLminN, where CLmaxS is the critical load of acidity expressed in terms of sulphur (S) only (i.e. where nitrogen (N) deposition is zero), CLmaxN is the critical load of acidity expressed in terms of N only (i.e. where S deposition is zero) and CLminN represents the long-term N removal processes (e.g. N uptake and immobilisation), i.e. the amount of N that can be deposited before it will begin to contribute to surface water acidification (see Henriksen and Posch (2001) for more detail). Because both S and N compounds can contribute to the exceedance of the acidity critical load, these three 'acidity' critical load parameters are required to determine the combinations of deposition that will not exceed the critical load. In general, the lower the values of CLmaxS and CLmaxN, the more sensitive the sites are to acidification from atmospheric deposition. In summary, critical loads have been successfully developed and implemented in Europe and other areas for assessing the impacts of air pollution on essential ecosystem services and for informing public policy for reductions in air pollution emissions.

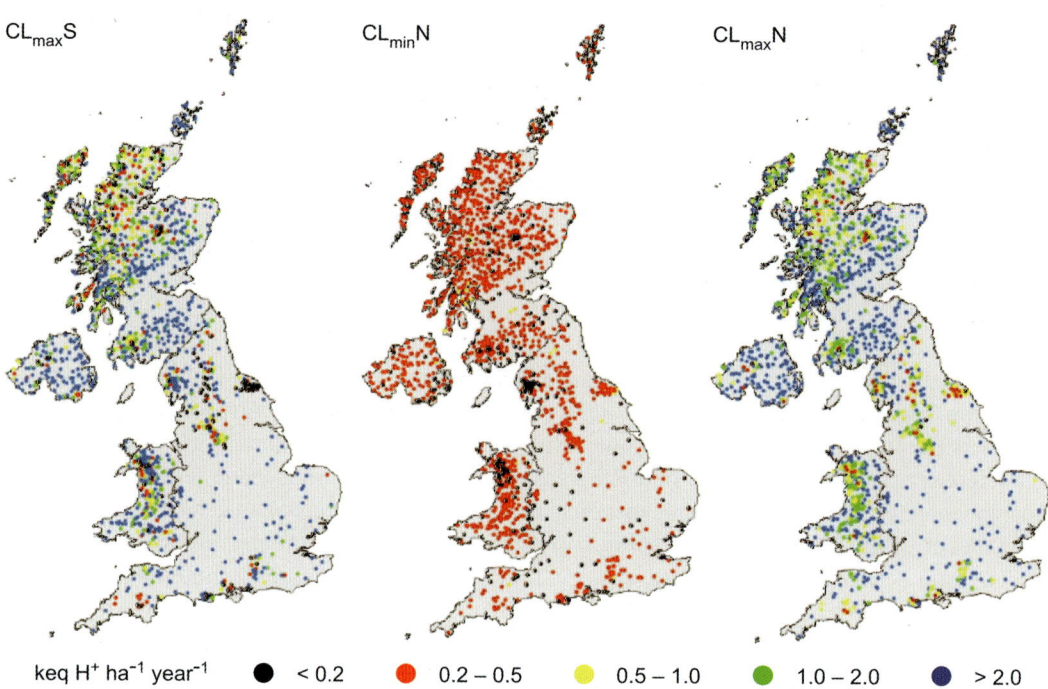

Figure 4.15 Critical load exceedance maps for the UK freshwaters in 2004 calculated using the first-order acidity balance (FAB) model.
Source: Centre for Ecology and Hydrology, (http://cldm.defra.gov.uk).

Both the UK AWMN and ICP Waters Programme reported an increase in surface water DOC concentrations over the period 1990–2005 (**Figure 4.9c**; Freeman *et al.*, 2001; Skjelkvåle *et al.*, 2005), with concentrations doubling across the UK AWMN (Evans *et al.*, 2006). This trend has raised concerns about the provision of safe drinking water (Holden *et al.*, 2007) and the stability of soil carbon stores (Freeman *et al.*, 2001; Bellamy *et al.*, 2005). A range of different hypotheses have been proposed to explain this trend, including those linked to climate change (e.g. increase in temperature, changes in distribution and volume of rainfall and elevated atmospheric carbon dioxide (Freeman *et al.*, 2001, 2004)), changes in land management (e.g. Yallop and Clutterbuck, 2009) and recovery from acid rain (Evans *et al.*, 2006). Using data from the ICP Waters Programme and UK AWMN, Monteith *et al.* (2007) demonstrated that the spatial variation in DOC trends could be explained by a simple model based on changes in the chemistry of atmospheric deposition (predominantly SO_4 and Cl). This study provided strong support for the hypothesis that reduction in the emissions of SO_2 over the last 20 years is the primary cause of the increasing trend in surface water DOC and, therefore, implied that these systems may simply be returning to their preindustrial conditions. Thus, rising DOC concentrations in freshwaters may to a large extent reflect recovery from the effects of acid deposition, rather than ecosystem degradation in response to climate or land use change.

While sulphur deposition has declined rapidly over the last 20 years, atmospheric deposition of reactive N compounds has shown much smaller changes; reduced N and wet deposition of nitrate have declined by only 10% in the UK (Fowler et al., 2005). Where N deposition exceeds the biological demand of the terrestrial ecosystem (termed N saturation), the excess N is removed in drainage water. This increased N leaching can lead to acidification of surface waters, when the NO_3 is accompanied by Al^{3+} or H^+, and eutrophication, where N is the limiting nutrient. However, identifying the precise impact of increased N leaching on aquatic ecosystems is complicated by the variety of chemical forms in which N may exist in atmospheric deposition, terrestrial biomass and surface waters. In addition, a variety of terrestrial processes separate N deposition from its effects and reduces the ability to attribute observations of aquatic effects to known rates of deposition. The ratio of $NO_3:(NO_3 +SO_4)$ in surface waters provides an index of the influence of NO_3 on chronic acidification status, assuming that both anions are derived from anthropogenic acid deposition. A value of greater than 0.5 indicates that NO_3 has a greater influence on the acidification of surface water than SO_4. Ratios of $NO_3:(NO_3 +SO_4)$ in lakes and streams of acid sensitive regions of Europe and North America in the 1980s and 1990s were generally low (< 0.3), with typically < 15% of the acidity explained by NO_3 (Chapman and Edwards, 1999). However, Curtis et al. (2005) showed that as S deposition decreases, and if N deposition remains static and the ability of a catchment to store atmospheric N is exceeded, increased NO_3 leaching at the UK AWMN may counterbalance any reduction in SO_4 due to reduction in emissions of SO_2, resulting in little change or even a decrease in the acidity of surface waters over the longer term.

G THE FUTURE

Over the last century different strategies have been developed to overcome the conflict between water quality deterioration due to human activities and water use that include a combination of:

1 wastes treated before discharging to a water body
2 waters treated in order to meet specific water quality requirements (e.g. drinking water)
3 water quality determined and procedures adopted that prevent deterioration (e.g. land management options).

In most countries these strategies have been accompanied by the development of laws and regulation to help to prevent and reduce water pollution. For example, in the United States there are the 1972 Clean Water Act and the 1974 Safe Drinking Water Act. However, one of the biggest problems with water pollution is its transboundary nature. Many rivers cross countries, so pollution discharged in one country can cause water quality problems in neighbouring countries. While environmental legislation in one country can make it tougher for people to pollute, to be really effective these laws need to operate across national and international borders (see Chapter 11). The European Union has water-protection laws (known as directives) that apply to all of its member states. They include, for example, the 1976 Bathing Water Directive (updated 2006), which seeks to ensure the quality of the waters that people use for recreation. More recently the European Union has introduced the Water Framework Directive (WFD), which has the following key aims:

- expanding the scope of water protection to all waters, surface waters and groundwater

> **REFLECTIVE QUESTION**
>
> What methods can be adopted to reduce pollution from agriculture, urban areas and mine drainage?

- for all inland waters to achieve 'good ecological status' by a set date
- water management to be based on river basins
- 'combined approach' of emission limit values and quality standards
- getting the price right
- getting the citizen involved more closely
- streamlining legislation.

The Directive requires the establishment of river basin districts, some of which will traverse national boundaries, and for each of these a 'River Basin Management Plan' that will be updated every six years. There are four distinct elements to the river basin planning cycle: characterisation and assessment of impacts on river basin districts; environmental monitoring; the setting of environmental objectives; and the design and implementation of the programme of measures needed to achieve them. However, in England and Wales, it was estimated in 2008 that approximately one-fifth of all river water bodies are at risk or probably at risk of failing to reach 'good ecological status' by 2015, due to pollution from point sources, while over 85% of surface water bodies are currently at risk or probably at risk of failing to reach 'good ecological status' because of diffuse pollution. While the WFD has set stringent targets for achieving 'good ecological status' of all surface waters, it is important to remember that surface waters have been used and degraded by man over decadal and centennial timescales, and thus the 2015 WFD targets need to be considered against these longer timescales of system response.

To date, studies investigating the impact of climate change on the global water cycle have concentrated on water resources and water supply. Hence there is currently much uncertainty about the potential impacts of climate change on surface water quality. In their comprehensive review of the potential impact of climate change on river water quality, from a UK perspective, Whitehead *et al.* (2009) make the following key points:

1 The IPCC Fourth Assessment Report (IPCC, 2007) had a relatively short section on the potential impacts of climate change on river water quality.
2 The most immediate response to climate change is expected to be an increase in river and lake water temperatures that will result in changes in water quality (e.g. concentration of dissolved oxygen, and ecology – see Chapter 6).
3 Changes in the timing, intensity and duration of precipitation will lead to changes in flow regime that will affect water quality and ecology. However, there is a lot of uncertainty about the likely impacts due to changes in regional precipitation.
4 Lower river flows will result in less volume available for dilution of pollution downstream of point discharges.
5 Increased storm events, especially in summer, would lead to more frequent incidences of combined sewer overflows, discharging highly polluted waters into receiving water bodies.
6 More intense rainfall and flooding could result in increased suspended solids, sediment yields and associated contaminant metal fluxes.
7 Nutrient loads are expected to increase.
8 There is a need for models to represent the whole ecosystem and the uncertainty in model prediction to be quantified.

Overall, climate change studies, especially in relation to water quality and ecology, are at fairly early stages and the outcomes are subject to considerable uncertainty (Whitehead *et al.*, 2009). Hence more research is required in this area.

Some people believe that global changes in population and economic development will have a much larger impact on water supply and quality than will changes in climate over the next few decades (e.g. Vörösmarty *et al.*, 2000 and see Chapter 2). With more people now living in cities than in rural areas, and with an estimated growth in the urban population of 1.5 billion over the next 20 years (UNPD, 2007), providing safe drinking water to urban inhabitants is a major challenge.

In cities with large upstream populations, the risk of drinking water containing high concentrations of nutrients and bacteria is high. A recent study showed that the number of people living in cities with an upstream population density of 19 people per hectare was 53 million, all in Africa and Asia; and that 890 million people live in cities with an upstream population density greater than 5.5 people per hectare (McDonald et al., 2011). Where these cities have adequate resources, they are able to treat the polluted water before distribution to the urban population. However, most cities in the developing world are unable to afford water treatment and are reliant on international aid or loans to build water treatment plants and sewage treatment works (McDonald et al., 2011). The lack of sewage systems in villages and towns upstream of cities has serious consequences for drinking water quality downstream. Other ways of improving water quality in rivers entering cities is to change land use or management upstream. To obtain a better understanding of how surface water quality will change in the future, an integrated approach is required that brings together climate change, water and land users and socio-economic communities. There is also a need to mobilise financial resources, particularly in the developing world, to support integrated approaches to improving water quality (see Chapter 8).

> **REFLECTIVE QUESTION**
>
> How might climate change impact upon water quality?

H SUMMARY

Rivers and lakes are vital natural resources; they provide drinking water, habitats for many different types of flora and fauna, and are an important resource for industry and recreation. However, due to human activities, a significant proportion of them are now polluted or under threat from pollution (e.g. Vörösmarty et al., 2010). Protecting and improving the aquatic environment is an important part of achieving sustainable development and is vital for the long-term health and well-being of humans and wildlife.

The hydrological pathway by which water travels through a watershed has a large control on the solute concentration. For example, water that passes rapidly over or through the watershed has a lower solute content than water that has moved slowly through the watershed. Temperature, redox potential and pH have a strong control on the form and oxidation state of different solutes. Global and regional patterns in surface water chemistry are predominantly controlled by differences in geology and climate, with soil, vegetation and land use being more important controls at the local scale. Surface water chemistry also displays temporal patterns at a range of scales from diurnal to seasonal to long term. These temporal patterns arise due to the processes controlling the supply of solutes available for leaching varying over time, and the hydrological processes that generate runoff within a catchment are dynamic and vary with climate and season.

Surface water chemistry is directly threatened by human activities. River and lake chemistry is affected by widespread land cover change, irrigation, urbanisation, industrialisation and its legacy, engineering schemes like reservoirs, and pollution from atmospheric deposition. Legislation, population growth and climate change may all influence surface water chemistry in the future.

FURTHER READING

Campbell, N., D'Arcy, B., Frost, A., Novotny, V. and Sansom, A. 2005. *Diffuse pollution: theory, control measures, practical experiences.* IWA Publishing; London.

This is an introductory text covering the nature, causes and significance of urban and rural diffuse pollution. Best management practices to tackle the problems are examined, as are the ways in which the adoption of such practices may be brought about. It makes use of case studies, from an EU perspective, to examine the strengths and weaknesses of various approaches, including SUDS.

Haygarth, P. and Jarvis, S. 2002. *Agriculture, hydrology and water quality.* CABI Publishing; Wallingford, Oxfordshire, UK.

This book contains a collection of review articles of global problems of diffuse water pollution from agriculture, which affects the water quality of surface waters and the oceans.

Issues in Ecology (http://www.esa.org/science_resources/issues_ecology.php).

These reports present the consensus of a panel of scientific experts on issues relevant to the environment and are written in a language understandable by non-scientists. Relevant reports include 'Nonpoint pollution of surface waters with phosphorus and nitrogen' and 'Impact of atmospheric pollution on aquatic ecosystems'.

Kelly, M. 1988. *Mining and the freshwater environment.* Elsevier Applied Science; London.

Although this book concentrates to a significant extent on the impacts of metal mining, it also provides good coverage of the general principles and processes that lead to acid mine drainage and the extent of their impacts on the freshwater environment.

Lawson, E.A. 2000. *Aquatic pollution: an introductory text.* John Wiley and Sons; Chichester.

This book introduces the basic concepts and issues in aquatic pollution. It describes the effects of pollution associated with urban runoff, acid rain, sewage disposal, pesticides, oil spills, nutrient loading, and many more, on lakes, streams, groundwater and oceans.

Novotny, V. 2002. *Water quality: diffuse pollution and watershed management (2nd edition).* John Wiley and Sons; New York.

A comprehensive book that covers the sources and control of a wide range of pollution, from urban, industrial and highway pollution to agricultural pollution. It also covers management and restoration of streams, lakes, and watershed management techniques.

Novotny, V. (ed.) 1995. *Nonpoint pollution and urban stormwater management.* Technomic Publication Co; Lancaster, PA.

This book covers the most important topics and solutions of the diffuse pollution problem from a US perspective, with emphasis on urban sources and abatement.

Classic papers

Hasler, A.D. 1947. Eutrophication of lakes by domestic drainage. *Ecology* 28: 383–395.

A paper which showed how agricultural and urban pollution could lead to eutrophication, and hence lake ecosystems required good land management practice.

O'Connor, D.J. 1960. Oxygen balance of an estuary. *Journal of the Sanitary Engineering Division*, ASCE 86: Proceedings Paper 2472.

The foundation of modern water quality modelling for surface waters.

PROJECT IDEAS

- Measure the solutes and sediment concentrations of two tributaries just upstream of a confluence and also measure the same things immediately downstream of the confluence. Have there been chemical changes at the confluence or has the system just mixed?

- In an agricultural catchment, sample several tributaries and the main river channel at different points for key nutrient concentrations (e.g. nitrates and phosphates). Can you relate differences to land use?

- Test some water quality variables immediately upstream and downstream of a BMP/SUDS device and evaluate its performance.

REFERENCES

Armitage, P.D., Moss, D., Wright, J.F. and Furse, M.T. 1983. The performance of a new biological water quality score system based on macroinvertebrates over a wide range of unpolluted running-water sites. *Water Research* 17: 333–347.

Arora, K., Mickelson, S.K., Helmers, M.J. and Baker, J.L. 2010. Review of pesticide retention processes occurring in buffer strips receiving agricultural runoff. *Journal of the American Water Resources Association* 46: 618–647.

Battaglia, M., Hose, G.C., Turak, E. and Warden, B. 2005. Depauperate macroinvertebrates in a mine affected stream: clean water may be the key to recovery. *Environmental Pollution* 138: 132–141.

Battarbee, R.W., Mason, J., Renberg, I. and Talling, J.F. 1990. *Palaeolimnology and lake acidification.* ENSIS Publishing; London.

Batty, L.C., Atkin, L. and Manning, D.A.C. 2005. Assessment of the ecological potential of mine-water treatment wetlands using a baseline survey of macroinvertebrate communities. *Environmental Pollution* 138: 412–419.

Bellamy, P.H., Loveland, P.J., Bradley, R.I., Lark, R.M. and Kirk, G.J.D. 2005. Carbon losses from all soils across England and Wales 1978–2003. *Nature* 437: 245–248.

Bundschuh, M., Hahn, T., Gessner, M.O. and Schulz, R. 2009. Antibiotics as a chemical stressor affecting an aquatic decomposer-detrivore system. *Environmental Toxicology and Chemistry* 28: 197–203.

Burns, D.A., McHale, M.R., Driscoll, C.T. and Roy, K.M. 2006. Response of surface water chemistry to reduced levels of acid precipitation: comparison of trends in two regions of New York, USA. *Hydrological Processes* 20: 1611–1627, doi: 10.1002/hyp.5961.

Burt, T.P., Arkell, B.P., Trudgill, S.T. and Walling, D.E. 1988. Stream nitrate concentrations in a small stream in Southwest England over a 15 year period (1970–1985). *Hydrological Processes* 2: 155–163.

Burt, T.P., Howden, N.J.K., Worrall, F., Whelan, M.J. and Bieroza, M. 2011. Nitrate in United Kingdom rivers: policy and its outcomes since 1970. *Environmental, Science and Technology* 45: 175–181.

Campbell, N., D'Arcy, B., Frost, A., Novotny, V. and Sansom, A. 2005. *Diffuse pollution: theory, control measures, practical experiences.* IWA Publishing; London.

Chadwick, J.W., Canton, S.P. and Dent, R.L. 1986. Recovery of benthic invertebrate communities in Silver Bow Creek, Montana, following improved metal mine wastewater treatment. *Water, Air and Soil Pollution* 28: 427–438.

Chapman P.J. and Edwards, A.C. 1999. The impact of atmospheric N deposition on the behaviour of N in surface water. *In:* Langan, S.J. (ed.) *Impacts of nitrogen deposition on natural and semi-natural ecosystems.* Kluwer Academic Publishers; The Netherlands, 153–212.

Chapman, P.J., Reynolds, B. and Wheater, H.S. 1993. Hydrochemical changes along stormflow pathways in a small moorland headwater catchment in mid-Wales, UK. *Journal of Hydrology* 151: 241–265.

Clements, W.H. 1994. Benthic invertebrate community responses to heavy metals in the upper Arkansas River basin, Colorado. *Journal of the North American Benthological Society* 13: 30–44.

Colwill, D.M., Peters, C.J. and Perry, R. 1984. Water quality of motorway runoff. *Supplementary Report 823.* Transport Road Research Laboratory; Buckinghamshire.

Curtis, C.J., Evans, C.D., Helliwell, R.C. and Monteith, D.T. 2005. Nitrate leaching as a confounding factor in chemical recovery from acidification in UK upland waters. *Environmental Pollution* 137: 73–82.

Davies, J.J.L., Jenkins, A., Monteith, D.T., Evans, C.D. and Cooper, D.M. 2005. Trends in surface water chemistry of acidified UK freshwaters, 1988–2002. *Environmental Pollution* 137: 27–39.

DeNicola, D.M. and Stapleton, M.G. 2002. Impact of acid mine drainage on benthic communities in streams: the relative roles of substratum vs. aqueous effects. *Environmental Pollution* 119: 303–315.

Diamond, J.M., Bower, W. and Gruber, D. 1993. Use of man-made impoundment in mitigating acid mine drainage in the North Branch Potomac River. *Environmental Management* 17: 225–238.

Driscoll, C.T., Schaefer, D.A., Molot, L.A. and Dillion, P.J. 1989. Summary of North America data. *In*: Malanchuk, J.L. and Nilsson, J. (eds) *The role of nitrogen in the acidification of soil and surface waters.* Nordic Council of Ministers; Sweden.

Du Preez, L.H. et al. 2005. Seasonal exposures to triazine and other pesticides in surface waters in the western Highveld corn-production region in South Africa. *Environmental Pollution* 135: 131–141.

EC. 1991. *Council Directive 91/676/EEC concerning the protection of waters against pollution by nitrates from agricultural sources.* OJ L375. European Commission; Brussels.

EC. 2011. *Chemicals and the Water Framework Directive: draft environmental quality standards, diclofenac.* European Commission; Brussels.

Edwards, A.C. and Withers, P.J.A. 2008. Transport and delivery of suspended solids, nitrogen and phosphorus from various sources to freshwaters in the UK. *Journal of Hydrology* 350: 144–153.

Ellis, J.B. and Mitchell, G. 2006. Urban diffuse pollution: key management issues for the Water Framework Directive. *Journal of the Chartered Institution of Water and Environmental Management* 20: 19–26.

Evans, C.D., Chapman, P.J., Clark, J.M., Monteith, D.T. and Cresser, M.S. 2006. Alternative explanations for rising dissolved organic carbon export from organic soils. *Global Change Biology* 12: 2044–2053.

Fenn, M.E. et al. 2011. Setting limits: using air pollution thresholds to protect and restore US ecosystems. *Issues in Ecology* 14.

Ferreira da Silva, E. et al. 2009. Heavy metal pollution downstream the abandoned Coval da Mo nine (Portugal) and associated effects on epilithic diatom communities. *Science of the Total Environment* 407: 5620–5636.

Fowler, D., Smith, R., Muller, J.B.A., Hayman, G. and Vincent, K.J. 2005. Changes in the atmospheric deposition of acidifying compounds in the UK between 1986 and 2001. *Environmental Pollution* 137: 15–25.

Freeman, C., Evans, C.D., Monteith, D.T., Reynolds, B. and Fenner, N. 2001. Export of organic carbon from peat soils. *Nature* 412: 785.

Freeman, C. et al. 2004. Export of dissolved organic carbon from peatlands under elevated carbon dioxide levels. *Nature* 430: 195–198.

Garrod, G.D., Garratt, J.A., Kennedy, A. and Willis, K.G. 2007. A mixed methodology framework for the assessment of the voluntary initiative. *Pest Management Science* 63: 157–170.

Gibbs, R. 1970. Mechanisms controlling world water chemistry. *Science* 170: 1088–1090.

Gray, L.J. and Ward, J.V. 1983. Leaf litter breakdown in streams receiving treated and untreated metal mine drainage. *Environment International* 9: 135–138.

Gray, N.F. 1997. Environmental impact and remediation of acid mine drainage: a management problem. *Environmental Geology* 30: 62–71.

Gray, N.F. 1998. Acid mine drainage composition and the implications for its impact on lotic systems. *Water Research* 32: 2122–2134.

Gumiere, S.J., Le Bissonnais, Y., Raclot, D. and Cheviron, B. 2011. Vegetated filter effects on sedimentological connectivity of agricultural catchments in erosion modelling: a review. *Earth Surface Processes and Landforms* 36: 3–19.

Hasler, A.D. 1947. Eutrophication of lakes by domestic drainage. *Ecology* 28: 383–395.

Hatch, D., Goulding, K. and Murphy, D. 2002. Nitrogen. *In*: Haygarth, P.M. and Jarvis, S.C. (eds) *Agriculture, hydrology and water quality.* CABI Publishing; Wallingford, 8–27.

Haygarth, P. and Jarvis, S. 2002. *Agriculture, hydrology and water quality.* CABI Publishing; Wallingford.

He, M., Wang, Z. and Tang, H. 1997. Spatial and temporal patterns of acidity and heavy metals in predicting the potential for the ecological impact on the Le An river polluted by acid mine drainage. *The Science of the Total Environment* 206: 67–77.

Henriksen, A. and Posch, M. 2001. Steady-state models for calculating critical loads of acidity for surface waters. *Water, Air and Soil Pollution* 1: 375–398.

Henriksen, A., Kamari, J., Posch, M. and Wilander, A. 1992. Critical loads of acidity: Nordic surface waters. *Ambio* 26: 304–311.

Hinton, M.J., Schiff, S.L. and English, M.C. 1997. The significance of storms for the concentration and export of dissolved organic carbon from two Precambrian Shield catchments. *Biogeochemistry* 36: 67–88.

Holden, J. *et al.* 2007. Environmental change in moorland landscapes. *Earth Science Reviews* 82: 75–100.

Hooda, P.S., Edwards, A.C., Anderson, H.A. and Miller, A. 2000. A review of water quality concerns in livestock farming areas. *Science of the Total Environment* 250: 143–167.

IPCC. 2007. *Climate Change 2007: Impacts, Adaptation and Vulnerability. Contribution of Working Group II to the Fourth Assessment Report of the Intergovernmental Panel on Climate Change* (Parry, M.L. *et al.* (eds)). Cambridge University Press; Cambridge.

Issues in Ecology (http://www.esa.org/science_resources/issues_ecology.php).

Jackson, A.R.W. and Jackson, J.M. 2000. *Environmental science: the natural environment and human impact.* Pearson Education; Harlow.

Jarvie, H.P., Withers, P.J.A., Hodgkinson, R., Bates, A., Neal, M., Wickham, H.D., Harman, S.A. and Armstrong, L. 2007. Influence of rural land use on streamwater nutrients and their ecological significance. *Journal of Hydrology* 350: 166–186.

Jarvis, A.P. and Younger, P.L. 1997. Dominating chemical factors in mine water induced impoverishment of the invertebrate fauna of two streams in the Durham coalfield, UK. *Chemistry and Ecology* 13: 249–270.

Jeffries, D.S., Clair, T., Couture, S., Dillon, P., Dupont, J., Keller, W., McNicol, D., Turner, R., Vet, R. and Weever, R. 2003. Assessing recovery of lakes in southeastern Canada from acidic deposition. *Ambio* 32: 176–182.

Kalin, M. 2004. Passive mine water treatment: the correct approach? *Ecological Engineering* 22: 299–304.

Kasprzyk-Hordern, B., Dinsdale, R.M. and Guwy, A.J. 2008. The occurrence of pharmaceuticals, personal care products, endocrine disruptors and illicit drugs in surface water in South Wales, UK. *Water Research* 42: 3498–3518.

Kay, P., Blackwell, P.A. and Boxall, A.B.A. 2004. Fate of veterinary antibiotics in a macroporous tile drained clay soil. *Environmental Toxicology and Chemistry* 23: 1136–1144.

Kay, P., Edwards, A.C. and Foulger, M. 2009. A review of the efficacy of contemporary agricultural stewardship measures for addressing water pollution problems of key concern to the UK water industry. *Agricultural Systems* 99: 67–75.

Kay, P., Grayson, R., Phillips, M., Stanley, K., Dodsworth, A., Hanson, A., Walker, A., Foulger, M., McDonnell, I. and Taylor, S. 2012. The effectiveness of agricultural stewardship for improving water quality at the catchment scale: experiences from an NVZ and ECSFDI watershed. *Journal of Hydrology* 422–423: 10–16.

Kelly, M. 1988. *Mining and the freshwater environment.* Elsevier Applied Science; London.

Kladivko, E.J., Vanscoyok, G.E., Monke, E.J., Oates, K.M. and Pask, W. 1991. Pesticide and nutrient movement into subsurface tile drains on a silt loam soil in Indiana. *Journal of Environmental Quality* 20: 264–270.

Kleinmann, R.L.P., Crerar, D.A. and Pacelli, R.R. 1981. Biogeochemistry of acid mine drainage and a method to control acid formation. *Mining Engineering* 33: 300–305.

Koryak, M. and Reilly, R.J. 1984. Vascular riffle flora of Appalachian streams: the ecology and effects of acid mine drainage on *Justicia Americana* (L.) Vahl. *Proceedings of the Pennsylvania Academy of Science* 58: 55–60.

Laine, D.M. and Dudeney, A.W.L. 2000. Bullhouse minewater project. *Transactions of the Institution of Mining and Metallurgy Section A* 109: A224–227.

Lampkin, A.J. III and Sommerfield, M.R. 1982. Algal distribution in a small, intermittent stream receiving acid mine drainage. *Journal of Phycology* 18: 196–199.

Lawson, E.A. 2000. *Aquatic pollution: an introductory text.* John Wiley and Sons; Chichester.

Letterman, R.D. and Mitsch, W.J. 1978. Impact of mine drainage on a mountain stream in Pennsylvania. *Environmental Pollution* 17: 53–73.

Likens, G.E., Bormann, F.H., Pierce, R.S., Eaton, J.S. and Johnson, N.M. 1977. *Biogeochemistry of a forested ecosystem.* Springer; New York.

Littlewood, I.G. 1992. *Estimating contaminant loads in rivers: a review.* Report No. 117, Institute of Hydrology; Wallingford.

McDonald, R., Douglas, I., Revenga, C., Hale, R., Grimm, N., Gronwall, J. and Feketa, B. 2011. Global urban growth and the geography of water availability, quality and delivery. *Ambio* 40: 537–446.

Meybeck, M. 1981. Pathways of major elements from land to oceans through rivers. In: *River inputs to ocean systems*, UNEP/UNESCO Report. 18–30.

Meybeck, M., Kuusisto A., Makela, A. and Malkki, E. 1996. Water quality. *In:* Bartram, J. and Balance, R. (eds) *Water quality monitoring: a practical guide to the design and implications of freshwater quality studies and monitoring programmes.* Chapman and Hall; London, UK.

Mitchell, G. 2001. *The quality of urban stormwater in Britain and Europe: database and recommended values for strategic planning models.* Technical Report, School of Geography; University of Leeds.

Mitchell, M.J., Driscoll, C.T., Kahl, J.S., Likens, G.E., Murdoch, P.S. and Pardo, L.H. 1996. Climatic control of nitrate loss from forested watersheds in the Northeast United States. *Environmental Science and Technology* 30: 2609–2612.

Monteith, D.T. and Evans, C.D. 2005. The United Kingdom's Acid Water Monitoring Network: a review of the first 15 years. *Environmental Pollution* 137: 3–13.

Monteith, D.T., Reynolds, B. and Evans, C.D. 2000. Are temporal variations in the nitrate content of UK upland freshwaters linked to the North Atlantic Oscillation? *Hydrological Processes* 14: 1745–1749.

Monteith, D.T., Stoddard, J.L., Evans, C.D., de Wit, H.A., Forsius, M., Hogasen, T., Wilander, A., Skjelkvale, B.L., Jeffries, D.S., Vuorenmaa, J., Keller, B., Kopacek, J. and Vesely, J. 2007. Dissolved organic carbon trends resulting from changes in atmospheric deposition chemistry. *Nature* 450: 537–539.

Morgan, J.J. and Stumm, W. 1965. The role of multivalent metal oxides in limnological transformations, as exemplified by iron and manganese. *Advances in Water Pollution Research, Proceedings of the Second International Conference on Water Pollution Research* 1: 103–118.

Mullaney, J.R., Lorenz, D.L. and Arntson, A.D. 2009. *Chloride in groundwater and surface water in areas underlain by the glacial aquifer system, Northern United States.* U.S. Geological Survey Scientific Investigations Report 2009–5086.

Neal, C., Jarvie, H.P., Neal, M., Hill, L. and Wickham H. 2006. Nitrate concentrations in river waters of the upper Thames and its tributaries. *The Science of the Total Environment* 365: 15–32.

Nilsson, J. and Grennfelt, P. (eds) 1988. *Critical loads for sulphur and nitrogen.* UNECE/Nordic Council workshop report, Skokloster, Sweden. Nordic Council of Ministers; Copenhagen.

Novotny, V. (ed.) 1995. *Nonpoint pollution and urban stormwater management.* Technomic Publication Co. Lancaster, PA.

Novotny, V. 2002. *Water quality: diffuse pollution and watershed management (2nd edition).* John Wiley and Sons; New York.

O'Connor, D.J. 1960. Oxygen balance of an estuary. *Journal of the Sanitary Engineering Division*, ASCE 86: Proceedings Paper 2472.

Øygarden, L., Kvaerner, J. and Jenssen, P.D. 1997. Soil erosion via preferential flow to drainage systems in clay soils. *Geoderma* 76: 65–86.

Posch, M., Hettelingh, J.P., de Smet, P.A.M. and Downing, R.J. 1997. *Calculation and mapping of critical thresholds in Europe.* RIVM Report No. 259101007. National Institute of Public Health and the Environment; The Netherlands.

Pretty, J.N., Brett, C., Gee, D., Hine, R.E., Mason, C.F., Morison, J.I.L., Raven, H., Rayment, M.D. and van der Bijl, G. 2000. An assessment of the total external costs of UK agriculture. *Agricultural Systems* 65: 113–136.

Reynolds, B., Emmett, B.A. and Woods, C. 1992. Variations in streamwater nitrate concentrations and nitrogen budgets over 10 years in a headwater catchment in mid-Wales. *Journal of Hydrology* 136: 155–175.

Rule, K.L., Comber, S., Ross, D., Thornton, A., Makropoulos, C.K. and Rautiu, R. 2006b. Sources of priority substances entering an urban wastewater catchment – trace organic chemicals. *Chemosphere* 63: 581–591.

Schindler, D.W. 1988. Effects of acid rain on freshwater ecosystems. *Science* 239: 149–157.

Schmiermund, R.L. and Drozd, M.A. 1997. Acid mine drainage and other mining-influenced waters (MIW). *In* Marcus, J.J. (ed.) *Mining environmental handbook.* Imperial College Press; London, 599–617.

SEPA. 1996. *State of the Environment Report.* Scottish Environmental Protection Agency.

Singer, P.C. and Stumm, W. 1970. Acidic mine drainage: the rate-determining step. *Science* 167: 1121–1123.

Skjelkvåle, B.L. et al. 2005. Regional scale evidence for improvements in surface water chemistry 1990–2001. *Environmental Pollution* 137: 165–176.

SNIFFER. 2006. *Retrofitting sustainable urban water solutions.* Report UE3(05)UW5, Scottish and Northern Ireland Forum For Environmental Research [online]. Available from: www.sniffer.org.uk.

Sprague, L.A., Hirsch, R.M. and Aulenbach, B.T. 2011. Nitrate in the Mississippi River and its tributaries, 1980–2008; are we making progress? *Environmental Science and Technology* 45: 7209–7216.

Stoddard, J.L., Jeffries, D.S., Lükewille, A., Clair, T.A., Dillon, P.J., Driscoll, C.T., Forsius, M., Johannessen, M., Kahl, J.S., Kellogg, J.H., Kemp, A., Mannio, J., Monteith, D., Murdoch, P.S., Patrick, S., Rebsdorf, A., Skjelkvåle, B.L., Stainton, M.P., Traaen, T., van Dam, H., Webster, K.E., Wieting, J. and Wilander, A. 1999. Regional trends in aquatic recovery from acidification in North America and Europe. *Nature* 401: 575–578.

Stutter, M.I., Chardon, W.J. and Kronvang, B., 2012a. Riparian buffer strips as a multifunctional management tool in agricultural landscapes: introduction. *Journal of Environmental Quality* 41: 297–303.

Stutter, M.I., Dunn, S.M. and Lumsdon, D.G. 2012b. Dissolved organic carbon dynamics in a UK podzolic moorland catchment: linking storm hydrochemistry, flow path analysis and sorption experiments. *Biogeosciences* 9: 2159–2175.

Tremaine, S.C. and Mills, A.L. 1991. Impact of water column acidification on protozoan bacterivory at the lake sediment–water interface. *Applied and Environmental Microbiology* 57: 775–784.

Tyler, C.R., Lange, A., Paull, G.C., Katsu, Y. and Iguchi, T. 2007. The roach (Rutilus rutilus) as a sentinel for assessing endocrine disruption. *Environmental Sciences* 14: 235–253.

Tyler, S.J. and Ormerod, S.J. 1992. A review of the likely causal pathways relating the reduced density of breeding dippers *Cinclus cinclus* to the acidification of upland streams. *Environmental Pollution* 78: 49–56.

UNPD. 2007. *World Urbanization Prospects: The 2007 Revision.* United Nations Population Division; New York.

US Senate. 2002. Federal Water Pollution Control Act, as amended through P.L. 107–303, November 27, 2002.

Van Damme, P.A., Hamel, C., Ayala, A. and Bervoets, L. 2008. Macroinvertebrate community response to acid mine drainage in the rivers of the High Andes (Bolivia). *Environmental Pollution* 156: 1061–1068.

Vörösmarty, C.J., Green, P., Salisbury, J. and Lammers, R.B. 2000. Global water resources: vulnerability from climate change and population growth. *Science* 289: 284.

Vörösmarty, C.J., McIntyre, P.B., Gessner, M.O., Dudgeon, D., Prusevich, A., Green, P., Glidden, S., Bunn, S.E., Sullivan, C.A., Reidy Liermann, C. and Davies, P.M. 2010. Global threats to human water security and river biodiversity. *Nature* 467: 555–561.

Walling, D.E. and Webb, B.W. 1985. Estimating the discharge of contaminants to coastal waters by rivers: some cautionary comments. *Marine Pollution Bulletin* 16: 488–492.

Walling, D.E. and Webb, B.W. 1986. Solutes in river systems. In: Trudgill, S.T. (ed.) *Solute processes.* Wiley and Sons Ltd; Chichester, 251–327.

Webb, B.W., Phillips, J.M., Walling, D.E., Littlewood, I.G., Watts, C.D. and Leeks, G.J.L. 1997. Load estimation methodologies for British rivers and their relevance to the LOIS RACS(R) programme. *The Science of the Total Environment* 194: 379–389.

Webb, J.S., McGuiness, S. and Lappinn-Scott, H.M. 1998. Metal removal by sulphate-reducing bacteria from natural and constructed wetlands. *Journal of Applied Microbiology* 84: 240–248.

Wetzel, R.G. 2001. *Limnology lake and river ecosystems.* Academic Press; New York.

White, I. and Alacon, A. 2009. Planning policy, sustainable drainage and surface water management: a case study of Greater Manchester, UK. *Built Environment* 35: 516–530.

Whitehead, P.G., Wilby, R.L., Battarbee, R.W., Kernan, M. and Wade, A.J. 2009. A review of the potential impact of climate change on surface water quality. *Hydrological Sciences Journal* 54: 101–123.

Wiseman, I.M., Edwards, P.J. and Rutt, G.P. 2003. Recovery of an aquatic ecosystem following treatment of abandoned mine drainage with constructed wetlands. *Land Contamination and Reclamation* 11: 221–229.

Worrall, F., Spencer, E. and Burt, T.P. 2009. The effectiveness of nitrate vulnerable zones for limiting surface water nitrate concentrations. *Journal of Hydrology* 370: 21–28.

Wren, C.D. and Stephenson, G.L. 1991. The effect of acidification on the accumulation and toxicity of metals to freshwater invertebrates. *Environmental Pollution* 71: 205–241.

Yallop, A.R. and Clutterbuck, B. 2009. Land management as a factor controlling dissolved organic carbon release from upland peat soils 1: spatial variation in DOC productivity. *The Science of the Total Environment* 407: 3803–3813.

Younger, P.L. 1997. The longevity of minewater pollution: a basis for decision making. *The Science of the Total Environment* 194/195: 457–466.

Younger, P.L. 2000. Holistic remedial strategies for short- and long-term water pollution from abandoned mines. *Transactions of the Institution of Mining and Metallurgy, Section A. Mining Industry* 109: 210–218.

CHAPTER FIVE

Groundwater

L. Jared West and Noelle E. Odling

LEARNING OUTCOMES

After reading this chapter you should be able to:

- describe the role of groundwater in the hydrological cycle
- understand factors influencing recharge and water balance, and the potential impacts of groundwater abstraction and injection
- appreciate which characteristics of rocks and soils determine whether they are groundwater aquifers
- understand causes of groundwater movement and the factors that affect flow rates below the ground surface, including the important concepts of hydraulic head, hydraulic gradient and Darcy's Law
- appreciate that groundwater flow can be represented using hydrogeological maps and numerical simulations
- explain the basics of groundwater abstraction/injection techniques and their associated problems
- describe the factors that influence the natural chemical constituents of potable groundwater
- describe key groundwater contaminants, the major threats to groundwater quality and the main factors influencing groundwater vulnerability to pollution.

A INTRODUCTION

Around 22% of the world's population rely solely on groundwater. An estimated 10% of global groundwater consumption is unsustainable in that water is being extracted faster than it is being replenished, and 10% of global food supply is based on such unsustainable groundwater exploitation (USAID, 2011). Given growing demand from population growth and development, coupled with reductions in sustainable yields due to climatic change, the twenty-first century is likely to see increased conflicts and other undesirable socio-economic consequences arising from insufficient and dwindling groundwater resources.

This chapter provides an introduction to groundwater resources and the physical and chemical processes which influence their availability. Beginning with the nature and occurrence of groundwater, and concepts of recharge, discharge and water balance, the idea of groundwater management is introduced. Then, the physical processes of groundwater flow are dealt with in more detail, in order to explain the distribution of extractable groundwater resources (**aquifers**). Flow processes that occur in aquifers, how they are represented using maps or numerical models, and factors affecting groundwater abstraction or re-injection are explained (the physical construction of wells is briefly described). The second part of the chapter deals with the chemical processes in groundwater, including natural constituents and how they are derived, and groundwater pollutants. The concept of groundwater vulnerability to pollution and well protection are described and the factors which influence these are discussed.

B GROUNDWATER RESOURCES

1 Recharge and discharge to groundwater

Groundwater constitutes an important component of the hydrological cycle, shown schematically in **Figure 5.1**, and is principally derived from precipitation, surface runoff and standing water bodies such as lakes and reservoirs. The addition of water to groundwater reserves is known as **recharge**, and humans also make a contribution to this via surplus **irrigation**, pipe, sewer and canal leakage, and deliberate aquifer augmentation schemes (artificial recharge; see section E5). In the upper layers of soil and rock there may be pores and voids which are not filled with water, but with air. This is known as the unsaturated or **vadose zone**. In this zone, water tends to flow downward under the force of gravity (although it may also be diverted laterally: West and Truss, 2006). In the saturated (or **phreatic**) zone beneath this, all pores and voids are water filled (except for occasional small pockets of trapped gas) and groundwater tends to flow laterally, typically following the topographic gradient. Groundwater may eventually intercept the ground surface, forming springs or seeps, or enter stream beds where it contributes to stream **baseflow**. Groundwater can also emerge as springs on the sea bed. Artificial abstraction, for irrigation or domestic and industrial supply, also constitutes an important means of groundwater discharge.

2 Rainfall recharge

The absorption of surface water into the ground is called **infiltration** (see **Figure 5.2**). Not all of the water that is present on the surface infiltrates:

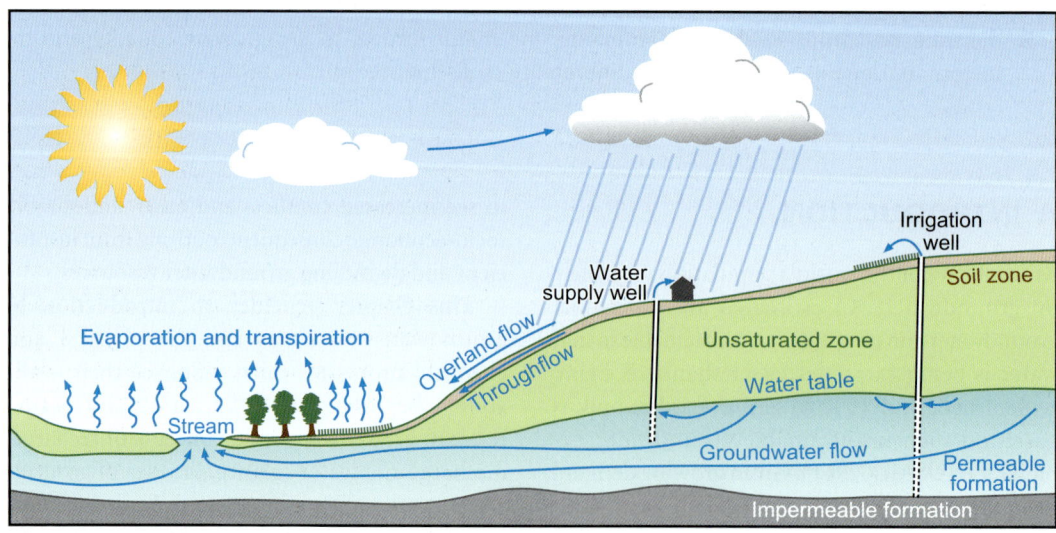

Figure 5.1 Role of groundwater in the hydrological cycle.

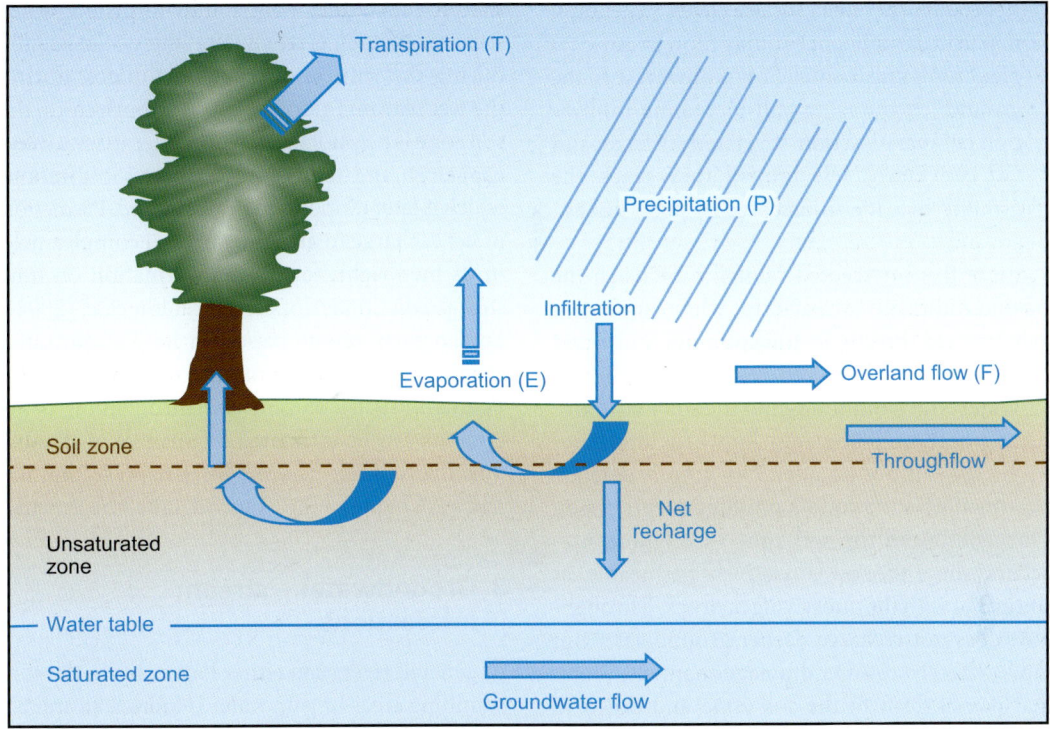

Figure 5.2 Factors influencing rainfall recharge to groundwater.

some runs off as **overland flow**, and some evaporates (see Chapter 3). Of the water that does infiltrate, some is drawn back to the surface by evaporation and some is drawn back to the surface by plants. Excess water from this process is then transpired by the plant leaves. A mature oak tree can draw water from as much as 10 m deep into the ground. The maximum rate at which infiltration can occur for a particular medium is known as its **infiltration capacity**. If the rate of precipitation exceeds infiltration capacity, then surplus water may lead to overland flow. There are a number of factors which determine how much water contributes to overland flow compared to that which infiltrates or is otherwise removed (e.g. by evaporation and transpiration):

1. The properties of the ground surface, for example, its porosity (void space) **permeability** (ability to allow water to flow through or permeate, see Section D).
2. Rainfall intensity and duration. The more intense the rainfall is, the more likely it is to exceed the infiltration capacity of the ground. This is why tropical regions, which commonly experience high-intensity rainfall, have a high percentage of **infiltration-excess overland flow** (see Chapter 3) compared to temperate climates, which have less intense rainstorms. Duration of precipitation and degree of ground saturation are also important; if the upper soil layer reaches its maximum water-holding capacity, then no further infiltration will be able to occur. Surplus water will then run off from the surface as **saturation-excess overland flow**.
3. Gradient of the ground. Steeper ground will encourage greater overland flow.

4   Vegetation (or other) surface cover. Vegetation in particular influences infiltration because (i) plant roots create small flow pathways into the ground, encouraging and increasing infiltration compared to non-vegetated surfaces, and (ii) conversely, less rainfall may reach the ground since it is intercepted by plant leaves.

Some of this intercepted rainfall may reach the ground indirectly by leaf-drip, while some of it may be used directly in **transpiration** and never reach the ground.

Water which infiltrates into the soil may move laterally as **throughflow** (also known as **interflow**) within the soil. See Chapter 3 for a more detailed treatment of such runoff production processes. Throughflow in the soil zone (see **Figure 5.2**) occurs where the soil is relatively permeable, as compared with the underlying bedrock. Throughflow does not recharge deeper groundwater but moves laterally to enter drainage channels. Water that passes through the soil zone and down to the phreatic zones contributes to groundwater recharge. In temperate climates this occurs mainly during the winter; in tropical climates, during the wet season. In drier seasons in both cases, the soil zone is below its water-holding capacity (**field capacity**), and tends to absorb any infiltration, which is later evaporated or transpired. Prediction of net recharge to groundwater is a complex field in its own right; for more information on how such prediction is done, see Scanlon *et al.* (2002). For an early classic paper on how evapotranspiration is estimated from meteorological data, see Penman (1948). Runoff and throughflow components can be estimated by analysis of streamflow hydrographs (see **Figure 5.3** and section B3; and see Chapter 3 for a more detailed treatment).

3 Groundwater – stream interactions

In general, rivers can either lose water to the surrounding areas or gain water (**Figure 5.3a** and **b**).

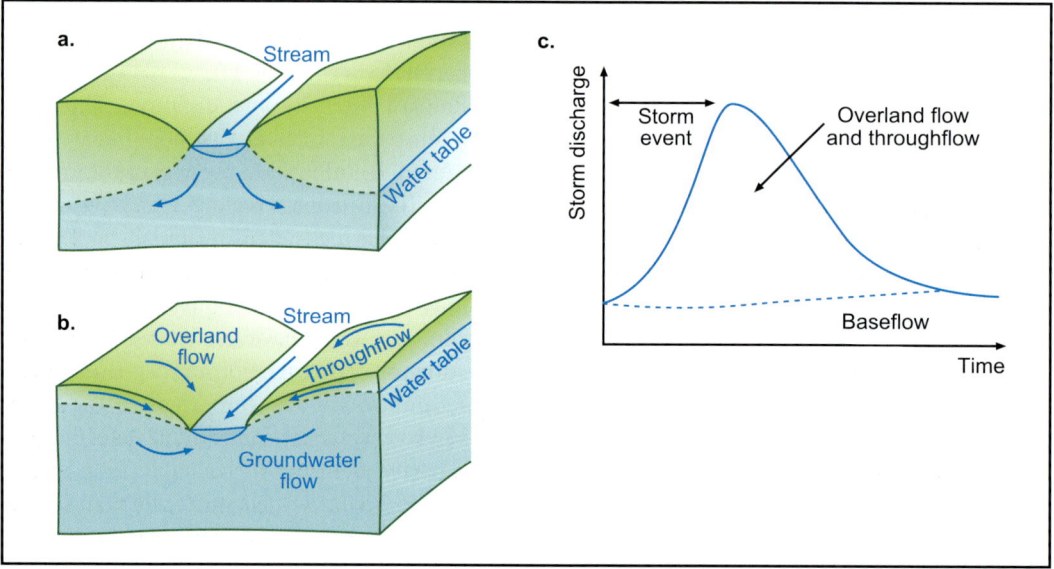

Figure 5.3 Interactions between streams and groundwater: (a) losing stream; (b) gaining stream; (c) the stream hydrograph response showing the baseflow component due to groundwater discharges.

The direction of these flows depends on the difference between the stream level (**stage**; see Chapter 3) and the groundwater level. During a flood period, increased stream stage may induce infiltration of stream water into the aquifer; subsequent declines in stream stage cause a reverse motion of infiltrated water. The interaction between groundwater and stream water will be dependent on the geology and the topographic characteristics of the area. Streams on permeable catchments in temperate countries often show net gain of water from groundwater, and hence are called 'gaining streams'. Ephemeral streams (dry for some parts of the year) in arid climates more often lose water to groundwater and may be the main source of groundwater recharge in some areas. An increasing body of literature exists on methods of measuring exchange between streams and aquifers (e.g. Jones and Mulholland, 2003). One approach is hydrograph separation, which can be a useful tool to determine groundwater contributions at different flow stages during a flood event (**Figure 5.3c** and see also Chapter 3).

4 Groundwater management

The simplest approach to evaluation of sustainable groundwater resources is to determine inputs and outputs to groundwater in a defined area known as a groundwater catchment (or historically, a groundwater basin) and by doing so to conduct a water balance evaluation. Groundwater catchments represent defined areas from which the in- and out-flows of water can be specified, i.e. they are conceptually similar to the idea of a surface water catchment or drainage basin. Hence they vary from just a portion of a geological formation, where this can be isolated by the presence of groundwater divides, to several geological units, where these are hydraulically interconnected. A groundwater catchment may be confined to a portion of a stream drainage catchment, i.e. only where permeable rocks are present. However, in extensive areas of highly permeable rocks such as limestones, surface and groundwater catchments may have completely different configurations so that groundwater divides bear little relation to the topographic divides that control surface drainage. Definition of catchments is now regarded as an essential first step in the holistic management of water resources (see section E).

Evaluation of water balance allows management of groundwater resources via prediction of the impact of a change, such as increased abstraction, or climatic change, on other abstractions and the aquatic environment (e.g. river baseflows). For example, any increase in abstraction must ultimately reduce one of the other outputs from groundwater (e.g. river baseflow), or increase one of its inputs (e.g. by pulling water in from an adjacent groundwater body or, worse, a salt-water saturated formation). In order to evaluate the water balance, the change in volume of groundwater stored over a given time period (typically a year) is found by:

$$\text{Change in storage} = (P - E - T - F - I) + G - L + A, \quad [5.1]$$

where P is precipitation, E is evaporation, T is transpiration, F is overland flow, I is throughflow (interflow) (see **Figure 5.2**), G is gains from surface water, L is losses to surface water, and A is artificial abstraction (where negative) or injection (where positive), with each item having units of volume per unit time. If the change in storage is negative, a reduction in groundwater storage will take place. However, in non-arid climates, other terms will adjust to make the change in storage zero again. For example, the baseflow component of a gaining stream may be reduced. What needs to be decided in groundwater management is whether the environmental consequences of such changes are acceptable. In order to implement groundwater management, it is usual for regulatory bodies to license groundwater abstractions. It is also increasingly common to consider groundwater and surface water resources in a holistic way, called integrated catchment management. This reflects the interaction of

BOX 5.1 CASE STUDIES

The Nubian Sandstone Aquifer System

The Nubian Sandstone Aquifer System (NSAS) underlies the northeastern area of the Sahara Desert in Egypt, Libya, Chad and Sudan, covering an area of 2 million km^2 (**Figure 5.4**). It is the largest known 'fossil' aquifer and is estimated to contain some 50,000 km^3 of water. The area of the NSAS forms a flat plateau, with mountains in the south and east. The aquifer system is composed of rocks of two different ages. The older rocks (the Nubian Aquifer System – NAS) which crop out in the south are Late Jurassic to Cretaceous in age (160–100 million years old) and are medium- to coarse-grained sandstones. The younger rocks (the Post Nubian System – PNAS) which crop out in the north are Palaeocene in age (around 56 million years old) and are sandstones and carbonates (limestones). These are separated by mudstones which partially limit communication between the two systems. The aquifers have a **transmissivity** (see section D2 below) in the range 2 to 72 m^2 per day with saturated thicknesses from 100 to 650 m. The aquifer contains two major basins: the Kufra Basin, covering southeastern Libya, northeastern Chad and northwestern Sudan, and the

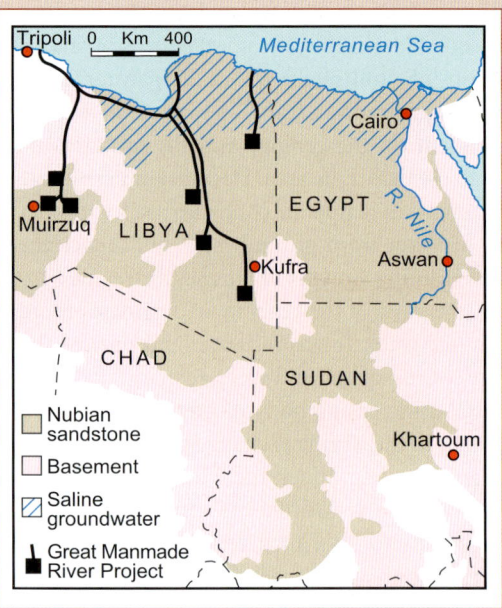

Figure 5.4 Map of the NSAS showing country boundaries, direction of groundwater flow, Kufra and Dakhla Basins and location of Al Kufrah and the GMRP.

Dakhla Basin in Egypt and eastern Libya. Groundwater flow is very slow (around 2 m per year) in a northerly direction. Close to the Mediterranean coast the groundwaters are highly saline.

The average rainfall of the area is less than 5 mm per year and there is thought to be negligible recharge to the aquifer. The groundwater is older than 20,000 years and some waters in the deeper parts are estimated, from radioisotope dating techniques, to be 1 million years old (Patterson *et al.*, 2005). The groundwater is of atmospheric origin and fell as rain during wet phases in the Pleistocene (11,700 years ago to 1.8 million years ago). The area has been **hyper-arid** for the last 3500 years. Since recharge is very low, the NSAS is a non-renewable source of water and, like an oil reservoir, will become depleted through exploitation in time.

The region of the NSAS has a total population of around 762,000 and the aquifer is the main source of water for many. Up until 2001, some 3.8 km^3 of water had been extracted but in recent times utilisation of the aquifer has been growing rapidly. Extraction rates are today estimated at around 3 km^3 per year, most of which is used for irrigation. In the 1970s, Libya started major exploitation of the NSAS to support irrigation at Al Kufrah in southeast Libya, using central-pivot sprinkler systems with a radius of 1120 m. This was followed in the mid-1980s with the Great Man-Made River Project (GMRP). Today some 2.37 km^3 of groundwater is extracted per year from over 1000 wells and transported through 5000 km of pipeline to the Libyan coast. When the project is complete, it is planned that

continued

6 km³ per year will be abstracted, mostly for agricultural purposes. Egypt also has plans to develop the Western Desert utilising NSAS groundwater for domestic and agricultural use (**Figure 5.5**). Sixteen cities are under development and a further 41 planned to provide for 20% of the population and to expand Egypt's agricultural land by 37% by 2017.

In order to manage the water resources of the NSAS, the Joint Authority for the Management of the NSAS System was created by Libya and Egypt in 1992 and was joined by Sudan and Chad in 1999. In the on-going 'Nubian Project', the four NSAS countries (Egypt, Libya, Chad and Sudan) with the IAEA (International Atomic Energy Agency), UNDP (United Nations Development Programme), GEF (Global Environment Facility) and UNESCO (United Nations Educational, Scientific and Cultural Organization) are working to establish a rational and equitable management of the NSAS for sustainable socio-economic development of the region. The Nubian project aims to fill knowledge gaps about the NSAS for a better understanding of the aquifer system in order to inform future management policies both within and across countries.

Figure 5.5 Well drilling in Egypt.
Source: Photo courtesy of Mahmoud Al-Araby Hussein.

streams and groundwater outlined above. In the European Union such an approach is now mandatory as a result of the Water Framework Directive (European Commission, 2000) and the Groundwater Directive (European Commission, 2006). A different approach is required for 'fossil' aquifer systems where water is effectively being mined as a non-renewable resource, such as the Nubian Sandstone Aquifer System (**see Box 5.1**). In such cases, sustainable usage is not possible, so demand needs to be managed in order to provide supply for a finite period.

REFLECTIVE QUESTION

What is the difference between the surface drainage basin area and the groundwater drainage basin area and what are the main reasons the two areas might be very different?

C AQUIFERS

1 Aquifers and confining layers

The term 'aquifer' refers to a geological stratum which contains water and which allows the movement of water within it. It is porous, in that it has space represented by pores or fractures which can be occupied by water, and permeable (see section D), in that the water can move through this space. An example would be sand or sandstone. Other types of geological formation that make good aquifers are discussed in sections C2 to C4. If a well were drilled into such strata and the water extracted, the well would keep on refilling. We refer to strata with relatively low permeability, such as silt or siltstone, as confining layers (**Figure 5.6**). In well A, shown in **Figure 5.6**, drilled in the permeable formation, the water will rise to the level of the **water table** (uppermost part of the saturated zone) and the formation is referred to as an unconfined, or phreatic, aquifer. In well B, drilled through the low-permeability confining layer into the aquifer beneath, the water will rise to a greater elevation than the top of the aquifer, because this layer contains water under pressure. The aquifer is referred to as a confined aquifer at this point. The water in well B rises towards a **piezometric** (or potentiometric) surface; this is an imaginary surface determined by the pressure in the aquifer. Because the piezometric surface at well C is above the ground surface, water from this well would overflow naturally. Such a well is known as an **artesian well** and the aquifer in this area is called an artesian aquifer.

Low-permeability strata above the water table may produce a locally elevated water table, known as a perched aquifer. Perched aquifers may be of limited lateral extent, and sometimes may exist only in response to particularly high infiltration (after a prolonged and heavy rainstorm, for instance). They may also be of considerable lateral extent and thickness, and be occasionally mistaken for permanent aquifers. Because of their variability, perched aquifers do not provide reliable long-term water supply sources.

2 Igneous and metamorphic rock aquifers

Most of the best aquifers around the world are found in sediments and **sedimentary rocks**. **Igneous rocks** and **metamorphic rocks** are commonly poor aquifer rocks. This is largely because of the mode of formation of these broad rock groups. Igneous rocks form from the solidification of magma, a process which usually creates a texture of interlocking crystals. This produces very little primary porosity (void space produced at the time of rock genesis), except for occasional **vesicles** in vesicular rocks. However, some volcanic rocks such as basalt have a well-developed fracture network formed during cooling, and this provides good permeability. For example, the Columbia Plateau Basalts in the northwest USA are one of the highest-yielding aquifers in the world (Foxworthy, 1983). Other extrusive igneous

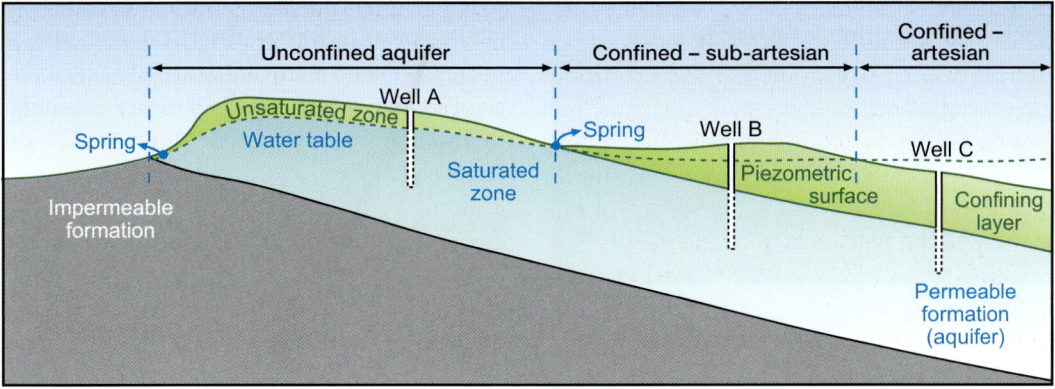

Figure 5.6 Confined and unconfined aquifers.

rocks, such as pumice, may have very high porosities, but due to poor connectivity between void spaces, permeability is low. Metamorphic rocks form from the alteration of existing rocks by the application of heat and/or pressure. This tends to reduce their porosity and permeability below that of the **parent rock**.

Secondary porosity may later develop in some igneous and metamorphic rocks, due to alteration and decomposition of minerals during weathering. Tectonic activity and stress-release processes may also lead to the generation of fractures which can act as conduits for groundwater. This is the means by which some of these rocks can be developed, usually locally and on a small scale, as fractured aquifers (**Box 5.2**). Assessment of the yield of such aquifers, however, is costly because changes in rock properties due to weathering and other secondary processes are extremely variable. However, these aquifers are vitally important in many developing countries, where they are a useful local supply of water (Wright and Burgess, 1992).

3 Unconsolidated aquifers

On a global scale, as much as 90% of developed aquifers are in unconsolidated materials (Todd, 1980) such as sediments in river valleys and floodplains, buried valleys and inter-montane valleys. These consist predominantly of sands and gravels (clays can have a high porosity, but since they have very fine pores, permeability is low and they make poor aquifers). The high permeability commonly found adjacent to water courses generally gives very high aquifer yields, and this is often augmented by inflow from streams. Floodplain deposits also provide high yields, but they are more variable and can be thin. The High Plains Aquifer in the USA is a very important water source and consists of fluvial sands and gravels stretching from South Dakota to Texas (Miller and Appel, 1997). In western USA, inter-montane valleys contain aquifers. They comprise deep valley deposits derived from erosion of adjacent rock walls and can be very productive. Groundwater in such aquifers is replenished from streams seeping into **alluvial fans** at the mouth of the valley.

4 Sedimentary rock aquifers

Sedimentary rocks form from the deposition of particles or sediments derived from the weathering and erosion of other rocks. Sediments initially become consolidated and cemented, which usually reduces porosity, though later **diagenesis**, fracturing and weathering may increase the porosity and the permeability. Most sandstones and **conglomerates** will yield some water. Those that

Figure 5.7 Sedimentary rock aquifers in the United Kingdom: (a) Permo-Triassic Sandstone; (b) Cretaceous Chalk.

make the best aquifers are usually poorly cemented, such as the Permo-Triassic Sandstones in Northern Europe (**Figure 5.7a**; Allen *et al.*, 1997) and the Nubian Sandstone Aquifer System in North Africa (**Box 5.1**). Limestones, which are composed of carbonate rather than silicate minerals, have special properties as aquifers because of their high solubility in rainwater (in other words, the limestone dissolves in water). Limestone areas may contain large solution cavities, and dissolution along joints can lead to large fissure-flow permeability. Connectivity between such fissures and cavities is critical, but it is difficult to predict whether a borehole sunk at random will intercept these. Also, because the cavities in limestone are relatively large, the rock tends to drain relatively quickly after rainfall, and so it is an unreliable water source. Chalk is a type of limestone with relatively closely spaced fractures (**Figure 5.7b**) and actually makes a very good aquifer, although it can be more susceptible to drought than sandstone aquifers, for the reasons just given.

REFLECTIVE QUESTION
What types of aquifer are there and which rocks or sediments make the best aquifers?

D GROUNDWATER FLOW PRINCIPLES

1 Fluid potential and hydraulic head

M. King Hubbert of the US Geological Survey investigated the flow of water in porous media and defined, theoretically, the potential of water to flow (see classic paper: Hubbert, 1940). He argued that the potential for water flow was a measurable, physical quantity and that it should also be some quantity whose properties were such that water would flow from regions where it was high to regions where it was low. This quantity was referred to as fluid potential (by analogy with electrical potential, or voltage, which causes electricity to flow). The fluid potential of groundwater is seen by the elevation to which water will rise in a well, which is called **hydraulic head** (units of length, L). Given constant fluid density, water will generally flow from regions of high hydraulic head to regions of low hydraulic head. The absolute value of the hydraulic head is always arbitrary because it depends on the datum from which well-water level is measured (ground surface or mean sea level, for example). However, this does not matter, because it is the difference in hydraulic head, rather than its absolute value, which causes water to flow.

2 Darcy's Law

In 1856, Henri Darcy published the results of some experimental work in which he was able to calculate the amount of flow through sand (Darcy, 1856). The purpose of the work was to determine the size of sand filter beds for public water supply. The measured rule that he discovered turned out to be very widely applicable for quantifying flow of fluids in both porous and fractured media (Holden, 2005). Darcy's experiment is illustrated in **Figure 5.8**; Darcy's Law can be written as

$$q = \frac{Q}{A} = -K \frac{\Delta h}{\Delta l}, \quad [5.2]$$

where Q is total discharge [units of length^3 and time^{-1}, L^{3}T^{-1} e.g. m^3 day^{-1}], q is **specific discharge** [LT^{-1}] (volume of flow rate divided by cross-sectional area), A is cross-sectional area [L^2], Δh is the difference in hydraulic head between the measurement points [L], Δl is the distance between measurement points [L] and K [LT^{-1}] is a constant of proportionality representing permeability, called **hydraulic conductivity**. The negative sign in Equation [5.2] indicates that water flows from high to low hydraulic head. The value of K depends on the properties of the porous medium and fluid properties (e.g. viscosity and density). However, in hydrogeology the fluid concerned is usually water at environmental temperatures, so hydraulic conductivity can be considered to be a property of the rock or soil. Although the original Darcy experiment was on sand, fractures also show a linear relationship between hydraulic gradient and flow (**Box 5.2**). For typical values of hydraulic conductivity for various geological media, see Domenico and Schwartz (1998). Another useful parameter in hydrogeology is the transmissivity, T, of an aquifer system (units L^{2}T^{-1}); this is the product of the hydraulic conductivity K, and the saturated thickness of an aquifer b. It represents the overall capacity of the aquifer system to transmit water. Typical values for productive aquifers range from a few tens to greater than 10,000 m^2 per day.

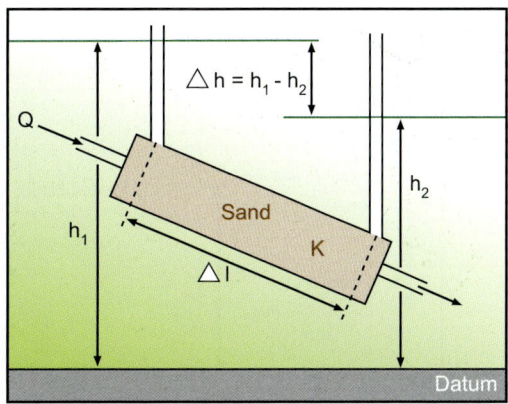

Figure 5.8 Column experiment that can be set up in the laboratory illustrating Darcy's Law for flow in sand.

3 Hydraulic gradient and specific discharge

In order for water to flow in an aquifer, it has to be in response to a gradient in hydraulic head. Horizontal hydraulic gradient in an unconfined aquifer is given by the slope of the water table (**Figure 5.9a**). In a confined aquifer it is similarly given by the slope of the piezometric surface. In well A in **Figure 5.9a**, the water table is at height h_A above a datum, and in well B, at a horizontal distance Δl from well A, the water table is at height h_B above the same datum. The hydraulic gradient is given by

$$\frac{h_A - h_B}{\Delta l} = \frac{\Delta h}{\Delta l} = i \quad [5.3]$$

Where the water table is flat, there is no horizontal hydraulic gradient and hence no horizontal flow. However, hydraulic gradients also occur in the vertical direction. In areas where surface and rain water is recharging groundwater (**Figure 5.9b**), vertical hydraulic gradients are downward. In areas where groundwater discharges (typically areas of relatively low topography, **Figure 5.9c**),

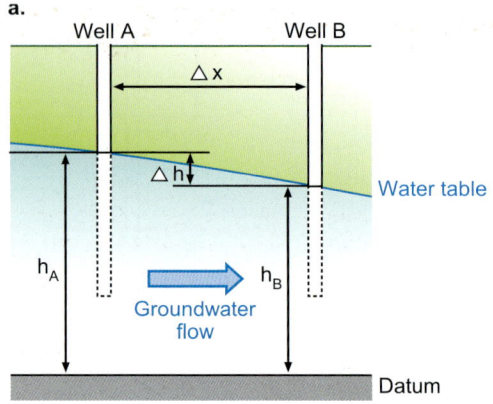

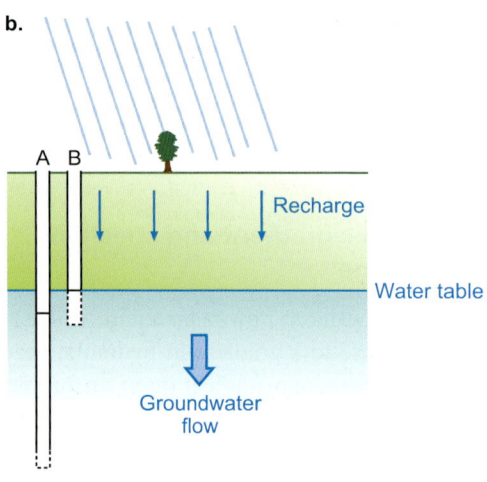

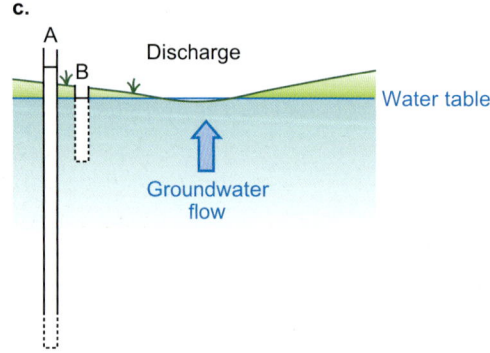

Figure 5.9 Hydraulic gradients in unconfined aquifers: (a) horizontal gradient causing lateral flow; (b) downward gradient in recharge zone; (c) upward gradient in discharge zone (dashed well casing indicates where wells are open to surrounding aquifer).

vertical hydraulic gradients are upward. Very strong upward hydraulic gradients (e.g. exceeding values of one) can produce dangerous quicksand conditions in unconsolidated sediments, because the force exerted by the moving water on the sediment grains overcomes the intergranular contact forces. In order to estimate the volumetric flow rate per unit area in an aquifer, sometimes called the specific discharge, it is simply necessary to multiply the hydraulic gradient given by Equation [5.3] with the hydraulic conductivity (Equation [5.2]). A wide range of hydraulic conductivies are found in formations used as aquifers (10^{-10} m s^{-1} to 1 m s^{-1}: Singhal and Gupta, 1999). With a typical hydraulic gradient for a lowland aquifer of around 10^{-3} (i.e. 1 m hydraulic head change per km) these yield specific discharges of 10^{-13} to 10^{-3} m s^{-1}. These are smaller than typical stream-specific discharge values, but as the cross-sectional areas of aquifers are often much larger than those of streams, there are broadly similar amounts of water flow within them. This can certainly be expected in permeable catchments in temperate climates, where much river baseflow originates from groundwater.

4 Hydrogeological maps

Hydrogeological maps can be constructed from seasonally averaged groundwater heads, in order to show groundwater flow patterns and available resources. An example of a hydrogeological map in **Figure 5.12** shows a section of the River Wylye, which flows over the outcrop of the unconfined Cretaceous Chalk aquifer in southern England. In this case, the hydraulic head contours represent the seasonally averaged water table elevation. As flow generally occurs in the direction of the hydraulic gradient, we can infer that in this example flow is exiting the aquifer to the River Wylye, because the contours indicate that flow is towards the river. Hence, the River Wylye is an example of a 'gaining stream' (see section B3), as groundwater flows from the aquifer into the stream. In permeable catchments like this one,

BOX 5.2 TECHNIQUES

Fractured aquifers

Fractures exist on all scales from microscopic (microfractures) to those which span the thickness of the Earth's brittle crust (such as the San Andreas Fault) and they are known to influence groundwater flow throughout this scale range. Open fractures along which groundwater can flow are found in relatively strong rocks such as crystalline, volcanic and well-**lithified** sedimentary rocks. The fractures which can influence flow include joints (in which displacement is perpendicular to the fracture walls) and faults (which show **shear displacement** within the plane of the fracture). Joints are very common throughout the Earth's crust, whereas faults tend to occur in zones of intense tectonic activity, mainly close to tectonic plate boundaries.

Fractured aquifers are important worldwide. Crystalline and volcanic rocks crop out over 40% of the land surface and carbonate rocks (limestones) account for a further 12% of the land surface. In these rocks the fractures are the sole source of moving water and water storage. Carbonate aquifers are particularly important worldwide and some 25% of the world's population depend on water from aquifers in carbonate rocks. The chief mineral constituent of carbonate aquifers is calcite (calcium carbonate), which is slightly soluble in water. Thus, water flowing through fractures in carbonate aquifers dissolves the fracture walls and enlarges the fracture **apertures** (distance between fracture walls), creating rapid flow pathways for water. Areas where carbonate rocks crop out often form **karst** landscapes in which water flows in subsurface systems of fractures and caves. Groundwater is the only supply of water for domestic and industrial use in karst areas.

In nature, fracture walls are rough and thus the distance between fracture walls (fracture aperture) is highly variable. This gives rise to flow channelling with the plane of the fracture. For characterising the impact of a fracture on groundwater flow, it is useful to think in terms of the **equivalent hydraulic fracture** with an equivalent aperture as a smooth-walled channel that conducts the same flow as the natural rough fracture.

Discharge per unit length of fracture, Q_f is governed by the Cubic Law (**Figure 5.10**), which is related to Darcy's Law as follows:

$$Q_f = -\frac{b^3}{12\mu}\rho g i \qquad [5.4]$$

where b is the equivalent hydraulic aperture [L], ρ is the density of water [M L^{-3}], g is the gravitational constant [LT^{-2}], μ is the dynamic viscosity of water [ML^{-2}T^{-1}], and i the hydraulic gradient in the plane of the fracture (dimensionless). The Cubic Law shows that flow in a fracture is proportional to the cube of the fracture aperture. Thus flow in fractured aquifers is highly sensitive to fracture aperture.

The hydraulic conductivity of fractured aquifers ranges from very low (10^{-10} m s^{-1}) to high (10^{-3} m s^{-1}), overlapping with the wider range shown by

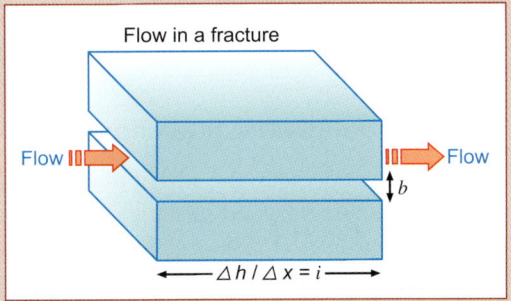

Figure 5.10 Flow in a fracture of aperture b and hydraulic gradient i, is described by the Cubic Law.

sedimentary aquifers (10^{-10} m s^{-1} to 1 m s^{-1}) (Singhal and Gupta, 1999). Fractured aquifers differ most from sedimentary aquifers in that the porosity which controls groundwater flow is much smaller than in typical sedimentary aquifers (Figure 5.11). Typical porosities for sedimentary aquifers are 10–30%, whereas porosities in fractured aquifers typically range from 0.001 to 1%. The effect of porosity on flow rate can be visualised by comparing the flow in two pipes, one of small diameter and one of large diameter. If the volumetric flow rate (in m^3 s^{-1}) is the same in both pipes, the water in the small pipe must flow faster. Thus for a given discharge rate the velocity of water is inversely related to porosity and in a fractured rock water must travel much faster to achieve the same discharge rate as water through a highly porous sedimentary aquifer. This has a dramatic effect on the rate of contaminant movement in fractured aquifers, where contaminants tend to travel much faster than in sedimentary aquifers. This is an important consideration when assessing vulnerability to contamination or designing remediation schemes to clean up contaminated aquifers.

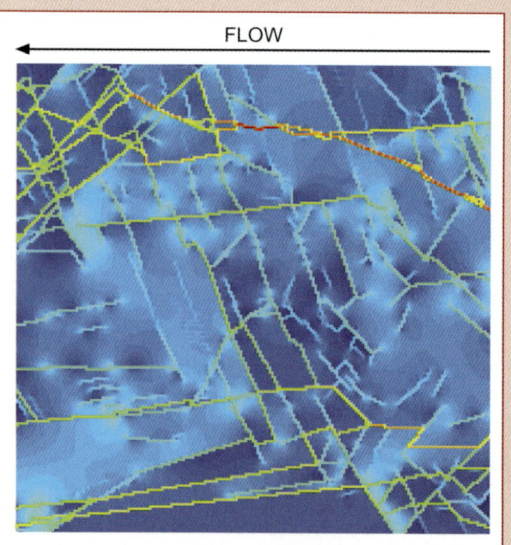

Figure 5.11 Two-dimensional computer simulation of flow through a fractured rock where the rock matrix is itself permeable and porous. Blue shows low flow rates and red high flow rates. The fractures show higher flow rates (yellow, red) than the matrix (blue) and flow rates in the matrix are enhanced where fractures converge to form 'bottle-necks' in the flow pathways (green).

the groundwater component forms an important contribution to flow and ensures that river flow continues throughout the year. Unfortunately, it also means that river flows are highly sensitive to groundwater elevations. If groundwater resources are depleted by over-abstraction, there can be a highly detrimental effect on aquatic ecosystems by reducing minimum river flows (see Chapter 6).

5 Groundwater flow models

In recent decades, the advent of digital computing resources has led to static representations of hydrogeological maps being largely replaced by more sophisticated representations of groundwater flow models (e.g. see Anderson and Woessner, 1992). Such models can be used to make predictions about the effects of changing the stresses (e.g. recharge rates, pumping rates, river stage) on the system. Changes in water levels and, hence, changes in the amount of water stored in aquifers are simulated numerically. The explicit modelling of changes in hydraulic head with time allows seasonal fluctuations in water levels, for example, to be simulated, or they can be used to provide information on the timescale of impacts due to climate or land-use change. However, such models require much more data in order to be calibrated properly (e.g. monthly water level measurements and recharge estimates). An example of a groundwater modelling output showing the effects of adjusting groundwater abstraction

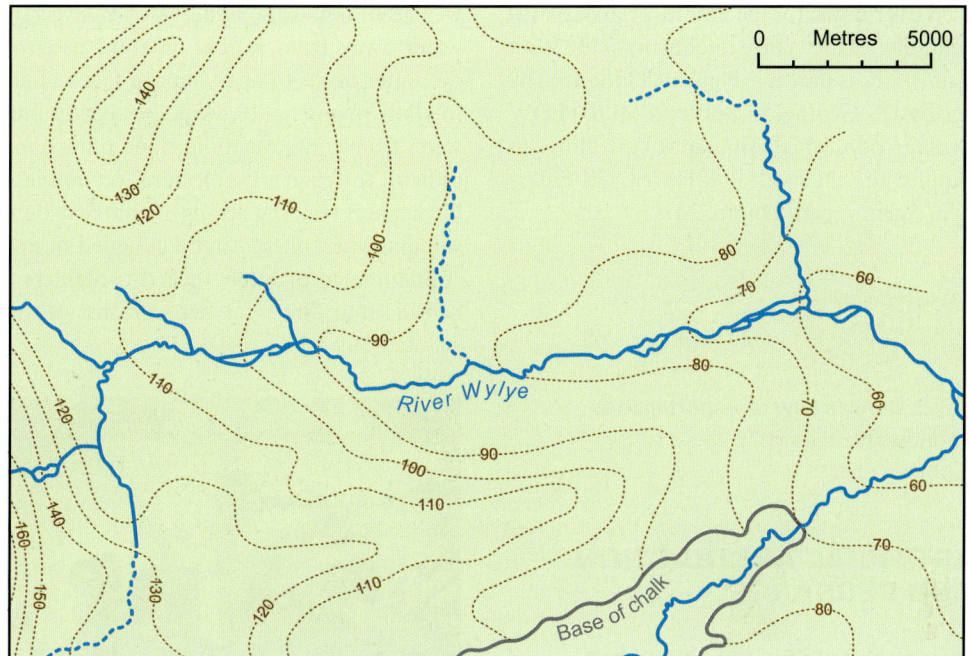

Figure 5.12 Hydrogeological map showing the River Wylye flowing over unconfined Cretaceous Chalk aquifer in southern England. Contours are water table elevations in metres above mean sea level.

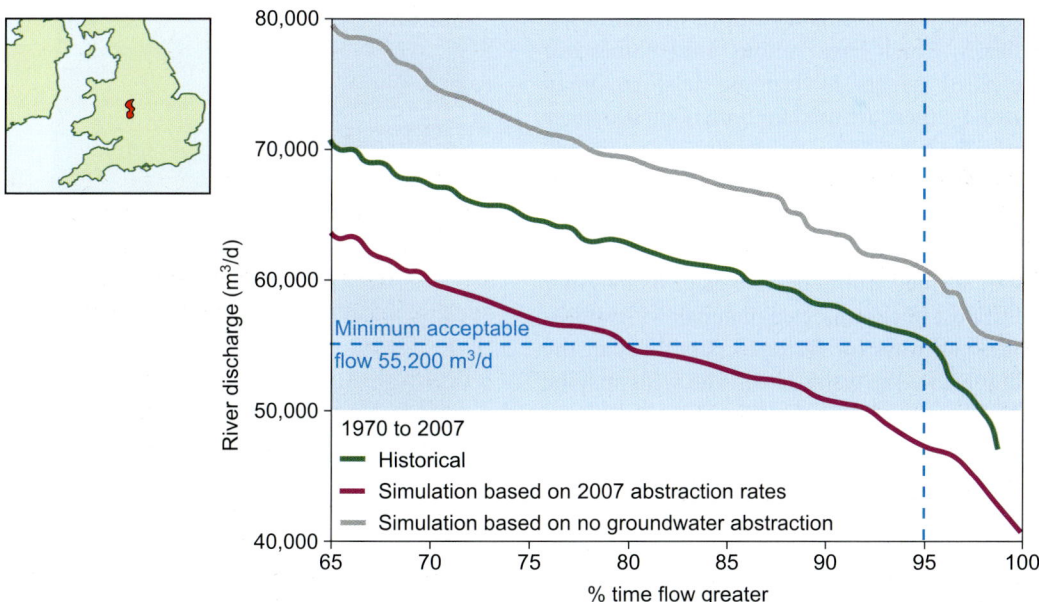

Figure 5.13 Impact of groundwater abstraction on flow at a gauging station in the River Worfe, West Midlands, UK, based on groundwater flow modelling of the underlying Permo-Triassic Sandstone aquifer (data supplied by the UK Environment Agency). Note that the difference between 'historical' and 'simulation using 2007 abstraction rates' arises because actual abstraction rates rose over the period 1970 to 2007.

rates on river flows, for the Permo-Triassic Sandstone aquifer of the Worfe catchment in the West Midlands, UK, is shown in **Figure 5.13** (derived by using the US Geological Survey's MODFLOW modelling code, which simulates both flows in the aquifer and the river: Rolf Farrell, UK Environment Agency, pers. comm. 2011).

> **REFLECTIVE QUESTION**
>
> Why is Darcy's Law of importance to groundwater science?

E ARTIFICIAL ABSTRACTION AND RECHARGE

1 Low-technology groundwater abstraction

Groundwater has been exploited for supply and irrigation for many thousands of years, via hand-dug wells, and via sloping tunnels or infiltration galleries, which have various local names but are called 'qanats' in Iran and 'falaj' in Oman (**Box 5.3**); these are used for both domestic water supply and irrigation. In developing countries, hand-dug wells are still perhaps the most important sources of relatively clean drinking water (**Figure 5.14**). The lining of these wells is commonly formed from concrete rings or masonry to support unconsolidated material, with appropriate pathways for water to pass through the lining below the water table. The wells may be unlined at depth in consolidated rock aquifers. They tend to have relatively large diameters (1 to 1.5 m) to accommodate the well diggers, and also to allow the well to act as a reservoir in low-yielding aquifers so that peak demand (i.e. cooking and washing times) can be accommodated (Misstear *et al.*, 2006). The annulus (ring) behind the well lining should be filled with cement grout to prevent surface water contamination and the

well head should also be designed to divert spilled water away from it. The method used to lift the water needs to be of appropriate technology to allow maintenance, while protecting the well water from contamination. Hand pumps may be difficult to repair in some rural economies, and bucket and winding arrangements may be more suitable where these can be designed to prevent contaminated privately owned containers from coming into direct contact with the well water.

Figure 5.14 Hand-dug wells: (a) a shallow, manually drilled borehole, equipped with an Afridev hand pump, at Chiaquelane, near Chokwe, Mozambique; (b) a timber-lined dug well in the village of Malii Anzas, southern Siberia, Russia.

Source: Photos courtesy of © David Banks, Holymoor Consultancy Ltd.

An alternative approach to low-technology groundwater water supply for small communities is to develop gravity-driven pipework from natural springs which are topographically above the village concerned. Here, a major design consideration is ensuring that the route of the pipework across the landscape produces an appropriate topographic gradient to drive flow.

2 Modern groundwater well construction

Modern groundwater abstraction wells are commonly boreholes drilled using either percussive or rotary drilling methods, or methods which combine both actions. The 'intake section' of the well may be unlined in stable rock, but will be supported with permeable casing called

BOX 5.3 CASE STUDIES

The qanats of Iran

The qanats of Iran are carefully constructed tunnel systems (or infiltration galleries) that are used to transport groundwater under gravity from mountain regions to the adjacent plains. In this way, water is transported to arid and semi-arid areas where it is used to support human settlement and agriculture. Qanats are thought to have originated over 3000 years ago, with the earliest known historical reference dating from the seventh century BC (Wulff, 1968). From ancient Persia, qanat technology spread east to China, south to the Arabian Peninsula, west to the Mediterranean and north to Afghanistan and Azerbaijan.

In the construction of a qanat, a mother well is first dug to the water table, usually in an alluvial fan at the margin of a mountain range where the aquifer is recharged by rain and snow melt and the

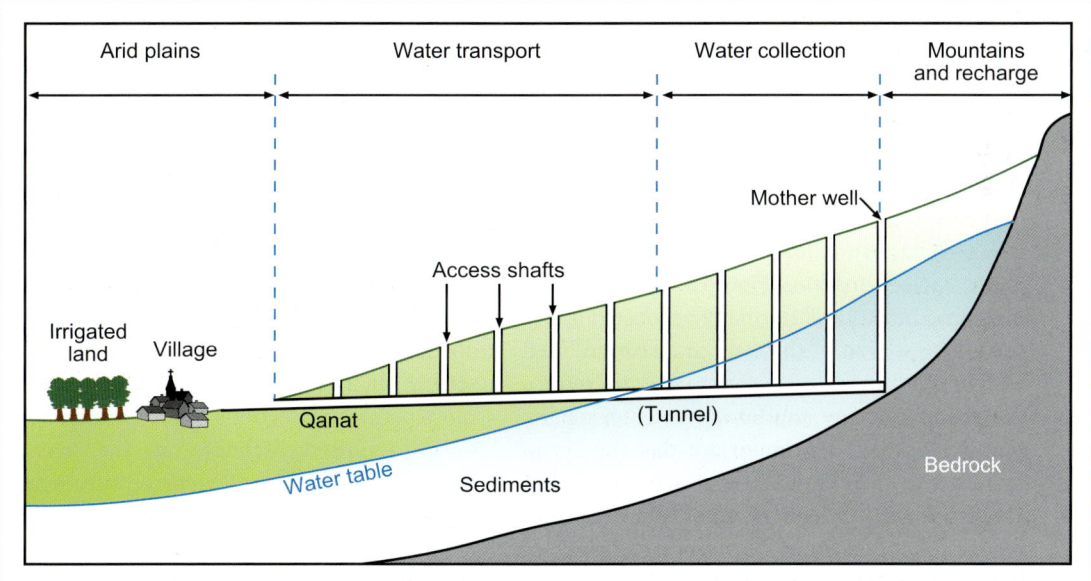

Figure 5.15 Cross-section illustrating the construction of a qanat showing relative positions of the mountain range and adjacent plains, water table and qanat.

continued

water table is high. Mother wells are typically less than 50 m deep but may be up to 300 m. A qanat tunnel is then dug, starting at the down-slope end and working towards the mother well, with vertical shafts at 50 to 100 m intervals for ventilation and removal of spoil (**Figure 5.15**). Qanats are typically dug in alluvial sediments and have an elliptical cross-section of around 1.5 m in height. They are supported in loose sediments by baked-clay hoops. The qanat tunnel has a gradient of between 1:1000 to 1:1500 – too steep a gradient would risk erosion by the flowing water and tunnel collapse. The length of a qanat is controlled by the slope of the ground surface and the water table. Most qanats are around 5 km in length but they can be up to 50 km long, such as some in central Iran. Qanat discharge is typically 60 m^3 per hour but can be up to 300 m^3 per hour. A discharge of 60 m^3 per hour is equivalent to 108,000 m^3 over the five-month growing season and is enough to irrigate 10 to 20 hectares of land. Qanats can be seasonal or flow all year round, and unused winter flow is often stored in nearby dams.

Qanats were crucial in ancient times for enabling cultivation in arid to semi-arid terrains in Iran. Up until the 1960s qanats supplied water for half of the cultivated land in Iran and there were some 25,000 qanats still in use in Iran in the 1980s (Beaumont *et al.*, 1989). However, since the 1950s, there has been increased development of pumped vertical wells, accompanied by expansion of newly cultivated land. This has resulted in over-exploitation of groundwater and falling water tables, which has significantly reduced qanat flow in many cases. Tehran was supplied with water by 12 qanats from the Albortz mountains, 40 km to the north, until the 1930s. In the 1940s, Tehran began to expand, and its population increased from a few thousand to over 8 million in 2010. Demand for water rapidly outstripped the capacity of the qanat system and in the early 1960s dams were built on two rivers (the Karaj and Jaj rivers) to supplement the water supply. This is, however, no longer sufficient and now there are plans to exploit groundwater in the alluvium of the Tehran plateau.

It seems certain that the role played by qanats in Iran and many other countries will decline in future, due to the cost of construction and maintenance and falling water tables from increased groundwater exploitation. However, qanats still play an important role in supplying water for irrigation in some parts of the world such as Oman, Yemen and China.

well-screen, surrounded by permeable pea-gravel, in less stable formations (**Figure 5.16**). The pump (or pump intake) will normally be located above the intake section of the well, in a section lined with impermeable casing, in order to reduce the chances of any aquifer solids entering the pump. However, it is important that the pump is sufficiently far below the resting water level to avoid running dry when the water level in the well reduces due to seasonal fluctuations or abstraction. As with hand-dug wells, the well head needs to be protected from ingress of contaminants; this is achieved by using concrete headworks which direct surface water away from the well head, and by filling the annulus around the impermeable, upper section of the well casing with cement grout. The diameter and depth of the abstraction well will depend on the geological setting, such as the depth to the water table and most-permeable sections of the aquifer, but also on the required discharge rate (i.e. maximum pumping rate). More extensive site investigation will be appropriate for design of a deep, drilled well than for a shallow, hand-dug well, in order to ensure that the economic investment of well construction is worthwhile. The maximum discharge rate will determine the size of pump required, and hence the minimum casing

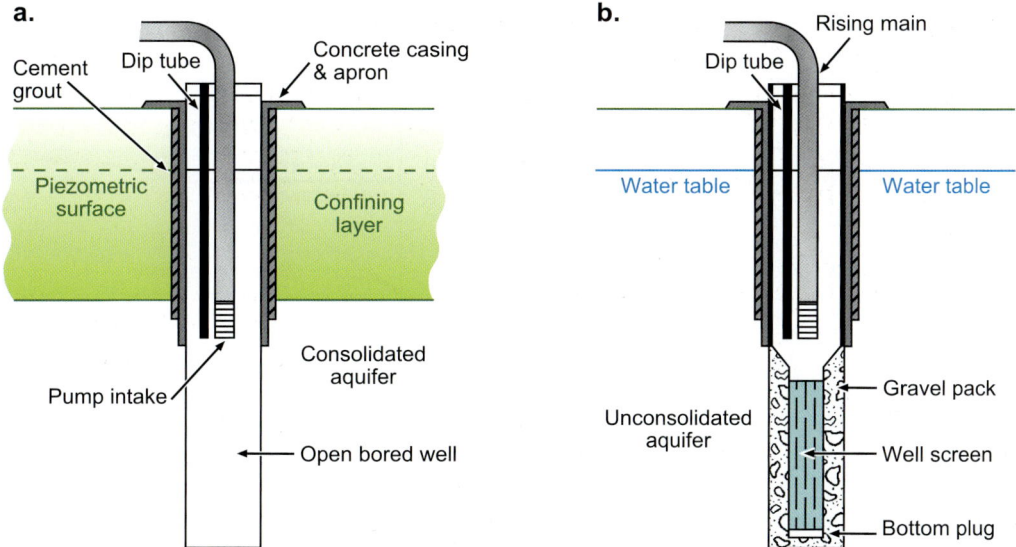

Figure 5.16 Modern groundwater abstraction wells: (a) well with self-supporting open section in a consolidated confined aquifer; (b) well with screen and gravel pack in an unconsolidated unconfined aquifer.

diameter for the section of the well containing the pump. However, the diameter of the intake section of the well may need to be larger than this, in order to ensure that the well is hydraulically efficient, enabling the water to enter the well from the surrounding formation at the required rate. Low well efficiency will mean that the water level in the well falls excessively during pumping, despite there being sufficient water in the surrounding formation.

3 Well testing

There are various reasons for testing groundwater abstraction wells, but the main reason from a water supply perspective is to ensure that the well performs as expected. In other words, can the well yield the required flow rate without the water level in the well falling excessively (this water table fall is referred to as 'drawdown')? There are two components to total well drawdown – they are drawdown in the aquifer near the well, and the additional drawdown between the aquifer and the water level in the well (**Figure 5.17a**); the latter is called 'well loss'. The efficiency of the well is defined as the ratio of the aquifer drawdown to total well drawdown; efficiencies of >80% are considered desirable. Generally, larger diameter wells are more efficient, but efficiency also reduces (i.e. well loss increases) as the pumping rate increases. Hence, there will be some upper limit to the pumping rate above which well drawdown will become large and well efficiency will reduce sharply (**Figure 5.17b**). In order to measure this limit, and define the yield characteristics of the well, it is usual to perform a 'step pumping test', both immediately after well construction and periodically to check that well performance has not deteriorated (Clark, 1977). This involves pumping the abstraction well until the drawdown has stabilised at a range of (usually increasing) pumping rates. The initial portion of the drawdown versus pumping rate plot, where well losses are small, is sometimes assumed to be linear in order to define the **specific capacity** of the well, which is the yield per unit drawdown (units L^2T^{-1},

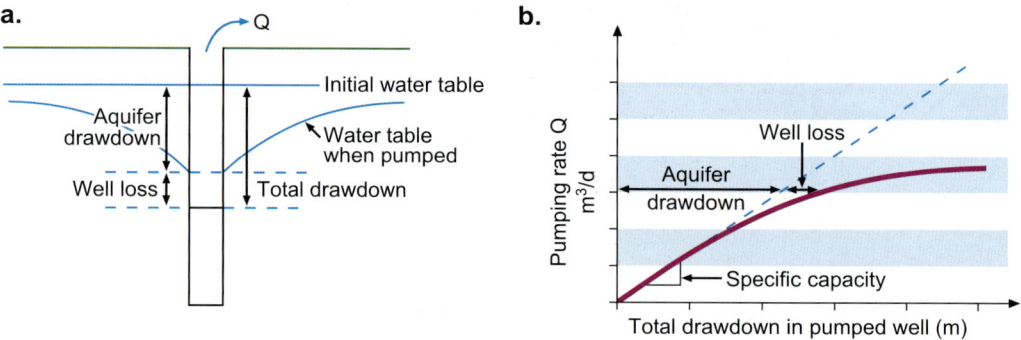

Figure 5.17 Well loss and well efficiency: (a) cross-section illustrating well loss; (b) example plot showing variation of well drawdown with pumping rate. Well losses become large, and efficiency reduces sharply as pumping rates become high.

e.g. m^3 s^{-1} per m of drawdown). As the well losses are relatively small for low pumping rates, the specific capacity under such conditions depends primarily on the properties of the aquifer such as its hydraulic conductivity (see section D2). To an approximation, if well losses are small, the specific capacity of the well can be estimated as 1.22 Kb where K is hydraulic conductivity and b is the saturated thickness of aquifer penetrated by the open section of the well (Misstear et al., 2006: 321). Note that Kb is the effective transmissivity intersected by the well (see section D2).

As well as the response of the abstraction well itself, it is also necessary (especially for larger-scale abstractions) to predict the impact of groundwater abstraction on the surrounding water table and any other wells or water bodies (such as rivers) in the vicinity. The response of the aquifer surrounding an abstraction well to pumping is complex, as the drawdown will depend on not only the properties of the aquifer and the pumping rate, but also the duration of pumping, and any surrounding hydrological features such as rivers, lakes and other abstraction wells, and recharge from rainfall. Understanding and predicting these responses requires longer-term pumping tests (typically at least a week in duration), where the water level is monitored not only in the pumped well itself but also in surrounding 'observation' wells. The theoretical responses of an idealised confined aquifer to pumping an abstraction well were described in a seminal paper by Charles Theis of the US Geological Survey (see classic papers: Theis, 1935) and still forms the basis of predicting effects of abstraction and injection of groundwater via wells today.

4 Problems with groundwater wells

Several physical, chemical and microbiological processes can interfere with the performance of groundwater wells and cause a reduction in efficiency and water quality over time. Physical processes include the clogging of the aquifer, gravel pack, well screen or the well itself, due to the ingress of fine particles which have been removed from their original positions in the aquifer by high groundwater velocities induced by pumping. In addition to reducing the efficiency of the well, such fine particles can also cause abrasion of the pump. Chemical fouling can result from the precipitation of minerals caused by a change in the chemical conditions between the aquifer and water in the well, for example, caused by degassing of carbon dioxide from the water in the well which is at a lower pressure than in the aquifer, or mixing of shallower, oxygenated groundwater

Figure 5.18 Buildup of iron bacteria on the intake of a submersible pump at Great Heck Borehole, Triassic Sherwood Sandstone, East Yorkshire, England. The rising main is 15 cm in diameter. The biofilm has accumulated in the form of large compact mass of gel.
Source: Photo courtesy of Gerd Cachandt, Arup.

with deeper, anoxic groundwaters within the well. Chemical processes can also cause corrosion of metallic well casing, leading to leaking casing joints which may allow ingress of shallower, contaminated water. Microbiological processes include the formation of biofilms (**Figure 5.18**) and the precipitation of biominerals such as ferric hydroxide, which can lead to biofouling of gravel packs, well screens and pumps. The above factors all relate to the performance of the well and the pump system, and hence can be addressed by replacing parts, cleaning operations (for example, using chlorine to treat biofouling) or, in the worst instance, decommissioning and the construction of a new well. More serious are problems related to the aquifer itself. Wells can go dry as a result of regional groundwater-level falls, which may be caused by over-abstraction, or reductions in aquifer recharge related to climate change. Water quality can decline regionally as a result of excessive use of agrochemicals, salt-water intrusion and ingress of surface waters. The latter effects are themselves often related to over-abstraction. For example, arsenic contamination of groundwaters in Bangladesh and West Bengal is thought to be driven by dissolution of arsenic-bearing iron oxy(hydr)oxide minerals, caused by ingress of surface water from rice paddy fields, which itself results from groundwater abstraction (Nickson *et al.*, 1998) (see section D1 of Chapter 8). In order to identify and deal with problems associated with groundwater abstractions, Misstear *et al.* (2006) recommend that abstraction wells should be monitored for water level (either manually or automatically using pressure sensors and data recorders), discharge rate (i.e. with a flow meter), water chemistry and microbiology, via a sampling tap or line in the well head. Periodic CCTV inspection of the borehole, and step pumping tests, are also recommended.

5 Artificial recharge and conjunctive use of groundwater and surface water

Artificial or enhanced recharge represents an attempt to increase the amount of rainfall recharge or surface water recharge to groundwater, for example, by modifications to topography to encourage infiltration rather than runoff, or by direct injection via pumping wells. Several reasons exist for artificially recharging groundwater, including replenishment of supplies on a long-term basis, or shorter-term storage of water underground (known as Managed Underground Storage, MUS, or Aquifer Storage and Recovery, ASR). For water storage, regions can be used where the ground contains non-**potable water**, as well as natural groundwater aquifers, in order to store water in times of excess (for example, high river flows) for later extraction and usage. Use of groundwater as temporary storage facilitates **conjunctive use** of groundwater with surface water resources, which means coordinated operation of both resources to meet demand. For example, surface water supplies can be used to meet supply requirements in most years, with some water being transferred to groundwater storage via artificial recharge. In years of low precipitation where surface water supplies are inadequate, stored groundwater can be used to supplement supply (Todd, 2005). However,

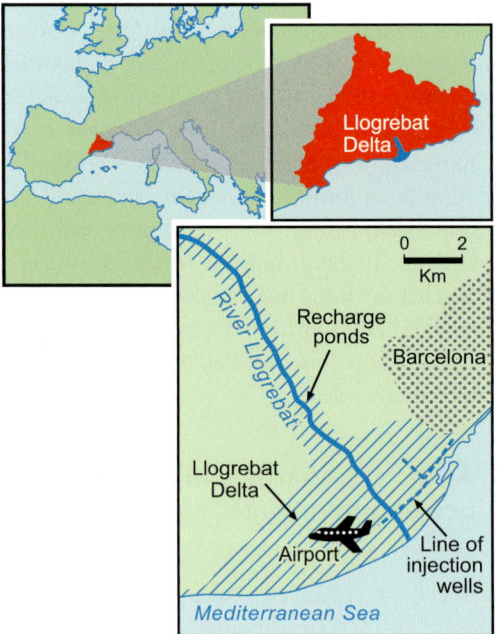

artificial recharge is also used in other contexts, for example, to prevent or limit the spread of saline intrusion (see Section G3). For example, freshwater from the Llogrebat River near Barcelona, Spain is artificially recharged to a deltaic aquifer, to limit the spread of saline water from the Mediterranean Sea, which has previously been inducted into the aquifer as a result of over-abstraction for municipal water supply (**Figure 5.19a**). Two approaches are used: water is encouraged to infiltrate the Logrebat Delta aquifer during periods of high flow via recharge ponds (**Figure 5.19b**), and is also injected directly into the deltaic aquifer via wells, to form a freshwater 'barrier' (**Figure 5.19a**).

Figure 5.19 Artificial recharge in the Llobregat Delta Aquifer, Barcelona, Spain: (a) (*left*) map showing location of river, recharge ponds, and injection wells; (b) (*below*) recharge pond.

Various issues arise that are relevant to the design and management of artificial recharge schemes. Firstly, an appropriate source of recharge water is required: this can be rainfall, storm runoff, river water, or treated wastewater. The variety of methods for enhancing recharge or injecting water into shallow or deeper aquifers, some of which have been in use since the early nineteenth century, are described by Todd (1980). One common issue is clogging of the recharge basins or wells with fine particles or biofilms, which reduces their efficiency over time; this can be dealt with by removing the finer particles by settling prior to injection, periodic drying of the recharge pond and scarification of its base; injection wells can be cleared by backflushing, and/or via chemical treatments. Another issue concerns the quality of recharge water; while some degradable pollutants (e.g. coliform bacteria) are effectively removed via filtration in granular aquifers, this process is less efficient in fracture flow systems and ineffective in karstic systems. There is increasing concern about the impact on groundwater quality of recalcitrant and emerging pollutants such as pharmaceuticals, most notably endocrine system disruptors (e.g. estrogens) (see Chapters 6 and 8). These are not removed by conventional wastewater treatments and may survive for long timespans in the environment (National Research Council, 1999); their threat to groundwater may be made much worse by artificial recharge, especially of treated wastewater. Issues specific to the relatively new technology of Aquifer Storage and Recovery (ASR) include mixing and diffusion effects with *in situ* waters, which may be non-potable, and water–rock interactions (e.g. release of arsenic and fluorine from aquifer minerals); both of these effects tend to degrade water quality and reduce the volume of usable water below that injected. The ratio of usable water to that injected in ASR is called the **recovery efficiency**, and most working ASR schemes have shown increased efficiency with each cycle of water injection and abstraction. Matrix-flow systems such as poorly cemented sandstones have generally been found to give higher recovery efficiencies than fracture flow aquifers. For further information concerning the ASR/MUS techniques, see Pyne (2005) and National Research Council (2008).

> **REFLECTIVE QUESTION**
>
> What are the key considerations when constructing a well?

F GROUNDWATER CHEMISTRY

1 Groundwater quality

Groundwater used for public supply is often extremely clean (high quality) compared with surface water, and therefore represents a valuable resource. Most of this 'clean' groundwater is 'meteoric' water (supplied from the atmosphere) from relatively shallow depths of up to a few 100 metres. Rocks at greater depth usually contain saline brines which are unsuitable for drinking. The reason for this is that if water remains in contact with rocks for very long time periods, the minerals in the rock dissolve and increase the **salinity**. Unfortunately, over-abstraction of potable groundwater can cause upwelling and mixing of these brines with fresher groundwaters (e.g. see Bottrell *et al.*, 2006). Therefore it is very important to be able to characterise groundwater quality and ensure that groundwater abstraction and land-use management prevent practices which lead to deterioration, for example, over-abstraction or inappropriate application of agro-chemicals such as nitrate.

2 Natural groundwater chemistry

As most groundwater that is used for water supply is derived from rainfall, its chemistry is a result of the chemistry of rainwater and the soils and rocks through which the water has passed, and the residence time in the aquifer. As residence time increases, groundwater tends to pick up more and more dissolved solids as a result of mineral dissolution (or weathering) reactions. Some of these reactions give waters their distinctive tastes, characterised, for example, by the recent trend to market 'natural' mineral waters as healthy. However, many mineral dissolution reactions result in a deterioration of water quality. Similarly, meteoric derived waters can mix with old, **connate water** (deposited within sedimentary rock, and perhaps saline), or magmatic waters produced by volcanic activity, or natural brines present in deeper crustal rocks. It is useful to be able to identify the sources for a particular water composition, as this tells us something about where the groundwater originates, and to distinguish anthropogenic pollutants from natural groundwater constituents.

3 Rainfall and soil water chemistry

Some constituents of groundwater are derived from rainfall and soil water. Rainfall typically contains less than 20 mg L^{-1} of total dissolved solids (TDS) (see Chapter 4), although it is less in rural areas away from industrial centres, which are sources of industrial atmospheric emissions. It also contains dissolved gases from the atmosphere, notably oxygen (O_2) and carbon dioxide (CO_2). Dissolved CO_2 dissociates to form bicarbonate **ions** (HCO_3^-) and hydrogen ions, i.e. acidity (H^+):

$$H_2O + CO_2 \text{ (g)} = H_2CO_3 \text{ (aq)}$$
$$= H^+ \text{ (aq)} + HCO_3^- \text{ (aq)} \quad [5.5]$$

which reduces the pH below neutral (pH 7). The equilibrium pH of rainwater in contact with CO_2 at current atmospheric levels is pH 5.7 (Freeze and Cherry, 1979). Dissolved oxygen makes the rainwater an oxidising solution as well as being acidic. When a well-developed soil horizon is present, bacteria in the soil produce CO_2 and increase the amount that is dissolved in the water, reducing the pH further. Where the atmosphere contains oxides of sulphur as a result of industrial emissions or volcanic activity, these also dissolve in rainwater, producing more H^+ ions, which can cause the pH to fall to between 3 and 4; this is known as acid rain (see Chapter 4). As a result of its dissolved constituents, water recharging aquifers from the soil zone is both acidic and oxidising and it is these elements which drive many of the mineral dissolution reactions described below.

4 Mineral dissolution (weathering) reactions

The composition of groundwater derived from rainfall varies according to the lithology of the aquifer and the residence time, due to mineral dissolution reactions (also known as mineral weathering reactions). Such reactions result in a characteristic suite of dissolved ions and lead to different TDS for groundwaters, according to the geology. Both TDS and ion concentrations are given on branded mineral waters and it is an interesting exercise to try to relate these to the geology of the area from which the water originates. Another property of groundwaters that results from geology is their hardness, which reflects the content of cations Ca^{2+} and Mg^{2+} that can precipitate and cause mineral scaling, for example, as a result of heating or reaction with soaps. Scaling causes problems with pipe blockages as well as producing a characteristic 'water mark' or 'scum line' in baths and sinks when soaps are used. Hard waters can be recognised in domestic use because they require far more soap in order to produce suds (e.g. the instructions on washing powder indicate that larger amounts are required for hard water). Hardness is usually expressed as mg L^{-1} equivalent dissolved $CaCO_3$, with waters containing < 60 mg L^{-1} classified as soft, 60–150 mg L^{-1} as medium and > 150 mg L^{-1} as hard; hard waters often have to be treated to reduce their hardness before they can be used in industrial processes, although they are quite potable and may represent a useful source of dietary calcium and magnesium. The key mineral dissolution reactions that control the major components of most natural groundwaters are described below.

4.1 Carbonate dissolution

$$CaCO_3(s) \text{ (calcite)} + H^+ \text{ (aq)}$$
$$= Ca^{2+} \text{ (aq)} + HCO_3^- \text{(aq)} \quad [5.6]$$

Where there are carbonate minerals such as calcite ($CaCO_3$) or dolomite ($MgCa(CO_3)_2$) within the aquifer (even a small amount) they tend to dominate the groundwater chemistry. This is because these weathering reactions are relatively fast. Resulting groundwaters are called 'calcium-magnesium-bicarbonate' waters because the major ions are calcium and/or magnesium and bicarbonate. Usually, the final bicarbonate concentration in such waters is between 170 and 260 mg L^{-1}, giving TDS between ~300 and 700 mg L^{-1}; the higher concentrations result where atmospheric carbon dioxide can easily reach the groundwater, which allows more H^+ to be produced as it is consumed by reaction 5.6 (Freeze and Cherry, 1979). As the dominant cations are calcium and magnesium, which precipitate with bicarbonate on heating, these waters are classified as hard. This type of hardness can be reduced by boiling the water, and hence is known as temporary hardness. Groundwaters dominated by carbonate dissolution are common in the shallow subsurface and in groundwaters with short residence time, because of the ubiquitous occurrence of carbonate minerals in soils.

4.2 Gypsum and halite dissolution

Where rock salts such as gypsum ($CaSO_4.2H_2O$) and halite (NaCl) are present these will dominate the groundwater chemistry, e.g. gypsum dissolution:

$$CaSO_4.2H_2O \text{ (s) (gypsum)}$$
$$= Ca^{2+} \text{ (aq)} + SO_4^{2-} \text{ (aq)} + 2H_2O \quad [5.7]$$

As such salts are very soluble, this can lead to very high concentrations of ions in groundwater; TDS of up to 5000 mg L^{-1} or more, which makes these 'calcium-magnesium-sulphate' waters unpotable. However, highly soluble salts will not be present in active flowing zones in groundwater because they will have already been dissolved out. It is more common to encounter the effects of such salts where over-abstraction causes water from less permeable units to enter aquifers. Given their high content of cations such as calcium, these waters are classified as hard in that they precipitate with soaps. However, unlike 'temporary hardness', where calcium/magnesium ions are balanced by bicarbonate ions, here they are balanced by sulphate and choride ions, so this hardness cannot be removed by boiling the water and is known as permanent hardness. Despite the name, permanent hardness can be removed by passing the water through ion exchange resins known as water softeners.

4.3 Silicate weathering

In rocks comprised of silicate minerals (e.g. sedimentary rocks such as quartz sandstones, and most igneous rocks), silicate mineral weathering reactions become important controls on groundwater chemistry (Garrels and Mackenzie, 1971):

Aluminosilicates:

$$2NaAlSi_3O_8 \text{ (s) (albite, Na-feldspar)}$$
$$+ 2H^+ \text{ (aq)} = Al_2Si_2O_5(OH)_4 \text{ (s)}$$
$$\text{(kaolin clay)} + 2Na^+ \text{ (aq)}$$
$$+ 2H_4SiO_4 \text{ (aq)} \quad [5.8]$$

Non-aluminous silicates:

$$Mg_2SiO_4 \text{ (s) (olivine)} + 4H^+ \text{ (aq)}$$
$$= 2Mg^{2+} \text{ (aq)} + H_4SiO_4 \text{ (aq)} \quad [5.9]$$

Ions derived from silicates will be present in the resulting groundwater (e.g. depending on the type of silicate, Na^+, K^+, Mg^{2+} or Ca^{2+}), but often in relatively low concentrations because these reactions are relatively slow. In the absence of other reactions, the major anion present is bicarbonate HCO_3^- derived from rainfall and soil zone processes. Where silicate weathering reactions are limited, giving overall low TDS of < 100 mg L^{-1}, the resulting types of water can be marketed as 'pure' and represent a high proportion of branded mineral waters; these waters are very soft. However, in many silicate rocks such as volcanic rocks and sandstones the groundwater may have longer residence times, producing intermediate TDS values and hardness; many also contain sulphide or sulphate minerals which can produce much higher TDS values (see reaction 5.11 below).

4.4 Cation exchange

$$ClayNa\ (s) + Ca^{2+}\ (aq)$$
$$= Clay_2Ca\ (s) + 2Na^+\ (aq) \qquad [5.10]$$

Cation exchange reactions are relatively important in sedimentary rock aquifers such as sandstones, which contain some clay minerals. As these rocks were deposited in the ocean basins, such clays have their exchangable surface sites saturated with sodium (Na^+) and potassium (K^+) ions because these predominate in seawater. As meteoric water ingresses, cations like Ca^{2+} and Mg^{2+} derived from, for example, carbonate weathering in the soil zone, are progressively exchanged onto clay minerals, and Na^+ and K^+ are released in their place as the groundwater moves deeper into the aquifer. This process is called **water softening** because the 'sodium/potassium waters' produced are relatively soft (i.e. these cations do not react with soaps). A good example of this sequence of reactions is described by Price (1993) for the Cretaceous Chalk aquifer of Northern Europe, and results in the groundwater chemistry illustrated in **Figure 5.20**. As cation exchange reactions

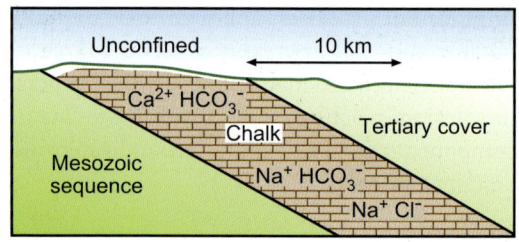

Figure 5.20 Schematic diagram of dominant chemistry in the Cretaceous Chalk aquifer of southern England. Calcium and bicarbonate ions in the unconfined aquifer arise from carbonate dissolution. In the confined aquifer calcium ions are exchanged for sodium and potassium ions present on clay minerals. Deeper within the confined aquifer connate waters derived from seawater are present.

simply replace one ion with another, they do not themselves increase the TDS load of groundwaters, which will therefore depend on other mineral dissolution reactions.

4.5 Oxidation reactions

As infiltrating water contains dissolved oxygen, some minerals such as sulphides of iron (Fe) or manganese (Mn) in the soil or deeper aquifer can be oxidised by bacteria, to produce sulphate and acidity:

$$FeS_2\ (s)\ (pyrite) + 3.5O_2\ (aq) + H_2O$$
$$= Fe^{2+}\ (aq) + 2SO_4^{2-}\ (aq)$$
$$+ 2H^+\ (aq) \qquad [5.11]$$

Iron (II) produced is oxidised to iron (III), which, if oxygen is still present, is then precipitated as a hydroxide:

$$Fe^{2+}\ (aq) + 0.25O_2\ (aq) + 2.5H_2O$$
$$= Fe(OH)_3\ (s) + 2H^+\ (aq) \qquad [5.12]$$

Where carbonate minerals are present, the acidity produced is neutralised by carbonate dissolution, leading to groundwater dominated by calcium, sulphate and bicarbonate ions; this may

Figure 5.21 Acid mine drainage showing precipitation of ferric hydroxides, emerging from the Kalavassos Copper Mine, Adit, southern Cyprus.

Source: Photo courtesy of Rachael Spraggs.

remain potable but will contain both temporary and permanent hardness. Where no carbonate is present the groundwater may become acidic, with the pH dropping as low as 4. This can produce unpotable waters because of the tendency to dissolve trace elements such as arsenic (Nordstrom *et al.*, 2002). Where the oxygen in the water is all used up, such that the groundwater becomes anoxic, iron will remain in solution as Fe^{2+}, but this will then precipitate as iron (III) hydroxide where such waters come into contact with oxygen. An extreme case of this occurs where acidic mine drainage emerges (**Figure 5.21**; see also **Figure 4.13**). The resulting iron (III) hydroxide precipitated can smother benthic organisms, destroying stream ecosystems. Where put into supply, Fe and Mn precipitation can stain porcelain, so waters have to be oxified by treatment. A good example illustrating the effects of oxygen consumption on groundwater chemistry of a carboniferous sandstone aquifer is given by Banks (1997). Sulphide weathering reactions typically result in groundwaters with TDS of 400 to 1000 mg L^{-1}, but can ultimately produce TDS values of several thousand mg L^{-1} where rocks are exposed to excess oxygen via mining processes and produce acid mine drainage (see Chapter 4, section F3 for a more detailed treatment of acid mine drainage).

> **REFLECTIVE QUESTION**
>
> Why does the chemistry of groundwater naturally vary from one place to another?

G GROUNDWATER POLLUTION

1 Groundwater pollutant properties

There are a very wide range of groundwater pollutants, including those that can harm humans, animals or plants and those that have detrimental effects on natural ecosystems where water emerges into rivers. Pollutants enter groundwater as a result of human activities such as mining, industry, waste and sewage disposal, agriculture, or where excessive water is pumped out for supply purposes, causing influx of 'naturally' contaminated water (e.g. salt-water from the sea or from deeper, saline groundwaters).

Two important characteristics influence the extent of detrimental effects of groundwater contaminants. These are the extent of interaction and attachment to the solid matrix (known as **sorption**) and the rate of transformation of contaminants into less harmful substances, for example, by biological processes. The extent of sorption depends on the mineralogy of the aquifer and on the type of contaminant. Good sorbents include clay minerals and iron minerals such as those responsible for the red coloration of many sandstones, as well as the natural organic matter that is present within soils and rocks. Generally, positively charged ionic contaminants, for example, toxic metal ions such as cadmium, are strongly adsorbed by the negatively charged surfaces of clay minerals and iron oxides; neutral contaminants such as organic chemicals may be adsorbed by natural organic matter. Negatively charged anionic contaminants such as nitrate, chromate and arsenate ions are typically not adsorbed and migrate freely in groundwater. Strongly adsorbed contaminants are often effectively immobilised because the vast majority of the contaminant is attached to the solid phase. However, re-mobilisation of such adsorbed contaminants can occur as a result of chemical changes such as changes in the concentration of dissolved oxygen in the groundwater, for example, arsenic sources from the natural minerals within aquifers in Bangladesh (Nickson *et al.*, 1998) and radioactive technetium released from nuclear industry-contaminated sites in the UK (Burke *et al.*, 2005).

Transformation of organic contaminants into less harmful substances is usually via biological processes, although radioactive decay is also important for radioisotopes. Substances that decay rapidly are less of a problem than those that persist for long periods or indefinitely because contaminants may be unable to reach groundwater wells before decay takes place. For example, many human pathogens cannot survive for very long in groundwater and, provided wells are sited far enough away from sources such as manure spreading and latrines, passage through the subsurface acts as an effective water treatment. Unfortunately, some types of contaminants, such as the anti-knock additive in unleaded petrol, methyl-tertiary butyl ether (MTBE), are not transformed quickly enough and so can reach supply wells. Biological decay processes are often dependent on local chemical conditions, such as the presence or absence of oxygen or nutrients (Spence *et al.*, 2001), which can result in highly variable or unpredictictable decay rates.

2 Contaminant source characteristics

It is useful to classify sources of groundwater contamination according to their spatial characteristics. Spatially, the size of a source of contamination varies hugely, from a few metres for a leaking septic tank, for example, to several hundred square km for a large area of agricultural chemical application to farmland, but generally we can speak of point sources (relatively small compared to the thickness of the aquifer) and diffuse sources, which are large compared with this thickness. Point sources give rise to 'plumes' of contamination which migrate away and spread out, whereas diffuse sources give rise to very large areas of groundwater contamination (**Figure 5.22**). Generally, point sources may lead to the shutdown of individual pumping wells, but diffuse sources can cause whole aquifers to be damaged.

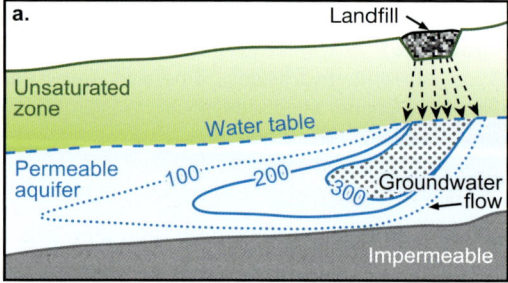

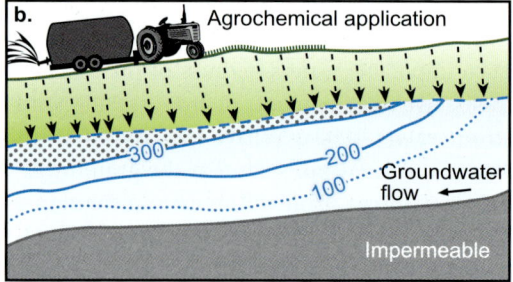

Figure 5.22 Contaminant source characteristics: (a) landfill site representing point source; (b) agrochemical application representing diffuse source. The numbers on contour lines are contaminant concentrations in mg L^{-1}.

3 Types of groundwater contaminant

The following broad groups of groundwater contaminants can be recognised.

3.1 Radionuclides

Elements such as uranium, plutonium, and some isotopes of strontium, caesium and technetium are subject to radioactive decay. They are produced mainly by the nuclear industry in mining and processing uranium ore, fuel reprocessing, and waste and weapons disposal. As the radioactive decay process produces ionising radiation it is especially damaging where the elements get inside living cells by consuming contaminated groundwater. Some radioisotopes, such as those of strontium (i.e. $^{90}Sr^{2+}$), are positively charged and are thus subject to sorption reactions. However, many others, such as uranium, form negative ions (e.g. uranyl UO_4^-), which are highly soluble and can thus migrate for long distances in groundwater.

3.2 Trace metals

This group includes 'metalloid' elements such as arsenic (As), as well as metals such as lead (Pb) and copper (Cu) which are toxic in relatively low concentrations in groundwater. The major sources are mining, industrial wastewater streams, runoff from industrial solid wastes (e.g. slagheaps) and agriculture. However, some of the most widespread groundwater contamination from these types of contaminants is natural, but made worse by groundwater abstraction, such as arsenical contamination of groundwaters in Bangladesh and West Bengal (Nickson *et al.*, 1998) (see Chapter 8). The main problem with trace metals such as Pb is their ability to **bioaccumulate** within organisms.

3.3 Nutrients

This group includes ions and compounds containing nitrogen (N) and phosphorus (P), such as nitrate (NO_3^-) and ammonium (NH_4^+). The main sources of these contaminants are

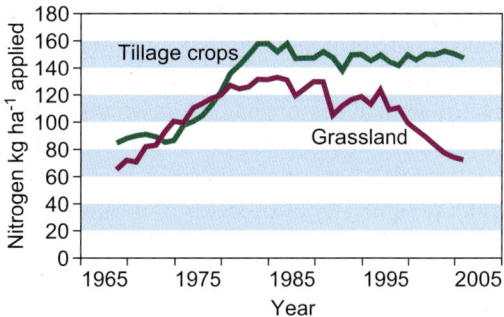

Figure 5.23 Trends in nitrate fertiliser application in the United Kingdom.
Source: Data from National Fertiliser Survey, after Allshorn, 2008.

agriculture (artificial fertiliser application and cultivation of virgin soils) and organic wastes such as sewage (leaking sewers and sewage disposal) and landfill sites. The main health effect of nitrate in drinking water is methemoglobinemia (Schwartz and Zhang, 2003) – 'blue baby syndrome' (see Chapter 8). Nitrate is currently threatening many large aquifers in developed countries as a result of the extensive increases in mineral fertiliser applications to arable land (**Figure 5.23**) in the latter part of the last century. The World Health Organization recommends a maximum concentration in drinking water of 45 mg L^{-1}. Many wells have had to be shut down for exceeding this concentration. Denitrification of abstracted groundwater is possible, but it is expensive. As an alternative, many large water suppliers currently pursue a policy of 'blending' abstracted waters from various sources in order to reduce the concentration below the limit. However, the sustainability of this policy is unclear, given the continuing rising trends seen for nitrate in some aquifers.

3.4 Major ions (salts)

This group includes ions such as Ca^{2+}, Mg^{2+} and Na^+, SO_4^{2-}, Cl^-. Extremely high concentrations of these make the water unfit for human consumption, and for many industrial uses. The health effects of consuming such water are usually not so serious or long term. However, very high concentrations of these substances give the water a bad taste, and ultimately make it unpotable. The main sources of major ion contaminants to groundwater are industrial or domestic waste (in landfills), natural saline waters being drawn into the aquifer by over-abstraction, and mixing with deep groundwater (brines). Where saline waters are drawn into groundwater aquifers from the sea this is known as marine incursion (see Ergil, 2000, and the example of the Llogrebat Delta in Barcelona illustrated in **Figure 5.19**). Many coastal aquifers have been seriously damaged by this process. However, salinisation can also occur as a result of evaporative recycling of irrigation water, where water is continuously extracted, used for irrigation and recharged to groundwater (Milnes and Renard, 2004). Salt concentrations increase progressively because each irrigation results in a proportion of the water being evaporated or transpired. Salinisation problems are among the greatest threats to groundwater worldwide, and their effects are likely to have massive socio-economic consequences as pressures on groundwater increase as a result of climate changes over the coming century.

3.5 Organic compounds

Organic compounds (those containing carbon, hydrogen and sometimes oxygen, sulphur and nitrogen) represent a complex group of groundwater contaminants. There are a huge range of sources of organic compounds: they include fossil biomass such as petroleum and coal, as well as biological decay of modern biomass (e.g. in landfills). Two main subgroups can be recognised according to solubility: compounds that are relatively insoluble in water, which thus form a separate liquid phase within the subsurface, called Non Aqueous Phase Liquids (NAPLs), and organic compounds that are soluble in water. The behaviour of NAPLs depends on whether their specific gravity is greater or less than that of water. Those that are less dense – Light Non Aqueous Phase Liquids (LNAPLs) – float on the water table; those that are more dense than water – Dense Non Aqueous Phase Liquids (DNAPLs) – sink through the groundwater to the base of the aquifer (**Figure 5.24a** and **b**). As even NAPLs are slightly soluble, they progressively dissolve (**Figure 5.24c**), but this process can result in the contamination of huge volumes of groundwater and take many years.

Organic contaminants can be also subdivided, depending on their chemical structure, into:

- Aliphatic hydrocarbons – these are chains of carbon atoms with hydrogen attached. They tend to have low solubility and are less dense than water (i.e. they form LNAPLs). The smallest molecules (e.g. methane CH_4 and ethane C_2H_6) are gases at 20°C, middle-sized molecules such as octane (in petroleum) are volatile liquids and the heavier molecules are tarry non-volatile liquids. The main sources of these compounds are leaks and spills at oil refineries, petrol stations, airfields and from pipelines.
- Aromatic hydrocarbons – these compounds contain at least one benzene ring (a ring structure of six carbon atoms with specific properties). They are more soluble in water than aliphatic compounds and are extremely toxic. The most common groundwater contaminant compounds in the group are benzene, toluene, ethylbenzene and xylene – these are called the BTEX compounds. These are slower to degrade in groundwater than aliphatic compounds, and so are often the most significant contaminants. Another set of aromatic compounds that are often found in groundwater are the Poly-Aromatic Hydrocarbons (PAHs), which are hydrocarbons containing several benzene rings, often derived from coal processing. These compounds are very toxic to humans and aquatic ecosystems.
- Halogenated hydrocarbons – where some of the hydrogen atoms have been replaced by chlorine, bromine or fluorine atoms, are denser than water and often form DNAPLs, although they

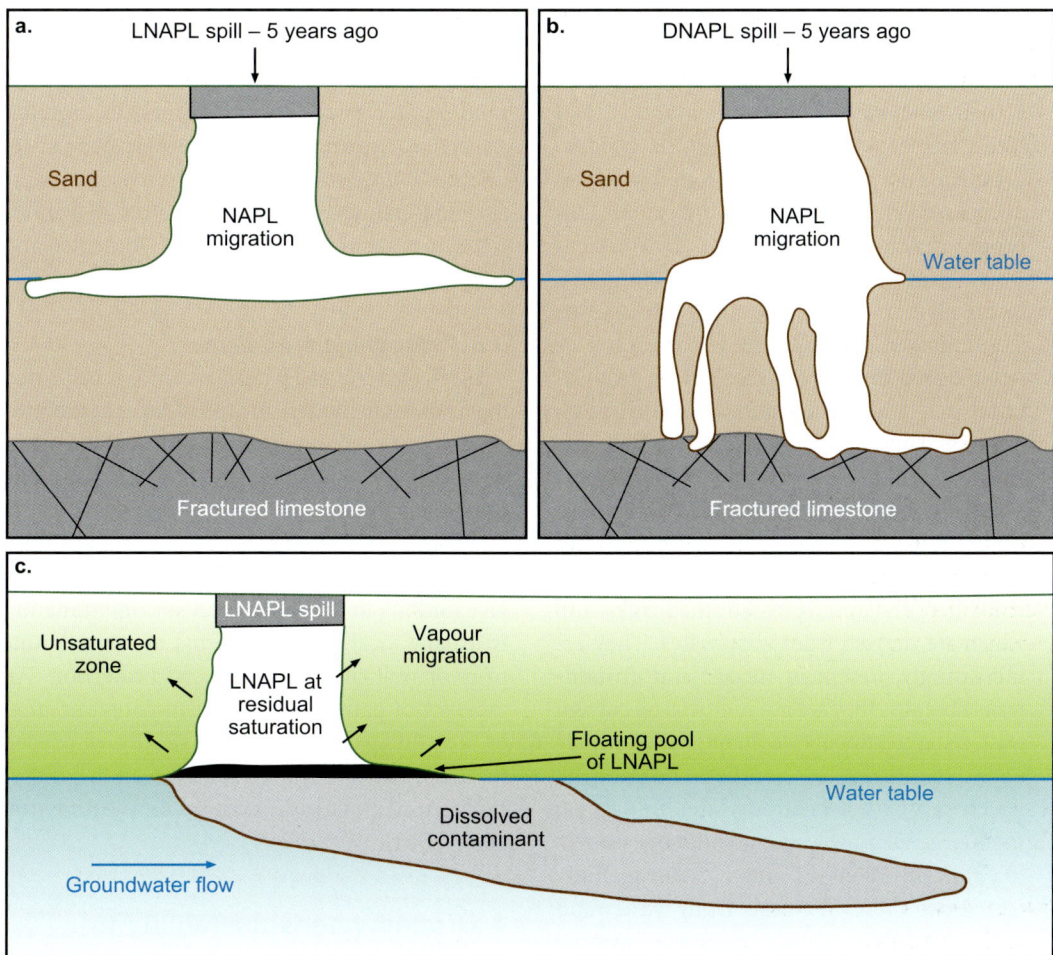

Figure 5.24 Behaviour of Non Aqueous Phase Liquids (NAPLs) in groundwater aquifers: (a) dense NAPLs; (b) light NAPLs; (c) dissolution of LNAPL to form a plume of contaminated groundwater.

may be soluble. Halogenated aliphatic compounds such as trichloroethane (TCE) are used as solvents for dry cleaning and as cleaning agents in metal manufacturing. TCE is quite soluble in water and is the biggest cause of groundwater abstraction well shutdown in developed countries such as the USA. Other important groundwater contaminants include halogenated aromatic compounds (e.g. chlorobenzene), which is extensively used for agricultural purposes, and polychlorinated biphenyls (PCBs), which are chlorinated hydrocarbons containing two benzene rings and are extremely toxic. These have often leaked into groundwater from large-scale electrical equipment such as transformers present at old industrial sites (they have now been phased out of use, due to their high toxicity).

- Biological degradation products – compounds produced by the degradation of biological materials, including organic acids characterised by the COOH (carboxylate) group, alcohols, characterised by the OH (hydroxyl) group and esters (which contain the C-O-C linkage), as well as more complex molecules such as sugars and proteins. These compounds are quite water soluble; one environmental effect is their **biochemical oxygen demand** (BOD) (see Chapter 4), which means that microorganisms use them as a food source, and use up the oxygen present in groundwaters. This produces a chemically reducing environment which can cause inorganic contaminants to dissolve or change into more toxic forms (e.g. sulphates to sulphides). The main source of biological decay products entering groundwater are landfill sites and sewage disposal lagoons, slaughterhouses and tanneries.

- Environmentally persistent pharmaceutical pollutants – these are relatively new chemical compounds which include pharmaceutical and personal care products, such as antibiotics, painkillers and antidepressant medicines, and which are derived from wastewater. They are increasingly present in surface and groundwaters because they are not removed by current wastewater treatment (see Chapter 4). Many are endocrine disruptors (synthetic hormones), which can have a relatively large impact on microorganisms, wildlife and humans, even in low concentrations $<< 1mg\ L^{-1}$. Our inability to remove these chemicals from wastewater streams and their unpredictable environmental effects are likely to be a key challenge in future water management. Section D3 in Chapter 8 provides more details about these new pollutants and potential health impacts.

The extent to which organic compounds are adsorbed or undergo biological decay once they are dissolved in the groundwater is very variable and this group of compounds have presented great difficulties where spills from industrially contaminated sites have degraded large volumes of groundwater and impacted upon surface water or groundwater abstraction wells up to several kilometres distant. A good example where this occurred is the Hyde Park Landfill Site in Niagara, New York State, where chemical wastes containing chlorinated aromatic hydrocarbons were dumped in an unlined landfill site in 1954. These DNAPLs and LNAPLs migrated downward into a glacial sand aquifer underlain by a limestone aquifer. As well as creating problems for housing subsequently built near the site, these contaminants also migrated several hundred metres in groundwater to reach the Niagara River (Domenico and Schwartz, 1998).

3.6 Pathogenic organisms

Viruses, bacteria and parasites can be transported in groundwater and there are instances of groundwater contamination producing most of the major diseases that are carried in water (typhoid, cholera, polio, hepatitis) (see Chapter 8). The sources are usually human or animal sewage or landfills. Fortunately, pathogenic microbes cannot survive for long in the subsurface, so contamination is usually local – and incidents are usually due to poor well siting or well head protection. An incident of pathogenic organism contamination of a municipal groundwater supply which led to 2300 cases of illness and seven deaths at Walkerton, Ontario is reported by Worthington *et al.* (2002).

4 Groundwater vulnerability to pollution

Groundwater vulnerability to pollution is the likelihood of contaminants reaching groundwater after introduction at the ground surface. Vulnerability to contamination from the surface is an increasingly important issue as more stringent regulations on water quality and the need to protect groundwater resources clash with the economic incentives to use plant-protection products (fertilisers, pesticides, herbicides) in the agricultural industry. Vulnerability is controlled by complex interactions between anthropogenic activities, natural stresses (rainfall, runoff) and intrinsic characteristics of the aquifer system. The intrinsic groundwater vulnerability is a function of the following parameters: the nature and

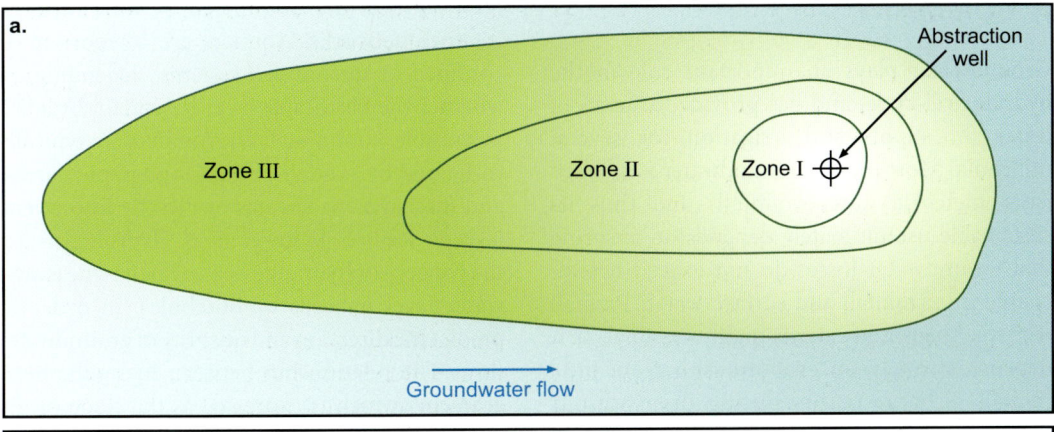

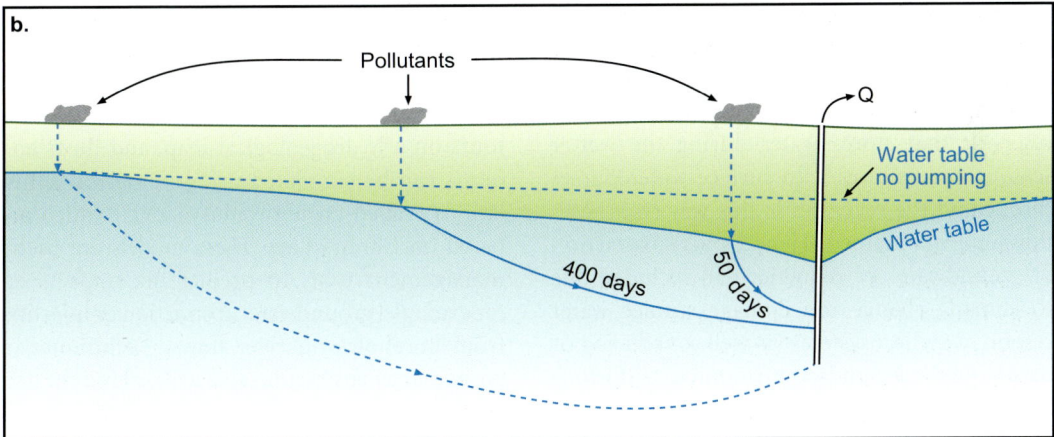

Figure 5.25 Well-head or source protection zones as defined by the Environment Agency, England: (a) plan view; (b) cross sectional view. Zone I: 50 day travel time to well; Zone II: 400 day travel time to well; Zone III: well catchment area. Zones are used for land use planning such as waste disposal site licensing.

thickness of any overlying confining layer, depth to groundwater and characteristics of the aquifer materials. Generally, fractured aquifers are relatively vulnerable, as compared with systems where flow is mainly via the matrix pores. Approaches to groundwater vulnerability assessment include semi-quantitative empirical (index-based) and statistical methods on large scales, usually used to produce maps of intrinsic groundwater vulnerability, and numerical groundwater flow and solute transport modelling methods to predict vulnerability of specific wells (Robins, 1998; Pavlis et al., 2010). This allows well head or source protection zones to be defined around groundwater wells (Frind et al., 2006); potentially polluting land uses such as petrol stations and landfills are then restricted within these zones. **Figure 5.25** illustrates how source protection zones can be defined, based on the ease with which contaminants can reach the supply well.

REFLECTIVE QUESTION

What are the main types of groundwater contaminants and what influences their detrimental effects?

H SUMMARY

Groundwater plays an important role in the hydrological cycle, and has provided sources of water for supply and irrigation for several millennia. Most potable groundwater is sourced from meteoric water (rainfall) and thus its sustainable use ultimately depends on ensuring that groundwater losses do not exceed net recharge from rainfall and surface water. Rainfall recharge to groundwater is highly seasonal, as it depends strongly on evapotranspiration; little rainfall recharge occurs during drier seasons in temperate climates. In arid climates, recharge from temporary surface water bodies during the wet season is the main source. As well as acting as a direct supply, groundwater also acts as a reservoir that stores water during the wetter seasons and releases it to surface watercourses throughout the year, and in this way river baseflows are maintained. Hence, over-abstraction of groundwater or reductions in recharge due to climatic change can damage surface water resources and ecosystems, as well as reducing or damaging the groundwater resource. Unfortunately, many of the world's groundwater aquifers are currently being over-abstracted, and combined with reductions in recharge due to climatic change in some areas, this is likely to lead to severe future ecological, sociological and humanitarian consequences.

The geographical distribution of groundwater depends on the physical properties of the ground as well as the availability of hydrological recharge. Generally, hydrologists distinguish aquifers and non-aquifers according to whether useful amounts of groundwater can be abstracted from wells or springs. The main characteristic of an aquifer is that it must be sufficiently permeable in order to allow groundwater to be extracted. Generally, sediments of relatively large grainsize (sands and gravels) are suitable but silts and clays are not. With consolidated rocks, fracture state and the resulting fracture connectivity are relatively more important than grainsize, and many rock aquifers are dominated by flow through fracture networks. Aquifers are categorised as confined or unconfined systems, depending on whether the permeable layer is overlain by a less permeable, confining layer. Aquifers are typically recharged by rainfall in high topographic areas, and discharge via springs or directly into rivers, lakes or sea beds in lower areas. Hydrologists use the concept of hydraulic head, which is measured water level in wells or boreholes, in order to predict the direction and quantity of groundwater flow. The relationship between hydraulic head gradient (which is expressed as the slope of the water table in unconfined aquifer systems) and flow is linear in most subsurface systems; this relationship is known as Darcy's Law.

Hydraulic heads can be represented as contours on a hydrogeological map, and flows and heads can be simulated using numerical simulations (called groundwater models) which are based on Darcy's Law. These models are useful management tools to predict the impacts of, for example, groundwater abstraction or injection from boreholes on river flows. Techniques of groundwater abstraction and artificial re-injection (recharge) have been developed over many years and hence involve a range of technologies. Older forms which are still appropriate in rural and developing country contexts use manual technologies and gravity-driven flow; modern approaches more typically use wells which are made from drilled boreholes with electrical pumps. The efficiency of a well depends on its geometry (diameter and screen length) as well as the pumping rates; excessive pumping rates for a given size lead to poor efficiency.

Groundwater derived from meteoric water is often of relatively high quality compared to surface water, as a result of the natural filtration capacity of the subsurface. However, it is important to recognise that such groundwater contains a range of natural substances from mineral dissolution. Which mineral dissolution reactions are operative depends on the mineralogy of the aquifer. While many reactions give bottled mineral waters their distinctive taste, others can make the water unpotable. Similarly, mixing with deeper waters which are not derived from rainfall (e.g. original waters present when the rock was formed) and induction of seawater into groundwater aquifers can also reduce quality. Furthermore, a range of natural and artificial substances are introduced into groundwater as a result of anthropogenic activities such as agriculture and industry, and increasingly via artificial recharge. These include heavy metals and radioisotopes, nutrients, major ions (salts), pathogens and organic compounds (including pharmaceutical products), some of which do not degrade and interfere with the operation of biological systems. Both intrinsic groundwater vulnerability to pollution and the vulnerability of individual sources (e.g. wells) depends on the properties of the aquifer and the overlying soil, with fracture flow systems being relatively vulnerable as compared with matrix flow systems, because of relatively rapid groundwater movement and poor filtration. A variety of approaches have been developed for characterising the vulnerability to pollution of groundwater aquifers and of specific well abstractions.

FURTHER READING

Anderson, M.P. and Woessner, W.W. 1992. *Applied groundwater modelling*. Academic Press; New York.

This book provides an excellent introduction to the technique of numerical simulation of groundwater flows as used in hydrogeological practice. Suitable for quantitative higher level undergraduate and master's level courses.

Freeze, R.A. and Cherry, J.A. 1979. *Groundwater*. Prentice-Hall; New Jersey.

Comprehensive textbook on physical and chemical hydrogeology; notable for natural groundwater chemistry coverage, which is not always included in more recent textbooks.

Misstear, B., Banks, D. and Clark, L. 2006. *Water wells and boreholes*. John Wiley and Sons; Chichester and New York.

Covers all aspects of water well drilling and well design; lots of good photographs.

Price, M. 2002. *Introducing groundwater (2nd edition)*. Nelson Thornes; Cheltenham.

This book provides a good basic introduction to physical hydrogeology for undergraduate students.

Pyne, D. 2005. *Aquifer storage recovery: A guide to groundwater recharge through wells (2nd edition)*. ASR Press; Florida.

Comprehensive coverage of the technical aspects of ASR from a United States perspective.

Schwartz, F.W. and Zhang, H. 2003. *Fundamentals of ground water*. John Wiley and Sons; Chichester and New York.

This book provides a balanced coverage of both physical and chemical hydrogeology from a United States perspective; suitable for undergraduate and master's level courses.

Todd, D.K.T. 2005. *Groundwater hydrology (3rd edition)*. John Wiley and Sons; Chichester and New York.

Thorough coverage of physical hydrogeology principles including well testing theory; suitable for quantitative higher level undergraduate and master's level courses.

Younger, P.L. 2007. *Groundwater in the environment*. Blackwell; Oxford.

Thorough coverage of groundwater hydrology fundamentals, catchment hydrology, interaction with surface water, and groundwater management.

Classic papers

Hubbert, M.K. 1940. The theory of ground-water motion. *Journal of Geology* 48: 785–944.

This seminal paper describes the application of Bernoulli's concept of 'fluid potential' to groundwater.

Penman, H.L. 1948. Natural evaporation from open water, bare soil and grass. *Proceedings of the Royal Society of London, A* 194: S120–145.

Highly influential paper on estimation of evaporation from meteorological data; gave rise to methods for assessing sustainable groundwater resources.

Theis, C.V. 1935. The relation between the lowering of the piezometric surface and rate and duration of discharge of a well using groundwater storage. *Transactions of the American Geophysical Union* 2: 519–524.

Origin of the main approaches used to measure aquifer properties from response to pumping.

PROJECT IDEAS

- Measure water level in a well or borehole through the year (you can use a dip-meter if available, but a weight attached to a measuring tape is a good substitute). Plot the water level against the date through the year and see if you can identify the main recharge period (i.e. when the water level is increasing). If more than one well/borehole/pond is available, compare the water level responses to see how they vary for wells in different geological deposits (identify these using a geological map of the area if available) and with the topography (normally, areas that are higher with respect to valley base level will show larger variations).

- Find a local aquifer using a geological/hydrogeological map. Identify, for example, from map information, its relationship to nearby surface water features (springs, wetlands, streams, lakes, the sea); i.e. is the aquifer likely to interact with these features? Produce a list of all possible environmental impacts of taking water from this aquifer via pumped wells. What strategies might be adopted to mitigate any adverse impacts of groundwater abstraction?

REFERENCES

Allen, D.L., Brewerton, L.J., Coleby, L.M., Gibbes, B.R., Lewis, M.A., MacDonald, A.M., Wagstaff, S.J. and Williams, A.T. 1997. *The physical properties of major aquifers in England and Wales.* British Geological Survey Technical Report WD/97/34.

Allshorn, S.J.L. 2008. Flow and solute transport in the unsaturated zone of the Chalk in East Yorkshire. Unpublished PhD thesis; University of Leeds.

Anderson, M.P. and Woessner, W.W. 1992. *Applied groundwater modelling.* Academic Press; New York.

Banks, D. 1997. Hydrogeochemistry of Millstone Grit and Coal Measures groundwater in South Yorkshire. *Quarterly Journal of Engineering Geology* 30: 237–256.

Bottrell, S.H., West, L.J. and Yoshida, K. 2006. Combined isotopic and modelling approach to determine the source of saline groundwaters in the Selby Triassic Sandstone aquifer, UK. *In:* Barker, R.D. and Tellam, J.H. (eds) *Fluid flow and solute movement in sandstones: the Onshore UK Permo-Triassic Red Bed Sequence.* Geological Society, London; Special Publications 263: 325–338.

Burke, I.T., Boothman, C., Lloyd, J.R., Mortimer, R.J.G., Livens, F.R., and Morris, K. 2005. Effects of progressive anoxia on the solubility of technetium in sediments. *Environmental Science and Technology* 39: 4109–4116.

Clark, L. 1977. The analysis and planning of step drawdown tests. *Quarterly Journal of Engineering Geology* 10: 125–143.

Darcy, H. 1856. *Les fontaines publiques de la ville de Dijon.* V. Dalmont; Paris.

Domenico, P. and Schwartz, F.W. 1998. *Physical and chemical hydrogeology.* John Wiley and Sons; Chichester and New York.

Ergil, M.E. 2000. The salinisation problem of the Guzelyurt Aquifer, Cyprus. *Water Research* 34: 1201–1214.

European Commission. 2000. *Water Framework Directive*. Directive 2000/60/EC. OJ L 327, 22.12.2000:1–73.

European Commission. 2006. *Groundwater Directive*. Directive 2006/118/EC. OJ L 372, 27.12.2006: 19–31.

Foxworthy, B.L. 1983. Pacific North-West region. *In*: Todd, D.K. (ed.) *Ground water resources of the United States*. Premier Press; California, 590–629.

Freeze, R.A. and Cherry, J.A. 1979. *Groundwater*. Prentice-Hall; Hemel-Hempsted.

Frind, E.O., Molson, J.W. and Rudolph, D.L. 2006. Well vulnerability: a quantitative approach for source water protection. *Ground Water* 44: 732–742.

Garrels, R.M. and Mackenzie, F.T. 1971. *Evolution of sedimentary rocks*. Norton; London.

Holden, J. 2005. Darcy's Law. *In*: Lehr, J.H. and Keeley, J. (eds) *Water encyclopedia, Volume 5*. John Wiley and Sons; New York, 63–64.

Hubbert, M.K. 1940. The theory of ground-water motion. *Journal of Geology* 48: 785–944.

Jones, J.B. and Mulholland, P.J. (eds). 2003. *Streams and groundwater*. Academic Press; New York.

Miller, J.A. and Appel, C.L. 1997. *Ground water atlas of the United States, Segment 3, Kansas, Missouri and Nebraska, Hydrologic Investigations Atlas, 730-D*. U.S. Geological Survey; Virginia.

Milnes, E. and Renard, P. 2004. The problem of salt recycling and seawater intrusion in coastal irrigated plains: an example from the Kiti aquifer (Southern Cyprus). *Journal of Hydrology* 288: 327–343.

Misstear, B., Banks, D. and Clark, L. 2006. *Water wells and boreholes*. John Wiley and Sons; Chichester and New York.

National Research Council. 1999. *Hormonally active agents in the environment*. National Academy Press; Washington, DC.

National Research Council. 2008. *Prospects for managed underground storage of water*. National Academy Press; Washington, DC.

Nickson, R., McArthur J., Burgess, W., Ahmed, K.M., Ravenscroft, P. and Rahmann, M. 1998. Arsenic poisoning of Bangladesh groundwater. *Nature* 395: 338.

Nordstrom, D.K. 2002. Public health – worldwide occurrences of arsenic in ground water. *Science* 296: 2143–2145

Patterson, L.J. *et al*. 2005. Cosmogenic, radiogenic and stable isotope constraints on groundwater residence time in the Nubian Aquifer, Western Desert of Egypt. *Geochemistry Geophysics Geosystems* 6: 1–19.

Pavlis, M., Cummins, E. and McDonnell, K. 2010. Groundwater vulnerability of plant protection products: a review. *Human and Ecological Risk Assessment* 16: 621–650.

Penman, H.L. 1948. Natural evaporation from open water, bare soil and grass. *Proceedings of the Royal Society of London, A* 194: S120–145.

Price, M. 1993. The Chalk as an aquifer. *In*: Downing, R.A., Price, M. and Jones, G.P. *Hydrogeology of the Chalk of NW Europe*. Oxford University Press; Oxford, 35–58.

Price, M. 2002. *Introducing groundwater (2nd edition)*. Nelson Thornes; Cheltenham.

Pyne, D. 2005. *Aquifer storage recovery: a guide to groundwater recharge through wells (2nd edition)*. ASR Press; Florida.

Robins, N.S. 1998. Recharge: the key to groundwater pollution and aquifer vulnerability. *In*: Robins, N.S. (ed.) *Groundwater pollution, aquifer recharge and vulnerability*. Geological Society; London, Special Publications 130: 71–76.

Scanlon, B.R., Healy, R.W. and Cook, P.G. 2002. Theme issue: groundwater recharge. *Hydrogeology Journal* 10.

Schwartz, F.W. and Zhang, H. 2003. *Fundamentals of ground water*. John Wiley and Sons; Chichester and New York.

Singhal, B.B.S. and Gupta, R.P. (eds). 1999. *Applied hydrogeology of fractured rocks*. Kluwer Academic Publishers; The Netherlands, 151–168.

Spence, M.J., Bottrell, S.H., Higgo, J.J.W, Harrison, I. and Fallick, A.E. 2001 Denitrification and phenol degradation in a contaminated aquifer. *Journal of Contaminant Hydrology* 53: 305–318.

Theis, C.V. 1935. The relation between the lowering of the piezometric surface and rate and duration of discharge of a well using groundwater storage. *Transanctions of the American Geophysical Union* 2: 519–524.

Todd, D.K. 1980. *Groundwater Hydrology (2nd edition)* John Wiley & Sons; New York.

Todd, D.K.T. 2005. *Groundwater hydrology (3rd edition)*. John Wiley and Sons; Chichester and New York.

USAID. 2011. *USAID website [online]*. [Accessed on 23 August 2011]. Available from: www.usaid.gov.

West, L.J. and Truss, S.W. 2006. Borehole time domain reflectometry in layered sandstone: impact of measurement technique on vadose zone process identification. *Journal of Hydrology 319:* 143–162.

Worthington, S.R.H., Smart, C.C. and Ruland, W.W. 2002. Assessment of groundwater velocities to the municipal wells at Walkerton. *In:* Stolle, D., Piggott, A.R. and Crowder. J.J. *Proceedings of the 55th Canadian Geotechnical and 3rd Joint IAH-CNC and CGS Groundwater Specialty Conferences, Niagara Falls, Ontario, October 20–23, 2002.*

Wright, E.P. and Burgess, W.G. (eds). 1992. *The hydrogeology of the crystalline basement aquifers in Africa.* Geological Society Special Publication No. 66. Geological Society; London.

Wulff, H.E. 1968. The Qanats of Iran. *Scientific American*, April 1968, 94–105. *In:* Beaumont, P., Bonine, M. McLachlan, A. (eds). 1989. *Qanat, kariz, and khattara: traditional water systems in the Middle East and North Africa.* Middle East & North Africa Studies Press Limited; Kent.

CHAPTER SIX

Aquatic ecosystems

Lee E. Brown, Colin S. Pitts and Alison M. Dunn

LEARNING OUTCOMES

After reading this chapter you should be able to:

- understand the main groups of aquatic organisms and their interactions
- evaluate some of the differences between river and lake ecosystems
- outline the key aspects of aquatic ecosystem functional processes
- recognise some major ways in which humans have altered aquatic ecosystems.

A INTRODUCTION

Rivers and lakes, the main focus of this chapter, constitute less than 0.01% of the world's water resources and cover only ~0.8% of the Earth's surface. Wetlands (areas where the water table reaches the surface and persists long enough to support aquatic plants) can also be considered as freshwater ecosystems (see Dobson and Frid, 2009) but these are not considered in this chapter. The study of any 'ecosystem' unavoidably requires some knowledge of both the living organisms and their effective environment. Where necessary, this chapter makes reference to the aquatic environment to place understanding of the biota into relevant context, but the reader is directed to Chapters 2 and 3 for relevant information on hydrology and Chapters 4 and 5 for a detailed consideration of water quality dynamics.

The terms 'stream' and 'river' are often used interchangeably when referring to flowing waters, because in reality there is no obvious distinction, although the latter term is typically used when describing a 'larger' running water course. Some ecologists use the term **lotic** to describe running water systems of any size but the term 'river' is used for consistency in this chapter. The world's longest river ecosystem is thought by many to be the 6695 km-long River Nile, which flows through North East Africa and reaches the Mediterranean Sea in Egypt. However, some Brazilian scientists recently proposed that the Amazon is longer at 6800 km (Roach, 2007). The disagreement stems in part from imprecise definitions of a river's ultimate source and the point at which the estuarine zone becomes 'sea'. Nevertheless, the Amazon River is the world's largest in terms of volume, with an annual average discharge of >175,000 m³ s⁻¹, and it is likely to have the highest biodiversity.

Rivers are extremely diverse in their geomorphological form (see Chapter 3) and physicochemical characteristics (see Chapter 4), which both influence the remarkable diversity of organisms that we find in flowing water ecosystems. Describing the characteristics of an individual river ecosystem can be difficult, owing to this diversity of characteristics. A widely used framework to aid understanding of nested river ecosystem geomorphological units is that of hierarchical organisation (Frissell *et al*., 1986). Spatial units include the whole catchment, river segments, **river reaches**, **mesohabitats** and **patches** (or microhabitats) (**Figure 6.1**). Streams and rivers are recognisable by their major unidirectional (upstream to downstream) movement of water, sediment and other material. This 'continuum' has formed the basis of many conceptual ideas about how river ecosystems are structured and function, most notably the River Continuum Concept, which suggested that changes in the river environment from source to sea were accompanied by a distinct zonation of stream invertebrates, fish and material processing (see Vannote *et al*., 1980). In general, the headwaters of river systems are typically shallow and steep, with fast velocities and coarse bed sediments (**Figure 6.2**). With distance, downstream width and discharge increase but gradient and sediment sizes decrease. The changing river environment can lead to notable zonation of aquatic biodiversity and ecosystem functioning along a river's course.

Lakes and ponds move much more slowly than rivers and are considered as still or standing bodies of water. They can be described as being **lentic**, from the Latin word meaning 'sluggish'. Lakes are much larger than ponds and, while there is not an internationally recognised standard, they

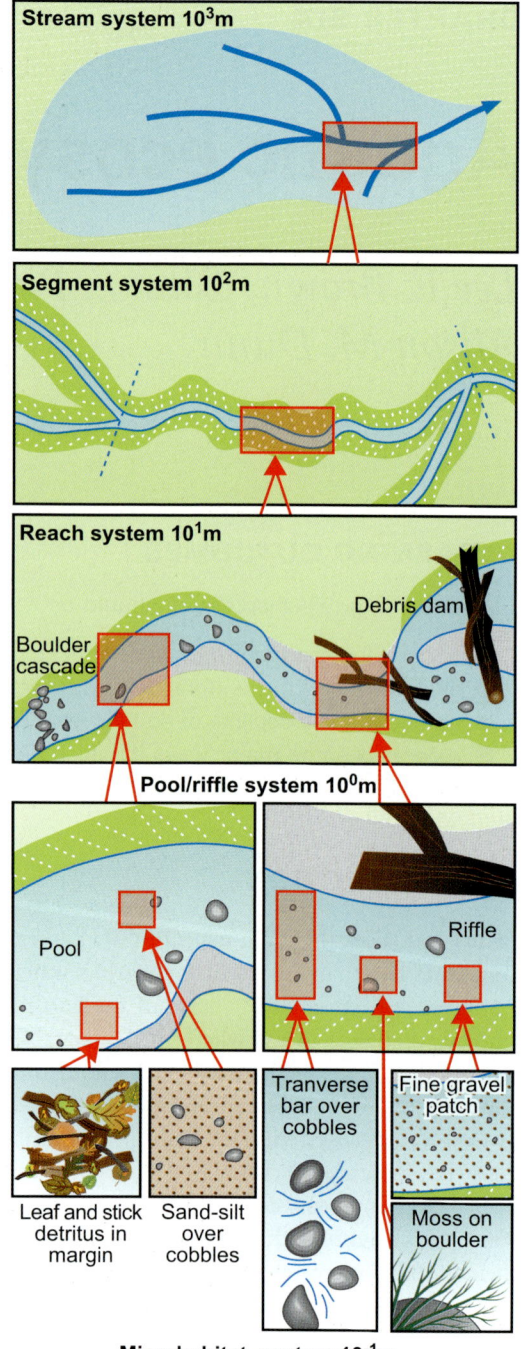

Figure 6.1 Hierarchical structure of river ecosystems.

Source: Reproduced with kind permission from Springer Science+Business Media: *Environmental Management*, A hierarchical framework for stream habitat classification: viewing streams in a watershed context v10, 1986, 199–214, Frissell, C.R., Liss, W.J., Warren, C.E. and Hurley, M.D., Figure 2.

Figure 6.2 Visible changes occur to the river environment from headwaters to lowland: (a) an open canopy UK moorland stream; (b) a meltwater-fed river in the French Pyrenees; (c) a densely forested stream in southeast Alaska; (d) a forested river in northwest North America; (e) a mid-reach on the River Aire, UK; (f) the River Danube in the lowlands of Austria; and (g) the braided Matukituki River in New Zealand.

Figure 6.3 Diversity of lake habitats in the Windermere catchment, northwest England: (a) the stony littoral zone of Rydal Water; (b) a fringing littoral reed bed in Windermere; (c) littoral reed swamp in Rydal Water; and (d) the expansive pelagic zone of Windermere.
Source: Photos (a)–(c) courtesy of Gary Rushworth.

are usually more than 2 ha in area and so are influenced strongly by wind turbulence. Lake Superior in the Great Lakes region of North America is the largest freshwater lake ecosystem, with a surface area of 82,414 km^2 and a maximum depth of 406 m. The largest freshwater lake by volume is Lake Baikal, southern Siberia, which is 31,500 km^2 in surface area and up to 1637 m deep. Lake Baikal is over 25 million years old and hosts over 2700 species of plant and animal, many of which are **endemic** (found nowhere else on the planet). Natural lakes can be created by tectonic processes, such as the fault line that is responsible for the creation of Lake Baikal. They may also be the result of the influence of repeated glaciations on the landscape, such as the Great Lakes of North America. Alternatively, they may arise from fluvial processes, in the case of oxbow lakes (see Chapter 3). Those of man-made origin are created by impounding a river behind a dam wall, creating an artificial lake in the river valley.

While lakes may appear to be more homogeneous than rivers, they can nevertheless possess a high diversity of habitat. The **littoral zone** of a lake is the area in the upper part of the water body that is marginal to the lake shore and receives abundant sunlight to the bed (**Figure 6.3**). This area also provides habitat for a wide variety of invertebrates and fish species. The **pelagic photic zone** is the area in the upper part of the water body

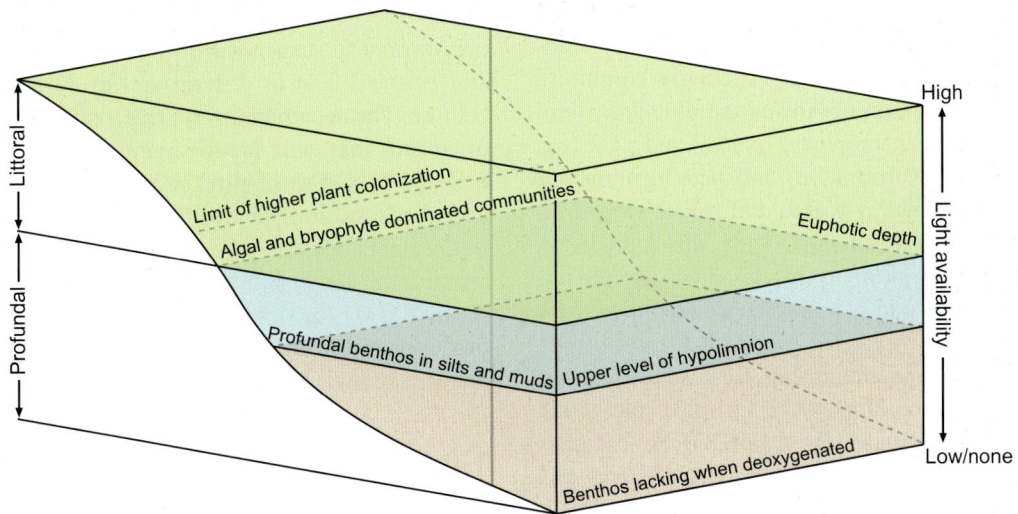

Figure 6.4 A general scheme of littoral and profundal habitats in a freshwater lake.
Source: After Moss, 1998. Reproduced with kind permission by John Wiley & Sons; Moss, B. 1998. *Ecology of freshwaters: man and medium, past to future.* Blackwell Publishing; Oxford.

that receives sunlight, but is open water and away from the shoreline. This is the area in which most primary production takes place. The **profundal zone** is in the lower and deep open water away from the shoreline. The **benthic zone** is the area at the bottom of the lake, at the transition between the water body and sediment and substrate (**Figure 6.4**). Ponds, being smaller, are occasionally overlooked in terms of their ecological importance. Nevertheless, both permanent and temporal ponds play an important role in sustaining the life cycles of a number of plants, insects and amphibians. In the case of temporary ponds there are many species that have developed adaptations and life history strategies to cope with changes in hydrological conditions. These include invertebrates that burrow into deeper, moist sediment and insects and amphibians which use water for part of the life cycle, before metamorphosis into their mature forms.

This chapter considers river and lake ecosystems in terms of their major biotic groups, including primary producers, consumers and parasites. The way in which these groups interact is considered with an introduction to food web structure. The chapter subsequently considers the role that aquatic organisms play in ecosystem processes, including organic matter production, respiration and nutrient cycling, and introduces the concept of ecosystem services. Finally the focus of the chapter turns to systems that have been altered significantly through the actions of humans, with a consideration of the effects of flow regulation, land use alteration, point-source pollution, eutrophication, climate change, invasive species and aquaculture. Boxed features throughout the chapter provide more detailed aquatic ecosystem methodological and management insights, including ways in which freshwater biota can be used to assess the condition of freshwater bodies.

B LIFE IN AQUATIC ECOSYSTEMS

While freshwaters cover only a small area of the Earth's surface (see section A1), they contain a disproportionately high diversity of plants and animals with at least 6% (or >100,000) of known species (Dudgeon *et al.*, 2006). There are >10,000 known fish species and some 90,000 species of invertebrates (with insects, crustaceans, molluscs and mites being particularly common). Other notable groups include amphibians, mammals, birds, macrophytes (plants) and algae (e.g. diatoms, phytoplankton), and there are numerous bacteria, fungi, protozoa and rotifer species inhabiting freshwaters. Despite this knowledge, understanding of freshwater diversity remains incomplete and new species are identified every year.

Organisms inhabiting freshwaters can be grouped according to their trophic role within aquatic **food webs** (see section B3). For example, **producers** (or autotrophs) are the plants and algae that synthesise biomass from inorganic compounds and light. Organisms that obtain organic matter as **consumers** are collectively called heterotrophs, and this group includes species that feed on producers (i.e. herbivores) or other consumers (i.e. carnivores or predators). A third group of consumers consumes dead organic matter and these are collectively known as **detritivores**. The diets of carnivores and detritivores can be heavily subsidised with organic materials from adjacent terrestrial ecosystems (e.g. terrestrial insects or leaf litter from trees, respectively). A fourth group are the **parasites**, and these exist on or in living organisms, benefiting at their host's expense. These groups are discussed in further detail in sections B1 and B2. These classifications are a convenient starting point, but it should be borne in mind that often many aquatic organisms have omnivorous diets, feeding on a variety of resources. Resources originating within the aquatic ecosystem and directly providing energy to the aquatic food web are termed **autochthonous**, while those originating from adjacent terrestrial ecosystems are termed **allochthonous**.

A variety of terms are used to explain where aquatic organisms spend the majority of their existence. Benthic organisms (or the **benthos**) live on, in or near the bed sediments of rivers or lakes. **Nektonic** organisms (or the **nekton**) is the term for organisms that can *actively* move around within the water column, whereas **planktonic** organisms live suspended in the water column and *passively* float or drift around in the water column. The **neuston** is a term reserved for those organisms that exist predominantly on, or just beneath, the surface of water bodies. Many terrestrial plants and animals are intricately linked with freshwater habitats, in particular existing in, on or around the bank habitats (the **riparian** zone; Naiman *et al.*, 2005), but in this chapter we focus predominantly on those organisms that spend the majority of their life cycle in aquatic habitats.

1 Primary producers

Primary producers are organisms that convert inorganic compounds into biomass through the process of **photosynthesis**. In rivers, there are two major groups of producers: macrophytes, which are large aquatic plants, and much smaller organisms, which can be collectively considered as algae. Aquatic macrophytes predominantly consist of flowering plants, mosses and liverworts (Allan and Castillo, 2007), and while as a group they are most successful in lakes and slower-flowing parts of rivers, some mosses and liverworts are often found in swift-flowing rivers. They have four different growth habits (**Figure 6.5**): (i) emergent plants with roots; (ii) floating plants with roots; (iii) floating plants without roots; and (iv) submerged plants with roots (Murphy, 1998). Angiosperms (flowering plants) are typical of slower, more nutrient-rich waters, while plants with emergent vegetation or floating leaves are found only in rivers if they are deep and slow flowing. Macrophytes play important roles in influencing aquatic habitats by slowing the velocity of water flow and encouraging the deposition of fine sediments. Macrophytes are found where light is able to reach the bed sediments, and so they are found in most areas of rivers except for the larger lowland stretches and estuaries that carry large quantities of suspended sediment (highly turbid rivers). In lakes aquatic macrophytes are a characteristic feature of the

The processes of shredding and/or physical breakdown due to the action of moving water and sediment break the coarse material into fine particulate organic matter (FPOM; < 1 mm). FPOM is easily maintained in suspension in rivers or lakes, and can be collected passively from the water column by organisms with filtering appendages, such as the invertebrate larvae of blackflies (Simuliidae) or those that collect food using nets (e.g. hydropsychid caddis flies). Other filterers, such as freshwater mussels (e.g. the zebra mussel *Dreissena polymorpha*), actively pump water and associated suspended particles through their mouth parts and in doing so remove the organic particles. Fine particulates can also be deposited on the bed of lakes and streams where flows are slower, and a third group of detritivorous consumers (collector-gatherers) feed on these deposits. Organic matter also forms part of a river's or lake's dissolved load, and the vast microbial component of aquatic food webs is fuelled largely by dissolved organic carbon (DOC) originating from soils, water and sediments (Meyer, 1994). Carbon and other nutrients in river water can be absorbed and assimilated by plants and microbes or adsorbed to sediments. When these nutrients are eventually released back to the water, some may be immediately recycled into the biota, or they remain as dissolved nutrients for later uptake by biota or abiotic mineralisation.

Parasitism is a major consumer strategy, although it is often neglected by community ecologists, despite an estimated 40% of all species being parasitic (Dobson *et al.*, 2008). Parasites exploit hosts at all trophic levels, and can affect host population dynamics as well as behavioural interactions with other species. In a study of estuarine salt marsh communities in North America, Kuris *et al.* (2008) reported 199 free-living animal species, 15 free-living vascular plants and 138 animal and plant parasite species. Parasite biomass made up ~1% of total biomass, which does not seem large. However, the biomass of parasitic trematode worms was, in fact, comparable to that of the top predators in the estuaries. In addition to exploiting the host, parasites provide food for species at higher trophic levels, either when they are eaten along with the host or as free-living life stages. For example, in the horn snail *Cerithidea californica*, the dominant invertebrate grazer in the estuaries, trematode parasites averaged 22% of the body mass of their host. Hence, when the host is consumed the parasites also provide a large resource for predators (Hechinger *et al.*, 2009).

Parasites can change the functional role of their hosts, and the strength of interactions of the host with other species in an ecosystem, either by reducing host population density or by changing the behaviour of the host (Hatcher and Dunn, 2011). These indirect effects of parasites can cascade throughout the trophic levels, leading to

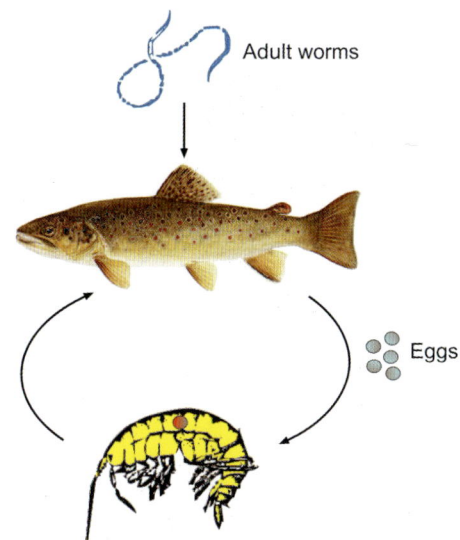

Figure 6.6 The life cycle of the trophically transmitted parasitic worm *Echinorhyncus truttae*. The adult worms live in the definitive fish hosts, which egest *E. truttae* eggs in the faeces. The intermediate amphipod host becomes infected when it eats the eggs, which subsequently develop into a bright orange cystacanth. The parasite makes the amphipod host more active and it seeks light. As a result, the host is at risk of predation by fish, and the parasite has an increased chance of trophic transmission back to a definitive fish host.

that graze on diatoms and algal growth, and fish such as the Grass Carp which feed on larger macrophytes and algae. Aquatic organisms that consume fruits can subsequently act as seed dispersers, including, for example, many fish species in the Amazon River, some of which have diets composed of up to 98% fruit (Reys *et al.*, 2009). Predatory (carnivorous) aquatic organisms include fish such as the pike and perch which feed on smaller fish, many fish and amphibians that feed on aquatic and terrestrial invertebrates, and invertebrate larvae such as some stoneflies and dragonfly that prey on smaller aquatic and terrestrial animals.

Detritivory is a major feeding strategy adopted by many aquatic organisms, with feeding possible on a variety of non-living organic-matter sources, including terrestrial plant litter, woody debris, pollen, excreted waste or dead animals. In particular, detritivores play major roles in processing organic matter that falls into water bodies from riparian (bankside) trees. In streams where plant leaves form a large component of the terrestrial plant input, leaves usually aggregate around obstructions, forming leaf 'packs'. Leaves are broken down in various ways that have been studied in detail (Benfield, 2006; Gessner *et al.*, 1999). For example, deciduous leaf litter that falls in autumn is largely nutritionally poor because the tree will have reabsorbed most of the soluble nutrients. However, the remaining leaf biomass still contains cellulose and lignin, which is partly consumed by fungi and bacteria. This consumption can serve to *condition* the litter, making coarse pieces of leaf litter more palatable to aquatic invertebrate consumers. Organisms that feed on this material have mouthpart adaptations that allow them to consume large pieces of plant material and small pieces of wood (>1 mm) known as coarse particulate organic matter (CPOM). CPOM is consumed predominantly by a group of organisms known as shredders, such as freshwater shrimp (*Gammarus*), louse (*Asellus*) and some stoneflies (e.g. Leuctridae; **Table 6.1**).

Table 6.1 Functional Feeding Groups (FFG) of invertebrates, their feeding mechanisms, food sources and typical size range of particles ingested

FFG	Feeding mechanism	Main food source	Particle size range (mm)
Collector-filterer	Suspension feeders – filtering particles from the water column	FPOM, detritus, algae, bacteria, fungi	0.01–1.0
Collector-gatherer	Deposit feeders – collecting deposited particles of FPOM or sediment	FPOM, detritus, algae, bacteria, fungi	0.05–1.0
Grazer	Graze surfaces of rocks, plants and wood	Producers such as attached algae, and associated detritus, bacteria and fungi	0.01–1.0
Predator	Capture prey and either engulf whole or pierce it and ingest body fluids	Living animals	>0.5
Shredder	Chewing	CPOM or live plant tissue	>1.0
Parasite	Microparasites absorb nutrients, often living within host cells	Producers and consumers	NA
	Macroparasites, endoparasites attach to and absorb nutrients from host gut or tissues. Ectoparasites feed on host blood or tissues (e.g. gills)		

Source: After Cummins *et al.*, 2005, reprinted with permission from Taylor and Francis.

Note: FPOM and CPOM = fine/coarse particulate organic matter, respectively.

littoral zone, and emergent macrophytes with root systems often emerge and grow upward above the water surface. Examples include the common reed *Phragmites* and bulrush *Typha*. Submersed and floating-leaved macrophytes, such as waterlilies *Nymphaea* and pondweed *Potomogeton* also inhabit these locations. Some lakes are shallow across their entire area and so macrophytes grow extensively. Slower-flowing stretches of river may also be colonised by macrophytes. Macrophytes serve as food sources and habitat for some animals, as well as providing substrate for some epiphytic algae (i.e. those that grow on living plants).

Algae lack true tissues and multi-cellular gametangia, and they contain chlorophyll *a* (Lowe and LaLiberte, 2006). Most aquatic algae in rivers grow attached to, or on, inorganic substrate and are termed epilithic (Wehr and Sheath, 2003). Others can grow on fine sediments (epipelic), sand (episammic) or wood (epixylic). A major group of the aquatic algae is the Bacillariophyta (diatoms); these are microscopic organisms that construct a shell (frustule) from silica. Owing to their diversity and varied responses to environmental conditions, they are widely used in water quality assessments. Other groups include green algae (Chlorophytes), yellow-brown algae (Chryophytes), red algae (Rhodophyta) and blue-green algae (Cyanobacteria). Algal distribution in rivers is strongly influenced by light availability as well as water chemistry, water temperature, current velocity and the abundance of aquatic herbivores. Some algae are free living in the water column in slow-flowing parts of rivers, and these are collectively known as phytoplankton. However, this group of algae is more common in lakes than in rivers.

The open water habitat of lakes contains a complex assortment of phytoplankton species, but many of these are invisible to the naked eye except for where they form immense colonies or blooms. Phytoplankton range in size from 1 μm (picoplankton), < 5 μm (ultraplankton), 5–60 μm (nanoplankton) and >60 μm (net plankton). This group includes a hugely diverse array of bacteria, green, yellow-green and golden-yellow algae, diatoms, cryptophytes, euglenoids and dinoflagellates (Moss, 1998). The growth of phytoplankton is strongly influenced by light and dissolved nutrients such as nitrogen and phosphorous. Periphytic algae such as *Cladophora* can be found attached to the substrate in the shallow littoral areas. Floating macrophytes, such as duckweed *Lemna*, can be found in the deeper profundal waters. However, free-floating algae, known as phytoplankton, carry out most of the primary production in this part of a lake. These include diatoms such as *Tabellaria* and *Asterionella*, desmids such as *Staurastrum*, dinoflagellates such as *Peridinium*, and blue-green algae such as *Microcystis* and *Anabaena*. These are grazed upon by small, free-floating animals known as zooplankton, a group which includes a variety of cladocerans, copepods and rotifers.

2 Consumers

Consumers are organisms that are heterotrophic, obtaining their energy from the consumption and processing of organic matters originating from either primary producers or other heterotrophs. This broad group has a diverse array of feeding modes, including herbivory (feeding on plant matter), frugivory (fruits), detritivory (non-living matter), carnivory (animal matter) and parasitism. The latter consumption strategy occurs at the expense of the 'prey' or host, but, unlike in the other feeding modes, the prey often remains alive for a prolonged period. Aquatic consumers play key roles in processing organic matter in aquatic ecosystems and in linking the energy and biomass created by primary producers, or input from adjacent terrestrial systems, to consumers higher up the food web (either in aquatic systems or even back to those in the terrestrial ecosystem; see section B3).

Herbivorous aquatic organisms include zooplankton which feed on phytoplankton in lakes and slow-flowing rivers, macroinvertebrates

Figure 6.5
Examples of aquatic producers: (a) emergent reeds (*Phragmites*) and floating rooted waterlilies; (b) close-up view of the rooted waterlily with floating leaves and flowers; (c) floating duckweed; (d) filamentous algae; and (e) rock with a thin covering of diatoms and green algae, visible below the water line.

changes in both productivity and community structure. For example, detritus processing by the isopod crustacean *Caecidotea communis* was reduced by >40% as a result of infection by the acanthocephalan parasite *Acanthocephalus tahlequalhensis* (Hernandez and Sukhdeo, 2008a). Parasites have also been shown to affect grazing indirectly, initiating so-called *trophic cascades*. For example, in some streams in Michigan, USA, parasite outbreaks caused crashes in populations of caddis flies (*Glossoma nigrior*), dominant grazers of periphyton. The resulting reduction in grazing led to increased periphyton biomass, as well as increased densities of the other organisms which compete with the caddis flies for food. This in turn led to increased densities of predatory caddis flies and stone flies (Kohler, 2008).

Parasites can have indirect effects on predator–prey interactions because many use both members of the relationship as hosts, and they can manipulate anti-predator behaviour of their intermediate host to enhance trophic transmission to the next host. For example, *Gammarus pulex* (amphipod crustaceans) that are infected by the acanthocephalan parasite *Echinorynchus truttae* show increased activity and move towards light (uninfected amphipods seek out shade to avoid predation); these changes in behaviour increase the risk of predation by trout, and hence increase the rate of parasite transmission (MacNeil et al., 2003). The strength of predator–prey linkages can also be affected by parasites. *G. pulex* that are infected by *E. truttae* (**Figure 6.6**) were found to eat 30% more isopods than did uninfected hosts, probably as a result of the metabolic demands of the parasite and of their increased activity (Dick et al., 2010). Conversely, white-clawed crayfish (*Austropotomobius pallipes*) that harboured the

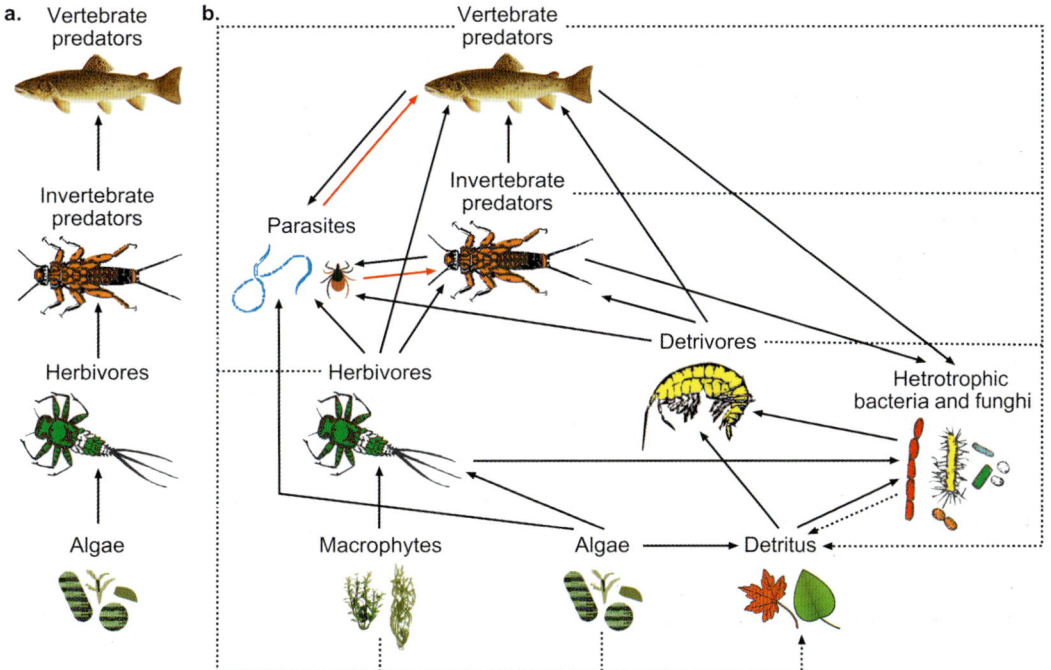

Figure 6.7 Schematic outline of (a) a simple aquatic food chain, and (b) a simple aquatic food web. Solid lines represent feeding links and broken lines represent decomposition pathways. Red arrows denote energy flows from parasites to predator when consumed along with the host.

Table 6.2 Food web summary statistics for rivers, estuaries and lakes

Study locations (and no. published food webs from that location)	No. of species	No. of links between species	Connectance (links per (species)^2 to give proportion of all possible links observed)	Mean food chain length
Rivers				
Afon Hirnant, UK	33	112	0.10	
Alaska, USA (n = 4)	38–45	150–193	0.10–0.11	1.65–3.10
Appalachian Mountains, USA (n = 4)	35–41	125–200	0.10–0.15	2.45–2.97
Artificial streams, Dorset, UK (n = 4)	61–71	320–492	0.09–0.11	2.03–2.17
Bere Stream, UK	142	1383	0.07	
Broadstone Stream, UK (n = 11)	24–128	90–721	0.04–0.21	4.88–5.38
Duffin Creek, Canada (n = 7)	31–39	101–146	0.09–0.11	
Muskingham Brook, USA (n = 8)	33–48	97–171	0.07–0.09	2.86
Southeast USA (n = 4)	58–105	126–343	0.03–0.04	1.52–3.13
South Island, New Zealand (n = 20)	49–113	110–950	0.03–0.08	1.56–4.42
Tadnoll Brook, UK	59	170	0.05	
Tai Po Kau Forest, Hong Kong	28	157	0.20	2.04
UK streams (n = 20)	19–87	56–1653	0.12–0.29	
Estuaries				
Chesapeake Bay, USA	31	68	0.07	3.99
St Mark's Seagrass, USA	48	221	0.10	4.8
Ythan Estuary, UK	92	409	0.05	5.91
Carpinteria salt marsh	134	1120–2313	0.06–0.13	2.2–3.1
Standing freshwaters				
Lake Tahoe, USA	172	3885	0.13	
Little Rock Lake, USA	92	1008	0.12	>10.0
Lochnagar, UK (n = 18)	45–81	270–831	0.14–0.38	
Mirror Lake, USA	586	14709	0.15	
North American lakes (n = 50)	10–74	17–571	0.06–0.17	4.04
Tuesday Lake, USA	56	269	0.09	4.2–4.6
Skipwith Pond, UK	35	276	0.23	6.22
Lake Tatvan	50	432	0.17	

Source: After Brown *et al.*, 2011; Carpinteria Salt Marsh data from Lafferty *et al.*, 2006; Lake Tatvan data from Amundsen *et al.*, 2009. Reproduced with kind permission by John Wiley & Sons; Brown, L.E., Edwards, F.K., Milner, A.M., Woodward, G. and Ledger, M.E. 2011. Food web complexity and allometric-scaling relationships in stream mesocosms: implications for experimentation. *Journal of Animal Ecology* 80: 884–895. (c) 2011 The Authors. *Journal of Animal Ecology* (c) 2011 British Ecological Society, *Journal of Animal Ecology* 80, 884–895.

microsporidian parasite *Thelohania contejeani* have been seen to eat less prey than their uninfected competitors, probably as a result of muscle damage by the parasite (Hatcher and Dunn, 2011).

3 Food webs

Food chains are representations of interactions between consumers and resources, linking autotrophic producers to top predators via a series of intermediate consumers (**Figure 6.7a**). In reality, food chains do not exist in isolation, because consumers feed on multiple resources and thus form food webs, which are complex networks of species connected to one another by their feeding interactions (**Figure 6.7b**). The study of food webs is important because the complex interactions mean it is difficult to predict how ecosystems respond to environmental change by simply extrapolating from studies of individual populations or groups of species (Woodward *et al.*, 2010). Many of the best-described food webs originate from running waters, and structural characteristics have been catalogued from a variety of habitats, including upland and lowland rivers, forested and open-canopy river basins, and tropical, temperate and Arctic environments (Brown *et al.*, 2011). Numerous food webs have also been documented from lakes and river estuaries (**Table 6.2**), and comparative analysis of food webs available from different aquatic habitats shows comparable food web size (number of species, number of links) and complexity (connectance), but generally lower mean food chain length in rivers. Some key influences on aquatic food webs include flow regimes, river catchment

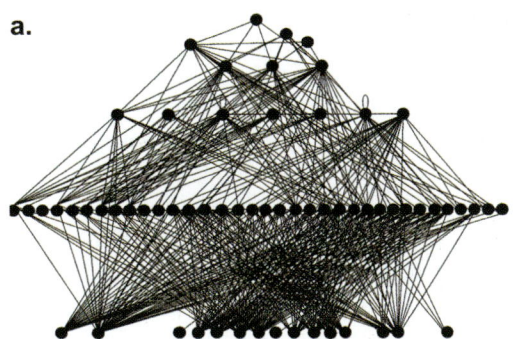

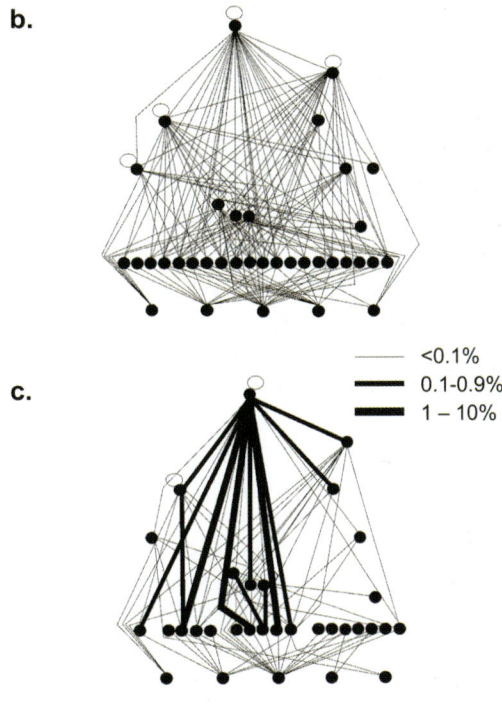

Figure 6.8 Connectance food webs from (a) Felbrigg Hall Lake, UK, and (b) Broadstone Stream, UK. Panel (c) shows the Broadstone Stream web with feeding links quantified based on numbers of prey eaten per capita 24 hr^{-1} (as a percentage of numbers m^{-2}). In each web, filled circles represent species or groups of species and lines represent feeding links. Food webs are arrayed hierarchically, based on species' trophic levels.

Sources: (a) After Rawcliffe *et al.*, 2010. Reproduced with permission of John Wiley & Sons: Rawcliffe, R., Sayer, C.D., Woodward, G., Grey, J., Davidson, T.A. and Jones, J.I. 2010. Back to the future: using palaeolimnology to infer long-term changes in shallow lake food webs. *Freshwater Biology* 55: 600–613. (b) and (c) After Woodward *et al.*, 2005. Reprinted from Woodward, G., Speirs, D.C. and Hildrew, A.G. Quantification and resolution of a complex, size-structured food web. *Advances in Ecological Research* 36: 85–135, copyright 2005, with permission from Elsevier.

size and disturbance regime, as well as ecosystem size, thermal regime, pH, invasion by new species and surrounding land use (e.g. Thompson and Townsend, 2005).

Connectance food webs are those which illustrate individual feeding links between species (e.g. **Figure 6.7** and **Figure 6.8a** and **b**), and they are constructed by sampling numerous individuals of constituent species, then examining their diet through microscopic analysis of gut contents to observe dietary links. Species can then be assigned trophic positions as basal species (those with consumers but no prey – i.e. primary producers), intermediate species (those with both consumers and prey) and top species (those with prey but no consumers – i.e. top predators). Food webs from rivers and lakes typically show trophic pyramids, whereby the number of species and their biomass decreases with food web height, due mainly to inefficiencies in energy transfer between trophic levels (i.e. not all energy synthesised or consumed by prey is available to their consumers, due to respiration and growth costs).

One major drawback with connectance food webs is that they assume that all feeding links are equal, because they do not provide any information on relative strength. Consequently, ever more detailed food webs are being developed with quantified links (**Figure 6.8c**) in attempts to develop more detailed knowledge of how aquatic food webs vary in space and over time. Furthermore, there are developing ideas about how aquatic food webs interact with other types of ecological networks such as host-parasite webs (Hernandez and Sukhdeo, 2008a; **Figure 6.9**), and the implications of these cross-overs for food web dynamics. Many food web studies of aquatic (and indeed terrestrial) ecosystems have tended to omit parasites, perhaps because of their relatively small biomass. However, parasitism is the most common consumer strategy (Kuris *et al.*, 2008), and parasites also influence the trophic interactions of their hosts (section B2). Inclusion of parasites in a stream food web (**Figure 6.9**) increased species richness (28 free-living and 10 parasitic species were recorded), linkage density and the number of possible and observed links (Hernandez and Sukhdeo, 2008b). Similarly, including parasites in a lake food web increased species richness (37 free-living and 13 parasitic species), food chain length and connectance (Amundsen *et al.*, 2009).

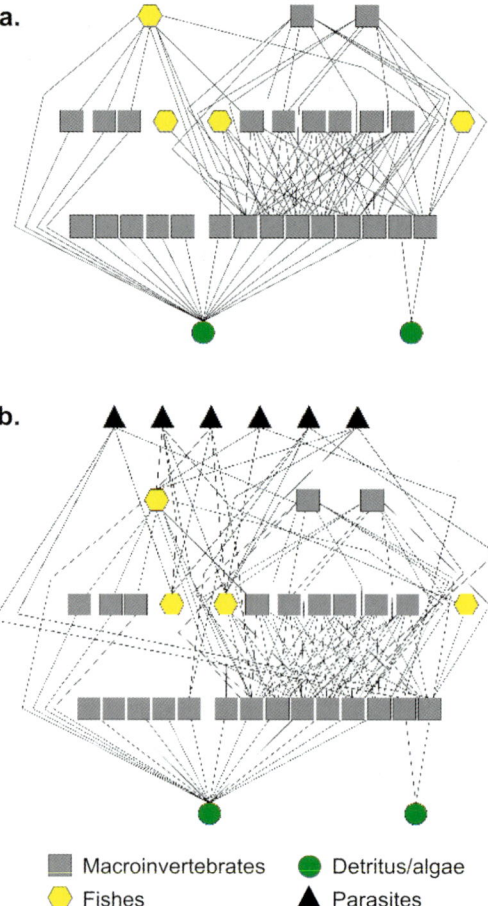

Figure 6.9 Food web structure of Muskingham Brook, New Jersey, USA, during fall 2003. Panel (a) shows the food web excluding parasites, and (b) shows the same food web with parasitic links included. Relative positions are not arrayed according to trophic level.

Source: After Hernandez and Sukhdeo, 2008a. Reprinted from Hernandez, A.D. and Sukhdeo, M.V.K. Parasite effects on isopod feeding rates can alter the host's functional role in a natural stream ecosystem. *International Journal for Parasitology* 38: 683–690, copyright 2008, with permission from Elsevier.

Aquatic food webs and those of the adjacent terrestrial ecosystem (or the riparian zone) are often strongly linked through the consumption of allochthonous leaf litter by shredders. Additionally, the diet of aquatic predators, such as fish and some invertebrates, can also be supported by terrestrial invertebrates which fall into the water. In this sense it can be said that the terrestrial ecosystem feeds, or subsidises, the aquatic ecosystem (Nakano et al., 1999), and some studies have shown that the diet of river-dwelling fish can be made up of >50% terrestrial invertebrates (Baxter et al., 2005). Research has also indicated reciprocal subsidies from aquatic to terrestrial ecosystems (Nakano and Murakami, 2001). Aquatic insects develop in the stream as larvae (nymphs), but towards the end of their life cycle they emerge as adult flies with a terrestrial aerial stage. The emergence of adult aquatic insects is typically concentrated into a small part of the year in temperate regions (peaking in early summer), whereas in tropical areas it may occur year round, providing food for terrestrial organisms such as spiders, bats, birds, amphibians and beetles. Some birds and amphibians also prey on aquatic invertebrate larvae.

> **REFLECTIVE QUESTION**
>
> What are functional feeding groups, and how might they vary along a river in a forested catchment?

C AQUATIC ECOSYSTEM FUNCTIONAL PROCESSES

1 Primary production

Primary production represents an important source of autochthonous energy to aquatic ecosystems. Primary productivity can be defined as the rate of formation of organic matter from inorganic carbon by photosynthesising organisms (Bott, 2006) and can be measured in a variety of ways (**Box 6.1**). Aquatic primary production is often thought of as being less important than allochthonous inputs of terrestrial plant litter, but this is not the case for open-canopy (non-forested) sites, some stretches of forested rivers where light penetration is relatively high, or large lakes (Murphy, 1998). Producer turnover (i.e. replacement of dead, consumed or eroded plants or algae) can be rapid compared to detritus stores, even though standing stocks of the latter may be much higher (Hershey and Lamberti, 1998). In rivers, primary production often shows a general increase with distance downstream into the middle reaches, particularly where the headwaters are forest covered and turbidity is relatively low. However, there can be high variation owing to the character of the river at different locations and over time (e.g. Vannote et al., 1980; Murphy, 1998). The edges and bottoms of lakes are considered to be relatively productive and diverse where light penetration is high and there are abundant nutrients in sediments and water (Moss, 2010).

The rate of aquatic primary production is influenced by a multitude of factors, many of which are common to running and standing waters. Regional and catchment-scale features such as climate, topography, geology and land use affect key proximate influences such as light availability, water chemistry/nutrient concentrations and temperature (Allan and Castillo, 2007). Light is vital for photosynthesis, and seasonal changes in primary production can be attributed to the annual solar cycle, which also drives thermal dynamics. The amount of light reaching the water surface is strongly affected by shading either from overhanging vegetation or from topography, while light reaching the bottom of rivers and lakes is a function of water depth and turbidity (a measurement of the amount of suspended particles in the water) (Wetzel, 2001). There can, however, be light-saturation effects, and thus other variables may serve to limit production (Murphy, 1998). Phosphorous (as dissolved inorganic P) and

BOX 6.1 TECHNIQUES

Measuring aquatic primary productivity

Primary production can be represented as:

$$GPP = NPP + R$$

where: GPP = Gross Primary Production; NPP = Net Primary Production; and R = Respiration. NPP represents the energy or biomass available to herbivorous consumers after the producer's respiration 'costs' are taken into account. Net Ecosystem Production (NEP) can also be determined by calculating the difference between NPP and the respiration of all heterotrophs in an aquatic system. Measurement of primary production can involve quantifying the accrual of periphyton on introduced substrate (e.g. unglazed tiles or microscope slides), determining chlorophyll a concentrations, measuring changes in dissolved O_2 or CO_2 concentrations *in situ* or in enclosed chambers, or adding the isotope ^{14}C and tracking its uptake into plants and algae (Howarth and Michaels, 2000; Bott, 2006; Figure 6.10). The biomass of primary producers is sometimes used as a measure of production but this can be misleading because, while some plants and algae may maintain only low standing stocks (e.g. perhaps due to high grazing pressures), they might also actually have a high turnover (e.g. constant and rapid replacement of lost biomass) and thus high production (Murphy, 1998). Standing stock measurements are also problematic because there is no way to differentiate the amount of production that has been respired or consumed.

Figure 6.10 Examples of the various methods used by freshwater ecologists to study primary production: (a) dual-station oxygen measurements using battery-powered dataloggers connected to dissolved-oxygen sensors; (b) from left to right, a photosynthetically active radiation sensor to measure incoming light, a thermistor used to measure stream temperature and a dissolved-oxygen sensor; (c) a 3 × 3 cm patch of algae has been sampled from a rock for later analysis of algal community composition and extraction of chlorophyll a; (d) unglazed 10 × 10 cm^2 quarry tiles attached to bricks to allow algal accrual over time.

nitrogen (N mainly as nitrate or ammonia) are often key limiting nutrients and are usually implicated in eutrophication, where excess algal growth causes severe water quality problems (see section E5). Silica is also important because it is required for the construction and maintenance of diatom frustules, and in some lakes spring diatom blooms can deplete silica concentrations significantly, such that other forms of algae start to become more abundant (Wetzel, 2001). Other important influences on the rate of primary production can include current velocity, substrate type and quality and grazing pressures from herbivores (Murphy, 1998; Allan and Castillo, 2007). In a study of 30 streams from 30–50°N, Lamberti and Steinman (1997) showed that gross primary production (GPP) ranged from 3.5 to 5400 g C m^{-2} y^{-1}, averaging 560 g C m^{-2} y^{-1}. Estimated average GPP for lakes worldwide is 260 g C m^{-2} y^{-1}, with 40 g C m^{-2} y^{-1} of this attributed to excess nutrient input and eutrophication (Lewis, 2011).

2 Secondary production

Secondary production involves the formation of heterotrophic biomass (microbial and animal) for a given area and over a given time, and it represents the processes of recruitment and growth, respiration, emigration and mortality (Benke, 1984; see **Box 6.2**). While the approach has been commonly used to calculate production of individual populations of heterotrophs (Huryn and Wallace, 2000), it is nevertheless possible to

---

BOX 6.2 TECHNIQUES

Measuring aquatic secondary production

Similar to producers, where only a fraction of GPP is converted to biomass, some of the material consumed by heterotrophs is not converted to biomass because of inefficiencies in assimilation and the use of energy and matter for respiration (Benke, 1984). Secondary production can thus be considered as:

$I = A + F$

where: I = ingestion; A = Assimilation, and; F = Faeces

Biomass that is assimilated (A) is then processed by the individual:

$A = P + R + E$

where: P = Production of biomass; R = Respiration, and; E = Energy lost during excretion

At the individual level, P can be considered as growth, while at the population or community level it represents the collective growth of all individuals. Measuring growth involves quantitative sampling of species cohorts over time (usually a minimum of 12 months) and calculating production from changes in species density and biomass. The mean biomass over the entire year (B) can be calculated for each species, and used to calculate P/B ratios to illustrate species turnover (Benke, 1984). P is determined by availability of food, but also how efficiently that food is converted to tissue, which is determined by assimilation efficiency (A/I), net production efficiency (P/A) and gross production efficiency (P/I). Indicative values of assimilation efficiency for aquatic invertebrates are usually in the order of 10% for vascular plant detritus, 30% for diatoms and algae, 50% for fungi, >70% for animals and 50% for microbes, with net production efficiencies of ~40% (Benke and Wallace, 1997; Benke, 2010).

calculate secondary production for entire trophic levels or assemblages, and to use these estimates to quantify energy flow through whole food webs (e.g. Hall *et al.*, 2000). The latter method is particularly important because it reveals that the 'weighting' of individual links in aquatic food webs is not equal, with most being weak but some being extremely strong. These differences in link weighting can have implications for the stability of ecosystems under disturbance conditions (Montoya *et al.*, 2009).

Reported values of secondary production for aquatic ecosystems are common for invertebrates, with values from across the world ranging between approximately 10^0 to 10^3 g m^{-2} y^{-1} (Huryn and Wallace, 2000). **Meiofauna** (invertebrates < 250 µm when fully grown) can be abundant in streams, and their intermediate role between microbial/detrital resources and larger organisms means that they can contribute significantly to secondary production (e.g. Tod and Schmid-Araya, 2009). In small, cold, stony open-canopy moorland streams, such as those in the North Pennines of the UK, fish production is low, ranging from just 1.02–3.5 g m^{-2} y^{-1} for trout and 0.48–7.43 g m^{-2} y^{-1} for bullhead (Crisp *et al.*, 1975). In contrast, in river systems where there are abundant prey production and favourable habitat conditions, trout production can reach >100 kg ha^{-1} y^{-1} (10 g m^{-2} y^{-1}; Huryn, 1996). Many estimates of trout production seemingly exceed estimates of prey production, a discrepancy which has been termed the Allen Paradox (Hynes, 1970). This situation arises because fish diet can be subsidised heavily from terrestrial (riparian) prey inputs, invertebrate production in the **hyporheic zone** (the sediments beneath the stream bed) or by cannibalism. All of these prey sources can be difficult to quantify.

3 Community respiration

Respiration is the biological process in which organic matter is utilised by organisms, typically in the presence of dissolved oxygen, to provide energy for metabolism, growth and reproduction. Both autotrophs (algae, macrophytes) and heterotrophs (animals, bacteria, fungi) respire organic matter, and together these account for community respiration. Quantifying community respiration (CR) is particularly important in studies that are also measuring primary production (P), because it allows the operator to examine ratios of P:R (or, more precisely, GPP:CR). Where GPP > CR there is a net addition of energy to an ecosystem and it can be considered net autotrophic. In contrast, the system can be said to be net heterotrophic where GPP < CR (Bott, 2006). Respiration is fuelled by primary production that occurs within the river (autochthonous production), or by allochthonous resources (e.g. leaf litter, dissolved organic carbon from soils) that enter the river from adjacent terrestrial ecosystems.

Community respiration of aquatic systems can be determined from measurements of dissolved oxygen over daily time periods. Other methods for monitoring the consumption of oxygen include enclosing surface stones in respiration chambers, enclosing bed sediments in respiration tubes followed by incubation, or collecting water samples and incubating for fixed time periods (e.g. Naegeli and Uehlinger, 1997). Smaller streams are dominated by benthic metabolism, whereas planktonic metabolism predominates in larger rivers and lakes (Thamdrup and Canfield, 2000).

4 Nutrient cycling

Advances in our understanding of instream nutrient supply, transformation and retention have developed around concepts of nutrient *spiralling*, in the sense that nutrient *cycles* in streams are inevitably coupled with the downstream movement of water, and are therefore better conceptualised as spirals (Newbold *et al.*, 1982). The approach, which is founded on experimental additions of nutrients to streams (**Figure 6.11**), followed by sampling at fixed locations downstream of the addition, allows estimates of: the distance required for nutrients to com-

Figure 6.11 A battery-powered peristaltic pump being used to deliver small quantities of dissolved NO_3^-, NH_4^+ and organic carbon from a tank into an Arctic tundra stream as part of an experiment to measure nutrient spiralling lengths and uptake velocity.

plete cycling through dissolved to particulate to consumer phases and then return to the water column (so called spiralling lengths); and the **bioavailability** (and/or demand) for nutrients within the ecosystem (uptake velocities). Ensign and Doyle (2006) reviewed >400 studies reporting spiralling lengths and uptake velocities for streams and the vast majority of these had been undertaken in small first- and second-order headwater streams (see glossary for explanation of **stream order**). More remains to be done to understand nutrient spiralling in larger rivers further down the river network. Nutrient limitation can also be assessed using diffusing substrates. In essence, nutrient mixtures (N, P, N+P) are dissolved in agar and then placed into small containers (**Figure 6.12**). The lid of the container is replaced with a porous substrate (usually a clay tile or wood, which enables comparative assessments of epilithic and epixylic biofilm growth, respectively) through which the nutrients can diffuse into the water column before being utilised by producers and heterotrophic microorganisms.

Figure 6.12 Nutrient-diffusing pots with wooden (brown) and clay (orange) substrate.

> **REFLECTIVE QUESTION**
>
> What are the key functional processes of aquatic ecosystems and how could changes in water quality impact upon these processes?

> **REFLECTIVE QUESTION**
>
> Thinking of a local river or lake – how do you benefit personally from that water body and, more widely, how does society benefit from the ecosystem services it provides?

D AQUATIC ECOSYSTEM SERVICES

Ecosystem services are the benefits that the human population is able to gain from a functioning ecosystem. The Millennium Ecosystem Assessment provides a framework which divides the various services into *provisioning*, *regulating*, *cultural* and *supporting* services (Millennium Assessment, 2005). The main provisioning services associated with freshwater ecosystems include food (fish and other edible biota production), fuel (wood and peat), biochemical (material from biota), genetic (resistant genes and medicine), and biodiversity (diversity of species and gene pools). Important regulating services include flow regulation (wetland storage and groundwater recharge, floodplain water storage), sediment transport (form and function of river channel and floodplain) and waste assimilation (pollutants and excess nutrient cycling). The main cultural services associated with aquatic ecosystems include recreation associated with angling, and amenity values (education, research). Supporting services include primary production, secondary production and nutrient cycling. Many freshwaters have been, and continue to be, managed poorly by humans (see section E), and in many locations this has led to degradation or loss of some ecosystem services at the expense of others (e.g. Maltby *et al.*, 2011). Balancing the competing demands for ecosystem services from multiple stakeholders is emerging as a major political, economic and scientific challenge. Other examples of ecosystem services provided by freshwaters and ecosystem service approaches are provided in Chapter 10.

E HUMAN MODIFICATION AND MANAGEMENT OF AQUATIC ECOSYSTEMS

1 Overview

Humans have severely altered aquatic eco-systems in a variety of ways. These include hydrological (e.g. water abstraction, dams), geomorphological (e.g. channelisation, culverts, dams, gravel extraction), water quality (e.g. acidification, organic pollution, nutrients, thermal pollution) and biological modifications (e.g. introduced species). Vörösmarty *et al.* (2010) conducted a global analysis of threats to aquatic biodiversity and concluded that 65% of freshwater habitats have moderate to high threats. Freshwater biodiversity is disproportionately at risk from habitat modification, because while freshwaters cover only 0.8% of the Earth's surface, they are home to an estimated 6% of all species (Dudgeon *et al.*, 2006). Additionally, aquatic ecosystems are rarely affected by individual modifications. In reality, multiple stressors can combine to exacerbate the threat to aquatic biodiversity (e.g. Ormerod *et al.*, 2010). For example, an urban stream may have been channelised to reduce flood risk, it can have a modified flow regime owing to runoff from impermeable surfaces and it may receive pollution from point sources associated with manufacturing plants or sewage treatment works. This section provides an overview of some of the most common human modifications of rivers and lakes and the biological consequences of such actions.

2 River flow regulation

Many river ecologists consider the natural **river regime** (see Chapter 3) as a 'master' environmental variable, due to its influence on a suite of physical, chemical and biological processes. Key aspects of the natural flow regime are the *magnitude* of flow extremes and the *rate of change* between different flow states, plus the *frequency*, *duration* and *timing* of a given flow magnitude (see Poff *et al.*, 1997). High flows of a natural flow regime (e.g. due to spring snow melt) can be important for cleaning river gravels prior to fish spawning, whereas lower flows are often necessary for juvenile fish to be reared (Acreman *et al.*, 2009). River systems and their constituent species have adapted to the natural flow regime over long timescales, and so alteration of flows by river regulation (e.g. dam construction, abstractions or diversions) can severely alter habitats by changing the thermal or sediment transport regime, impeding the migration of organisms and leading to large-scale changes in the structure and functioning of aquatic ecosystems (Lytle and Poff, 2004). Dams constructed for water storage constitute a major impact on river ecosystems globally (**Figure 6.13**), with an estimated 15% of river runoff held in large reservoirs (>15m height; Nilsson *et al.*, 2005). In the USA, there are an estimated 2.5 million structures controlling river flow, and only 2% of rivers are considered to be unimpacted (Lytle and Poff, 2004). A further 10% of global runoff is extracted through abstractions (Vörösmarty and Sahagian, 2000). These human interventions in the water cycle are thought to have contributed to major negative impacts on aquatic biodiversity and some ecosystem services (Vörösmarty *et al.*, 2010).

A detailed study by Poff and Zimmerman (2010) evaluated the links between river flow regulation and ecological response by considering 165 studies conducted worldwide across a range of different rivers. Negative responses of river ecosystems due to flow alterations were reported

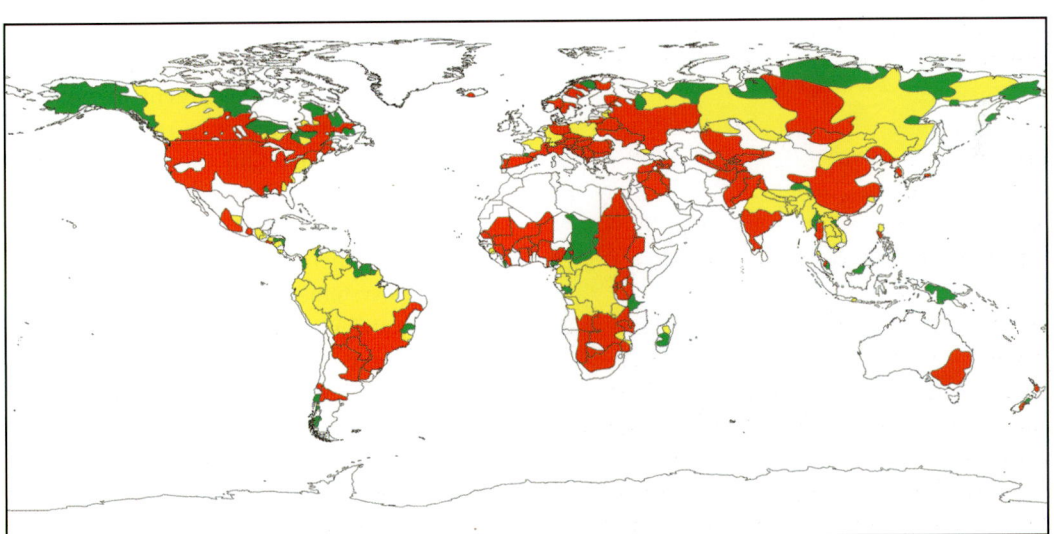

Figure 6.13 River channel fragmentation and flow regulation effects of dams on 292 of the world's major river systems. Green = unimpacted, yellow = moderately impacted and red = strongly impacted catchments. Uncoloured areas are not covered by large river systems but may also be strongly affected by river regulation.
Source: Modified from Nilsson *et al.*, 2005.

in 152 (92%) studies. Dams intercept large quantities of sediment, and sections of river downstream of impoundments can be characterised by coarsening of bed sediments, due to the removal of fines, or alternatively the loss of peak flow events may see a progressive infilling of the interstices with finer materials (Stanford and Ward, 2001; Poff *et al.*, 1997). Similar changes can occur when water is abstracted from rivers, and the effects can be particularly impactful to fish, whose eggs develop in stream bed sediments, or invertebrate larvae, which live, migrate and feed in and on bed sediments. Fish populations in particular can be seriously impacted upon by alterations to flow, with low flows and flow regulation structures impeding migration, altered thermal regimes seriously affecting physiology, and loss of prey as a result of changed aquatic and riparian food webs. Lowered flows can also reduce dissolved oxygen concentrations, due to reduced turbulence/re-aeration, and because of a lowered heat capacity (warmer waters can hold less dissolved oxygen). Changes to the flow regime can lead to major shifts in invertebrate and algal community composition, alterations to the cycling of nutrients and a depression of primary and secondary production (Ledger *et al.*, 2011). In contrast, there can be some 'positive' changes to ecosystems, including increases in non-native species cover, encroachment of plants onto river sediments that are no longer inundated (Poff and Zimmerman, 2010; see **Figure 6.14**, for example), increased abundance of small, short-lived invertebrates and enhanced growth of algae and bryophytes on stabilised river bed sediments (Armitage, 2006).

Recognition of the negative effects of river regulation on aquatic and riparian ecosystems has resulted in the development of policy to manage flows and attempt to restore degraded ecosystems and maintain those that are least affected (Petts, 1996; Arthington *et al.*, 2006). The approach involves regulation of water abstraction to ensure that there is enough water to maintain aquatic ecosystems. In some heavily regulated systems, water is released from dams to create artificial floods in an attempt to restore some aspects of the natural flow regime. For example, in Switzerland, periodic water releases on the Spöl River are made during the course of the year to raise the continuous baseline flow from ~1.6m^{3}s^{-1} to >40 m^{3}s^{-1} (**Figure 6.15**). Associated river ecosystem studies have shown that such interventions can cause shifts in the aquatic ecosystem, including the elimination of mosses from previously stabilised river sediments and the development of invertebrate communities

Figure 6.14 Regulated river channel below Digley Reservoir, West Yorkshire, UK. Vegetation encroachment is notable along the banks and on rocks in the centre of the river channel, owing to the loss of high-magnitude flood flows.

Figure 6.15 The Spöl River, Switzerland, (a) before and (b) during an artificial flood generated by water release from a reservoir.
Source: Photos courtesy of Christopher Robinson.

that are more resilient to flood disturbances (Robinson and Uehlinger, 2008). Other approaches being adopted in some developed regions of the world include the removal of smaller dams (e.g. Bednarek, 2001; Babbitt, 2002) so as to restore natural flows and allow free passage of migratory fish to upstream reaches. Short-term effects of dam removal include increases in fine sediment loads downstream, although some of these sediments may have associated pollutants, necessitating sediment remediation. Despite the known negative effects of regulation on river ecosystems, new large dams continue to be proposed and constructed each year in developing countries, including in Brazil on the Amazon, where aquatic ecosystems are particularly rich in biodiversity.

3 Land use alteration: urbanisation

River catchments worldwide have been altered significantly as a consequence of land use change (e.g. Harding *et al.*, 1998; Paul and Meyer, 2001; Allan, 2004), with major effects on water quality (see Chapter 4) and aquatic ecosystems. Some major examples of land use change which impact upon aquatic ecosystems include afforestation and deforestation, the turning over of land to agriculture, and the development of towns and cities and their associated infrastructure. The *urban stream syndrome* is a term used to describe the ecological impact that an urbanised environment has upon the watercourses that receive their run-off (Meyer *et al.*, 2005). The symptoms include: (i) hydrological change, with reduced base flow and enhanced peak flows; (ii) modified thermal regimes linked to i, in addition to changes in shading/exposure, point-source discharges and runoff from paved areas; (iii) altered water chemistry with enhanced nutrients and contaminants; (iv) transformed channel morphology (e.g. culverts, straightening of channels, reinforcement of banks) with reduced natural habitat (**Figure 6.16**); (v) altered ecosystem functioning (see section C); and (vi) a reduction of biodiversity.

Changes to ecosystem functional processes following urbanisation can include increased gross primary productivity and community respiration when compared to forested streams. Increased algal biomass has been recorded in urban streams with a prevalence of filamentous algae as compared to forested streams, which have a high proportion of diatoms (Taylor *et al.*, 2004). Heavily urbanised locations have P/R ratios of much less than 1, suggesting that heterotrophic metabolism is dominant (Meyer *et al.*, 2005). Nutrient uptake is also reduced in urbanised streams. Studies on aquatic macrophytes in urban watercourses are rare but they do appear to be affected by the loss of habitat, increased turbidity,

Figure 6.16 Channelised urban streams showing: (a) stone reinforcements of the bed and banks; (b) sheet piling reinforcement of banks; and (c) stone-filled gabion basket reinforcement of the left-hand bank and a brick wall forming the right-hand bank.

chemical contamination, increased velocities and coarser sediment profile that arise as a consequence of urbanisation.

Patterns of the impact of urbanisation on macroinvertebrate assemblages have shown consistent patterns worldwide. Robson *et al.* (2006) measured the degree of urbanisation of a catchment as the area covered by impermeable surface upstream of a series of monitoring sites. Chemical contamination increased with levels of impermeable surface cover, and that contamination was greatest during rainfall events. A reduction in macroinvertebrate diversity was also observed moving upstream to downstream. There was a decline in the numbers of taxa intolerant to disturbance, including Gammaridae (shrimp) and families from the orders of Ephemeroptera (mayflies), Plecoptera (stoneflies) and Trichoptera (caddis flies). In the degraded section of the river the macroinvertebrate assemblages were dominated by Asellidae (louse), Chironomidae (non-biting midges) and oligochaete worms. Walsh *et al.* (2005) demonstrated that even small levels of impermeable cover (< 10% of catchment area) can have overwhelming impacts on macroinvertebrates in streams. Fish communities are also impacted upon for a variety of reasons (Morgan and Cushman, 2005). Contaminants in both water and sediment can be directly toxic to fish. Channelisation and enhanced water velocity mean that there is a reduction in slow-flow zones and off-channel habitat for refuge (particularly important for juvenile fish) and spawning. Predator fish such as pike are compromised by the loss of tree and woody debris cover for concealment, and through indirect trophic links such as the decline in numbers of prey caused by chemical contamination.

4 Point-source pollution

Humans have a long history of discharging waste products from industrial processes and sewage treatment works through pipes or culverts into rivers, and such inputs are collectively termed 'point-source pollution' (see Chapter 4 for more detailed discussion). A major source of point-source pollution is a sewage treatment works. Waste from these sources can be rich in organic effluent, which leads to severe increases in **biochemical oxygen demand** (see Chapters 4 and 9), and thus dissolved oxygen concentration decreases in river reaches downstream of the outlet pipe, due to enhanced respiration rates (Mason, 2002). The severe lack of oxygen that can ensue following organic enrichment typically leads to losses of most invertebrates and fish, with the exception of a few tolerant groups such as worms and some midge larvae. The effects of aquatic point-source pollution have even been shown to propagate into riparian (terrestrial/river bank) ecosystems (Paetzold et al., 2011). However, in the last two decades there have been major improvements in river water quality in many parts of the developed world, linked to increased investment in enhanced sewage treatment technologies and recovery of aquatic ecosystems.

Problems still remain in rivers that receive sewage works effluent because treatment processes focus predominantly on reducing organic pollution by facilitating enhanced respiration in the treatment plant (see Chapter 9). These treatment processes are largely unable to remove the active compounds contained in human pharmaceuticals, recreational drugs, and personal care products such as toothpaste, shower gels and shampoo. Research in the last couple of decades has, for example, found that male roach (*Rutilus rutilus*) have developed female characteristics in many UK rivers, due to steroidal estrogens and chemicals that mimic estrogens reaching the river in sewage effluent (Tyler and Jobling, 2008). Other studies show reduced reaction speeds of fish when avoiding predators in water polluted by antidepressant drugs (Painter et al., 2009), changes to the breakdown of plant litter by microbes and fungi in the presence of antibiotics (Bundschuh et al., 2009) and some alterations to animal physiology and immunology, rendering them more susceptible to parasites (Morley, 2009). The full ecological effects of the many hundreds of synthetic chemicals released into rivers remain to be studied in detail. Chapter 8 provides more detail on the human health effects of these modern chemicals.

5 Eutrophication

Eutrophication is the process by which waters become enriched by nutrients, in particular nitrogen and phosphorus. The process can occur naturally but it is widely augmented by human activities (Moss, 2010). Enrichment causes increased algal production, turbidity and a change in the structure and functioning of lake ecosystems. Phosphorus is regarded as being the most important nutrient driving eutrophication because it usually limits primary production, due to its relative scarcity (Dillon and Rigler, 1974). The optimum atomic ratio of nitrogen to phosphorus for primary production is 16:1 (Redfield Ratio), but in most circumstances this ratio is exceeded, implying that phosphorus is the limiting nutrient. The origins of phosphorus from rural sources are largely due to erosion of soil particles from agriculture and forestry practices. The phosphates (PO_4^{3-}) transported on these sediments are usually bound with metals. The phosphate is released only at extremes of pH. However, the most important source of phosphorus to freshwaters is through treated sewage effluents and storm drainage, which can carry high concentrations of sodium tripolyphosphate, a compound used as a component of detergents.

Agricultural fertilisers are the main source of nitrogen in the form of nitrate (NO_3^-), which is highly soluble in water. The amount of artificial fertiliser used in agriculture has increased significantly in recent decades (see Chapter 4). Current global estimates of nitrate fertiliser production are around 100 million metric tonnes, compared to less than 10 million in 1950. Consequently, the environmental impacts of this escalating use are of increasing concern. Levels of nitrates in freshwaters tend to be at their lowest during the summer, when plants and crops are actively

Table 6.3 Critical thresholds of classification for trophic status

Trophic status	Mean annual total P concentration (µg L^{-1})	Mean annual chlorophyll *a* concentration (µg L^{-1})	Mean secchi disc depth (m)
Ultra-oligotrophic	< 4	< 2.5	> 6.0
Oligotrophic	< 10	2.5–8	> 3.0
Mesotrophic	10–35	8–25	3–1.5
Eutrophic	35–100	25–75	1.5–0.7
Hypertrophic	> 100	> 75	< 0.7

Source: After Dobson and Frid, 2009. By permission of Oxford University Press: *Ecology of aquatic systems, 2nd edition.*

growing. They become higher in winter, when plant decay and higher rainfall and reduced evapotranspiration allow nitrates to leach readily from soils into watercourses. Inputs of nitrogen also originate from animal manure in rural locations.

Lakes can be classified according to their nutrient concentrations and level of production (**Table 6.3**). Oligotrophic lakes are characterised by low nutrient status, high biodiversity and low levels of primary production. The enhanced levels of nitrogen and phosphorus in freshwaters disturb the ecological balance of production between aquatic macrophytes and phytoplankton, shifting systems to the eutrophic end of the spectrum. Algae are able to take advantage of the nutrients more readily than the macrophytes and a larger phytoplankton biomass leads to increased turbidity. This reduces light penetration and therefore further reduces the extent and presence of rooted macrophytes (Irvine et al., 1989).

As nutrient loadings increase, the nature of the phytoplankton communities begins to change from low-productivity types of diatoms and green algae to highly competitive, high-productivity types such as Cyanobacteria (also known as blue-green algae; **Table 6.4**). Blue-green algae have mechanisms that are able to provide them with a competitive advantage over other forms of phytoplankton. They are able to maintain optimum buoyancy for photosynthesis using internal gas vesicles which expand and contract to alter their density and position in the water column. They are able to store phosphorus in their cells as polyphosphates, and then utilise them when phosphate levels in the water become very low. In addition, in circumstances when phosphorus levels are very high and the Redfield Ratio is less than 16:1, then nitrogen availability becomes the limiting factor. Many types of blue-green algae are able to take advantage of these conditions and utilise nitrogen by fixation of N_2. Eutrophic conditions are characterised by frequent and intense algal blooms, high productivity, high turbidity and low visibility, and are dominated by species of blue-green algae.

Changes to the phytoplankton community also have effects on the composition of the zooplankton species which consume them. The mechanisms involved could include one or a combination of the following: the ability of the zooplankton species to graze on the algae; nutritional deficiencies of the phytoplankton; production of toxins by phytoplankton; and

Table 6.4 Examples of algal types associated with trophic status

Productivity	Main algal groups	Examples of typical algal genera
Oligotrophic	Diatoms	*Tabellaria, Cyclotella*
	Desmids	*Staurastrum*
Mesotrophic	Dinoflagellates	*Ceratium, Peridinium*
	Diatoms	*Asterionella, Cyclotella*
Eutrophic	Cyanobacteria	*Microcystis, Anabaena*
	Diatoms	*Asterionella, Fragilaria*

external factors, including other anthropogenic influences. The invertebrate community alters from one of high diversity and low biomass to one of low diversity and high biomass. In oligotrophic conditions, invertebrate species inhabit the macrophytes in lake shores, which provide refuge from predating fish. However, in eutrophic conditions, these are lost and the community becomes dominated by decomposers such as tubificids and chironomids in the benthic sediment.

Changes in fish communities have also been noted with the onset of eutrophication, due to a number of factors. The ability to prey upon different forms of zooplankton varies according to the foraging behaviour and feeding adaptations of each individual fish species. Consequently, changes in zooplankton species as prey will cause changes in fish species as predators. Furthermore, increased turbidity and loss of visibility, in combination with changes in the invertebrate community, means that there is a shift in predation from species using visual means to seek prey to those which forage among the benthos using olfactory means. The pattern of change among fish is from coregonids and salmonids in oligotrophic conditions to cyprinids such as bream in eutrophic conditions.

A series of wider issues are problematic in eutrophic aquatic systems. The high biomass of phytoplankton may become a problem during hours of darkness, when respiration is far in excess of photosynthesis. Severe oxygen depletion has resulted in fish kills around the world. Eutrophication gives rise to high production in the epilimnion in lakes with thermal stratification. This exacerbates the depletion of oxygen in the hyperlimnion, when the dead phytoplankton sink into the hypolimnion to decompose. Algal blooms have also been associated with the production of harmful toxins because the blue-green algae such as *Mycrocystis* and *Anabaena* produce hepatotoxins and neurotoxins (see section D1 of

Figure 6.17 Intense algal growth on a eutrophic water body next to a public footpath.
Source: Photo courtesy of Shutterstock/Kris Butler.

Chapter 8). Algal blooms can reduce the effectiveness of water treatment processes for supply by causing coagulation, sedimentation and filter blocking. They can also reduce the amenity and recreational value of freshwaters (**Figure 6.17**; see also **Figure 8.8** in Chapter 8).

6 Climate change (environmental warming and altered water availability)

The effects of climate change on freshwaters can be thought of as a complex amalgam of stressors, including acidification resulting from elevated atmospheric sulphate and nitrate concentrations, elevated dissolved CO_2 and the more obvious changes in water availability (see Chapter 2) and water temperature (Woodward et al., 2010). Freshwater ecosystems are likely to be highly vulnerable to climate change because they are relatively isolated and physically fragmented within a largely terrestrial landscape, and dispersal of organisms to other systems may be difficult, owing to a lack of hydrological connectivity and/or the poor dispersal abilities of many aquatic organisms. Freshwaters are also already heavily exploited by humans for the provision of 'goods and services' (see section D and Chapter 10) and they may be less resistant/resilient to change than might be the case if they were entirely pristine. A comprehensive discussion of how climate change can be expected to alter aquatic ecosystems could take up an entire book; therefore, this section refers only to selected studies to illustrate some possible responses. Those wishing to explore the effects of climate change on aquatic ecosystems in more detail can find good starting points in comprehensive review papers such as Heino et al. (2008), Palmer et al. (2009), Whitehead et al. (2009) and Woodward et al. (2010).

A major driver of aquatic ecosystem response to climate change will be alterations to river and lake thermal regimes. Evidence that aquatic ecosystems are warming is growing, with studies to date indicating significant warming of streams in the USA (e.g. Kaushal et al., 2010) and central Europe (Hari et al., 2006; Webb and Nobilis, 2007), and of various lakes around the world (Schneider and Hook, 2010). These thermal changes are expected to have strong effects on aquatic plants and animals because the **metabolism** and foraging behaviour of individual organisms is strongly influenced by temperature, and in freshwaters many organisms are ectothermic (i.e. unable to regulate their body temperature and therefore reliant on environmental temperature; see Woodward et al., 2010). For example, fish are particularly sensitive to changes in environmental temperature and some studies have forecast that up to 75% of local fish biodiversity could be lost by 2070, due to combined changes in flow and thermal regimes linked to climate change (Xenopoulos et al., 2005). Recent studies in experimental ponds have also illustrated how climate change might shift the metabolic balance of some freshwaters from net autotrophic (i.e. carbon sinks) to net heterotrophic (i.e. carbon sources) because under warming scenarios ecosystem respiration increases at a faster rate than primary production (Yvon-Durocher et al., 2010). In rivers, warming may lead to changes in the processing of accumulated terrestrial plant litter. This is because, as temperature increases, microbial breakdown of leaf litter increases but consumption by detritivores decreases (Boyero et al., 2011). Thus, while there should be no net difference in breakdown, the shift towards microbial decomposition could drive greater CO_2 production as opposed to carbon being bound up in detrital particles following detritivore consumption and processing.

Environmental warming has been shown to alter the structure of aquatic ecosystems in many studies. For example, O'Reilly et al. (2003) have observed 0.1°C warming per decade since 1913 in the upper water column of Lake Tanganyika, East Africa. This warming is associated with decreases in the primary productivity of the lake, and consequent declines in catches of fish. The latter organisms are a vital source of protein to humans

> **BOX 6.3 CASE STUDIES**
>
> ### Warming of Lake Washington, USA
>
> In a study of Lake Washington, northwest USA, Winder and Schindler (2004) documented warming trends since 1962. Lake Washington does not freeze and the water column mixes completely in winter but develops thermal stratification from around April to November, when warm water overlies a cold layer. Spring water temperatures were found to have risen, on average, by 1.4°C in the upper 10 m of the lake over the 40-year monitoring period. The researchers found that lake stratification occurred approximately 21 days earlier at the end than at the start of the study. This thermal change was linked with blooms of phytoplankton producers, which occurred earlier in the year in the latter part of the record, as compared with the earlier part of the record. However, the primary consumers of phytoplankton, the zooplankton *Daphnia*, did not respond accordingly and therefore the trophic link between these organisms was disrupted. This example illustrates how warming may produce ecosystem changes in ways that are difficult to predict, particularly because knowledge of the intricate direct and indirect links between species in aquatic systems remains relatively poorly understood.

living in countries surrounding the lake, and so further warming and ecosystem change could have major consequences for food supply in this region. **Box 6.3** provides an example of changes caused by lake warming in the USA.

In North Wales, UK, some of the longest continuous records of stream habitat data and aquatic invertebrates have been collected since 1981. By 2005, streams from open moorland had seen average temperature increases of 1.7°C, while forest streams increased by up to 1.4°C (Durance and Ormerod, 2007). The effects of this warming appear to be strongest in streams with circum-neutral pH, with reductions in the abundance of invertebrates and changes in the composition of communities observed. The authors hypothesised that for each future temperature increase of 1°C there could be a reduction of spring invertebrate abundance by 21%. Similar long-term studies of invertebrates in rivers of southeast Alaska have documented significant changes with widespread glacial retreat and subsequent stream ecosystem development since the late 1970s (Milner *et al*., 2008, 2011). More broadly, glacier retreat is occur-

ring across much of the world and is strongly linked to a warming climate. Further retreat is expected to lead to major shifts in the water sourcing of Arctic and alpine rivers, with glacier- and snow-melt reductions and changes in river hydrological, geomorphological and water quality dynamics. These changes could have significant, widespread consequences for plants and animals of river ecosystems, leading to changes in species diversity (potentially including some species extinctions) and abundance and the structure of food webs (Brown *et al*., 2007; Milner *et al*., 2009).

7 Invasive species

Biological invasions are defined as the spread of a non-indigenous species from the point of introduction to some substantial level of abundance, and aquatic ecosystems are particularly at risk of invasions by plants, invertebrates and vertebrates. Invasions are a major economic cost and rank second only to habitat destruction as drivers of biodiversity loss. Species that are deliberately introduced to a new habitat for aquaculture (see section E8) often escape and become invaders in the new habitat. Furthermore, accidental transportation and introduction by shipping, particularly as a result of ballast water exchange, is a major source of aquatic species introductions. Invasive species can change the physical properties of the habitat (ecosystem 'engineering'; Jones et al., 1994) as well as competing with, becoming prey for, preying on or parasitising native species. They can also drive changes in the productivity and biodiversity of aquatic ecosystems. Successive invasions have transformed freshwater communities; for example, the Rhine running through central and western Europe has suffered continual introductions of new species as a result of shipping traffic as well as the building of canals which have increased connections with other rivers in Europe. As a result, by 2006, invasive species were estimated to make up 90% of macroinvertebrate numbers (van Riel et al., 2006).

The water hyacinth (*Eichhornia crassipes*) has been described as the world's worst aquatic weed. It is an ecosystem engineer that affects the habitat structure of lakes and rivers, with knock-on effects on other species and great economic costs (Lowe et al., 2000). It originated in South America and has been introduced to new habitats as an ornamental plant. Once introduced to tropical or subtropical habitats, it can spread rapidly and populations can double in size in only 12 days. The water hyacinth impedes water flow, navigation and fishing and reduces biodiversity, as it deprives native plants of space, sunlight and oxygen. For example, the spread of this weed in Lake Victoria in Kenya has led to closures of a hydroelectric plant as well as creating habitats for mosquitoes and snails, vectors for the human diseases malaria and schistosomiasis. Invasive animals can also act as ecosystem engineers. The Chinese mitten crab (*Eriocheir sinensis*) damages banks (including flood defences) by burrowing into sediments, creating a network of tunnels that can eventually cause bank collapse. Banks along the River Thames in southeast England have receded up to 6 m as a result of mitten crab burrowing. The native range of this crab covers the Far East, but it was reported in Germany in 1912 and has spread through northern Europe (predominantly in ballast water) over the last 100 years. Significant populations are found in the Thames and Humber rivers in the UK, and across North America (Lowe et al., 2000). This invader affects both marine and freshwater ecosystems, as it is catadromous; the adults reproduce in saline estuaries and the juveniles then migrate upstream to freshwater, where they live for three to five years before migrating back to saline water as adults to breed.

For other invaders, their trophic interactions with other species may affect ecosystem structure and functioning. The amphipod *Dikerogammarus villosus* (aka 'the killer shrimp') invaded western Europe via the Danube–Main–Rhine canal, which opened in 1992. The rate of downstream movement has been estimated at 124 km per year, as a result of drift and transport on boats (van Riel et al., 2006). It has recently been recorded in the UK for the first time (MacNeil et al., 2010). In some habitats, there is evidence that the beds of the invasive zebra mussel, *Dreissena polymorpha*, facilitate invasion by the killer shrimp by providing habitat and food. The main impact of *D. villosus* stems from its predatory behaviour, which affects the abundance of a range of invertebrates and also has cascading effects through the ecosystem. Stable isotope analysis revealed the killer shrimp to be on the same trophic level as predatory fish (van Riel et al., 2006). It has been shown to eat more prey in the laboratory than does the native amphipod *G. pulex*, and in the field it may displace native amphipods both by outcompeting them and by preying on them (Bollache et al., 2008). Native amphipods are weaker predators, but process more detritus by shredding than does the killer shrimp. Hence, displacement of native amphipods by this invader could have knock-on effects on ecosystem functioning within invaded ecosystems (MacNeil et al., 2011).

Parasites (or a loss of parasites) can also affect invasions (Dunn, 2009). Animal and plant population growth is often regulated by parasites which cause increased mortality. However, invading species may lose their parasites either because, by chance, only uninfected hosts are introduced to the new habitat; or because parasitised (sick) individuals are more likely to die during translocation to, and during establishment in, the new range. This process is called 'enemy release' and may be one factor that explains the success of an invader and its ability to outcompete native species. For example, in Northern Ireland, UK, the native freshwater amphipod crustacean competes with three invasive amphipods: a European amphipod, *Gammarus pulex*, which is gradually replacing the native species, and two other species that arrived from North America via ballast water and the transport of ornamental pond plants. In Northern Ireland the native species harbours five different species of parasite which reduce its growth and survival, while the invaders are affected by only three (Dunn and Dick, 1998).

Some invaders are themselves parasites, brought to new habitats along with their invading hosts. These introduced parasites can go on to cause disease outbreaks in native hosts (emerging infectious diseases). For example, crayfish plague is a fungal pathogen which was accidentally introduced to Europe in the 1870s, when American signal crayfish were imported to fish farms. Signal crayfish carry the disease but are not killed by it; they act as reservoirs of infection, transmitting the parasite to endangered native crayfish. As a result of its competitive abilities and its ability to transmit plague to its native competitors, the signal crayfish has spread through large areas of Europe, driving native crayfish extinct (Holdich *et al.*, 2009). Similarly, spread of the fungal disease Chytridiomycosis by the American bullfrog is a key cause of worldwide decline in the diversity and abundance of frogs and toads (Garner *et al.*, 2006).

8 Aquaculture

Aquaculture is typically thought of as the production of fish under captive and controlled conditions (**Figure 6.18**). However, it also includes the production of aquatic plants, and animals such as crustaceans and molluscs, under similar circumstances. In this section most of the discussion will relate to fish production through aquaculture. Over 80% of global aquaculture is in Asia, with >50% occurring in China. The scale and intensity of aquaculture-based primary and secondary production may vary enormously, from low-intensity extensive ponds to highly controlled production on an industrial scale. In some situations the temperature, light and feeding regimes are completely optimised to ensure efficient production. Production through aquaculture methods has more than quadrupled since 1990 (**Table 6.5**). In comparative terms, the increase in levels of production by capture methods are much lower. This could be the result of overfishing, leading to the depletion of some wild stocks and the recognition that production under more controlled conditions is more reliable in terms of satisfying demand.

The environmental impacts of aquaculture can be complex and far-reaching. Aquaculture requires large volumes of water, and abstraction of water from the environment can lead to significant changes in hydrological regimes and so disturb ecosystems. The waste water from aquaculture operations will have high levels of ammonia and is also contaminated with excess feed and faecal material. This can lead to high levels of suspended sediments, elevation of biochemical oxygen demand and reduced levels of oxygen concentration. Camargo (1992) observed the impacts of a trout farm discharge on the macroinvertebrate community of the stream receiving the waste water. In the area immediately upstream and downstream it was found that species richness was reduced from 38 to 19 and the BMWP score (a measure of water quality; **Box 6.4**) was reduced from 168 to 19. The dense

Figure 6.18 Fish farm in a sheltered bay off the Isle of Skye, northern Scotland.

numbers of fish in same-age monocultures are also vulnerable to disease and parasite infestation. This may be exacerbated where populations have been manipulated genetically in order to improve productivity, but at the expense of the ability to resist diseases.

It is common for the feed used in aquaculture to be produced artificially, with high levels of nitrogen and phosphorus to improve the potential productivity. Enrichment of freshwaters with these nutrients in areas around rearing pens in lakes or in rivers receiving farm effluent can lead to issues of eutrophication (see section E5) if they are not controlled. The resultant algal blooms may produce toxins and, as they die and decompose, they can cause oxygen depletion. In addition, antibiotics and other therapeutic components are frequently added to artificial feed pellets. The toxic

Table 6.5 Comparison of aquaculture and capture production (tons) between 1990 and 2009

Production method	1990	2009
Global aquaculture (all species)	16,840,078	73,044,604
Global capture (fish species only)	74,788,090	76,014,632
Global freshwater aquaculture (all species)	7,630,758	33,837,426
Global freshwater aquaculture (fish species only)	7,056,194	29,976,476
Global freshwater capture (fish species only)	5,361,650	8,887,075

Source: United Nations Food and Agriculture Organisation.

BOX 6.4 CASE STUDIES

Biomonitoring and the European Water Framework Directive

The Water Framework Directive (WFD) is a piece of legislation developed by the European Union which uses ecosystem-based objectives to manage water resources (Kallis and Butler, 2001). It is intended to provide a common basis for the protection and sustainable use of water, in terms of quantity and quality, across the member states of the EU. The approach aims to rank the ecological status of rivers and lakes onto an ordinal scale of High, Good, Moderate, Poor and Bad. These status classifications are determined on the basis of observed versus expected values of biological groups, including diatoms, macrophytes, macroinvertebrates and fish. The classification system also considers hydrogeomorphological and water chemistry parameters. The lowest classification across the biological, hydrogeomorphological and water chemistry parameters determines the 'ecological' status of a water body. Expected values are determined from databases of high-quality reference sites. To ensure comparability between test and reference sites, a typology based on catchment geology, catchment size and altitude is used (Logan and Furse, 2002).

There are a number of different methods that have been developed to assess ecological status using various organisms and across a variety of aquatic habitats. One system that was developed and used widely prior to the introduction of the WFD focuses on macroinvertebrate benthic fauna and utilises the Biological Monitoring Working Party (BMWP) scoring system (Armitage *et al.*, 1983) to assess water quality. This approach has subsequently informed approaches such as the River Invertebrate Prediction and Classification System (RIVPACS; Wright *et al.*, 2000), which can be used to derive the type-specific 'expected' values that feed into the determination of ecological status classification. Similar approaches are used in Australia (Australian River Assessment System – AUSRIVAS). In the USA, individual states have different monitoring approaches, but some have adopted RIVPACS-type approaches.

The BMWP score for a location can be determined by taking a standard three-minute kick sample from the river bed, and identifying all the families of macroinvertebrates that are present. The families of macroinvertebrates are given scores between 1 and 10 in relation to their perceived tolerance to pollution (a score of 10 is given to organisms that are least tolerant to pollution, predominantly stoneflies, mayflies and caddis flies). A total BMWP score is then derived for the location from the sum of the scores of each family in the sample. An unpolluted location may have a BMWP score in excess of 100. Additional measures that can be calculated from the sample are the number of scoring taxa in the sample (*N* taxa) and the average score per taxon (ASPT). The ASPT is calculated as the total BMWP/*N* taxa, and is considered to be more robust to variations in sample size, sampling effort and seasonal fluctuations (Armitage *et al.*, 1983).

One of the main aims of the WFD is to prevent the deterioration of water bodies that are currently at good or better status, and to improve those that fall below this level. Member states are expected to consider ways to improve failing water bodies up to a minimum of good status by 2015, in the first instance. Recent evaluations have been undertaken to inform River Basin Management Plans for the EU WFD. Only 29% of English and Welsh water bodies (including rivers, lakes and groundwater) were classed as having 'good' status or above. In Scotland the picture was better, with 64% of all surface water bodies classified as 'good' or better (SEPA, 2009), while Romania had a slightly lower number

continued

(59%) in compliance with WFD targets. In France, the Rhine-Mass basin had 37% compliance whereas the Loire-Bretagne had only 25%. The WFD does allow member states to exempt some water bodies from the classification. Examples include surface waters where necessary improvements cannot reasonably be achieved by 2015, those where improvements in status are unfeasible or unreasonably expensive, if unforeseen or exceptional circumstances occur, or where new modifications are necessary to a water body (e.g. flood control).

impact of such chemicals released from uneaten pellets as they decompose is still to be properly assessed and understood in freshwater environments.

A much more significant issue that arises from aquaculture is the ecological and genetic impact on the environment of fast-growing alien species and **genotypes** that escape from captivity (Naylor *et al.*, 2001; see also section E7). The minimum impact that escaped organisms may have is to disturb the balance of the natural ecosystem through interactions with the communities that exist in their vicinity. This can be propagated on a longer term, perhaps even permanently, if the escapees are able to establish a sustainable breeding population. This has been a common problem associated with species of *Tilapia* which have originated in Africa but are now widespread through the continents around the tropics (De Silva *et al.*, 2004; Pullen *et al.*, 1997). Where aquaculture is used to increase the production of native species, then when individuals escape from the aquaculture system into the wild there can be large problems. The genetics of a cultured population of a species often exhibit a high level of inbreeding and abnormally high growth rates, and this can be at the expense of other important genetic traits. Consequently, these individuals compete aggressively with indigenous populations for food and habitat. Moreover, cross-breeding between captive and wild populations can lead to decreased fitness of the indigenous population because of the impact of the alien genotypes on the local gene pool (Ferguson *et al.*, 2007).

> **REFLECTIVE QUESTION**
>
> In what ways is it likely that humans have had an impact on the aquatic ecosystem of your nearest lake or river? What do you think the key impacts of warming and more extreme conditions (more droughts and more intense rainfall events) would be on the aquatic ecosystem in the river closest to your home?

F SUMMARY

This chapter has provided an introduction to some of the diverse habitats and biotic groups that can be found in rivers and lakes, and the differences between these two types of freshwater system. The trophic roles of various organisms have been considered, and the chapter has explored how freshwater ecosystem structure can be considered holistically through the study of food webs. The main functional processes that are studied in freshwater systems, namely primary production, secondary production, community respiration and nutrient cycling, have been explored. Some of the key methods used by aquatic scientists to study these processes have been outlined, and a consideration of the ecosystem services that these processes provide to humans has been introduced. The later sections of the chapter have provided overviews of some of the main ways in which human activities have led to the modification of aquatic ecosystems. It is important to remember that rivers and lakes are particularly diverse and well-studied ecosystems. The ideas discussed herein for the most part provide only introductions to general frameworks or ideas, rather than in-depth blueprints for how individual river or lake systems can be expected to work. The reader is encouraged to consult the recommended Further reading and key research papers/references to build up a more detailed understanding and knowledge of freshwater ecosystems.

FURTHER READING

Allan, D.J. and Castillo, M.M. 2007. *Stream ecology: structure and function of running waters.* Springer; Netherlands.

Provides a detailed review of most topics relevant to the study of stream ecology, including coverage of stream hydrology, water chemistry and human modifications of river systems.

Closs, G., Downes, B. and Boulton, A. 2004. *Freshwater ecology: a scientific introduction.* Blackwell Publishing; Oxford.

An introductory text to both rivers and lakes, aimed at those who have not previously studied freshwater ecology, including discussion of scientific methodology, key concepts and case studies. Each chapter is based around a key research question in aquatic ecology.

Hauer, F.R. and Lamberti, G. 2006. *Methods in stream ecology.* Elsevier; New York.

This book provides an overview of the scientific rationale and applications of most common methods used by stream ecologists, as well as more technical information on how to process data.

Mason, C.F. 2002. *Biology of freshwater pollution.* Pearson; London.

Discusses the multiple alterations of freshwater ecosystems such as through pollution, habitat modification and the introduction of invasive species.

Moss, B. 2010. *Ecology of freshwaters: a view for the twenty-first century.* Wiley-Blackwell; Chichester.

Mainly discusses lakes and wetlands but has some chapters on rivers. The book provides a detailed overview and is aimed at students who wish to gain an integrated view of freshwaters.

Wetzel, R.G. 2001. *Limnology. Lake and river ecosystems, 3rd edition.* Academic Press; California.

Provides a detailed review of lake ecosystems, with some comparative assessment of rivers in places. The text is more suited to advanced study.

Classic papers

Resh, V.H., Brown, A.V., Covich, A.P., Gurtz, M.E., Li, H.W., Minshall, G.W., Reice, S.R., Sheldon, A.L. and Wallace, J.B. 1988. The role of disturbance in stream ecology. *Journal of the North American Benthological Society* 7: 433–455.

An examination of how disturbance impacts on aquatic ecosystems.

Scheffer, M., Hosper, S.H., Meijer, M.L., Moss, B. and Jeppesen, E. 1993. Alternative equilibria in shallow lakes. *Trends in Ecology and Evolution* 8: 275–279.

Very important theory about non-linearity in lake systems which makes restoration of damaged lakes more difficult.

Schindler, D.W. 1974. Eutrophication and recovery in experimental lakes: implications for lake management. *Science* 184: 897–899.

An early but important paper on eutrophication and how to get lakes out of a eutrophic state.

Vannote, R.L., Minshall, G.W., Cummins, K.W., Sedell, J.R. and Cushing, C.E. 1980. The river continuum concept. *Canadian Journal of Fisheries and Aquatic Sciences* 37: 130–137.

Very important paper of relevance to the introductory material within this chapter.

PROJECT IDEAS

- Using the kick net sampling method, collect and identify samples of macroinvertebrates at several locations along your local river. At each site, measure variables such as dissolved oxygen concentration, pH, water temperature and electrical conductivity using handheld probes, and collect a sample of water for analysis of nutrients and metals in the laboratory. Hydrological and geomorphological measurements (e.g. width, depth, discharge) can also be measured. When analysing your datasets you could consider how the macroinvertebrates vary along the river, and how they are related to changes in hydrology, geomorphology and/or water quality.

- Contact your local environmental monitoring agency and request hydrological, water quality and/or ecological datasets for your local river/lake if they are available. Using such datasets you may be able to analyse whether there have been any notable changes over time in relation to changes in catchment land use or year-to-year weather conditions. Alternatively, you could compare your local river/lake to others in your region/country to assess what are the reasons for spatial differences.

- Collect freshly fallen leaves from one or more species of tree growing adjacent to your local river and then dry them out for a few weeks. You can then construct litter bags using known weights of leaf litter (a few grams per bag will suffice). Litter bags can be constructed from fine mesh to exclude invertebrates and gain a measure of fungal/microbial decomposition processes, and/or a coarse mesh which allows invertebrates to colonise, giving a measure of total breakdown. Litter bags can then be anchored to the bed of your local river for several weeks, during which time they will begin to decompose. You could leave all the bags in place for a fixed period of time, or remove some at various intervals to assess how litter decomposes over time. After retrieval of the litter bags from the river, return to the laboratory and carefully rinse off any sediment and attached invertebrates. Dry the remaining leaf litter, reweigh the material and subtract the value from the litter weight prior to placing them in the river, to assess how much decomposition has taken place.

REFERENCES

Acreman, M.C., Aldrick, J., Binnie, C., Black, A.R., Cowx, I., Dawson, F.H., Dunbar, M.J., Extence, C., Hannaford, J., Harby, A., Holmes, N.T., Jarrett, N., Old, G., Peirson, G., Webb, J. and Wood, P.J. 2009. Environmental flows from dams: the Water Framework Directive. *Engineering Sustainability* 162: 13–22.

Allan, J.D. 2004. Landscapes and riverscapes: the influence of land use on stream ecosystems. *Annual Review of Ecology, Evolution and Systematics* 35: 257–284.

Allan, J.D. and Castillo, M.M. 2007. *Stream ecology: structure and function of running waters.* Springer; Netherlands.

Amundsen, P.A., Lafferty, K.D., Knudsen, R., Primicerio, R., Klemetsen, A. and Kuris, A.M. 2009. Food web topology and parasites in the pelagic zone of a subarctic lake. *Journal of Animal Ecology* 78: 563–572.

Armitage, P.D. 2006. Long-term faunal changes in a regulated and an unregulated stream – Cow Green thirty years on. *River Research and Applications* 22: 947–966.

Armitage, P.D., Moss, D., Wright, J.F. and Furse, M.T. 1983. The performance of a new biological water quality score system based on macroinvertebrates over a wide range of unpolluted running-water sites. *Water Research* 17: 333–347.

Arthington, A.H., Bunn, S.E., Poff, N.L. and Naiman, R.J. 2006. The challenge of providing environmental flow rules to sustain river ecosystems. *Ecological Applications* 16: 1311–1318.

Babbitt, B. 2002. What goes up, may come down. *BioScience* 52: 656–658.

Baxter, C.V., Fausch, K.D. and Saunders, W.C. 2005. Tangled webs: reciprocal flows of invertebrate prey link streams and riparian zones. *Freshwater Biology* 50: 201–220.

Bednarek, A. 2001. Undamming rivers: a review of the ecological effects of dam removal. *Environmental Management* 27: 803–814.

Benfield, E.F. 2006. Leaf breakdown in stream ecosystems. *In*: Hauer, F.R. and Lamberti, G. A. (eds) *Methods in stream ecology*. Elsevier; New York, 579–590.

Benke, A.C. 1984. Secondary production of aquatic insects. *In*: Resh, V.H. and Rosenberg, D.M. (eds) *Ecology of aquatic insects*. Praeger; New York, 289–322.

Benke, A.C. 2010. Secondary production. *Nature Education Knowledge* 3: 23.

Benke, A.C. and Wallace, J.B. 1997. Trophic basis of production among riverine caddisflies: implications for food web analysis. *Ecology* 78: 1132–1145.

Bollache, L., Dick, J.T.A., Farnsworth, K.D. and Montgomery, W.I. 2008. Comparison of the functional responses of invasive and native amphipods. *Biology Letters* 4: 166–169.

Bott, T.L. 2006. Primary productivity and community respiration. *In*: Hauer, F.R. and Lamberti, G.A. (eds) *Methods in stream ecology*. Elsevier; New York, 263–290.

Boyero, L. *et al.* 2011. A global experiment suggests climate warming will not accelerate litter decomposition in streams but might reduce carbon sequestration. *Ecology Letters* 14: 289–294.

Brown, L.E., Hannah, D.M. and Milner, A.M. 2007. Vulnerability of alpine stream biodiversity to shrinking glaciers and snowpacks. *Global Change Biology* 13: 958–966.

Brown, L.E., Edwards, F.K., Milner, A.M., Woodward, G. and Ledger, M.E. 2011. Food web complexity and allometric-scaling relationships in stream mesocosms: implications for experimentation. *Journal of Animal Ecology* 80: 884–895.

Bundschuh, M., Hahn, T., Gessner, M.O. and Schultz, R. 2009. Antibiotics as a chemical stressor affecting an aquatic decomposer-detritivore system. *Environmental Toxicology and Chemistry* 28: 197–203.

Camargo, J.A. 1992. Temporal and spatial variations in dominance, diversity and biotic indices along a limestone stream receiving a trout farm effluent. *Water, Air and Soil Pollution* 63: 343–359.

Closs, G., Downes, B. and Boulton, A. 2004. *Freshwater ecology: a scientific introduction*. Blackwell Publishing; Oxford.

Crisp, D.T., Mann, R.H.K. and McCormack, J.C. 1975. The populations of fish in the River Tees system on the Moor House National Nature Reserve, Westmorland. *Journal of Fish Biology* 7: 573–593.

Cummins, K.W., Merritt, R.W. and Andrade, P.C.N. 2005. The use of invertebrate functional groups to characterise ecosystem attributes in selected streams and rivers in south Brazil. *Studies on Neotropical Fauna and Environment* 40: 69–89.

De Silva, S.R., Subasinghe, R., Bartley, D. and Lowther, A. 2004. *Tilapias as alien aquatics in Asia and the Pacific: a review*. FAO; Rome.

Dick, J.T.A., Armstrong, M., Clarke, H.C., Farnsworth, K.D., Hatcher, M.J., Ennis, M., Kelly, A. and Dunn, A.M. 2010. Parasitism may enhance rather than reduce the predatory impact of an invader. *Biology Letters* 6: 636–638.

Dillon, P.J. and Rigler, F.H. 1974. The phosphorous–chlorophyll relationship in lakes. *Limnology and Oceanography* 19: 767–773.

Dobson, A., Lafferty, K.D., Kuris, A.M., Hechinger, R.F. and Jetz, W. 2008. Homage to Linnaeus: How many parasites? How many hosts? *Proceedings of the National Academy of Sciences of the United States of America* 105: 11482–11489.

Dobson, M. and Frid, C. 2009. *Ecology of aquatic systems, 2nd edition*. Oxford University Press; Oxford.

Dudgeon, D., Arthington, A.H., Gessner, M.O., Kawabata, Z., Knowler, D.J., Lévêque, C., Naiman, R.J., Prieur-Richard, A.H., Soto, D., Stiassny, M.L.J. and Sullivan, C.A. 2006. Freshwater biodiversity: importance, threats, status and conservation challenges. *Biological Reviews* 81: 163–182.

Dunn, A.M. 2009. Parasites and biological invasions. *Advances in Parasitology* 68: 161–184.

Dunn, A.M. and Dick, J.T.A. 1998. Parasitism and epibiosis in native and non-native gammarids in freshwater in Ireland. *Ecography* 21: 593–598.

Durance, I. and Ormerod, S.J. 2007. Effects of climatic variation on upland stream invertebrates over a 25 year period. *Global Change Biology* 13: 942–957.

Ensign, S.H. and Doyle, M.W. 2006. Nutrient spiralling in streams and river networks. *Journal of Geophysical Research* 111: G04009.

Ferguson, A., Fleming, I., Hindar, K., Skaala, Ø., McGinnity, P., Cross T.F. and Prodöhl, P. 2007. Farm escapes. *In:* Verspoor, E., Stradmeyer, L. and Nielsen, J.L. (eds) *The Atlantic salmon: genetics, conservation and management.* Blackwell Publications Ltd; Oxford, 357–398.

Frissell, C.A., Liss, W.J., Warren, C.E. and Hurley, M.D. 1986. A hierarchical framework for stream habitat classification: viewing streams in a watershed context. *Environmental Management* 10: 199–214.

Garner, T.W.J., Perkins, M.W., Govindarajulu, P., Seglie, D., Walker, S., Cunningham, A.A. and Fisher, M.C. 2006. The emerging amphibian pathogen *Batrachochytrium dendrobatidis* globally infects introduced populations of the North American bullfrog, *Rana catesbeiana*. *Biology Letters* 2: 455–459.

Gessner, M.O., Chauvet, E. and Dobson, M. 1999. A perspective on leaf litter breakdown in streams. *Oikos* 85: 377–384.

Hall, R.O., Wallace, J.B. and Eggert, S.L. 2000. Organic matter flow in streams with reduced detrital resource base. *Ecology* 81: 3445–3463.

Harding, J.S., Benfield, E.F., Bolstad, P.V., Helfman, G.S. and Jones, E.B.D. 1998. Stream biodiversity: the ghost of land use past. *Proceedings of the National Academy of Sciences* 95: 14843–14847.

Hari, R.E., Livingstone, D.M., Siber, R., Burkhardt-Holm, P. and Guttinger, H. 2006. Consequences of climatic change for water temperature and brown trout populations in Alpine rivers and streams. *Global Change Biology* 12: 10–26.

Hatcher, M.J. and Dunn, A.M. 2011. *Parasites in ecological communities; from interactions to ecosystems.* Cambridge University Press; Cambridge.

Hauer, F.R. and Lamberti, G. 2006. *Methods in stream ecology.* Elsevier; New York.

Hechinger, R.F., Lafferty, K.D., Mancini III, F.T., Warner, R.R. and Kuris, A.M. 2009. How large is the hand in the puppet? Ecological and evolutionary factors affecting body mass of 15 trematode parasitic castrators in their snail host. *Evolutionary Ecology* 23: 651–667.

Heino, J., Virkkala, R. and Toivonen, H. 2008. Climate change and freshwater biodiversity: detected patterns, future trends and adaptations in northern regions. *Biological Reviews* 84: 39–54.

Hernandez, A.D. and Sukhdeo, M.V.K. 2008a. Parasite effects on isopod feeding rates can alter the host's functional role in a natural stream ecosystem. *International Journal for Parasitology* 38: 683–690.

Hernandez, A.D. and Sukhdeo, M.V.K. 2008b. Parasites alter the topology of a stream food web across seasons. *Oecologia* 156: 613–624.

Hershey, A.E. and Lamberti, G.A. 1998. Stream macroinvertebrate communities. *In*: Naiman, R.J. and Bilby, R.E. (eds) *River ecology and management: lessons from the Pacific coastal ecoregion.* Springer-Verlag; New York, 169–199.

Holdich, D.M., Reynolds, J.D., Souty-Grosset, C. and Sibley, P.J. 2009. A review of the ever increasing threat to European crayfish from non-indigenous crayfish species. *Knowledge and Management of Aquatic Ecosystems* 11: 394–395.

Howarth, R.W. and Michaels, A.F. 2000. The measurement of primary production in aquatic ecosystems. *In*: Sala, O.E., Jackson, R.B., Mooney, H.A. and Howarth, R.W. (eds) *Methods in ecosystem science.* Springer New York, 72–85.

Huryn, A.D. 1996. An appraisal of the Allen paradox in a New Zealand trout stream. *Limnology and Oceanography* 41: 243–252.

Huryn, A.D. and Wallace, J.D. 2000. Life history and production of stream invertebrates. *Annual Review of Entomology* 45: 83–110.

Hynes, H.B.N. 1970. *The ecology of running waters.* University of Toronto.

Irvine, K., Moss, M. and Balls, H. 1989. The loss of submerged plants with eutrophication II. Relationships between fish and zooplankton in a set of experimental ponds, and conclusions. *Freshwater Biology* 22: 89–107.

Jones, C.G., Lawton, J.H. and Shachak, M. 1994. Organisms as ecosystem engineers. *Oikos* 69: 373–386.

Kallis, G. and Butler, D. 2001. The EU water framework directive: measures and implications. *Water Policy* 3: 125–142.

Kaushal, S.S., Likens, G.E., Jaworski, N.A., Pace, M.L., Sides, A.M., Seekell, D., Belt, K.T., Secor, D.H. and Wingate, R.L. 2010. Rising stream and river temperatures in the United States. *Frontiers in Ecology and the Environment* 8: 461–466.

Kohler, S.L. 2008. The ecology of host–parasite interactions in aquatic insects. *In*: Lancaster, J., Briers, R. and Macadam, C. (eds) *Aquatic insects: challenges to populations*. CAB International; Wallingford.

Kuris, A.M., Hechinger, R.F., Shaw, J.C., Whitney, K.L., Aguirre-Macedo, L., Boch, C.A., Dobson, A.P., Dunham, E.J., Fredensborg, B.L., Huspeni, T.C., Lorda, J., Mababa, L., Mancini, F.T., Mora, A.B., Pickering, M., Talhouk, N.L., Torchin, M.E. and Lafferty, K.D. 2008. Ecosystem energetic implications of parasite and free-living biomass in three estuaries. *Nature* 454: 515–518.

Lafferty, K.D., Dobson, A.P. and Kuris, A.M. 2006. Parasites dominate food web links. *Proceedings of the National Academy of Sciences of the United States of America* 103: 11211–11216.

Lamberti, G.A. and Steinman, A.D. 1997. Comparison of primary production in stream ecosystems. *Journal of the North American Benthological Society* 16: 95–104.

Ledger, M.E., Edwards, F.K., Brown, L.E., Milner, A.M. and Woodward, G. 2011. Impact of simulated drought on ecosystem biomass production: an experimental test in stream mesocosms. *Global Change Biology* 17: 2288–2297.

Lewis, W.M. 2011. Global primary production of lakes: 19th Baldi Memorial Lecture. *Inland Waters* 1: 1–28.

Logan, P. and Furse, M. 2002. Preparing for the European Water Framework Directive: making the links between habitat and aquatic biota. *Aquatic Conservation: Marine and Freshwater Ecosystems* 12: 425–437.

Lowe, R.L. and LaLiberte, G.D. 2006. Benthic stream algae: distribution and structure. *In*: Hauer, F.R. and Lamberti, G.A. (eds) *Methods in stream ecology*. Elsevier; New York, 269–294.

Lowe, S.J., Browne, M. and Boudjelas, S. 2000. *100 of the world's worst invasive alien species* IUCN/SSC Invasive Species Specialist Group (ISSG). Auckland; New Zealand.

Lytle, D.A. and Poff, N. 2004. Adaptation to natural flow regimes. *TRENDS in Ecology and Evolution* 19: 94–100.

MacNeil, C., Dick, J.T.A., Platvoet, D. and Briffa, M. 2011. Direct and indirect effects of species displacements: an invading freshwater amphipod can disrupt leaf-litter processing and shredder efficiency. *Journal of the North American Benthological Society* 30: 38–48.

MacNeil, C., Fielding, N.J., Hume, K.D., Dick, J.T.A., Elwood, R.W., Hatcher, M.J. and Dunn, A.M. 2003. Parasite altered micro-distribution of Gammarus pulex (Crustacea: Amphipoda). *International Journal for Parasitology* 33: 57–64.

MacNeil, C., Platvoet, D., Dick, J.T.A., Fielding, N., Constable, A., Hall, N., Aldridge, D., Renals, T. and Diamond, M. 2010. The Ponto-Caspian 'killer shrimp', *Dikerogammarus villosus* (Sowinsky, 1894), invades the British Isles. *Aquatic Invasions* 5: 441–445.

Maltby, E. *et al.* 2011. *Freshwaters – open waters, wetlands and floodplains. In*: UK National Ecosystem Assessment. UNEP-WCMC; Cambridge.

Mason, C.F. 2002. *Biology of freshwater pollution*. Pearson; London.

Meyer, J.L. 1994. The microbial loop in flowing waters. *Microbial Ecology* 28: 195–199.

Meyer, J.L., Paul, M.J. and Taulbee, W.K. 2005. Stream ecosystem function in urbanizing landscapes. *Journal of the North American Benthological Society* 24: 602–612.

Millennium Ecosystem Assessment. 2005. *Ecosystems and human well-being: current state and trends: findings of the Condition and Trends Working Group* (eds. Hassan, R., Scholes, R. and Ash, N.). Island Press; New York.

Milner, A.M., Brown, L.E. and Hannah, D.M. 2009. Hydroecological effects of shrinking glaciers. *Hydrological Processes* 23: 62–77.

Milner, A.M., Robertson, A., Monaghan, K., Veal, A.J. and Flory, E.A. 2008. Colonization and development of a stream community over 28 years; Wolf Point Creek in Glacier Bay, Alaska. *Frontiers in Ecology and the Environment* 6: 413–419.

Milner, A.M., Robertson, A.L., Brown, L.E., Sonderland, S., McDermott, M. and Veal, A.J. 2011. Evolution of a stream ecosystem in recently deglaciated terrain. *Ecology* 92: 1924–1935.

Montoya, J.M., Woodward, G., Emmerson, M.C. and Solé, R.C. 2009. Press perturbations and indirect effects in real food webs. *Ecology* 90: 2426–2433.

Morgan, R.P. and Cushman, S.F. 2005. Urbanisation effects on fish assemblages in Maryland, USA. *Journal of the North American Benthological Society* 24: 643–655.

Morley, N.J. 2009. Environmental risk and toxicology of human and veterinary waste pharmaceutical exposure to wild aquatic host–parasite relationships. *Environmental Toxicology and Pharmacology* 27: 161–175.

Moss, B. 1998. *Ecology of freshwaters: man and medium, past to future*. Blackwell Publishing; Oxford.

Moss, B. 2010. *Ecology of freshwaters: a view for the twenty-first century*. Wiley-Blackwell; Chichester.

Murphy, M.L. 1998. Primary production. *In:* Naiman, R.J. and Bilby, R.E. (eds) *River ecology and management: lessons from the Pacific coastal ecoregion*. Springer-Verlag; New York, 144–168.

Naegeli, M.W. and Uehlinger, U. 1997. Contribution of the hyporheic zone to ecosystem metabolism in a prealpine gravel-bed river. *Journal of the North American Benthological Society* 16: 794–804.

Naiman, R., Décamps, H. and McClain, M. 2005. *Riparia*. Elsevier Academic Press; New York.

Nakano, S. and Murakami, M. 2001. Reciprocal subsidies: dynamic interdependence between terrestrial and aquatic food webs. *Proceedings of the National Academy of Sciences of the United States of America* 98: 166–170.

Nakano, S., Miyasaka, H. and Kuhara, N. 1999. Terrestrial–aquatic linkages: riparian arthropod inputs alter trophic cascades in a stream food web. *Ecology* 80: 2435–2441.

Naylor, R.L., Williams, S.L. and Strong, D.R. 2001. Aquaculture – a gateway for exotic species. *Science* 294: 1655–1656.

Newbold, J.D., O'Neill, R.V., Elwood, J.W. and Van Winkle, W. 1982. Nutrient spiralling in streams: implications for nutrient limitation and invertebrate activity. *American Naturalist* 120: 628–652.

Nilsson, C., Reidy, C.A., Dynesius, M. and Revenga, C. 2005. Fragmentation and flow regulation of the world's large river systems. *Science* 308: 405–408.

O'Reilly, C.M., Alin, S.R., Plisnier, P.D., Cohen, A.S. and McKee, B.A. 2003. Climate change decreases aquatic ecosystem productivity of Lake Tanganyika, Africa. *Nature* 424: 766–768.

Ormerod, S., Dobson, M., Hildrew, A.G. and Townsend, C.R. 2010. Multiple stressors in freshwater ecosystems. *Freshwater Biology* 55: 1–4.

Paetzold, A., Smith, M., Warren, P.H. and Maltby, L. 2011. Environmental impact propagated by cross-system subsidy: chronic stream pollution controls riparian spider populations. *Ecology*. 92: 1711–1716.

Painter, M.M., Buerkley, M.A., Julius, M.L., Vajda, A.M., Norris, D.O., Barber, L.B., Furlong, E.T., Schultz, M.M. and Schoenfuss, H.L. 2009. Antidepressants at environmentally relevant concentrations affect predator avoidance of larval fathead minnows (*Pimephales promelas*). *Environmental Toxicology and Chemistry* 28: 2677–2684.

Palmer, M.A., Lettenmaier, D.P., Poff, N.L., Postel, S.L., Richter, B. and Warner, R. 2009. Climate change and river ecosystems: protection and adaptation options. *Environmental Management* 44: 1053–1068.

Paul, M.J. and Meyer, J.L. 2001. Streams in the urban landscape. *Annual Review of Ecology and Systems* 32: 333–365.

Petts, G.E. 1996. Water allocation to protect river ecosystems. *Regulated Rivers – Research and Management* 12: 353–367.

Poff, N.L. and Zimmerman, J.K.H. 2010. Ecological responses to altered flow regimes: a literature review to inform the science and management of environmental flows. *Freshwater Biology* 55: 194–205.

Poff, N.L., Allan J.D., Bain M.B., Karr J.R., Prestegaard K.L., Richter B.D., Sparks R.E. and Stromberg, J.C. 1997. The natural flow regime: a paradigm for river conservation and restoration. *BioScience* 47: 769–784.

Pullen, R.S.V., Palomares, M.L., Casal, C.V., Dey, M.M. and Pauly, D. 1997. Environmental impacts of tilapias. *In:* Fitzsimmons, K. (ed.) *Tilapia aquaculture volume 2*. Northeast regional Agricultural engineering Service (NRAES) Cooperative Extension; New York, 554–570.

Rawcliffe, R., Sayer, C.D., Woodward, G., Grey, J., Davidson, T.A. and Jones, J.I. 2010. Back to the future: using palaeolimnology to infer long-term changes in shallow lake food webs. *Freshwater Biology* 55: 600–613.

Resh, V.H., Brown, A.V., Covich, A.P., Gurtz, M.E., Li, H.W., Minshall, G.W., Reice, S.R., Sheldon, A.L. and Wallace, J.B. 1988. The role of disturbance in stream ecology. *Journal of the North American Benthological Society* 7: 433–455.

Reys, P., Sabino, J. and Galetti, M. 2009. Frugivory by the fish *Brycon hilarii* (Characidae) in western Brazil. *Acta Oecologica* 35: 136–141.

Roach, J. 2007. Amazon longer than Nile River, scientists say. *National Geographic News*. June 18. Available from http://news.nationalgeographic.com/news/2007/06/070619-amazon-river.html.

Robinson, C.T. and Uehlinger, U. 2008. Artificial floods cause ecosystem regime shift in a regulated river. *Applied Ecology* 18: 511–526.

Robson, M., Spence, K. and Beech, L. 2006. Stream quality in a small urbanised catchment. *Science of the Total Environment* 357: 194–207.

Scheffer, M., Hosper, S.H., Meijer, M.L., Moss, B. and Jeppesen, E. 1993. Alternative equilibria in shallow lakes. *Trends in Ecology and Evolution* 8: 275–279.

Schindler, D.W. 1974. Eutrophication and recovery in experimental lakes: implications for lake management. *Science* 184: 897–899.

Schneider, P. and Hook, S.J. 2010. Space observations of inland water bodies show rapid surface warming since 1985. *Geophysical Research Letters* 37: L22405.

SEPA. 2009. *The river basin management plan for the Scotland river basin district 2009–2015.* Scottish Government; Edinburgh.

Stanford, J.A. and Ward, J.V. 2001. Revisiting the serial discontinuity concept. *Regulated Rivers: Research and Management* 17: 303–310.

Taylor, S.L., Roberts, S.C., Walsh, C.J. and Hatt, B.E. 2004. Catchment urbanisation and increased benthic algal biomass in streams: linking mechanisms to management. *Freshwater Biology* 49: 835–851.

Thamdrup, B. and Canfield, D.E. 2000. Benthic respiration in aquatic sediment. *In:* Sala, O.E., Jackson, R.B., Mooney, H.A. and Howarth, R.W. (eds) *Methods in ecosystem science.* Springer; New York, 86–103.

Thompson, R. and Townsend, C.R. 2005. Energy availability, spatial heterogeneity and ecosystem size predict food-web structure in streams. *Oikos* 108: 137–148.

Tod, S.P. and Schmid-Araya, J.M. 2009. Meiofauna versus macrofauna: secondary production of invertebrates in a lowland chalk stream. *Limnology and Oceanography* 54: 450–456.

Tyler, C.R. and Jobling, S. 2008. Roach, sex, and gender-bending chemicals: the feminization of wild fish in English rivers. *BioScience* 58: 1051–1059.

van Riel, M.C., van der Velde, G., Rajagopal, S., Marguillier, S., Dehairs, F. and de Vaate, A.B. 2006. Trophic relationships in the Rhine food web during invasion and after establishment of the Ponto-Caspian invader *Dikerogammarus villosus*. *Hydrobiologia* 565, 39–58.

Vannote, R.L., Minshall, G.W., Cummins, K.W., Sedell, J.R. and Cushing, C.E. 1980. The river continuum concept. *Canadian Journal of Fisheries and Aquatic Sciences* 37: 130–137.

Vörösmarty, C.J. and Sahagian, D. 2000. Anthropogenic disturbance of the terrestrial water cycle. *BioScience* 50: 753–765.

Vörösmarty, C.J., McIntyre, P.B., Gessner, M.O., Dudgeon, D., Prusevich, A., Green, P., Glidden, S., Bunn, S.E., Sullivan, C.A., Liermann, C.R. and Davies, P.M. 2010. Global threats to human water security and river biodiversity. *Nature* 467: 555–561.

Walsh, C.J., Fletcher, T.D. and Ladson, A.R. 2005. Stream restoration in urban catchments through redesigning stormwater systems: looking to the catchment to save the stream. *Journal of the North American Benthological Society* 24: 690–705.

Webb, B.W. and Nobilis, F. 2007. Long-term changes in river temperature and the influence of climatic and hydrological factors. *Hydrological Sciences Journal* 52: 74–85.

Wehr, J.D. and Sheath, R.D. 2003. *Freshwater algae of North America. Ecology and classification.* Academic Press; California.

Wetzel, R.G. 2001. *Limnology. Lake and river ecosystems, 3rd edition.* Academic Press; California.

Whitehead, P.G., Wilby, R.L., Battarbee, R.W., Kernan, M. and Wade, A.J. 2009. A review of the potential impacts of climate change on surface water quality. *Hydrological Sciences Journal* 54: 101–123.

Winder, M. and Schindler, D.E. 2004. Climate change uncouples trophic interactions in an aquatic ecosystem. *Ecology* 85: 2100–2106.

Woodward, G., Perkins, D. and Brown, L.E. 2010. Climate change and freshwater ecosystems: impacts across multiple levels of organisation. *Philosophical Transactions of the Royal Society B* 365: 2093–2106.

Wright, J.F., Sutcliffe, D.W. and Furse, M.T. 2000. *Assessing the biological quality of fresh waters: RIVPACS and other techniques.* Freshwater Biological Association; Cumbria.

Xenopoulos, M.A., Lodge, D.M., Alcamo, J., Märker, M., Schulze, K. and van Vuuren, D.P. 2005. Scenarios of freshwater fish extinctions from climate change and water withdrawal. *Global Change Biology* 11: 1557–1564.

Yvon-Durocher, G., Woodward, G., Jones, J.I., Trimmer, M. and Montoya, J.M. 2010. Warming alters the metabolic balance of ecosystems. *Philosophical Transactions of the Royal Society B* 365: 2117–2126.

CHAPTER SEVEN

Water demand planning and management

Adrian T. McDonald and Gordon Mitchell

> **LEARNING OUTCOMES**
>
> After reading this chapter you should be able to:
>
> ■ describe the key methods for measuring and estimating water demand
> ■ explain the key drivers of water demand and trends in water demand
> ■ outline methods of water management through resource development
> ■ outline methods of water management through efficiency.

A INTRODUCTION

This chapter concerns water demand and the challenges of the water gap, which arises where there is a greater demand for water than available supply in the forms in which it is demanded. Demand for water grows because of demographic change (rising population, or fall in average household size), increased needs for agricultural produce, urbanisation, and the supply of manufactured products. The Middle East and North Africa is an area which, broadly, is considered to have poor water resources (see Chapter 1) and is also where there have been very large reductions in water resources over the past 60 years (Jones, 2010). In other parts of the world, however, technological advances, drives to reduce demand through water efficiency, and reduced activity of water-consuming industry have resulted in declining demand (e.g. Germany). Water demand management involves not only large-scale water resource inventories (groundwater and surface water accounting, usage rates, replenishment rates, etc.) in order to try to secure resources to meet demand but, importantly, it also involves working with people to alter attitudes and behaviour so as to reduce water demand.

B SOURCES OF DEMAND

Just as water is not a simple, single product (floodwaters are very different from **potable water**, which in turn differs from urban wastewater effluents), so the demand for water varies, depending on the nature of the water involved. The demand for raw water is quite different from the demand for potable water. Indeed raw water for potable consumption is simply one of the – often minor – raw water demands. Chapter 1 outlines the proportion of global and developed/developing world water uses for different purposes (see e.g. **Figure 1.9**).

The major demand for water globally arises from irrigation requirements for agriculture. Thus, this demand will be influenced by moisture deficit and by the market for agricultural products. Soil moisture alters in the short term with weather variability, in the medium term with seasonal crop cycles and inter-annual variability (e.g. El Niño Southern Oscillation; see Chapter 2), and over longer periods with the influences of climate change. The demand for agricultural products is influenced by a range of factors, such as population growth, lifestyle expectations, world market prices, food and energy sourcing policies, and more. So a key point is to understand how the demand for water interacts with energy and food requirements and circumstances. For example, if more land is taken up to grow crops that are used as **biofuels**, then that is an additional water demand to that required for food provision. **Box 7.1** presents some interactions which the UK Chief Scientist has characterised as giving the possibility of the 'perfect storm'.

Demand for water abstraction (known as off-stream use) arises from industry, agriculture, and domestic water supply requirements, but water is also demanded without the need for abstraction (known as in-stream use; see Chapter 1). It is needed in river channels for power generation – in past times primarily for direct mechanical power but today also for various types of hydropower – head fall, run-of-the-river, micro-hydro. Water is also demanded within channels for navigation for shipping. The maintenance of water habitats and water quality is also an in-stream requirement. The most obvious and current example, at least in Europe, is the explicit requirement for sufficient water to maintain or achieve 'good ecological status' in all reaches of the river, as required by the EU Water Framework Directive (see Chapters 4 and 6). However, there has also been an *implicit* requirement in many countries that there must be sufficient water to dilute pollutants – summarised in the phrase 'the solution to pollution is dilution'. In the UK this was formalised in the 1912 Royal Commission on Sewage Disposal, where it was suggested that there should be a river flow of eight times the volume of any pollutant discharge (and that such river flow should be of fair

---

BOX 7.1 CONTEMPORARY CHALLENGES

The perfect storm

The Chief Scientist in the UK (at the time of writing), Professor Sir John Beddington, has alerted water managers and scientists, and policy makers and politicians, to the possibility of a 'perfect storm' whereby demand for energy and food in the BRIC economies (Brazil, Russia, India and China), driven by lifestyle and demographic change, when allied to climate change, drives up prices on global markets. This may encourage some countries which are currently net importers of food to try to grow more food locally. For example, global price increases would force increased irrigation-driven food and energy cropping in the UK. Should this arise in a period of drought it would be a challenge to provide the water and prioritise supply. Zheng Chunmiao, a Professor of Water Resources at Peking University, has noted that groundwater resources in China for agricultural irrigation are unsustainable. He told the UK's *Guardian* newspaper in 2012 that 'The government must adopt a new policy to reduce water consumption. The main thing is to reduce demand. We have relied too much on engineering projects. We must reduce food production. It would be more economical to import.'

See Project ideas at the end of the chapter for some work you could undertake on this topic.

quality at least – an often forgotten requirement). It follows, then, that the demand for water in stream is interconnected with the multiple off-stream uses via abstractions and discharges that occur on many river systems. Such demand therefore has to be evaluated for every river reach. This lies at the heart of abstraction licensing in those countries where legal processes exist to deal with licensing for abstraction (e.g. the USA, Australia, Germany, UK). Even in those countries some rivers are now over-licensed for abstraction, in order to satisfy demand.

> **REFLECTIVE QUESTION**
>
> Why is growing biofuels potentially both good and bad for society?

C MEASURING CURRENT POTABLE DEMAND

1 Water metering

At its core, demand is an economic concept (see Chapter 10) – how much of a particular good a population will choose to buy for a set price. Water is a rather more problematic resource to consider in that style, for two reasons; firstly, because it is not a matter of pure choice, as water is essential to life, and secondly, because demand is not always directly measured but must often be estimated. If a country has universal water metering, it is in theory a simple matter of aggregating some millions of meter readings to provide a measure of potable water demand. However, these meters will be 'dumb' meters, read at different times and themselves representing an aggregation of different time periods (e.g. perhaps read once a year or once every six months), and so deriving a credible and useful figure for water demand is still not trivial. **Box 7.2** provides details of more innovative and useful smart meters.

In some countries, however, not all homes have a water meter. Surprisingly, this is not just restricted to the developing world. The UK is one example, with only 40% of the country metered. This figure also varies markedly in different regions – in 2009 Anglian Water, in the east of England, had 80% of metered customers, while Scottish Water had a mere 530 metered customers. This remarkably low number for Scottish Water, 0.02% of the 2.3 million customers, was a reduction of over 10% from the previous year's figure. Customers with a meter tend to pay on the volume of water consumed, although there may also be a standing charge in some cases. In the UK customers without a meter pay a single charge based on the size/value of house they occupy. This is usually on a fairly simple scale related to independent assessments of value which are used in determining local government taxes. However, the problem remains as to how you estimate demand, and thus plan the supply needs, when there are no or few meter records.

2 Demand estimation

There are two main approaches to demand estimation based on either domestic consumption monitors or district meters (known in the industry as the bottom-up and top-down approaches, respectively). The bottom-up method aggregates estimates of demand made at the individual household level, while the top-down method disaggregates from an area measurement.

2.1 Bottom-up

A domestic consumption monitor or domestic water use survey is a sample of about 2000 consumers (at most, but often many fewer) who pay on a standard tariff (i.e. based on not having a meter). Establishing, and thereafter maintaining, a monitor is difficult, although the information that such a monitor provides can be invaluable to the industry. The sample of customers is

BOX 7.2 THE FUTURE OF WATER

Intelligent metering

Water meters have typically been installed in a meter box, often in the garden. They have to be physically read. The set-up in Figure 7.1a is not atypical. However, the gas and, particularly, the electricity industries in the developed world have adopted smart meters that measure continuously and so can allow smart tariffs and excellent customer information. Automatic meter reading can be carried out remotely. It often uses wireless technology to connect to a local station, which might be a van passing down the street to read all the meters. The frequency of the reading is the key to whether it is just a more convenient dumb meter or a much smarter device. A true smart meter reads the flows 'continuously' (e.g. Figure 7.1b), reports the information to the water company, analyses the information and presents various summaries to the consumer. So, in addition to the meter there is likely to be a consumer wireless display unit. A smart meter may be able to provide extremely valuable water-demand information which enables water providers to deliver services more efficiently. For example, knowing if there are areas or households with particular patterns in water demand (e.g. religious or cultural patterns) may enable water pressures to be altered during the day to suit those areas, thereby also reducing any volumes of water leaking from underground pipes at other times of the day.

Figure 7.1 (a) A dumb meter, which simply records accumulated usage, and (b) a smart meter, which provides continuous data on usage through time.

usually drawn at 'random' from the company's billing computer, but that does not mean that just entering the list of customers at random and printing the next several thousand names would be correct – start at K and you might get too many Khans, start at M and there might be too many Macs, and these have implications for properly representing cultural water-use differences, which we will discuss later. But let us assume for the moment that the initial sample was near random. The people on the list are contacted and invited to join the survey, but only a small proportion of those invited will choose to enter the survey and so they are self-selected, and this starts to introduce bias. People may decide to join the survey because they think they are low water users and hope to become a metered customer if the survey bears out their perception; others may be reticent to join a survey because of their own real or perceived educational limitations; while yet others may simply wish to avoid contact with bodies with 'authority'. Table 7.1 lists the range of known biases. All domestic consumption monitors contain bias, some of which are present from the outset while others build up during the maintenance and upkeep of the monitor. These biases can be partially counteracted by careful correction measures in the recruitment of the sample and the interpretation of the results, but such surveys will never give a precise measure and when used to report demand to fractions of a litre, as is sometimes required by regulators, give a false impression of precision. The challenge for demand analysis is to determine the overall direction of each bias and to allocate an estimate of the extent of bias so that a correction can be entered into the final demand estimate.

If the sample is really representative of the household population as a whole we can simply take the average. Unfortunately, in many countries water supply bodies are required by legislation to report not just the household water demand but the average per capita demand, and this introduces significant complexities. In particular, it requires that the size of every household in the sample must be known. Unfortunately

Table 7.1 Biases in domestic consumption monitors (DCM)

Bias	Description
Self-selection	When a water company wishes to create a DCM the people who choose to join are self-selected. They may join because they believe they are low water users, or because they are 'good citizens', or because they like being part of an experiment.
Staff	To get a DCM started it is possible that a company may encourage its own staff to join – both to get it underway and to iron out initial difficulties. However, such members will be more informed and are more likely to behave 'correctly'.
Exclusion	Not everyone feels able or is allowed to join a survey. In the early years of UK private water companies, local authorities in some areas, in a mistaken view of their own authority and because they had political disputes with the concept of 'privitisation', told council house tenants that membership of a water survey was not allowed. The inevitable bias which carries through to type of property, family size and several other factors has been difficult to eradicate. People can also self-exclude if, for example, they have concerns about language and competence.
Financial advantage	People are compensated for the time it takes to complete forms and record information. For some this is an income source, but it is likely to be biased towards those with time available and need for supplemental income.
Hawthorne effect	When people are surveyed they form an opinion about the purpose of the survey and tend to make an (unknown to them) effort to perform in the direction expected. If using less water or being more water efficient is perceived as being an appropriate outcome they will be biased in that direction.

Table 7.2 A very simplified domestic consumption monitor table

	Household consumption (litres per day)	Number of residents per capita (litres per day)	Water demand
	400	4	100
	150	1	150
	280	2	140
	360	3	120
Total	1190	10	119 or 127.5

Note: In a real table there are likely to be 1000 rows (households) and perhaps hundreds of columns, with each column holding a weekly or monthly average demand (each of those derived, in many cases, from 15-minute demand data), together with data on location, socio-economic information, and characteristics of the property. One outcome (first number, bottom right) is derived by dividing the sum of column 1 by the sum of column 2. The second outcome is derived by averaging the per capita demand of the sample households (second number, bottom right).

household sizes change frequently, due to births and deaths, lodgers, students away from home, and so on, and households typically move in their entirety every few years on average (prior to the marked slowdown in the housing market), so in, say, just three years, the sample information on which to calculate per capita demand could have drifted at least 25% into error. Thus, for the bottom-up approach water supply bodies must resurvey on at least an annual basis, with both cost and enhanced bias impacts.

Although the water industry seeks a large sample (e.g. 2000 in a domestic water use survey), for many supply bodies that figure falls significantly, as data returns may be incomplete and recruitment is difficult because people frequently leave the survey. Whatever the sample size, the water-demand analyst eventually finishes with a table similar to that shown in **Table 7.2**. Each row provides the result showing the water demand for a household. However, **Table 7.2** shows two alternative values. The first is derived from the averaging of the individual household per capita demands, and the second is arrived at by summing all the daily demands and dividing by the aggregate population.

Starting from an even more fundamental base, the demand for water can be calculated from the product of the population of interest multiplied by the per person demand. The total population can be segmented by household characteristic to perhaps gain a more sophisticated measure. The per person demand can be determined by the use of OVF (Ownership, Volume and Frequency) calculations, sometimes known as micro-components. Components of demand are such things as toilet flushing, garden watering, tap use, dishwasher use, etc. For each component the average ownership is multiplied by the average volume per use and by the average frequency of use, and all component use is summed to derive the use per person or per household. This is a very mechanistic process, the averaging hides a great deal of variability, and the values required are very difficult to determine with confidence. Clearly, the ownership of three baths does not result in the owner having triple the average number of baths, and so working with OVF determinations requires considered views of both the data and the outcomes. However, in some countries this is a mandatory approach to demand measurement. The measurement of micro-components is outlined in Ball *et al.* (2003).

2.2 Top-down

Water in a distribution system is monitored by a system of area meters. Top-down demand estimation is derived from these meters. Each meter covers, typically, a minimum of several hundred households, but the number is often much larger.

The meters measure total water passing through the pipe, irrespective of final use. Area meter-based determination of the household demand is therefore more complex and less direct. An area meter gives the water use in a relatively large area. It is therefore the sum of many demands, few of which are known. Initially the nightline is determined. This is the water demand in the early hours of the morning and assumes that household use is at a minimum. From the nightline an allowance for legitimate night use is removed, as is metered domestic usage and commercial and industrial use (also metered) for the night period; the remainder is deemed leakage. From the average demand on the area meter, commercial and metered demand is subtracted, as is the now 'known' leakage. Further allowances for estimated operational use, estimated theft, estimated emergency use and the lowered leakage under the lower-pressure daytime regime in the metered area are also entered into what is simply a mass balance. The result is average demand of the unmetered population. This can now be divided by the number of households to derive per household average unmetered water demand and by the population (although this too is often not a readily available, up-to-date, accurate value) to gain average per capita unmetered domestic demand.

The two values of bottom-up and top-down demand estimation will not be the same, but if they lie within 5% of each other little further action is deemed necessary. If they lie outside this band, then some form of outcome reconciliation is required which may either be an arithmetical or an investigative correction.

> **REFLECTIVE QUESTION**
>
> What methods can be used to estimate domestic and industrial water demand and how can we estimate leakage rates?

D TRENDS IN WATER DEMAND

Water demand is driven by population, lifestyle and economy. Of course lifestyle is dependent on age, affluence, education, culture, family size, and several other more minor factors, few of which are directly measurable. Water demand is trending upwards both globally and in most countries, and will continue to do so for the next 20 years because the population forecasts for that period are well understood. Thereafter the water demand is less certain, reflecting the uncertainty in the population forecasts. The components of the trend also change over time. In less developed countries irrigation is the major water demand, but, as countries become more industrialised, so commercial and domestic water requirement becomes more important. The trend depends on how much water is already demanded, and expected population growth. Globally, water demand is increasing at a 2% compound annual growth rate, and by 2030 a major gap between water resources demanded and sustainable resource availability is expected. This is illustrated in **Figure 7.2**, taken from a report by the global consulting firm McKinsey & Company, which aggregates the views and research of several major global companies.

The implication of this water-balance forecast is that, without significant interventions, there will be water shortages. In some countries, typically in the developed world, this might be addressed by demand reduction *and* resource development, but in many less developed nations resource development is the only option, assuming that the water resource is available for exploitation, a situation that does not always exist. The deeper implications of this last statement are troubling: drought, crop failure, inability to develop, water-related diseases and ultimately conflict (see Chapter 11) over access to water and, if the conflicts are across national boundaries, war. In the words of Ban Ki-moon:

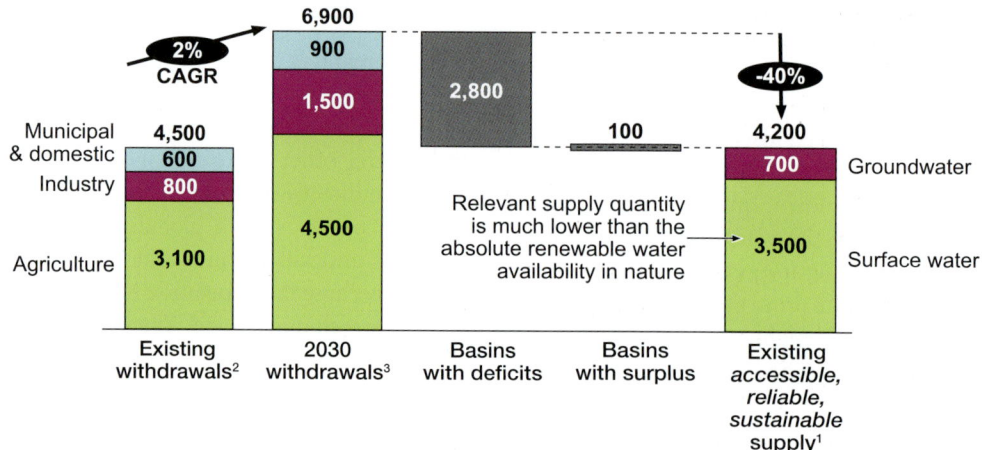

Figure 7.2 Current and forecast global water demand and resource availability.
Source: McKinsey & Company, 2010.
Notes: [1]Existing supply which can be provided at 90% reliability; [2]Based on 2010 agricultural production analysis; [3]Based on GDP, population and agricultural projections and considers no water productivity gains to 2030.

All are places where shortages of water contribute to poverty. They cause social hardship and impede development. They create tensions in conflict-prone regions. Too often, where we need water we find guns. [...] There is still enough water for all of us – but only so long as we keep it clean, use it more wisely, and share it fairly.

Ban Ki-moon, UN Secretary General (United Nations, 2012)

The OECD has forecast that the population suffering from *severe* water stress will rise by a further 1 billion to 3.9 billion by 2030 (**Figure 7.3**). Water stress is the least severe of the categories of reduced water availability. The OECD suggests that on a quantitative basis **water stress** occurs when the water resource falls below 1700 m³ per person per year, **water scarcity** occurs when the resource drops below 1000 m³ per person per year and **absolute water scarcity** occurs when the water resource falls below 500 m³ per person per year. The UN Water for Life decade also provides a qualitative definition of water scarcity:

Water scarcity is the point at which aggregate impact of all users impacts on the supply or quality of water under prevailing institutional arrangements to the extent that demand cannot be satisfied fully.

So, in the above definition scarcity occurs when the water resource availability is significantly below that which we have become used to and which we have anticipated being available in the future. The FAO World Resources Institute has mapped water scarcity (**Figure 7.4**), and since it is based on a quantitative measure that addresses the balance between water resources and population, the scarcity is not just focused on the desert countries. See also section E1 in Chapter 1 for a brief discussion of water scarcity.

Demand varies greatly in different countries (**Figure 7.5**). Of particular note are the very significant differences between countries at broadly the same state of economic development. The US per capita water demand is quadruple that of the UK. How much faith to place in such aggregate figures is, however, open to question; Germany is recorded in **Figure 7.5** as using more water than

WATER DEMAND PLANNING AND MANAGEMENT 211

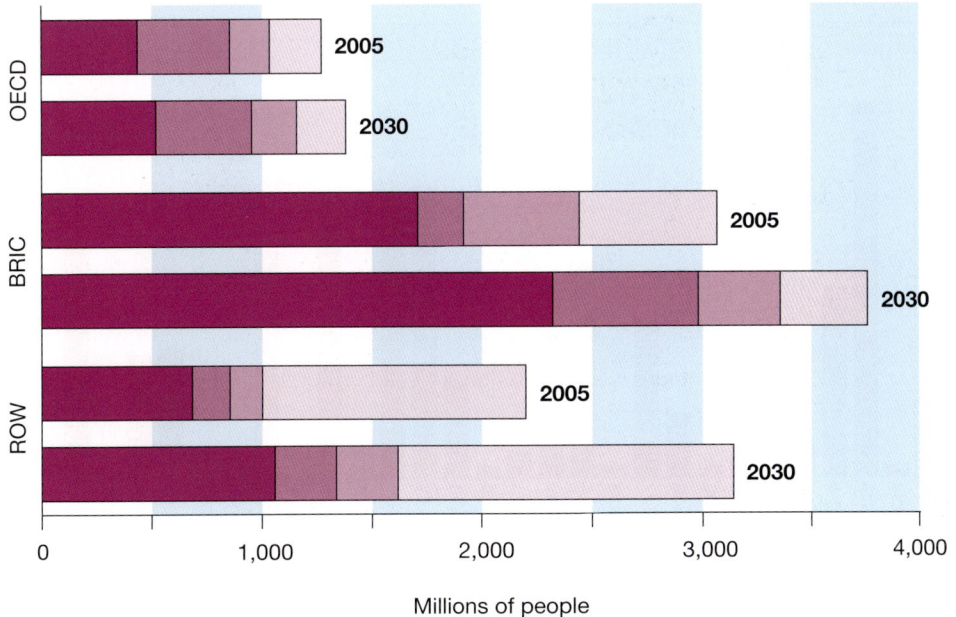

Figure 7.3 Populations living in areas of water stress.
Source: Kindly reproduced with permission from OECD (2008), *OECD Environmental Outlook to 2030*, OECD Publishing, doi: 10.1787/9789264040519-en.
Notes: OECD countries = 34 developed countries; BRIC countries = Brazil, Russia, India and China; ROW = rest of the world.

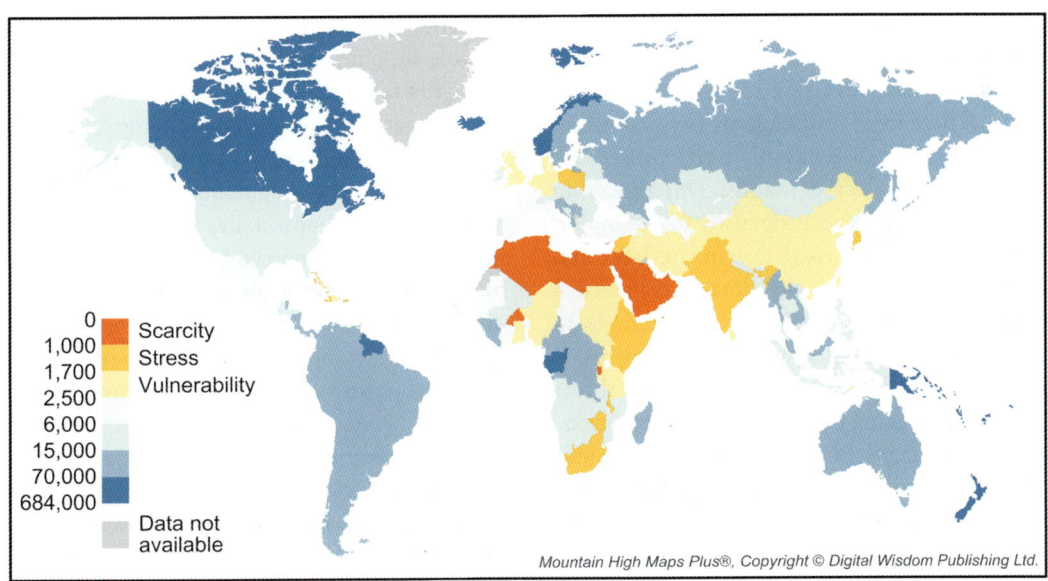

Figure 7.4 Global water scarcity distribution map.
Source: Reprinted with permission. Copyright © 2008, United Nations Environment Programme. No further use of this publication may be made for resale or for any other commercial purpose whatsoever without prior permission in writing from the United Nations Environment Programme.

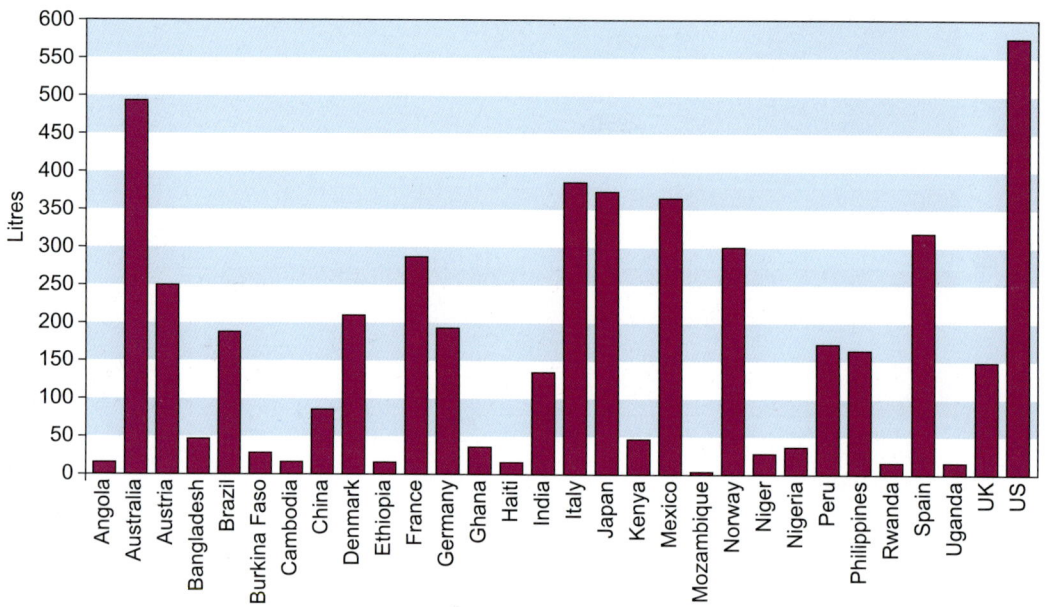

Figure 7.5 Water demand in different countries – average water use per person per day.
Source: Data from UNDP, 2006.

the UK but is quoted elsewhere as being more water efficient. Alternative tabulations of water use per head, for example, as reported by the EEA (n.d.) show Germany using 125 litres per person per day. There are of course reasons why countries at the same level of economic development can have different water demands. Differences in average household size, propensity to live in flats in Europe and in grassed lots in the USA and differences in the way the data is collected – including or excluding leakage is one example. In reality, all of the above as well as price drivers are responsible for the differences. **Figure 7.6** shows household water use indexed against water prices. At lower water prices a range of water demands exist: a Canadian household using triple the water of a Hungarian household. But as price increases above \$3 per m^3, the gap between highest and lowest water use diminishes, and demand is fairly consistent around the 50 m^3 per capita per year.

Water demand by industry can be for both raw and potable water. For example, drinks, canning or food processing companies such as Coca Cola, Crown Technologies or Birds Eye will require potable supplies, while steel making will require primarily raw water. However, many industries, such as energy processing and refining, that might be expected to require raw water in fact use considerable quantities of potable water.

Almost all potable water demands made by industry and commercial premises are metered in the developed world. Some companies have a diverse range of industrial users, and in times of economic 'tranquillity' it is possible simply to use a time trend to forecast future industrial water demand. However, in some cases the industrial demand is focused, for the most part, on a single industry, and so variations in the activity of that industry caused by markets or policy can cause dramatic changes in water demand. In these cases there is a need to track the factors that influence the demand for that particular industry's products or services. The water demand of a specific industry is driven by outputs, and these

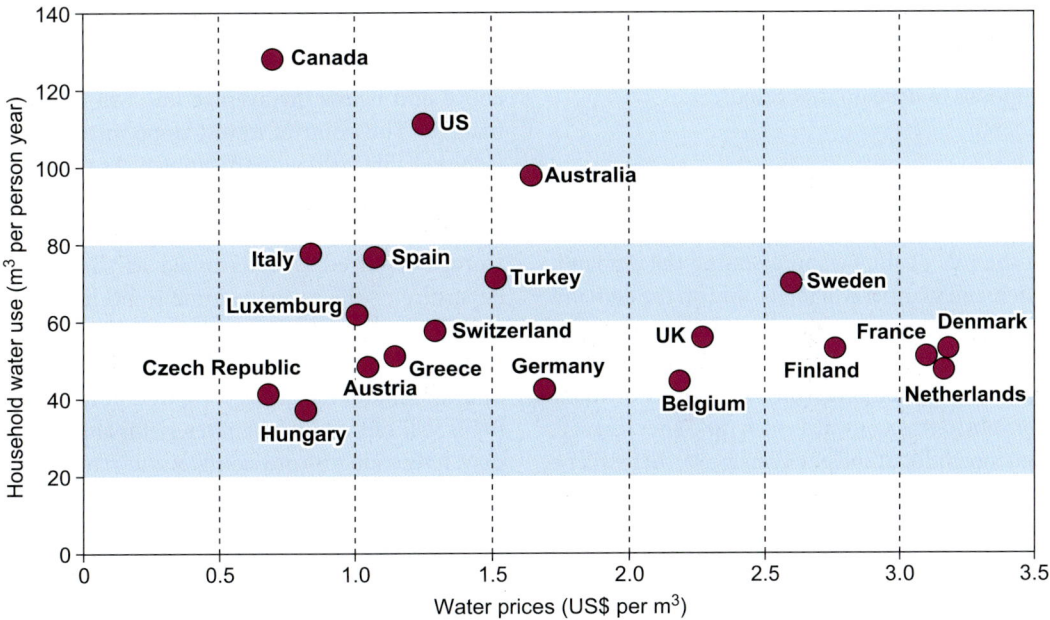

Figure 7.6 Household water use indexed against water prices across a range of countries.
Source: Data based on values reported in OECD, 1999.

can be measured either by material finished product metrics, by sales or by the labour force employed (albeit an input measure). This is modified slightly by water price (industry appears to be more responsive to small price changes than is domestic metered demand) and by water stress, as measured by regional soil moisture deficit. Since these drivers cannot be determined with certainty for the future, forecasts have to be made on the basis of a range of scenarios. Additionally, as we go into the future, the water-saving strategies of each industry need to be determined. These will depend on the technologies thought to be available and the rate of uptake of the technology. This in turn is likely to depend on the size of the individual company and the cost of water used, relative to other inputs (Mitchell *et al.*, 2000).

E TACKLING RISING DEMAND THROUGH WATER RESOURCE PLANNING AND WATER EFFICIENCY

Historically, the approach to water demand management has been to 'predict and provide' (*predict* the future water demand and devise and implement plans to *provide* sufficient supply for that demand). To a great extent this remains true today. Water supply organisations predict the major elements of demand for some decades ahead and present the forecasts of available water across the same time period, including any planned resource developments, to identify when resource availability will be insufficient to meet demand. This is then used as the basis

> **REFLECTIVE QUESTION**
>
> What is the water availability in your region in m³ per person per day and how does it compare to water stress/water scarcity measures?

of proposals to achieve a secure water balance with sufficient surplus to deal with supply perturbations. Historically, this would simply be proposals to develop new resources.

1 Water resource development

What constitutes water demand is very dependent on the role of the person assessing the demand. A demand planner will focus only on the demand made by domestic and commercial customers. The resource planner, on the other hand, needs to determine all the other uses and losses of water from abstraction onwards such that, after leakage, theft, operational and emergency use, there will be sufficient water to provide for the needs of the customer demand. The scale of the difference in these two interpretations depends on the nature of the water resource system and how it has been developed.

Water systems start with abstractions from ground or surface water (there are small exceptions, for example, rainwater harvesting). Seawater is effectively unlimited, but for all other sources the rate of exploitation determines whether the operation is sustainable. If the rate of exploitation of groundwater far exceeds recharge rate it is effectively the mining of a finite resource and should be analysed from that perspective. A surface water, say a river, could be exploited, to the level determined by single mass curve analysis, assuming there is no storage (to exploit further, to the long-term yield of a river, will require storage so that periods of above-average yield are stored to satisfy demand in periods of below-average yield). A single mass curve is a basic water resource analysis tool. It is a plot of the cumulative flow at a point in a river over time. It can be derived from daily data, but weekly or monthly total flow volumes are usually satisfactory for water resources purposes. The longer the period of record, the more faith the analyst can have in the outcomes. A straight line drawn from the final point to the origin is the long-term yield (LTY) of the river. In a temperate river the cumulative plot will depart only slightly from the LTY, but in a markedly seasonal river subject to Mediterranean, monsoonal or snowmelt drivers the annual departures above and below the average line will be more marked. The point of lowest slope in the curve is the reliable yield without storage. As storage is increased, so the river can be used in greater volume until the LTY is reached. The volume of storage required is determined as the largest departure of the actual volume trend line from the LTY line, but of course this assumes that the period is representative of the longer-term conditions. Additional storage volume beyond that point will not add to the river yield, although it may have value in providing a safety margin, a recreational resource, or the ability to deliver above-average volumes if required at critical periods – say, to generate power in high energy-demand periods.

Dams are by far the most common method of providing storage. The WWF (2012), drawing information from the World Commission on Dams and the International Commission on Large Dams (ICOLD), notes that of the 48,000 large dams (i.e. >15 m high) about half are in China and two-thirds in just four countries (China, Iran, Japan and Turkey). Half the dams are for irrigation and 20% for hydropower. In the twentieth century about $2 trillion was spent on these structures. **Figure 6.13** in Chapter 6 provides a map showing how river discharges around the world are regulated by dam structures. The ICOLD database provides extensive statistics on many aspects of dams, which you can interrogate at http://www.icold-cigb.org/GB/World_register/world_register.asp.

Raw waters are transferred to a treatment works by aqueduct, pipes and siphons, depending on local circumstances and the age of the system. Following treatment and disinfection (see Chapter 9 for the techniques by which this is achieved), which will vary according to the raw water source (upland waters tend to be cleaner than lowland river abstractions because of the long storage in a large reservoir and the maintenance of a relatively pristine catchment) the now potable water is measured into the water distribution network system to be held in service reservoirs, from which it is distributed by a bifurcating pipe system to individual homes and businesses. Area meters will record flows to different areas. The last few metres of the water supply pipe are owned by the customer and water meters may be located at the property boundary, external or internal to the house. After distribution to the house, the water system is covered by building regulations.

Abstracting, storing, treating and distributing water requires major investment in infrastructure and has impacts on ecology, landscape and energy. In the early industrial countries such as the UK and some parts of Western Europe, the infrastructure is now ageing and requires a major intervention and replacement programme. Reliance on a Victorian legacy, no matter how well built, has a finite time limit. Investment in infrastructure replacement in the UK has accelerated markedly in the last 20 years, but it will still take about 100 years to replace all the water assets and thus, at the end of the period, assets replaced today will again have become legacy assets. However, this is still a pleasing picture compared to the 800-year asset replacement rates of the 1970s. With climate change, however, it will be necessary to consider additional assets if periods of extreme flood and drought are to be managed successfully.

In many third world countries rapid urbanisation has resulted in shanty towns. Here infrastructure, as perceived in the western world, is non-existent, and investment in infrastructure development is not sufficient to match the rate of urban expansion. It may well be that water must be supplied to the shanty perimeter and informal local systems are encouraged to distribute locally.

All of the approaches discussed above are, to a greater or lesser extent, conventional. There are many less widely used techniques such as **cloud seeding**, **desalination** and iceberg sourcing. The latter two suffer from relatively high energy costs, as the water is provided at sea level and has therefore to be pumped up into the distribution system. Given that, even with more energy-efficient desalination systems and energy-efficient large iceberg towing, considerable energy has gone into the provision of the water. It is unlikely, therefore, that such sources will be widely exploited until water prices have risen considerably. Cloud seeding is rather different. Initially explored in the 1950s, cloud seeding promotes rainfall release from existing clouds and is fraught with problems in this more litigious world. Capturing water from clouds that had customarily deposited rain elsewhere; creating major flood-producing rainfalls; generating rain in the wrong place, all limit this technique – and of course there have to be clouds to seed in the first instance.

Considerable effort goes into sourcing, cleaning and distributing high-quality water, so unaccounted-for water, primarily leakage that does not reach the customer, is unwelcome. Many systems leak. Electricity leaks as heat from transmission lines. Universities 'leak' students who give up and prisons 'leak' prisoners. However, the water industry suffers leakage of around 10% in the better systems and closer to 25% in poorer systems. Leakage occurs as systems age and decay, as pressure increases and as ice, traffic, roots and drought shift the ground. Finding and fixing leaks is not trivial. Major bursts that reach the surface are reported by the public, but more modest leaks have to be traced by discrepancies in the local water balance and by locating the leak by the noise it makes, using either a skilled operator with a 'listening stick' or a correlator, in essence a software-supported microphone, to do the same task. Once a leak is located the ground can be excavated and the leak made good – but this is highly disruptive to traffic and residents, and so many companies now seek 'no-dig' solutions, by lining existing pipes with an inner plastic pipe, for example.

2 Water efficiency

Today, in some countries, the approach to water resource planning is different and is called 'twin track', achieving a water balance by addressing resource development *and* demand reduction. However, this approach introduces uncertainty into what is already an uncertain equation. Demand reduction is ultimately in the hands of the individual consumer, so there can be only limited confidence in: (i) the scale of demand reduction achievable, (ii) the longevity of that reduction, and (iii) the sustainability of the reduction in the face of changing societal circumstances and attitudes. Of course there is uncertainty in resource development as well – the water industry has no contract with its major supplier – rainfall – and so holds a reserve of water, called '**headroom**', in its water balance forecast accounts. In more extreme droughts headroom can prove insufficient and water supply organisations may ask customers to be more economical with their water use. If they respond, the reduced demand helps to stave off more aggressive water controls such as hosepipe bans, closure of water-using commercial companies such as car-wash firms, suspension of abstraction licences, rotational cut-offs, the *threat* of standpipes in the street and, in the final step, the cessation of household supply and the substitution of standpipes in the streets (effectively taking the consumer back to the early days of the Industrial Revolution).

Water supply organisations may have already promoted water efficiency and choose to use (or have been so instructed by a government regulator) the lowered water use simply to defer investment on major water development projects, and perhaps to plan for a permanently different demand. In such cases restraint and efficiency cannot now be called upon in time of drought. So the availability of this informal headroom (the ability of the public to be more water efficient in times of crisis) is important in offsetting climate, and thus resource, uncertainty. We need therefore to think through water efficiency in a more analytical way. In particular, we need to differentiate between embedded water efficiencies and behavioural efficiencies. The former should be encouraged, perhaps through the building regulations and the equipment and industry standards regimes. But the latter might best be retained as informal headroom until such time as there is a credible national plan to share headroom in the event of drought.

Both domestic and industrial water users are being encouraged to use less water. Groups such as 'Water use it wisely' in the USA or 'Waterwise' in the UK have been formed to encourage such reductions in water use. Industry, since its water use is metered, will be persuaded by economic arguments and will invest in new technologies once the saving is sufficient to overcome the many initial barriers and inertia that exist in all organisations. If a commercial company can gain some assurance that it will be less likely to be called upon to curtail operations because it is more water efficient, it is probable that risk reduction will be a greater driver than simple savings. However, in many countries water efficiency has risen up the political agenda. In the UK, for example, water efficiency is now tax deductible *in a single year* (as opposed to the norm of a small amount per year over several years), as this quote from UK Business Link reveals:

> The Enhanced Capital Allowance (ECA) scheme enables you to deduct the whole cost of your investment in water-saving technologies and products from your profits in the tax year that you make the purchase. The scheme is available to businesses that pay UK corporation tax or income tax, and that have enough profits for the allowance to be written off against.
>
> (HMRC, 2013)

All companies seek to minimise waste in their inputs and raw materials. However, unless a company is a large water user, the focus of waste minimisation will be on energy and raw materials. This is perhaps changing as companies, particularly in retail and companies with a direct link to customers, seek to market their good environmental credentials. A larger water-using company will seek to adapt its product design and manufacturing processes to use less water, and the particular approaches adopted, while often supported by environmental auditing and life cycle assessment of products and processes, will be driven by least-cost analysis with water simply as one cost.

Similarly, demand management can be practised in the agricultural sector. Here, water conservation relies on minimising evaporation and runoff. This can be achieved through a variety of practices, including replacing flood irrigation with more targeted overhead irrigation (sprinkler) systems or drip irrigation. These systems are costly to install, however, so conservation seeks to ensure maximum productivity by reducing soil compaction or creating furrows (to reduce runoff), and by more precisely understanding crop water requirements, through the use of evaporation pans and satellite remote sensing. As climate changes, water stress may lead to reduced agricultural water use through changing of crops, possibly including use of more drought-tolerant plants.

Reducing domestic demand is much more challenging. Raising prices is not an option for unmetered customers, while increasing prices for metered customers often raises concerns for low-income households, who may struggle to pay the higher prices, and so reduce usage to levels which are potentially detrimental to health and well-being. **Table 7.3** shows the water savings that are possible through behaviour change; these represent significant potential savings, but are difficult to achieve, as water companies must influence the attitudes and behaviours of the many different types of consumer that make up the population of water-using device owners. Technological devices can be built into the home to make them more efficient, a process that may be encouraged for new homes by building regulations. Technology, assumed to 'embed' savings more firmly than encouraging behavioural change through education and conservation campaigns, includes pressure-limiting valves, spray taps, low-flow showers, smaller baths and toilet cisterns, and dual-flush systems. However, home owners might not choose to replace a faulty spray tap with a similar device if they have found it inconvenient, a process known as regression. In older homes people can be encouraged to install water-capture devices such as the 'drought plug' (which facilitates using bath water for garden irrigation) or displacement devices such as 'hippo bags' (which ensure that each toilet flush is reduced by about a litre). Consumers now have more choice of water-efficient domestic appliances, although they must still choose the more water-efficient option. Influencing the behaviour of the consumer to install, retain and use water-saving devices and to modify lifestyles to use less water remains the key.

Local rainwater harvesting by capturing roof water and the recycling of grey water from baths and showers for use in toilet flushing are much more costly than any of the options discussed earlier and appear not to be cost-effective, particularly if retrofitted. More problematic are the potential health impacts associated with toilet aerosols containing potential pathogens from

Table 7.3 Potential water savings through change of habit proposed by Northumbrian Water and Essex & Suffolk Water, UK (reproduced with kind permission)

Change of habit	Saving per property per day, litres
Turn off the tap when brushing your teeth	30.7
Shower rather than bathe	25.0
Use a bowl when hand-washing dishes	17.5
Take shorter showers	17.0
Use a basin when personal washing	11.8
Repair dripping taps	8.9
Use full loads in washing machines	8.3
Use a bowl when washing food	6.5
Use full loads in dishwashers	3.8
Use water resourcefully in the garden	2.0

bird droppings or the accidental filling of, say, a child's paddling pool from a tap providing grey or roof water.

3 Resilience

Population growth and, particularly, concentration in urban areas, expectations of improving environmental quality, climate change and a more risk-averse public are fuelling expectations of resilient water systems that recognise risk and uncertainty and that can adapt to unexpected alterations in the supply–demand balance, whether this arises through system failure or through drought. Traditionally this has been achieved by holding sufficient 'headroom', a strategic reserve between expected supply and demand. Such a reserve is, however, at a cost and is planned never to be fully used. It is therefore likely that increasing the size of any reserve in an individual water supply organisation in many areas of the developed world, such as the UK, even in the face of growing uncertainty, will be difficult. There is therefore a need to share headroom between adjacent companies where possible. In effect this is a form of increased centralisation that can be achieved by informal agreements, formal water transfer agreements or by the acquisition of adjacent companies.

Altering the supply side of the supply–demand balance has greater certainty than managing down demand. Further, there is an additional loss of resilience in the simplistic reduction in demand. When demand is constrained by education, tariffs, information or habit this could be used to increase the gap between supply and demand, effectively increasing headroom and making the system more resilient to outside pressures, such as a prolonged drought. However, such a course of action is often unlikely and instead investment in new resources will be deferred. However, when an inevitable drought does occur the water managers can no longer call on behavioural changes and simple technical adaptations to constrain demand because these have already been employed. Demand management, naively pursued, can reduce resilience. What is required is, firstly, a more sophisticated selection of demand-management techniques so as to implement those efficiencies that cannot be delivered by voluntary behaviour change, but to retain for use in drought responses those adaptations that can be delivered quickly by customers – in effect, to retain the informal distributed 'headroom' that is customer-driven reduced demand in response to supply problems. Secondly, more robust decision making, as advised by the RAND think-tank below, is required:

> Consider large ensembles of plausible futures, not a single best estimate forecast.
> Seek robust strategies that perform well across many plausible futures, not a strategy optimal for one particular view. Employ adaptive strategies, ones that evolve over time in response to changing conditions. Use the computer as a 'prosthesis for the imagination,' not a calculator.
>
> (Lempert *et al.*, 2003)

This robust decision making might direct strategy towards (what some might see as) a sub-optimal solution, but it is a solution that is flexible across several futures and so is resilient (see Lempert *et al.* (2003, 2009) for more on resilient methods). This is important, as most water conservation opportunities are geographically decentralised, and exist across a diverse range of uses, and, hence, water companies tend to have relatively little direct control over demand (leakage control being an obvious exception). Under such circumstances, it is no surprise that water resource planners have traditionally sought to bridge the water gap through supply augmentation, rather than demand management. However, the impacts of increasing supply (economically, socially, environmentally) are now widely recognised in water resource planning, and 'supply-fix' is no longer automatically seen as the solution. Indeed, often the opportunities to increase supply through increasing storage, or further abstraction, do not exist. Thus integrated, adaptable strategies that may seek to increase supply, reduce demand, or both, but which are sensitive to place and water users in those places, and which recognise the limits and uncertainties associated with the various water management instruments, are required. In adopting this more adaptable approach water companies are more likely to continue to achieve promised standards of service in the face of the uncertainty inherent on both sides of the supply–demand equation.

> **REFLECTIVE QUESTION**
>
> What mechanisms can be used to tackle rising demand?

F SUMMARY

Balancing water demand and supply, and protecting or enhancing environmental quality at the same time, is challenging under conditions of growing uncertainty. Water resource development takes a long time, and so demand forecasting is required. Such forecasting requires a strong demographic element, and the information base on water-demand fluctuations at the household level is weak. Water supply also needs to account for a range of other demands on resources, including leakage, theft, operational use and emergency use. Water efficiency in both domestic and industrial arenas has a significant role, but sophistication in precisely what efficiencies are promoted is vital. Decisions need to be robust and to reflect uncertainty, if they are to contribute to a resilient system.

FURTHER READING

Butler, D. and Memon, F.A. (eds) 2006. *Water demand management.* International Water Association; London.

    A detailed review of water demand management.

Jones, J.A.A.(ed.). 2010. *Water sustainability: a global perspective.* Hodder Education; London.

    Describes the drivers of demand, and a range of technical and institutional approaches to closing the water gap.

Maksimovic, C., Butler, D. and Memon, A. (eds). 2003. *Advances in water supply management.* Balkema Publishers; Lisse/Abington/Exton/Tokyo.

    An edited book covering most aspects of water supply, with every chapter jointly authored by practitioners and scientists.

Postel, S. 1992. *Last oasis: facing water scarcity.* Worldwatch Institute Series. W.W. Norton and Company Limited; London.

    Describes the causes of water scarcity, the consequences of shortages, techniques for promoting water efficiency, and the policies and laws required to promote sustainable water use.

Classic papers

Bowden, C. 1977. *Killing the hidden waters.* University of Texas Press; Texas.

    This book charts the slow destruction of the water resources of the American South West. It has an indigenous people perspective and is relevant to other parts of the world today.

Clarke, G., Williamson, P., McDonald, A. and Kashti, A. 1997. Estimating small area demand for water: a new methodology. *Journal of the Institution of Water and Environmental Management* 11: 186–192.

Herrington, P. 1996 *Climate change and the demand for water.* HMSO; London.

Both of these papers were very influential in supporting policy or water company best practice.

> **PROJECT IDEAS**
>
> - Draw up additional evidence for the perfect storm on energy, food and water supply by searching for information on the internet. Possible questions to look up include: How many days' food supply does the world hold in reserve? How many droughts has your country suffered in the last 20 years? What proportion of US cropping is for energy?
>
> - Try to obtain some water meter data at area and household level and see if you can estimate the level of leakage which might be occurring in your area. If you have your own meter or can ask colleagues to keep records from theirs you could try plotting usage rates over time to identify patterns (daily, weekly, seasonal, etc.).
>
> - Search on the internet for household water use and water price information for different countries and add this to the graph on **Figure 7.6**; update data for countries already on the figure and add new countries to it. Are there any interesting patterns?

REFERENCES

Ball, A., Styles, M., Stimson, K. and Kowalski, M. 2003. Measuring microcomponents for demand forecasting. *In*: Maksimovic, C., Butler, D. and Memon, A. (eds) *Advances in water supply management*. Balkema Publishers; Lisse/Abington/Exton/Tokyo, 673–682.

Bowden, C. 1977. *Killing the hidden waters*. University of Texas Press; Texas.

Butler, D. and Memon, F.A. (eds). 2006. *Water demand management*. International Water Association; London.

Clarke, G., Williamson, P., McDonald, A. and Kashti, A. 1997. Estimating small area demand for water: a new methodology. *Journal of the Institution of Water and Environmental Management* 11: 186–192.

EEA. (n.d.). *Household water use*. European Environment Agency Website [online]. Available from: http://www.eea.europa.eu/data-and-maps/figures/household-water-use-1.

Herrington, P. 1996. *Climate change and the demand for water*. HMSO; London.

HMRC. 2013. www.hmrc.gov.uk/capital-allowances/fya/water.htm [accessed March 2013].

Jones, J.A.A. (ed.). 2010. *Water sustainability: a global perspective*. Hodder Education; London.

Lempert, R., Popper, S. and Bankes, C. 2003. *Shaping the next one hundred years: new methods for quantitative, long-term policy analysis*. RAND report MR 1626 [online]. Available from: http://www.rand.org/international_programs/pardee/methods.html.

Lempert, R., Scheffran, J. and Sprinz, D. 2009. Methods for long-term environmental policy challenges. *Global Environmental Politics* 9: 106–133.

Maksimovic, C., Butler, D. and Memon, A. (eds). 2003. *Advances in water supply management*. Balkema Publishers; Lisse/Abington/Exton/Tokyo.

Mitchell, G., McDonald, A., Williamson, P. and Wattage, P. 2000. A 'Standard Industrial Classification' coded strategic planning model of non-household water demand for UK regions. *Journal of the Chartered Institution of Water and Environmental Management* 14: 226–232.

OECD. 1999. *Household water pricing in OECD countries* [online] [accessed June 2012]. Available from: http://tinyurl.com/af4tg45.

OECD. 2008. *Environmental outlook to 2030* [online] [accessed June 2012]. Available from: http://www.oecd.org/dataoecd/29/33/40200582.pdf.

Postel, S. 1992. *Last oasis: facing water scarcity*. Worldwatch Institute Series. W.W. Norton and Company Limited; London.

UNDP. 2006. *Human development report. Beyond scarcity: power, poverty and the global water crisis*. UNDP; New York.

United Nations. 2012. *UN Department of Economic and Social Affairs [online]* [accessed June 2012].

Available from: http://www.un.org/waterforlife decade/scarcity.shtml. World Resources Institute (WRI) in collaboration with United Nations Development Programme.

WWF. 2012. *Dam facts and figures [online]* [accessed June 2012]. Available from: http://wwf.panda.org/what_we_do/footprint/water/dams_initiative/quick_facts/.

CHAPTER EIGHT

Water and health

Rebecca J. Slack

LEARNING OUTCOMES

After reading this chapter you should be able to:

- understand the important links between water and human health
- describe key water-related health issues and distinguish between biological, chemical and physical health effects
- describe the historical context of improved public health through enhanced sanitation and water supply
- recognise the different types and scales of health issues related to water around the world today
- appreciate the role of research and development in maintaining the health of growing global, particularly urban, populations.

A INTRODUCTION

All life on Earth requires water to survive. However, water can also be harmful to life. Chemical and physical hazards can threaten the health of individuals, populations and even whole ecosystems through changes in water quality (the characteristics of freshwater that make it suitable to sustain life) and water quantity (the availability of freshwater resources). Physical hazards such as drought and flooding, along with chemical hazards such as saline intrusion or pollution events, impact on the health of many organisms including humans. It is also true that many pathogenic organisms use water as a pathway to infect host organisms; for human hosts, water-related diseases that result from such infection make significant contributions to morbidity and mortality.

Ensuring a reliable and safe supply of water and removal and treatment of wastewater via adequate sanitation is essential for all human populations. If the quantity or quality of water is insufficient, impacts on health can be considerable. This chapter considers the dual issues of water quantity and water quality in relation to human health. The key causes of disease and ill-health associated with water will be discussed and, where appropriate, measures used to improve the water–health interface will be examined.

B THE HISTORY OF WATER AND HEALTH

The association between water and health has a long history. Human habitations have always been associated with freshwater, whether springs or wells, streams, rivers, or lakes. However, as populations increased rapidly in the nineteenth century, water and sanitation conditions in most cities became a major public health concern. There were frequent cases of contamination of the water supply by sewage, particularly as watercourses were used for both abstraction and wastewater disposal. While there were exceptions, notably the cities of Japan, whose economic, cultural and religious attitudes resulted in more sanitised environments, most cities suffered from high death rates, particularly of young children. In the UK, the work of social reformers such as Edwin Chadwick (1800–1890) led to the Public Health Act of 1848, which sought to improve sanitation and, hence, the health of the poor. This was some of the earliest legislation that dealt with public health and the environment, but the scientific understanding of the causes of disease was still very unclear.

There were many theories about the spread of infectious disease, with one of the most popular being the **miasmatic theory of disease**, prevalent at the time of the Public Health Act of 1848. This theory linked foul smells and stagnant air with the cause of disease. As miasmas were supposedly a product of environmental factors such as vapours from cesspits and open sewers, it was believed that diseases could not be passed from person to person, and could be prevented by purifying the air. While this did lead to some improvements in sanitary conditions and an attempt to clean up many towns and cities (as dirty water frequently smells unpleasant), there were still significant incidences of disease. The **germ theory of disease** challenged the miasmatic theory but started to gain momentum only in the 1850s when the work of Dr John Snow linked cholera with contaminated water supplies (see **Box 8.1**; Smith 2002). However, it was not until the 1860s and 1870s that the work of Louis Pasteur, Robert Koch and Joseph Lister confirmed that microorganisms were responsible for many diseases and the miasmatic theory was finally laid to rest. Since then, public health has improved considerably in the developed world. In particular, the discovery of penicillin in 1928 initiated the identification and application of antibiotics to the treatment of all bacterially associated water-related disease.

The development of water and wastewater treatment and distribution systems resulted in the most significant impact on public health and quality of life. The first recognised water treatment system was the slow sand filter-bed devised by James Simpson (1799–1869) and still used today. Wastewater treatments had been used sporadically from the early 1800s, when sewage sludge was cited to be an effective agricultural fertiliser. Subsequent sewage collection and treatment developments, including the London sewer network designed by Joseph Bazalgette, were a result of the recognition that faecal contamination of the water supply led to foul smells and disease (Halliday 2001). By the late nineteenth and early twentieth centuries, most towns and cities in the developed world had some form of sewer system and safe public water supply. Further development of water treatment systems included rapid sand filters, used extensively in the USA from the 1890s; disinfection of water supplies from the 1900s; and, for wastewater, biological wastewater treatment from the 1880s and activated sludge in the 1920s (see Chapter 9).

> **REFLECTIVE QUESTION**
>
> What were the key discoveries in understanding the links between water and health?

WATER AND HEALTH

BOX 8.1 CASE STUDIES

John Snow and cholera

Cholera spread from the Indian subcontinent in the early nineteenth century, rapidly becoming responsible for the deaths of millions worldwide. In 1849, Dr John Snow published a pamphlet called *On the Mode of Communication of Cholera* which stressed water as the medium for the spread of cholera. It was not met with much enthusiasm – the miasmatic theory of disease was still prevalent in most scientific circles. It was not until an 1854 outbreak of cholera in the Soho area of London that

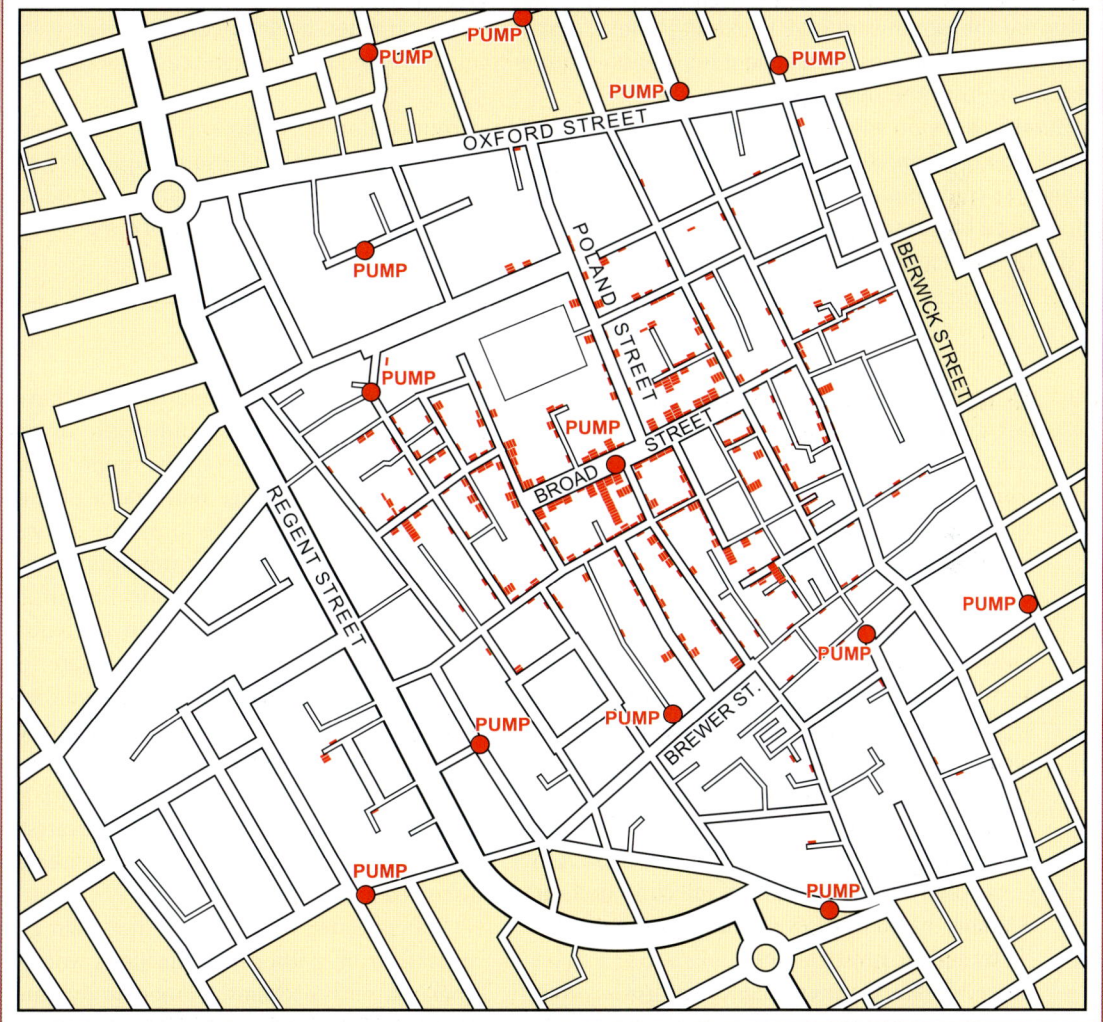

Figure 8.1 John Snow mapped the area surrounding the Broad Street pump, representing each cholera death by a bar. The greatest number of bars can be seen clustering around the infected pump.

continued

Snow could put his theory to the test. Snow mapped the locations of the homes of those who had died in the outbreak and determined that many cases were centred on a water pump in Broad Street (**Figure 8.1**). Despite some anomalies, including no cases reported in a nearby workhouse and workers at the local brewery escaping infection, but two cases in distant parts of London, Snow successfully petitioned the authorities to remove the handle of the pump and hence disable it from use. The number of cases of cholera, and subsequent deaths, rapidly fell, confirming the link to the pump. It later became apparent that the workhouse and brewery had private wells and the two deaths at a distance from the pump were a result of the householders preferring the taste of water from the Broad Street pump over their local water supplies. Further examination of the pump revealed that a few days before the first cases of disease, the water was reported to be cloudy and foul smelling. Water was supplied to the pump by the Southwark and Vauxhall Water Company, who extracted sewage-contaminated water from the River Thames; subsequent legislation required that all water companies in the London region take water from the river upstream of the city.

Snow's work on the 1854 cholera outbreak not only demonstrated how cholera was spread and how it could be prevented, it also provided support for the germ theory of disease and founded the science of **epidemiology**, which is the cornerstone of public health research. It also demonstrated the importance of water quality for human health. More information about John Snow's work is provided in the biographies by Vinten-Johansen and colleagues (2003) and Hempel (2007).

C INFECTIOUS DISEASES

There are five categories of infectious diseases associated with water. These are categorised by method of distribution rather than by microorganism type:

- water-borne diseases – infections spread by water
- water-carried diseases – diseases acquired by the accidental ingestion of, or exposure to, contaminated water
- water-washed diseases – infections spread by lack of water (for personal hygiene) and insanitary conditions that may not directly involve water as a means of infection transfer
- water-based diseases – infections transmitted by aquatic invertebrates
- vector-based water-related diseases – infections spread by insects that need water to complete their life cycle.

All categories, described in **Table 8.1**, have the potential to cause illness and death anywhere around the world. However, the resilience of the water supply and sewage network of developed countries ensures that the impact of all water-related diseases is kept to a minimum through the supply of high-quality water in sufficient quantities for high standards in hygiene and sanitation. Most developed countries are found in temperate regions where water-based diseases such as schistosomiasis and vector-borne diseases such as trypanosomiasis are not reported. Developing countries in tropical regions with poor water supply and sanitation are at the greatest risk of developing water-related infectious diseases, as discussed in subsequent sections.

Water-related infections are the main causes of death from pathogenic infections worldwide (malaria, schistosomiasis, amoebic dysentery). Although there is an emphasis on human health and infectious disease, most pathogens also affect a wide number of mammals, including cattle and sheep (typhoid and cholera are, however, human

Table 8.1 Water-related diseases: pathogens, treatment and prevention strategies

Infection type/ Sub-category	Disease	Pathogen	Treatment regime	Prevention
Water-borne				
Bacterial infections	E. coli infection	*Escherichia coli*	Antibiotics	Good personal hygiene Improved water quality
	Cholera	*Vibrio cholerae* (serogroups O1 and O139)	Oral rehydration therapy Antibiotics	Improved water quality Improved sanitation Vaccination
	Vibrio illness	*Vibrio vulnificus, Vibrio alginolyticus, Vibrio parahaemolyticus*	Antibiotics	Improved water quality Improved sanitation Good personal hygiene
	Typhoid fever	*Salmonella enterica* serovar Typhi (called *S. typhi*)	Antibiotics	Improved water quality Improved sanitation Good personal hygiene Vaccination
	Paratyphoid	*Salmonella paratyphi A, S. paratyphi B* (or *S. schottmuelleri*), and *S. paratyphi C* (*S. hirschfeldii*)	Antibiotics	Improved water quality Improved sanitation Good personal hygiene Vaccination
	Salmonellosis	*Salmonella* spp	Antibiotics	Good personal hygiene
	Leptospirosis	*Leptospira* spp	Antibiotics	Avoiding contact with water in likely areas of infection; covering cuts; preventing splashing to mucous membranes
	Bacillary dysentery	*Shigella dysenteriae*	Antibiotics Oral rehydration therapy	Good personal hygiene
	Botulism	*Clostridium botulinum*	Antibiotics	Good personal hygiene
	Campylobacteriosis	*Campylobacter jejuni*	Antibiotics	Good personal hygiene
	M. marinum infection	*Mycobacterium marinum*	Antibiotics	Good personal hygiene
	Tularaemia	*Francisella tularensis*	Antibiotics – streptomycin	Good personal hygiene
	Diarrhoeal disease	*Shigella flexneri, Shigella sonnei*	Antibiotics	Good personal hygiene

Table 8.1 continued

Infection type/Sub-category	Disease	Pathogen	Treatment regime	Prevention
Protozoan infections	Giardiasis	*Giardia lamblia*	Self-limiting Antibiotics	Improved water quality Better hygiene practices
	Amoebiasis (amoebic dysentery)	*Entamoeba histolytica*	Amoebicides (e.g. diloxanide)	Improved water and sanitation Good personal hygiene
	Balantidiasis	*Balantidium coli*	Antibiotics (e.g. tetracycline)	Improved water and sanitation Good personal hygiene
	Cryptosporidiosis	*Cryptosporidium parvum*	Supportive (e.g. oral rehydration therapy, antibiotics, etc.)	Effective water treatment/quality Avoiding contact with water in likely areas of infection
	Cyclosporiasis	*Cyclospora cayetanensis*	Cotrimoxazole	Effective water treatment
(Fungal infection)	Microsporidiosis	Microsporidia: *Enterocytozoon bieneusi*, *Encephalitozoon cuniculi*, and *Encephalitozoon intestinalis*	Albendazole and supportive care	Good personal hygiene Avoiding contact with water in likely areas of infection
Viral infections	Adenovirus infection	Adenovirus	Supportive	Chlorination of water Avoiding contact with infected water
	Gastroenteritis	Astrovirus, Calicivirus, Enteric adenovirus, Parvovirus, Rotavirus, Norovirus	Self-limiting: supportive care Oral rehydration therapy, antibiotics	Vaccination (e.g. rotavirus) Good personal hygiene Improved water quality
	SARS	Coronavirus	Supportive	Improved sanitation Good personal hygiene
	Hepatitis	Hepatitis A	Supportive: diet, hydration, therapeutic drugs	Vaccination Good personal hygiene Improved sanitation
	Poliomyelitis	Poliovirus	Supportive	Vaccination
	Polyomavirus infection	JC virus, BK virus, SV40	Supportive only	Good personal hygiene and sanitation
Parasitic infections	Ascariasis	*Ascaris lumbricoides*	Ascaricides	Good personal hygiene Improved sanitation
	Taeniasis	*Taenia solium*, *T. saginata*	Praziquantel and/or antihelminthics	Good personal hygiene Improved sanitation

	Hymenolepiasis	*Hymenolepis nana, H. diminuta*	Praziquantel and/or anti-helminthics	Good personal hygiene Improved sanitation
	Echinococcosis	*Echinococcus granulosus, E. multilocularis, E. vogeli*	Surgery with albendazole and/or mebendazole	Good personal hygiene Improved sanitation
	Coenurosis	*Taenia multiceps, T. serialis, T. brauni, T. glomerata*	Praziquantel and/or antihelminthics	Good personal hygiene Improved sanitation
	Enterobiasis	*Enterobius vermicularis*	Albendazole and/or mebendazole	Good personal hygiene
Water-carried				
	Otitis externa	Various bacteria, (e.g. *Pseudomonas aeruginosa*)	Antibiotics	Preventing water entering the ear canal
	Pseudomonas dermatitis/folliculitis	*Pseudomonas aeruginosa*	Self-limiting Antibiotics	Chlorination of recreational facilities Good personal hygiene
	Legionellosis	*Legionella* ssp. (e.g. *L. pneumophila*)	Antibiotics	Good general hygiene
	Pontiac fever	*Legionella* ssp. (e.g. *L. pneumophila*)	Antibiotics	Good general hygiene
Water-washed				
Skin and eye infections	Nosocomial infections	*Staphylococcus aureus*, methicillin resistant *Staphylococcus aureus* (MRSA), *Pseudomonas aeruginosa, Acinetobacter baumannii, Stenotrophomonas maltophilia, Clostridium difficile*	Isolation and care	Good personal hygiene
	Conjunctivitis	Not necessarily caused by pathogen (see Microsporidiosis)	Water/eye drops Self-limiting	Water/eye drops Good personal hygiene
	Leprosy	*Mycobacterium leprae, M. lepromatosis*	Antibiotics: rifampicin, clofazimine, dapsone	Good personal hygiene (although not very contagious)
	Scabies	*Sarcoptes scabiei*	Insecticide creams – permethrin	Good personal hygiene and general cleaning
	Tinea	Tinea species: *T. pedis, T. cruris, T. corporis, T. capitis*	Antifungal cream: miconazole, clotrimazole	Good personal hygiene
	Trachoma	*Chlamydia trachomatis*	Antibiotics: azithromycin, erythromycin, doxycycline/tetracycline	Good personal hygiene Improved water supply and sanitation
	Yaws (Frambesia tropica)	*Treponema pallidum*	Antibiotics	Good personal hygiene

Table 8.1 continued

Infection type/ Sub-category	Disease	Pathogen	Treatment regime	Prevention
Biting invertebrate	Typhus	*Rickettsia* species	Antibiotics: azithromycin, doxycycline/tetracycline	Good personal hygiene
	Relapsing fever (typhinia)	*Borrelia recurrentis*	Antibiotics: tetracycline	Good personal hygiene
	Spotted fevers	*Rickettsia* species and *Orientia* species	Antibiotics: azithromycin, doxycycline/tetracycline	Good personal hygiene
	Trench fever	*Bartonella quintana*	Antibiotics: doxycycline/tetracycline	Good personal hygiene
	Tularaemia	*Francisella tularensis*	Antibiotics: streptomycin	Good personal hygiene
	Bubonic plague	*Yersinia pestis*	Antibiotics: streptomycin	Good personal hygiene and general cleaning
Water-based				
Skin penetration	Schistosomiasis	*Schistosoma* spp.: *S. mansoni, S. intercalatum, S. haematobium, S. japonicum, S. mekongi*	Praziquantel and/or antihelminthics	Molluscicides; Prevent contact with infected water; Improved water quality and sanitation
Intestinal penetration	Dracunculiasis (Guinea Worm Disease)	*Dracunculus medinensis*	Surgical or traditional-winding removal of worm	Improved water supply (filtration/boiling)
	Clonorchiasis	*Clonorchiasis sinensis*	Chloroquine (partially effective)	Avoid uncooked fish
	Diphyllobothriasis	*Diphyllobothrium latum*	Diatrizoic acid or praziquantel	Avoid uncooked fish
	Fasciolopsiasis	*Fasciolopsis buski*	Praziquantel and/or anti-helminthics	Improved water supply and cooking raw vegetation
	Paragonimiasis	*Paragonimus westermani*	Praziquantel and/or anti-helminthics	Avoid undercooked crustaceans
Vector-based				
Breeding in water	Malaria	*Plasmodium falciparum, P. vivax, P. ovale, P. malariae,* and *P. knowlesi*	Mefloquine, doxycycline and the atovaquone/proguanil	Vector-control – insecticides, mosquito nets, covering or managing potential mosquito breeding pools

	Disease	Pathogen/Vector	Treatment	Prevention
	Dengue fever	Flavivirus – Dengue fever virus (spread by Aëdes aegypti, A. albopictus, A. polynesiensis, A. scutellaris)	Oral/intravenous rehydration therapies	Vector-control – insecticides, mosquito nets, covering or managing potential mosquito breeding pools
	Yellow fever	Flavivirus – Yellow fever virus (spread by Aëdes aegypti, A. albopictus, A. africanus, Haemagogus spp. Sabethes spp.)	Anti-viral drugs	Vaccination Vector-control – insecticides, mosquito nets, covering or managing potential mosquito breeding pools
	Chikungunya	Alphavirus – spread by Aëdes spp.	Chloroquine	Vector-control – insecticides, mosquito nets, covering or managing potential mosquito breeding pools
	Viral encephalitis/ viral haemorrhagic fever (other)	Arboviruses: Asfivirus, Phlebovirus, Orthobunyavirus, Nairovirus, Flavivirus, Coltivirus, Orbivirus, Alphavirus, Uukuvirus	Antivirals	Vector-control
	Lymphatic filariasis	Wuchereria bancrofti, Brugia malayi, B. timori	Albendazole/ivermectin Diethylcarbamazine/ albendazole	Vector-control – insecticides, mosquito nets, covering or managing potential mosquito breeding pools
	Subcutaneous filariasis	Loa loa, Mansonella streptocerca, Onchocerca volvulus, Dracunculus medinensis	Albendazole/ivermectin Diethylcarbamazine/ albendazole	Vector-control – insecticides, mosquito nets, covering or managing potential mosquito breeding pools
	Serous cavity filariasis	Mansonella perstans, M. ozzardi	Albendazole/ivermectin Diethylcarbamazine/ albendazole	Vector-control – insecticides, mosquito nets, covering or managing potential mosquito breeding pools
	Onchocerciasis (river blindness)	Onchocerca volvulus – Wolbachia pipientis	Ivermectin Doxycycline	Vector-control – insecticides, mosquito nets, covering or managing potential mosquito breeding pools
Bite near water	African trypanosomiasis (sleeping sickness)	Trypanosoma brucei	Pentamidine, suramin, eflornithine, melarsoprol, nifurtimox	Vector-control Medical intervention

Figure 8.2 Faecal contamination of water supplies is one of the commonest routes for the spread of water-related disease; farmed animals are a common source of such contamination in rural locations.
Source: Photo courtesy of Shutterstock/risteski goce.

specific): disease outbreaks can therefore have important consequences for the livelihoods of farming communities, particularly in non-industrialised countries. Livestock can also be a source of contamination, particularly in rural areas where surface water runoff to drinking water sources can be contaminated by faecal matter (**Figure 8.2**).

1 Water-borne disease

For centuries, outbreaks of typhoid and cholera were responsible for the deaths of hundreds of thousands of people. Improvements in water quality and the provision of separate supply and sanitation networks have eradicated incidences of these diseases in many countries. The last major outbreak of cholera in the United States was in 1910–1911, but outbreaks still occur today, usually following natural disasters (cholera in Haiti in late 2010) or after conflict/civil unrest (the Democratic Republic of Congo cholera outbreak in 2011) (WHO 2011a). Small, localised outbreaks are frequent where appropriate sanitation facilities are lacking. Outbreaks of these two water-borne diseases can result in high mortality rates, but treatments are available and epidemiological approaches can usually identify the source of the outbreak to prevent re-occurrence.

There are 3 to 5 million cases of cholera per year worldwide, resulting in about 120,000 deaths. Spread of the disease is a result of faecal contamination of drinking water; the bacteria that cause cholera are relatively fragile and a large infective dose is required. The classic symptom of cholera is acute diarrhoea which, if left untreated, will result in death from dehydration. The large volumes of diarrhoea produced also increase the likelihood of the spread of infection if appropriate disposal facilities are not available (e.g. in refugee camps) (WHO 2011b). The disease is **endemic** to the Indian subcontinent, Central and South America and sub-Saharan Africa. Infection can be reduced by implementing relatively simple measures to prevent faecal contamination of water supplies, such as bunds to prevent surface water drainage into wells, separating sanitation and water facilities, provision of hand pumps, etc. Treatment of the disease with oral rehydration therapies is very effective and relatively cheap. Antibiotics can be used to lessen the severity of the disease, but the spread of antibiotic-resistant mutations is raising concern. In areas where there is the greatest risk of infection, vaccination is being trialled, such as in Dhaka, Bangladesh, or in some places is already well established, such as in Zimbabwe (WHO 2010a).

Typhoid fever resulted in many deaths in the nineteenth century in the United States and Europe but was largely eradicated through the provision of water and sanitation services. With an estimated 16 to 33 million cases resulting in over 200,000 deaths per year, typhoid and paratyphoid (the 'enteric fevers') remain a major public health concern around the world today, and are one of the main causes of death in young children (Crump *et al.* 2004). Typhoid and paratyphoid are caused by faecal or urinal contamination of drinking water, but only a small dose is required to cause infection. If the patient recovers, bacteria will still be present in faeces and urine and can cause infection of others via poor hygiene, contaminated food or contaminated water. Healthy people can act as typhoid carriers (bacteria remains in the gall-bladder of 2–5% of patients), as shown by Mary Mallon ('Typhoid Mary') in New York in the early twentieth century. Antibiotics are the main form of treatment, drastically cutting the number of people who die from the disease. Other preventive measures are associated with provision of clean water (typhoid bacteria are killed by chlorination and removed by most filtration methods), adequate sanitation and hygiene education (e.g. hand washing).

Leptospirosis can infect a range of mammals and is spread by water contaminated by infected animal urine (particularly rodents). Although rare, the disease occurs in both temperate and tropical climates (Levett 2001). The bacteria can be ingested or can penetrate the skin or mucous membranes, with infection starting with influenza-like symptoms and progressing to jaundice and fever leading to liver failure ('Weil's disease'). The number of cases correlates with warm and wet conditions leading to epidemics most recently in Nepal (2010), Ireland (2010), and the Philippines (2009). Leptospirosis may have been responsible for almost preventing the British colonisation of North America (Marr and Cathey, 2010). Prevention is aided by provision of clean water, general good hygiene and awareness or risks when engaging in recreational activities around natural water systems (e.g. water sports).

There are many other water-borne diseases associated with human/animal faecal contamination of drinking or recreational waters, or simply poor personal hygiene (Pruss 1998; Black *et al.* 1989). Cryptosporidiosis, usually an acute self-limiting infection causing diarrhoea and abdominal pains, is estimated to be responsible for over 50% of all water-borne disease. Unlike many other water-borne infections, it is not uncommon in developed countries and there are numerous reports of cases associated with public swimming pools and water supply sources, despite water treatment processes (Fayer *et al.* 2000). Certain strains of *Escherichia coli*, one of the coliform group of bacteria used to assess the quality of food and water and indicative of faecal contamination, can result in serious illness and death in humans, as demonstrated by an outbreak in Germany in May 2011. The protozoan disease giardiasis causes diarrhoea but is rarely fatal, despite infecting millions of people worldwide: in the United States alone, up to 2 million people are estimated to be infected at any one time (Regli *et al.* 1991). Amoebiasis or amoebic dysentery, however, is a more common cause of death, with approximately 50 million people infected worldwide and 70,000 deaths per year. Bacillary dysentery (shigellosis) and related bacterial infections resulting in severe diarrhoeal infections cause an estimated 160 million cases and 1 million deaths per year (Kotloff *et al.* 1999). Gastroenteritis is used to describe a number of bacterial infections associated with contaminated water, including *Shigella*, *Campylobacter jejuni*, *Salmonella*, *Staphylococcus*, *Clostridium* and *Vibrio* species. Viral gastroenteritis may also be water-borne, including the highly infectious norovirus, polio, hepatitis A and hepatitis E. Many of these microorganisms are also found in shellfish grown in waters contaminated with sewage and this is a frequent cause of 'food poisoning' in developed countries.

Ascariasis is caused by the ingestion of eggs of the roundworm *Ascaris* species. It is estimated that 25% of the world's population may be infected by *Ascaris* ssp., particularly in Latin America and Africa, and some cases can result in pneumonia and respiratory failure. Enterobiasis, another roundworm disease and one of the most common childhood parasitic diseases around the world (known as pinworm or threadworm), can be spread by contaminated water but is more likely to result from poor personal hygiene or contaminated food. Other diseases resulting from the ingestion eggs of parasites include taeniasis and, less commonly, hymenolepiasis, echinococcosis and coenurosis, all of which result from tapeworm infections. While water is one mode of infection, it is usually undercooked food or poor personal hygiene which lead to infection (Bath *et al.* 2010).

2 Water-carried disease

Legionnaires' disease and Pontiac fever are not associated with faecal contamination of water but are associated with bacteria that thrive in warm aquatic environments, usually industrial cooling systems, hot tubs, windscreen washer systems and ornamental pools, particularly if water is stagnant. Regular cleaning and disinfection of pipes and pools help to prevent infection.

Hot tub rash, *Pseudomonas* dermatitis or folliculitis, is a specific water-carried disease which occurs when skin is in contact with contaminated water for long periods of time. The bacteria are commonly found in the environment, particularly in water and other moist environments. This tends to be a self-limiting skin disease and can be prevented by regular disinfection of hot tubs or swimming pools, combined with showering to minimise infection.

3 Water-washed disease

Many water-borne and water-carried diseases may also be considered to be water-washed diseases, as they are also spread by unwashed hands during food preparation and human-to-human contact (Mackintosh and Hoffman 1984). Provision of adequate supplies of water to maintain good personal hygiene, such as hand washing before leaving the bathroom or washing of infected clothing, is essential to prevent infection. This is particularly important for many of the diarrhoea-causing bacteria, viruses and protozoa as well as tapeworm infections. Provision of adequate, clean water for washing and cleaning can also help to reduce the spread of infections more generally. Transmission of air-borne infections such as influenza can be reduced by frequent hand washing, as can the spread of hospital-acquired (or **nosocomial**) infections such as *Clostridium difficile* and methicillin-resistant *Staphylococcus aureus* (MRSA).

Other water-washed diseases are infections of the body surface, the skin and eyes. These can be fungal infections such as tinea, conditions caused by bacteria, including yaws and leprosy, or parasitic invertebrates such as the skin-burrowing mite causing scabies; all result in skin sepsis or ulcers. The eye disorder conjunctivitis is a particular problem in arid countries, where the eyes can become dry and so unable to 'self-cleanse'. However, trachoma is a much more debilitating disease, leading to blindness in infected individuals, with 41 million cases worldwide and

8 million suffering severe vision impairment, particularly in sub-Saharan Africa (Resnikoff *et al.* 2004). Infection rates for all diseases can be reduced through provision of improved access to water and sanitation facilities and good personal hygiene practices such as regular hand and face washing.

Biting arachnids and insects, principally mites, ticks, lice and fleas, are responsible for a number of vector-related infections that can largely be prevented through good personal hygiene, which is dependent on access to suitable water supplies. Epidemics of typhus and other louse- or flea-borne diseases have been reported at intervals around the world, especially at times of war or civil unrest or in slum communities when access to water for washing is restricted. Tularaemia is largely spread via tick bites but can also be spread by flies and, in about 5 to 10% of US cases, by contaminated water sources (Ellis *et al.* 2002). Symptoms of tularaemia are similar to bubonic plague, a disease spread by rat fleas and largely thought to be responsible for killing 30–60% of the European population in the fourteenth century (in an epidemic called the Black Death) and the European Great Plagues of the seventeenth century. Although now rare, isolated outbreaks of bubonic plague are still reported.

4 Water-based disease

Water-based diseases are all worm infections (flukes or trematodes) which can be spread by aquatic invertebrates, particularly water snails, with humans or other mammals acting as the definitive host. Such diseases are rarely life-threatening but can cause discomfort and have a negative impact on livelihood.

Schistosomiasis (or bilharziasis) ranks just behind malaria in public health importance, infecting 200 million in many tropical and subtropical countries (Gryseels *et al.* 2006). Although the mortality rate associated with the disease is low, it does result in chronic illness which can affect the livelihoods of infected individuals. The most common route of infection involves the larval parasite, released from water snails, penetrating the skin during contact with infected waters (for fishing, rice planting, recreation). The subsequent development of the worm may be in the liver, blood vessels, bladder or kidneys, from where eggs will be released to infect faeces or urine (**Figure 8.3**). Programmes of eradication, focusing on eliminating water snail populations and mass treatment of infected populations, have been initiated in a number of countries where the disease is endemic. Limiting human contact with untreated water through better water quality and sanitation facilities is a further approach. The building of dams and irrigation schemes in endemic countries has also resulted in new habitats for water snails, and hence levels of the disease close to such developments have increased, an issue that has led to the design of new irrigation systems and approaches to reduce/prevent water contact.

Other trematode infections, including fasciolopsiasis, clonorchiasis, the fish tapeworm infection diphyllobothriasis, paragonimiasis, and anisakis result from larval ingestion in drinking water, raw or undercooked fish or crustaceans, or raw aquatic plants.

Dracunculiasis is spread through drinking water contaminated with infectious larvae of the nematode guinea-worm, which mature in the human body. The adult female migrates to the skin, usually of the foot, to release larvae into any water that the person may enter, to repeat the life cycle. Prevention of infection through improved water quality by filtering or boiling water is very effective in African countries where it is endemic and the Indian subcontinent, as is protecting well supplies from surface water runoff (Esrey *et al.* 1991).

5 Water-vector disease

Disease vectors, rather than the pathogens, may require water to complete their life cycle. One example of this is the mosquito, which acts as a

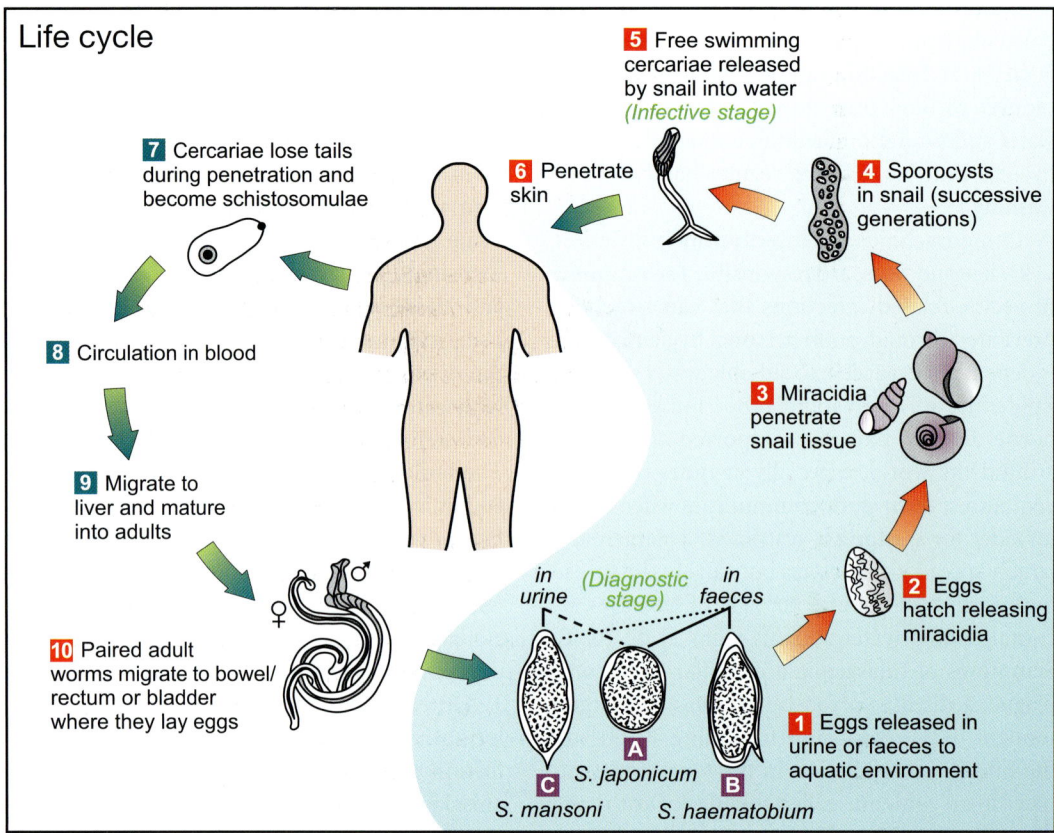

Figure 8.3 The life cycle of the *Schistosoma* spp. trematode worm, demonstrating infection of human hosts by the infective cercariae and subsequent release of eggs from human host to infect intermediate water snail hosts.
Source: Adaptation of an original figure by CDC/Alexander J. da Silva/Melanie Moser.

disease vector for malaria (**Box 8.2**). Malaria is the world's biggest public health concern, with over 250 million cases per year, resulting in almost 1 million deaths (2.23% of deaths worldwide) (WHO 2010b). While malaria is endemic in most tropical or subtropical countries (**Figure 8.4**), the majority of deaths are among young children living in sub-Saharan Africa. While treatments of malaria are available, preventive strategies focusing on the control of the mosquito vector, such as use of mosquito nets and spraying of insecticides, have been the target of recent attempts to reduce infection rates, but research is on-going in a number of areas (Nájera *et al.* 2011). Water devel-

opment projects can increase as well as decrease breeding habitats for disease-causing vectors. Man-made lakes and irrigation measures such as for rice cultivation may result in malaria epidemics; new dams in the Tennessee Valley in the 1930s led to an increase in malaria infections until control measures such as altering the water level were instigated (Hall *et al.* 1946).

Mosquitoes are responsible for the spread of other diseases, such as arboviruses and filariasis. Mosquito control measures are often the only effective disease management measures for these diseases. Culicine mosquitoes are the vector responsible for the spread of many arboviruses,

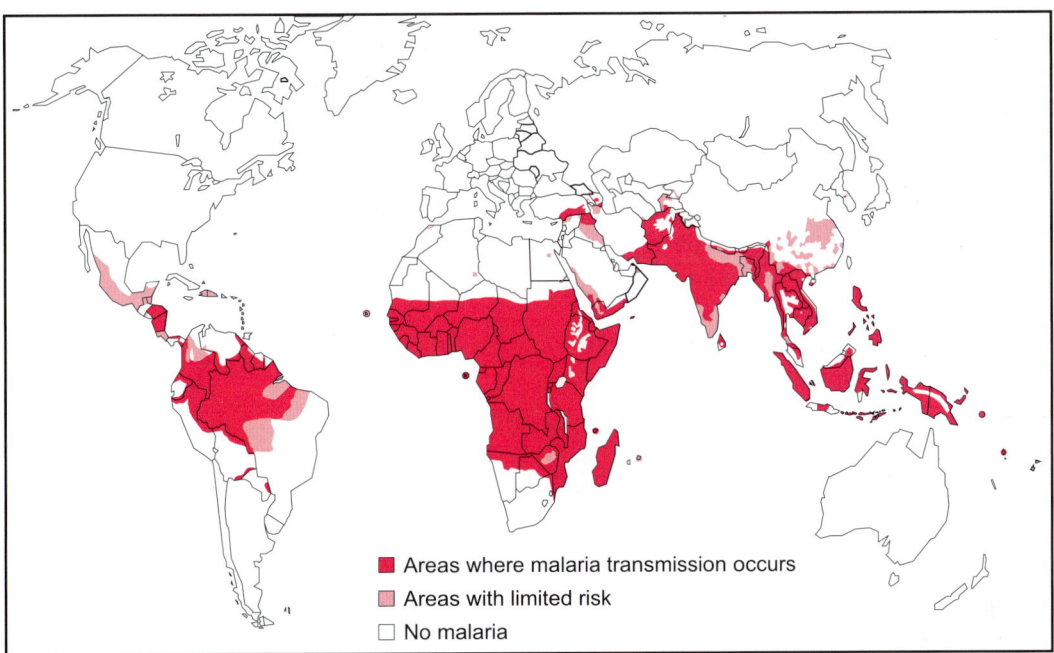

Figure 8.4 Map showing countries affected by malaria.

such as dengue virus and yellow fever. There are four types of dengue virus spread by mosquitoes of the *Aëdes* genus, principally *Aëdes aegypti*. About 5% of cases, usually a result of infection by more than one type of the virus (multiple infection), result in severe illness involving fever, joint and muscle pain (hence 'breakbone fever'), and bleeding which may lead to death from shock and haemorrhage. Of the 50 to 100 million infected across the tropics (Asia, Africa and Latin America), fewer than 25,000 deaths from dengue fever are reported (Gubler 1998). However, dengue fever has increased 30-fold over the last 50 years, due to deforestation (*A. aegypti* is an African forest mosquito), urbanisation and international travel. For yellow fever, a vaccine does exist although vaccination programmes do not exist in all the countries where the disease is endemic in Africa and South America. A haemorrhagic fever that causes jaundice in severe cases and can lead to death, yellow fever is also spread by the *A. aegypti* mosquito. It was a major cause of death in the nineteenth century, with epidemics across the United States and Europe; it is rarely reported in these countries today. It affects an estimated 200,000 people per year, with 30,000 deaths, but an effective vaccine does exist (Robertson *et al.* 1996). Other mosquito-vector diseases include lymphatic filariasis (LF), a chronic nematode disease which results in swelling in different parts of the body. Treatments exist and preventive strategies can be implemented, but it can take many years to diagnose, as clinical symptoms develop slowly. Other filariasis diseases include dracunculiasis (see section C4) and onchocerciasis.

Incidence of the nematode-bacteria disease onchocerciasis (river blindness), once a major cause of blindness worldwide, has been massively reduced in the last 20 years, due to international eradication programmes. Transmitted via the *Simulium* genus of blackflies, which breed in fast-flowing rivers in sub-Saharan Africa, Yemen and Latin America, the nematodes can persist in the

BOX 8.2 CONTEMPORARY CHALLENGES

Mosquitoes and malaria

The mosquito is the vector (an agent, usually an arthropod, which carries and transmits disease but remains unaffected itself) for malaria-causing *Plasmodium* species. There are almost 1 million deaths from malaria per year, with 85 to 90% of these occurring in sub-Saharan Africa. Many large organisations around the world are looking at methods to reduce the number of people who die or whose lives are blighted by this disease. The larval and pupal stages of the mosquito's life cycle are dependent on freshwater (Figure 8.5). By reducing the amount of standing water in malaria-affected regions, covering water-storage ponds to restrict access by mosquitoes, or use of biological controls in surface water, the number of mosquitoes can be reduced and rates of malaria infection lowered. Preventive medications are available but the high cost and long-term side-effects of these drugs prohibit their widespread use for all but the short-term traveller.

Malaria is spread by the *Anopheles* species of mosquito (Figure 8.6). While there are in excess of 400 species of Anopheles, only 30–40 species are common vectors for the malaria plasmodium. Of these, *Anopheles gambiae* is the disease vector for the most dangerous malarial plasmodium *Plasmodium falciparum*. Only the adult female mosquitoes feed on blood. Adult female mosquitoes become infected with *Plasmodium* sporozoites after feeding on the blood of an infected human. The sporozoites are carried in the salivary gland of the mosquito for infection of subsequent humans. Of the over 200 species of *Plasmodium*, only about 11 infect humans. Of these, *P. falciparum* causes the severest type of malaria and is responsible for 90% of deaths from malaria; *Plasmodium vivax*, *Plasmodium ovale*, *Plasmodium malariae* and *Plasmodium knowlesi* result in a milder version of the disease.

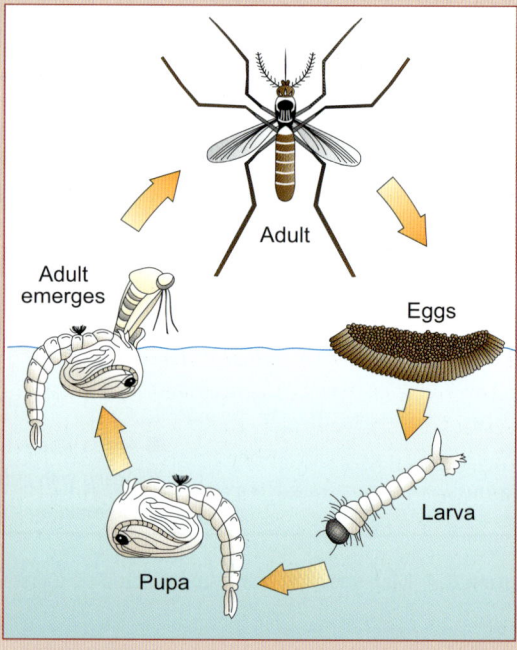

Figure 8.5 The life cycle of the mosquito is dependent on the presence of a water body.

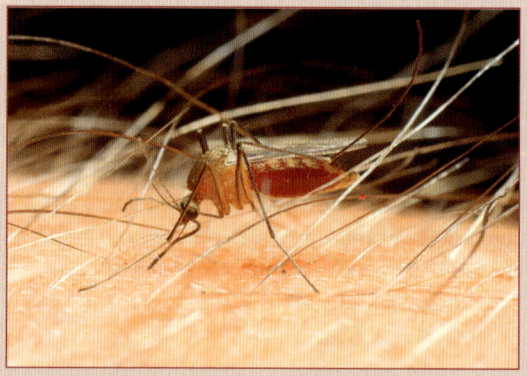

Figure 8.6 Adult female mosquitoes become infected after feeding on human blood.
Source: Photo courtesy of Shutterstock/Kletr.

The mosquito is still found in areas where malaria has been eradicated (e.g. Europe, United States), and hence its reintroduction is possible. Climate change may make northern latitudes more susceptible to malarial infection.

skin for many years, causing itching, swelling and tissue damage, particularly in the lower limbs and eyes. When the worms die, they release a **symbiotic** bacterium that increases the inflammatory response which, in chronic cases, can result in blindness. While the disease is not lethal, it can greatly alter the quality of life for the estimated 18 million people infected, and especially the 300,000 who are permanently blinded (Burnham 1998). Effective treatments for the disease are widely available, but blackfly control measures, particularly the use of insecticides, can greatly reduce transmission, as seen in the Onchocerciasis Control Programme in West Africa. For countries where vector control is not considered feasible or cost-effective, it has been suggested that water-engineering projects such as slowing flow may help to reduce fly numbers by changing the habitat of blackfly breeding sites. Such measures may be expensive and damaging to the wider environment and may also increase breeding sites for other disease vectors preferring slow or still water, such as mosquitoes.

African trypanosomiasis (sleeping sickness) and American trypanosomiasis (Chagas disease) are protozoan diseases spread by the tsetse fly and assassin bug, respectively. Neither disease vector is dependent on water for its life cycle but the tsetse flies associated with the spread of both chronic and acute sleeping sickness are the riverine tsetse flies of the subgenus *Palpalis*. Insecticide use and land management regimes have resulted in temporary reduction of tsetse fly populations (Kagbadouno *et al.* 2011), but medical and veterinary programmes may circumvent the eradication of the fly.

> **REFLECTIVE QUESTION**
>
> What are the main types of infectious disease associated with water and how can their incidence be reduced?

D CHEMICAL CONTAMINANTS

Water rarely contains just hydrogen and oxygen. Components that reflect the surrounding geology of a body of water (particularly if a groundwater resource); the interaction with aquatic organisms, sediment and atmospheric inputs; runoff from land; and anthropogenic inputs (**Figure 8.7**) can all determine the chemical load of water (see Chapter 4). **Box 8.3** provides guideline values for the many different compounds found in drinking water. Most of these substances are present at low levels and hence will have little impact on human health but there are incidences where levels exceed water quality guidelines and hence pose a threat to health. This section considers the main issues concerning the chemical content of water and its impact on human health.

1 Chemicals derived from the environment

Arsenic is found in many freshwater resources but high levels are recorded in groundwater-dependent areas with strata rich in arsenic rock deposits (Smedley and Kinniburgh 2002; Kavcar *et al.* 2009). High levels of inorganic arsenic have been reported in drinking water supplies in the Americas (Argentina, Chile, Mexico, United States) and Asia (India, China, Thailand), but it is in Bangladesh that the problem has reached epidemic proportions. In the 1970s, large numbers of deep tube-wells were dug to tap groundwater reserves relatively free of pathogens found in existing water sources and responsible for one of the world's highest child mortality rates from water-borne disease. However, from the early 1990s, it was apparent that a large number of these wells, an estimated 27% of all wells, were heavily contaminated with arsenic (at levels above 0.05 mg L^{-1} or 50 ppb). Around contaminated tube-wells, symptoms of arsenicosis were common: skin lesions (which, if infected, can become gangrenous), skin hardening, dark spots on limbs, swollen limbs and loss of feeling in hands and

Figure 8.7 Human sewage effluent can alter both the biological and chemical content of water. Other human activities, such as the disposal of solid waste to watercourses, as shown here, will also alter chemical water quality.
Source: Photo courtesy of Rachel Caffarate.

feet. Children may show impaired cognitive development, and long-term exposure increases the risk of cancers among the estimated 35 to 77 million people exposed to high arsenic levels. Various programmes have been initiated to prevent further exposure to arsenic-contaminated water in Bangladesh, including development of alternative water supply options and arsenic removal systems. In the United States, arsenic levels rarely exceed the 50 ppb level but some areas do exceed the MCL (**Box 8.3**), which was lowered in 2006 from 50 ppb to the WHO's guideline value of 10 ppb (0.01 mg L^{-1}), leading to the development of a range of water treatment options, from point of use to large-scale treatment (WHO 2008).

Fluoride occurs naturally in some water resources, often at levels exceeding recommended health levels of 0.7–1.2 parts per million (ppm) (0.7–1.2 mg L^{-1}); the WHO's guideline value is 1.5 mg L^{-1} (WHO 2008). While surface waters generally contain less than 0.5 ppm (0.5 mg L^{-1}), some groundwater reserves in Argentina, western states of the United States, northern Africa, eastern Africa, the Indian subcontinent and parts of China have fluoride levels exceeding 1.5 mg L^{-1}. High levels of fluoride in water can result in dental fluorosis, an unaesthetic brown mottling stain on the tooth surface, but a greater resistance to dental decay. As a consequence of lower levels of cavities, fluoride has been added at levels of

WATER AND HEALTH

> **BOX 8.3 THE FUTURE OF WATER**
>
> ## Maintaining water quality
>
> While we may initially be concerned with the biological quality of drinking water, particularly the microorganism load contained in water resources, chemical compounds found in water are also important considerations for human health and well-being. A number of the chemicals found in water may be derived from natural sources such as arsenic and fluoride. However, many chemicals occur in water as a result of human activities, particularly organic compounds such as hydrocarbons and pesticides. In all developed countries and many developing countries, water quality legislation or guidelines have established limits for these chemicals. The limits and guidelines are continuously reviewed and updated to ensure the safety of drinking water. The World Health Organization (WHO) has established the guideline values shown in **Table 8.2** for chemicals that are of health significance in drinking water (WHO 2008). National legislation in developed and many developing countries has set legally enforced limits for a number of chemicals based on available toxicological data, based on the WHO guideline values (**Table 8.3**). In the United States, National Primary Drinking Water Regulations (NPDWRs or primary standards) are legally enforceable standards that apply to public water systems. NPDWRs protect public health by limiting the levels of contaminants in drinking water, including organic chemicals, inorganic chemicals including heavy metals, radionuclides, disinfectants and disinfection by-products, as well as microorganisms. The legally enforced limits are called Maximum Contaminant Levels (MCLs) and represent the highest level of a contaminant allowed in drinking water. Maximum Contaminant Levels Goals (MCLGs) represent the level below which there is no appreciable risk to health. Including a margin of safety and forming the basis of MCLs, MCLGs are not enforceable but do represent a preferred contaminant limit for drinking water. The Drinking Water Directive applies across all European Union (EU) member states, ensuring harmonisation of high-quality water across the continent. Each EU member state has implemented these regulations in national legislation, as demonstrated by the Water Supply (Water Quality) Regulations 2000 in England. It is worth noting that a number of these regulatory instruments utilise the 'precautionary principle', which urges the adoption of cautious approaches when a risk is suspected but scientific consensus is still lacking (Principle 15 of the Rio Declaration on Environment and Development).

between 0.5 and 1.0 ppm to the water supply of a number of regions/countries, including the United States, Brazil and Australia, to reduce the public health problem of tooth decay. However, fluoridation of water supplies has been met with mixed support – often welcomed by public health authorities but viewed with suspicion by the public. In a number of European countries and Japan, water fluoridation has been discontinued. Although the negative health impacts from water fluoridation are not well reported, high-level fluoride exposure can lead to skeletal fluorosis, a reduction in bone strength caused by calcium displacement, which is seen in a number of countries with naturally occurring high levels of fluoride.

Radon is a naturally occurring radioactive gas formed during the decay of small amounts of uranium found in rock and soils, particularly granite and granite-derived alluvial soils. After tobacco smoking, radon is cited as the second most common cause of lung cancer. While levels in the environment are easily vented, levels in enclosed spaces such as homes or mines can build

Table 8.2 World Health Organization guideline concentrations for chemicals found in drinking water

Chemical	Guideline value (mg L^{-1})	Chemical	Guideline value (mg L^{-1})
Acrylamide	0.0005	Dimethoate	0.006
Alachlor	0.02	1,4-Dioxane	0.05
Aldicarb	0.01	Edetic acid	0.6
Aldrin and dieldrin	0.00003	Endrin	0.0006
Antimony	0.02	Epichlorohydrin	0.0004
Arsenic	0.01	Ethylbenzene	0.3
Atrazine and its chloro-striazine metabolites	0.1	Fenoprop	0.009
		Fluoride	1.5
Barium	0.7	Hexachlorobutadiene	0.0006
Benzene	0.01	Hydroxyatrazine	0.2
Benzo[a]pyrene	0.0007	Isoproturon	0.009
Boron	2.4	Lead	0.01
Bromate	0.01	Lindane	0.002
Bromodichloromethane	0.06	MCPA	0.002
Bromoform	0.1	Mecoprop	0.01
Cadmium	0.003	Mercury	0.006
Carbofuran	0.007	Methoxychlor	0.02
Carbon tetrachloride	0.004	Metolachlor	0.01
Chlorate	0.7	Microcystin-LR	0.001
Chlordane	0.0002	Molinate	0.006
Chlorine	5	Monochloramine	3
Chlorite	0.7	Monochloroacetate	0.02
Chloroform	0.3	Nickel	0.07
Chlorotoluron	0.03	Nitrate	50
Chlorpyrifos	0.03	Nitrilotriacetic acid	0.2
Chromium	0.05	Nitrite	3
Copper	2	N-Nitrosodimethylamine	0.0001
Cyanazine	0.0006	Pendimethalin	0.02
2,4-Db	0.03	Pentachlorophenol	0.009
2,4-DB	0.09	Selenium	0.04
DDTd and metabolites	0.001	Simazine	0.002
Dibromoacetonitrile	0.07	Sodium dichloroisocyanurate	50
Dibromochloromethane	0.1	Sodium (as cyanuric acid)	40
1,2-Dibromo-3-chloropropane	0.001	Styrene	0.02
1,2-Dibromoethane	0.0004	2,4,5-T	0.009
Dichloroacetate	0.05	Terbuthylazine	0.007
Dichloroacetonitrile	0.02	Tetrachloroethene	0.04
1,2-Dichlorobenzene	1	Toluene	0.7
1,4-Dichlorobenzene	0.3	Trichloroacetate	0.2
1,2-Dichloroethane	0.03	Trichloroethene	0.02
1,2-Dichloroethene	0.05	2,4,6-Trichlorophenol	0.2
Dichloromethane	0.02	Trifluralin	0.02
1,2-Dichloropropane	0.04	Trihalomethanes	1 (max)
1,3-Dichloropropene	0.02	Uranium	0.3
Dichlorprop	0.1	Vinyl chloride	0.0003
Di(2-ethylhexyl)phthalate	0.008	Xylenes	0.5

Table 8.3 Limits and guidelines set for drinking water for selected pesticides and other organic substances

Contaminants	mg L^{-1}			
	MCLG	MCL	WHO	EU
Pesticides				
Atrazine	0.003	0.003	0.002	0.0001
2,4-D	0.07	0.07	0.03	0.0001
Dichlorprop			0.1	0.0001
Glyphosate	0.7	0.7		0.0001
Malathion				0.0001
MCPA			0.002	0.0001
Mecoprop			0.01	0.0001
Simazine	0.004	0.004	0.002	0.0001
Cumulative pesticides				0.0005
Other organic pollutants				
Benzene	0	0.005	0.01	0.001
Ethylbenzene	0.7	0.7	0.3	
Toluene	1	1	0.7	
Xylene(s)	10	10	0.5	
1,1,1-trichloroethane	0.2	0.2		
1,2-dichloroethane	0	0.005	0.03	0.003
Dichloroethylenes	0.1	0.1	0.05	
Trichloroethylene	0	0.005	0.07	0.01
Tetrachloroethylene	0	0.005	0.04	0.01
Chlorobenzene	0.1	0.1		
Dichloromethane	0	0.005	0.02	
Pentachlorophenol	0	0.001	0.009	
DEHP	0	0.006	0.008	
Vinyl chloride	0	0.002	0.0003	0.0005
Polycyclic aromatic hydrocarbons	0	0.0002		0.0001
Polychlorinated biphenyls	0	0.0004		

Notes: MCLG = Maximum Contaminant Level Goal (US-EPA); MCL = Maximum Contaminant Level (US drinking water); WHO = World Health Organization guidelines; EU = European Union drinking water limits (applied to all member states).

up, due to lack of ventilation. For example, areas of the UK, particularly Cornwall and Derbyshire, and the USA, particularly Iowa and the Appalachian Mountains, demonstrate elevated levels of radon gas in homes well in excess of the WHO Action or Reference Level of 100 Bq m^{-3}, but generally within the WHO Upper Limit of 300 Bq m^{-3} (WHO 2008). Drinking water sourced from groundwater in radon-affected areas may contain elevated levels of radon. While risks from ingestion are low, radon can be volatilised during showering and other activities, and may contribute 1 to 5% of indoor airborne radon. In the USA, an MCL of 300 pCi L^{-1} (approximately 11,000 Bq m^{-3}) is used to restrict radon levels in drinking water. Water can be treated before entering homes if levels are particularly high.

Large-scale groundwater extraction in coastal

areas can lead to saline intrusion and threaten water supplies. When freshwater is rapidly drawn from groundwater reserves along the coast, the water table is drawn down, causing sea water from the coastal fringe to be drawn into the aquifer (see Chapter 5). Saline intrusion renders the water supply unfit for consumption and for use in irrigation. Problems occur around the world, with areas of North Africa, Mediterranean Europe, the Middle East, the United States, China and Mexico all reporting saline contamination of groundwater reserves. Careful management of groundwater water supplies and saltwater monitoring programmes can ensure that the risks of saline contamination are minimised. Coastal flooding may also temporarily increase the **salinity** of surface water sources, an issue that is mitigated by coastal sea defences but is likely to be a growing issue with climate change.

Cyanobacteria, often referred to as blue-green algae, occur in both freshwater and marine habitats (see section E5 of Chapter 6). When conditions are favourable – usually a combination of nutrient availability, sunlight and warm temperature – cyanobacteria can reproduce rapidly to form a bloom (**Figure 8.8**; see also **Figure 6.17** in Chapter 6). Between 30 and 50% of blooms are relatively harmless to human health. However, as certain cyanobacteria species produce toxins (the most common is called microcystin), the bloom may pose a health threat to humans and other organisms that drink or swim in water

Figure 8.8 Cyanobacteria blooms can have considerable impacts on water quality – and may release toxins into the water.
Source: Photo courtesy of Rachel Caffarate.

contaminated by cyanobacteria. As water sources used for drinking water abstraction in developed countries are generally managed to control for blooms, risks to humans are low, unless they are exposed through recreational activities such as wild swimming and water sports. Even exposure via these means is unlikely to result in death, although exposure to high levels of cyanobacterial hepatotoxins may result in liver damage, particularly in children, while neurotoxins may affect the nervous system. Generally, symptoms will start with skin and/or eye irritation, followed by nausea, abdominal pain, diarrhoea and fever. A number of deaths have been reported in South America and China, associated with cyanobacteria blooms in reservoirs and dams supplying drinking water (Chorus and Bartram 1999; WHO/UNICEF 2003). However, the effect on animals is greater, as they may not have access to managed drinking water resources, and there have been many reports of fatal poisonings (due to liver or nervous system damage) in wild and domestic animals that drink from lakes and reservoirs affected by blooms.

2 Classical anthropogenic chemicals

Aquatic contaminants derived from human activities, particularly activities associated with agriculture, have long been a problem. While high nitrate levels in groundwater and surface water may result from natural processes, most nitrate contamination is a result of human activities, particularly the over-application of inorganic fertilisers. Nitrate is also a degradation product of other nitrogen compounds derived from animal manure or sewage sludge; runoff from animal husbandry; or wastewater treatment or septic tank release. High levels of nitrate in drinking water can lead to the development of methaemoglobinaemia, or 'blue-baby syndrome', in infants under 6 months; after ingestion, nitrate is converted to nitrite, which reacts with haemoglobin to form methaemoglobin, resulting in significantly reduced oxygen-carrying capacity of the blood, and hence discoloration of the skin (Gupta *et al.* 2000). Elevated nitrate in the environment contributes to **eutrophication**, and hence affects the health of the aquatic ecosystem (see Chapter 6). Strict water quality guidelines for nitrate are in place in most developed countries as a safeguard against methaemoglobinaemia, and these are usually based on the WHO guideline value of 50 ppm (50 mg L^{-1}) of nitrate in drinking water (WHO 2008). While water and wastewater treatment can remove nitrate or reduce levels, measures to prevent nitrate entering water, such as catchment-sensitive farming in nitrate vulnerable zones (a legal requirement in the UK) and other land and water management practices can reduce the need for and therefore the cost of treatment. Other nutrients, particularly phosphates, benefit from land management strategies to reduce input to water resources, reducing levels of eutrophication and the need for treatment.

Pesticides are very commonly used in agriculture and horticulture to reduce yield loss due to pests and diseases. In 1962, Rachel Carson's book *Silent Spring* first voiced the concern that pesticide use was having a wider environmental impact than was initially intended (Carson 1962). Since then, most industrialised countries have instigated the monitoring of natural waters and drinking waters for pesticides, introduced legislation to control the types of pesticides that can be used, and set limits on the amount of pesticide residues permitted in drinking water. In the EU, individual pesticides must not be present in water above 0.0001 mg L^{-1} (limits are lower, at 0.03 µg L^{-1}, for aldrin, dieldrin and heptachlor), with a cumulative pesticide limit of 0.0005 mg L^{-1}; limits set elsewhere are based on the individual toxicity of each pesticide chemical (**Table 8.3**). Despite this, it has been reported that almost 90% of US surface water bodies are contaminated by pesticides (Gilliom *et al.* 2007). Many pesticides have been withdrawn from use, due to their persistent, bioaccumulative and toxic properties such as DDT (dichlorodiphenyltrichloroethane); DDT is still, however, used in a number of African countries to control mosquitoes. Although pesticides can affect the health of the environment, consequences for human health are not so well defined. Accidental ingestion, inhalation, or absorption through the skin of small amounts of pesticides during preparation or application can result in temporary ill-health, such as nausea or skin irritation; chronic exposure may lead to serious illness through increased risk of cancers, respiratory disease and neurological disorders (such as Parkinson's Disease) (Brown *et al.* 2005). Exposure to the herbicides 2,4-D and 2,4,5-T, which constituted the defoliant Agent Orange used by the US military during the Vietnam War (1962–1971), led to elevated levels of ill-health and birth defects, mainly due to contamination by 2,3,7,8-tetrachlorodibenzo-*p*-dioxin (TCDD) (Erickson *et al.* 1984). The consequences to human health through consumption of low-level contaminated water supplies are not well understood but are largely prevented, at least in the developed world, through strict drinking water quality limits, treatment regimes and pesticide control regulations.

A range of organic compounds associated with human activities can be discharged into the environment in wastewater if appropriate treatment processes do not exist. Most substances are associated with industrial effluent, particularly organic solvents used in manufacturing, including petrochemicals. Different organic chemicals impact upon the environment, and ultimately on human health, in different ways depending on the chemical, its concentration, whether or not it is discharged in a mixture with other chemicals, and the chemistry of the receiving waters. A common symptom of water pollution by organic substances is fish death, which usually results from oxygen depletion (**hypoxia**); hypoxia is also a water quality indicator for sewage/wastewater pollution which may also be discharged with effluent containing organic pollutants. Many organic compounds are priority pollutants, and hence their levels in water are closely monitored (**Box 8.3**). Effective removal during wastewater treatment, particularly using tertiary treatment such as ozonation and ultraviolet radiation (see Chapter 9), should prevent entry into receiving waters; water treatment for drinking water supplies offers a secondary treatment process. In developed countries, environmental exposure will be through accidental release, which will result in heavy fines and the enforcement of 'polluter pays' principles. While human exposure will be unlikely, recreational exposure to pollution incidences may result in gastrointestinal or neurological illness. In developing countries where wastewater and water treatments are less effective, environmental and health consequences can be more extreme. Recently, reports of high rates of birth defects in the Iraqi city of Fallujah have been suggested to be linked to chemical contamination of the city's water supply, although more recent studies suggest a link to depleted uranium (Alaani *et al.* 2011). Previously, it has been suggested in both developed and developing countries that perceived pockets of illness and disease, including birth defects, cancer clusters and increased infant mortality rates, may be linked to water and soil contaminated by waste disposal sites and other associated industrial activities, but for many cases the epidemiological evidence is inconclusive (Elliott *et al.* 2009). The most high-profile of all such cases is Love Canal in the USA (Vianna and Polan 1984; Gensburg *et al.* 2009; Austin *et al.* 2011), where a community living above a closed chemical waste landfill in the 1970s reported high levels of illness and birth defects. This incident resulted in new federal legislation in 1980 focusing on hazardous waste landfills (Comprehensive Environmental Response, Compensation, and Liability Act (CERCLA)). Overall, although human health impacts from exposure to contaminated water are not well defined, environmental consequences are well documented and the toxicological profile of most chemicals is well described, hence the limits imposed on drinking water.

Heavy metal contamination of water can be a consequence of natural processes such as weathering of rocks, or derived from human activities. Incidences of water contamination by heavy metals such as lead and chromium usually derive from human activities. Lead, a neurotoxin which has an adverse impact on the brain development of infants and young children, has been used to make water pipes for centuries, and in many older properties in the developed world (pre-1970 in the UK and much of Europe; pre-1986 for the USA) lead pipes may still be found. Generally, water utility companies have replaced older lead pipes with plastic pipes up to the point of entry to a household, but it is then generally the householders' responsibility to replace household lead water pipes with pipes made from copper or other materials. Households with lead pipes are recommended to run their taps to flush out standing water before using water for drinking or cooking, but should look ultimately to replace the pipework, particularly if young children are resident.

Chromium, unlike lead, is a water contaminant associated with industry. Runoff from contaminated land or poorly treated industrial effluent (particularly from dye works) can lead to elevated chromium levels in water sources, and ultimately in drinking water. In particular, hexavalent chromium (Cr(VI)) is a recognised human carcinogen, but its presence in drinking water is still common in many developed countries (Costa 1997). With an MCL of 0.1 mg L^{-1} applicable to total chromium (including Cr(VI)), levels in tap water across the USA and other countries can sometimes exceed this for Cr(VI) alone. Removal through ion exchange, reverse osmosis and coagulation/flocculation is effective (see Chapter 9).

Many metals and other substances can enter water as a consequence of water (or wastewater) treatment. Contamination incidences are rare and usually accidental, but have resulted in some significant public health issues. In the English town of Camelford in 1988, a large amount of aluminium sulphate, which is used in water treatment for coagulation/flocculation (see Chapter 9), entered the water supply and resulted in widespread health effects which varied from urinary and dermal complaints, diarrhoea and nausea to more long-term health issues that have been attributed to the pollution incidence but are still disputed. Key among the long-term effects were osteodystrophy and damage to brain function, leading to dementia and Alzheimer's disease-like symptoms (Ward *et al.* 1978). Incidences of chlorine leaks from water and wastewater treatment works have also been reported, often with considerable impact on the local environment.

3 Emerging contaminants

Groups of compounds have been identified over the last few decades which may pose a threat to human health and the health of the wider environment. Often, the evidence to support health links is still much disputed and it is far from clear what impact these contaminants have on health. Perhaps the most famous of these are endocrine disrupting compounds (EDCs). These compounds can mimic hormones (e.g. the female hormone oestrogen), inhibit the action of hormones (e.g. antiandrogenic effects), or alter the normal regulatory effects of the endocrine system and, potentially, the nervous and immune systems. EDCs include a wide range of chemicals, most of which have been identified in natural waterways, derived from a wide range of sources such as:

- dichlorodiphenyltrichloroethane (DDT) and other pesticides
- polychlorinated biphenyls (PCBs)
- bisphenol A
- polybrominated diphenyl ethers (PBDEs)
- phthalates
- alkylphenols
- perfluorooctanoic acid (PFOA)
- dioxins and furans
- various pharmaceuticals
- complex chemical mixtures such as industrial effluents, landfill leachate and municipal sewage.

EDCs have been cited as being responsible for a range of reported environmental consequences, including declining fertility in fish, birds and mammals; reduced hatching success and offspring survival in fish, amphibians, birds and reptiles; demasculinisation/feminisation or defeminisation/masculinisation across all vertebrate groups, particularly aquatic animals; and thyroid function and development problems in fish and birds. Laboratory analyses in particular demonstrate the endocrine disruption behaviour of individual chemicals or mixtures. The role of the toxic effect is still uncertain, with suggestions that EDCs have a direct impact on endocrine function or that the impact results indirectly from other stresses on other physiological functions. Evidence of impact on human health is less substantiated, with no causal relationship established between EDC exposure and health effects, despite suggestions that exposure to EDCs may increase rates of breast cancer and endometriosis in women, and testicular and prostate cancers in men; result in abnormal sexual development; reduce male fertility; alter pituitary and thyroid gland function; suppress the immune system; and have neurological effects. While levels of individual EDCs in the environment, and specifically in drinking water, may be very small (measured in ppm or ppb (parts per billion)) and hence may easily be buffered by a healthy adult (Kuch and Ballschmiter 2001), the impact of mixtures of these chemicals and the response of foetuses and new-born babies is still unclear (Crisp *et al.* 1998).

Certain pharmaceuticals and veterinary medicines demonstrate endocrine disrupting behaviour, which is essential for their therapeutic purpose, such as the female contraceptive 17-α-ethinyloestradiol. Chapter 6 describes some of the impacts of these chemicals observed on aquatic organisms. Low levels of pharmaceutical residues have long been detected in natural waterways and there is increasing evidence that these residues may also be found in drinking water, albeit at trace levels (see also Chapter 4). Such residues include a wide range of therapeutic drug categories, from analgesics to antibiotics and treatments for coronary and respiratory diseases. While levels are low, the health effects from long-term exposure to biologically active chemicals have not been fully realised (Kuemmerer 2009).

Chlorine is the most common method of drinking water disinfection worldwide and it is very effective at killing pathogens that remain in the water following filtration. While chlorine is a relatively cheap and effective disinfectant, it can react with natural organic compounds in water to produce the disinfection by-products trihalomethanes (THMs) and haloacetic acids (HAAs). THMs and HAAs are classified as carcinogenic agents, so exposure to these compounds in drinking water may increase cancer risk (Boorman *et al.* 1999). To reduce the amounts of THMs and HAAs produced in disinfected water and comply with regulatory requirements, water companies attempt to remove as much of the organic content of water (particularly dissolved organic carbon (DOC)) as possible. Removing high levels of DOC from water can be a costly (financially and energy-wise) process and many water utility companies in the developed world are involved in research aimed at reducing DOC prior to water treatment. While ozonation and UV disinfection do not lead to the generation of harmful residues, they tend not to be as effective, or are suitable for use only in certain situations.

Nanomaterials are increasingly used in consumer products such as clothes and cosmetics (e.g. silver nanoparticles added to socks to prevent foot odour and zinc oxide in sunscreen), as well as in water and wastewater treatment processes. The properties of nanomaterials offer very cost-effective methods of improving health and well-being through cleaner water. However, the environmental fate and behaviour of nanomaterials in treated sewage effluent, septic tanks, and as air-borne particles requires investigation to ensure that environmental exposure does not pose a health risk. Two factors in particular have raised concerns about the increasing use of nano-particles: their chemical properties and size. At nanoscales, the properties of many chemicals change, due to alterations to the surface area-to-volume ratio, increasing their solubility in water. Due to their small size, nanoparticles have the potential to be taken up by organisms via dermal absorption, inhalation or ingestion. However, there is no evidence that yet suggests that nanomaterials commonly used in consumer products and in the treatment of water or wastewater have a negative impact on environmental and human health (Moore 2006; Nowack and Bucheli 2007). Work is on-going in this area.

> **REFLECTIVE QUESTION**
>
> What are the main chemical contaminants and why is there a concern about EDCs?

E PHYSICAL WATER RISK

The previous sections have largely focused on water quality, including microbiological load and chemical contamination; water quantity has not been considered except with regard to water-washed disease and water availability for personal hygiene. Yet, water availability is vital to support health – directly through the provision of drinking water, but also for food production. This section evaluates some of the complex issues that can be considered to be physical risks to health from water. Between 1999 and 2008 over 2,200 million people were affected by approximately 2,445 water-related natural disasters, 83% of all natural disasters, leading to over 870,000 deaths (IFRC 2009).

1 Flooding

The physical process of flooding is described in section D of Chapter 3. Flooding can lead to direct loss of life through drowning, but can exert a greater influence on health through disease, loss of financial security and social cohesion, and famine. Flood waters can easily become contaminated with sewage, leading to the spread of water-borne diseases such as cholera and typhoid (**Figure 8.9**). For instance, during the monsoon season in India and Bangladesh, levels of cholera peak rapidly (Koelle *et al.* 2005; Akanda *et al.* 2011). In many countries with a well-developed water supply and wastewater network, this risk is well recognised and can be mitigated against; countries lacking this infrastructure are placed at a greater risk. An increase in standing water may also result in an increase in vector-borne diseases such as malaria, dengue fever and West Nile Fever.

The greatest impact that flooding has on health is associated with loss of shelter, civil unrest, crop failure, loss of transport links for movement of food and emergency aid, reduced access to healthcare professionals, loss of power, radiation leaks from damaged nuclear power plants, etc. (Watson *et al.* 2007). All can result in both physical and mental health consequences. Some, if not all, of these impacts were seen in the aftermath of Hurricane Katrina in New Orleans in 2005 and of the Japanese tsunami following the earthquake of 2011 (**Box 8.4**). The impact of floods can be reduced through the construction of suitable barriers (coastal and riverine), maintaining floodplains (free from development), development of flood strategies, and increasing awareness of populations living in flood-prone areas.

2 Drought

Where water is scarce, low-quality water resources may be consumed, increasing the likelihood of water-borne disease or illness from water contaminated by chemical residues. Drought is also linked to famine, conflict and civil unrest, and is a key factor in many social and economic disasters around the world, from the Dust Bowl of the mid-western states of the United States (Schubert *et al.* 2004) and Canada during the

Figure 8.9 Too much water can have health consequences – increasing the spread of disease.
Source: Photo courtesy of Shutterstock/think4photop.

1930s (see **Box 2.2** in Chapter 2) to famines and war in sub-Saharan Africa in more recent years. Reducing the impact of drought on human health and the environment can include the construction of water-storage ponds or containers (although this will also increase the number of mosquito breeding sites, and so elevate malaria risk), planting of drought-resistant crops, development of drought management strategies, water-sharing mechanisms, land management and aid programmes.

3 Food production

Food production is reliant on the provision of water for irrigation of crops (**Figure 8.13**), drinking water/food for cattle and the processing/cooking of food. Particularly in sub-Saharan Africa, drought is the main factor behind crop failure, but too much water can also have a catastrophic effect (as seen in the Irish potato famine in the 1840s, when wet and warm conditions provided the ideal environment for the spread of potato blight, *Phytophthora infestans*). Malnourishment further aids infection by waterborne diseases, increasing the risk of death, as compared to well-nourished individuals.

Many countries around the world are already suffering considerable water stress, which will be exacerbated by climate change and increasing population. To meet increasing demands for food production, and also for the production of biofuels for energy, more land will need to be taken into production and a greater reliance will be placed on irrigation. Agriculture is already responsible for the largest abstraction of water, with a global average of 70% of abstracted freshwater used directly in agriculture (see Chapter

BOX 8.4 CASE STUDIES

Hurricanes, earthquakes and tsunamis

Location: New Orleans, USA
Date: 29 August 2005
Event: Hurricane Katrina, Category 3 storm (landfall)
Result: The city of New Orleans is built on marshlands of the Mississippi River delta, with much of the city below sea level. Hurricane Katrina caused a storm surge of up to 9 m in height along a 32 km stretch of the Mississippi coastline, with New Orleans hit by a 6 m-high surge. The New Orleans surge caused over-topping of a number of levees built to protect the city, while other concrete floodwalls collapsed as safety margins were exceeded. Large areas of the city were flooded (Figure 8.10).

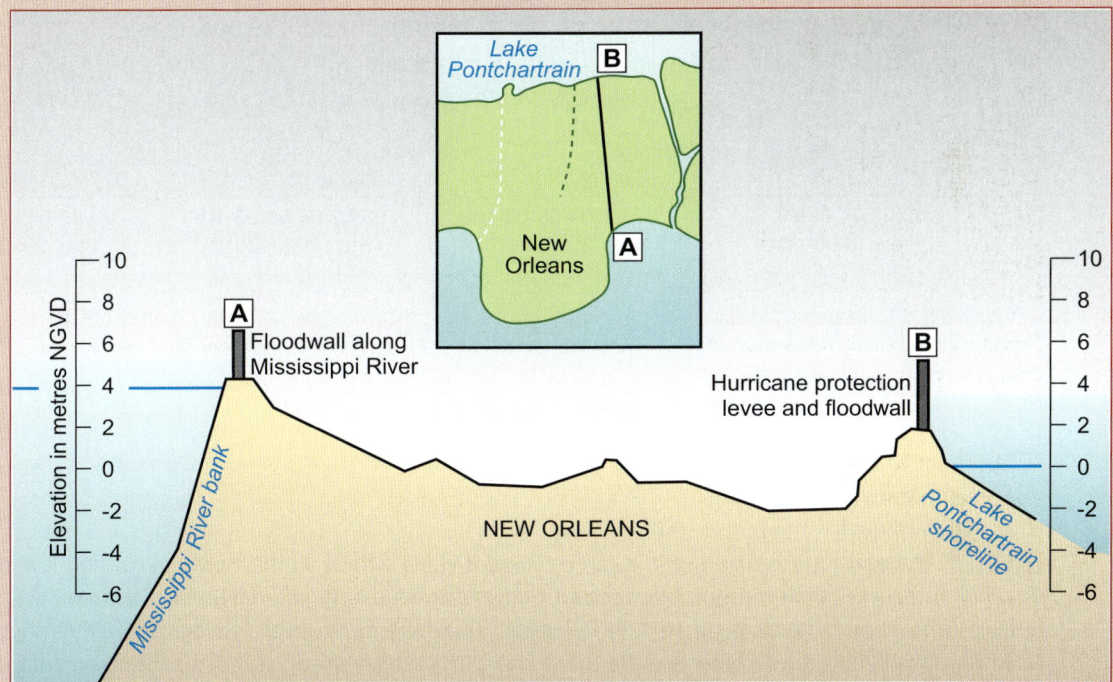

Figure 8.10 Cross-section through the city of New Orleans, showing dependence on levees and floodwalls.

Health impacts: While 1,836 deaths were associated with Hurricane Katrina, 700 were a result of the levee breaches in New Orleans. Over 1 million people from the central Gulf coast were relocated across the USA. About 300,000 homes were destroyed or made uninhabitable. Looting, including of food and water supplies, increased and crime rates generally rose.
Aftermath: New Orleans' population plummeted in the years following the flooding and has yet to fully recover (Figure 8.11). Unemployment doubled in the region, with economic loss as oil production shut down for 6 months following the hurricane, and crop loss from flood waters. It is estimated to have cost the US government in excess of $100 billion.

continued

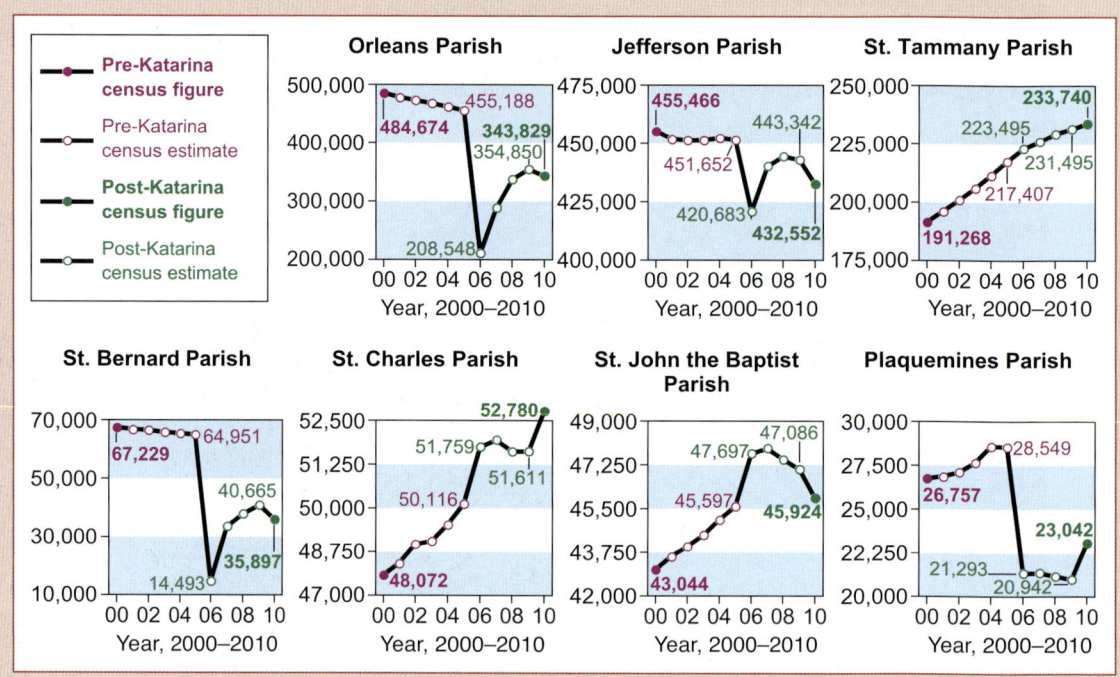

Figure 8.11 Population tracking and the impact of Hurricane Katrina on New Orleans. Even by 2010, many people had not returned to large areas of New Orleans affected by flood waters – a consequence of loss of homes and a rise in unemployment in the area.

Location: Tohoku, Japan
Date: 11 March 2011
Event: Earthquake, magnitude 9.0
Result: The earthquake resulted in a large tsunami wave which hit the northeast coast of Japan. The wave, in excess of 10 m high, overtopped many of the existing sea defences. This was exacerbated by a drop in land level of up to 1 m in places, due to ground-level movements resulting from the earthquake (Universitat Politècnica de Catalunya 2011). Large areas of the coastline were flooded and infrastructure, including houses and roads, was severely damaged as the wave travelled up to 10 km inland. At Fukushima I Nuclear Power Plant, three nuclear reactors went into meltdown following flooding of emergency generators and loss of power.
Health impacts: Of the 15,821 deaths recorded by April 2011, 93% were as a result of drowning. A number of towns were almost completely destroyed (**Figure 8.12**), and key services were lost for many more – 1.5 million households lost water supplies. Following the Fukushima nuclear disaster, residents within 20 km of the plant were evacuated, with many unlikely to return to their homes, due to the high levels of radioactivity released during the meltdown.
Aftermath: With 300,000 people displaced and living in temporary accommodation, the area has suffered considerably, both socially and economically. Industrial production was suspended following the earthquake and tsunami, and the cost of rebuilding is estimated to be in the region of $120 billion.

Figure 8.12 The immediate aftermath of the Japanese tsunami, March 2011.
Source: Photo courtesy of Shutterstock/yankane.

Final word: Both case studies feature incidents that affected areas of the developed world for which expensive flood prevention measures were in place (although not necessarily fit for purpose). Similar incidences in the developing world can result in greater levels of mortality and morbidity, with longer term health impacts resulting from lack of food, water and shelter. For instance, the Indian Ocean tsunami in 2004 is estimated to have killed almost 200,000 people and displaced over 1.5 million.

12), but the total climbs to 86% of abstracted freshwater if we include the processing of agricultural products (Hoekstra and Chapagain 2008). As rainfall becomes less predictable in more regions, a shift in agricultural practices towards more drought-tolerant plant varieties is needed to ensure a resilient food supply.

> **REFLECTIVE QUESTION**
>
> What are the key health risks associated with flooding and droughts?

Figure 8.13 Irrigation of crops may be needed to obtain food with optimal nutritional content – boosting general health. The quality of the water used may also have an effect on health.
Source: Photo courtesy of Shutterstock/Elena Elisseeva.

F SOCIAL AND ECONOMIC WATER RISK

Clean, safe water and good sanitation may be available, but not accessible to all in any given population. The barriers may be physical, for instance, not being connected to a water pipe or sewer (**Figure 8.14**), but will largely result from social and/or economic factors. There is a well-defined difference between water accessibility in developed countries as compared to developing countries.

In the developed world, all permanent homes have a water supply (either public or private) and sewerage connection (sewer network or septic tank), and householders pay for clean water and to dispose of waste water. Non-payment of water

Figure 8.14 Collecting water can often involve a walk of several kilometres. For children, this may mean they do not have time to attend school.
Source: Photo courtesy of Shutterstock/Tim Booth.

BOX 8.5 TECHNIQUES

Disease burden and socio-economic loss

Premature death, ill-health or disability can have significant impacts on the livelihoods of individuals, families, communities and populations. There are various measures that can be used to calculate the cost, both financial and non-financial, of disease, but increasingly most disease burden studies are using DALYs.

A DALY is a **D**isability **A**djusted **L**ife **Y**ear and is defined by the WHO as 'the sum of years of potential life lost due to premature mortality (years of life lost, YLL) and years of productive life lost through living in states of less than full health due to disease/injury (years of life lost due to poor health/disability, YLD)', so representing a common metric for mortality (death) and morbidity (poor health/disability) (see Equation [8.1]: the weight factor reflects the severity of the disease on a scale from 0 (perfect health) to 1 (death)). First developed by the World Bank, the DALY is now widely used as a disease burden measurement by a number of international organisations: one DALY represents the loss of the equivalent of one year of full health.

$$DALY = YLL + YLD \qquad [8.1]$$

where YLD = number of incident cases in that period × average duration of the disease × weight factor.

While most measures of disease burden (where 'disease' encompasses all factors influencing physical and mental health, including infectious agents, non-communicable illness, and physical factors such as flooding and earthquakes) consider mortality only, the DALY seeks to evaluate the consequences of a lifetime spent with poor health or a disability that may influence social and economic well-being, including, for example, the ability to earn money to feed and house a family. In 2004, the global average burden of disease was 237 DALYs per 1,000 population, with about 60% due to premature death and 40% to non-fatal health outcomes. For particular diseases, the difference between mortality and DALY may be small, while others show a considerable difference (Table 8.4). For water-related diseases, drowning always results in death (100% mortality), while diarrhoeal disease, the second-largest cause of global disease burden, has a mortality to morbidity ratio of 77%, and malaria a ratio of 68% (Table 8.4). Neuropsychiatric disorders, however, result in higher levels of morbidity (mortality:morbidity ratio of 16%), a significant disease burden that would rank much lower if considering mortality alone (Table 8.4).

Table 8.4 Global burden of disease for selected factors (based on data presented by WHO 2004)

Disease	Deaths (% of global total)	DALYs (% of global total)
Diarrhoeal disease	3.7	4.8
Malaria	1.5	2.2
Drowning	0.7	0.7
Neuropsychiatric disorder	2.1	13.1
(Unipolar depressive disorder)	(0)	(4.3)

DALYs provide an indication of where the greatest social, economic and health needs are, helping to prioritise intervention strategies such as improved sanitation or the provision of mosquito nets. Projections of 2004 global DALYs to 2030 suggest that diarrhoeal disease will fall from 4.8% of global DALYs, currently the second leading cause of burden of disease, to 1.6%, a ranking of 18 in the burden

continued

list. However, these projections are based on a sustained and additional effort to address Millennium Development Goals (MDGs) (MDG 7 addresses access to safe drinking water and basic sanitation) and the economic development of low-income countries: if MDGs are not attained or surpassed, and economic growth remains stagnant, the diarrhoeal disease burden may remain high.

DALYs are usually calculated on the basis of age and sex only, although geographic location may also be considered. However, not all ages are considered equally, with age weightings and time discounting frequently applied to the calculation of YLL and YLD. The WHO applies non-uniform age weights that give less weight to years lived at young and older ages (Figure 8.15) and a 3% discounting rate so that future years lost are not considered equally as future uncertainty increases (a 3% discount suggests that a weighted year of life saved next year is worth 97% of a year of life saved this year). Using discounting and age weights, a death in infancy corresponds to 33 DALYs, and deaths at ages 5–20 years to around 36 DALYs.

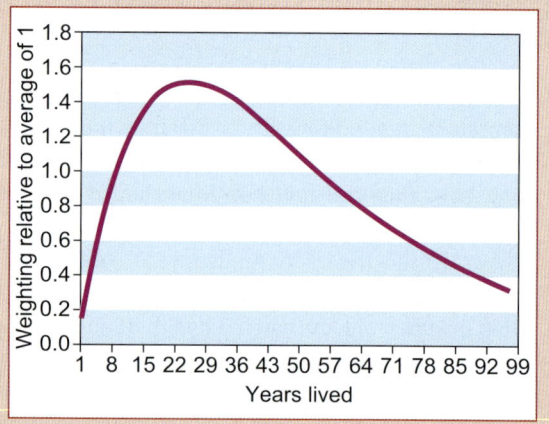

Figure 8.15 Age weightings applied to calculation of DALYs.

bills will rarely result in a home being disconnected from a public water supply and sewerage network, due to the health issues that would result. Social and economic exclusion from accessibility to water and sanitation is therefore rare. However, as water metering is increasingly adopted in many developed countries, poorer households may seek to reduce their water use, at a risk to hygiene and sanitation practices.

Developing countries may be at greater risk from social and financial barriers to water supply and sanitation. People living in slum areas or in rural locations will be particularly vulnerable, due to poor water/wastewater infrastructure (social exclusion). Similarly, householders may be unwilling or unable to pay for connections to existing networks (economic exclusion). Civil unrest, corruption, or lack of political will may also hamper access to water supply and sanitation, ultimately leading to impacts upon health and well-being. Death or chronic diseases resulting from waterborne infection or other water-related issues, such as crop failure due to too much/too little water, can also have a significant impact on the social and economic status of an individual, community or population (Box 8.5). Social and financial barriers to water accessibility are crucial to the consideration of water and health in developing countries, and Chapters 10, 11 and 12 discuss these issues in more detail.

REFLECTIVE QUESTION

What are the main differences in socio-economic water-related health risks in the developed and developing worlds?

WATER AND HEALTH

G HOW SAFE IS YOUR WATER?

Water and sewage distribution networks and treatment systems are a staple of all public health schemes in the developed world, and are discussed in more detail in Chapter 9. Stringent drinking water quality standards have been implemented in all industrialised countries, which usually adopt WHO guideline values (**Box 8.3**). There is an increasing focus on ecological water quality, as demonstrated by the EU's Water Framework Directive, which seeks to improve the chemical, biological and geomorphological status of all European water bodies (see also Chapters 4, 6,

---

BOX 8.6 TECHNIQUES

Minimising disease

As is apparent from the text, there are a number of strategies involving water that can be followed to improve health. Provision of clean water and sanitation facilities considerably reduces the transmission of water-related disease. Hand washing is one of the most important ways of controlling the spread of infections, especially those that cause diarrhoea and vomiting, and respiratory disease. The recommended method is the use of soap, water, and paper towels. Healthcare professionals are taught a standard procedure, as shown in **Figure 8.16** (Ayliffe *et al.* 1992), for maximising the effectiveness of hand washing to ensure that the whole hand is washed for the removal of transient microorganisms.

Hand washing technique

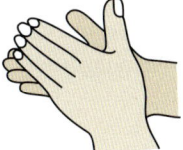

1 Palm to palm

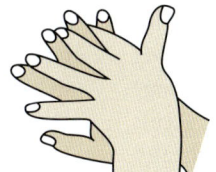

2 Left palm over right dorsum and right palm over left dorsum

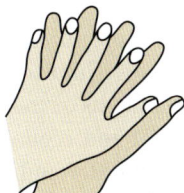

3 Palm to palm fingers interlaced

4 Backs of fingers to opposing palms with fingers interlocked

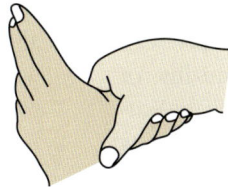

5 Rotational rubbing of left thumb clasped in right and vice versa

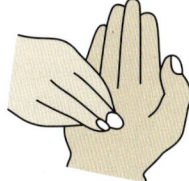

6 Rotational rubbing, backwards and forwards with clasped fingers of right hand in left palm and vice versa

Figure 8.16 An appropriate hand-washing technique to minimise the spread of disease.

and 11). With advanced treatment facilities available for the last 100 years for the removal of microorganisms from water, there has been a greater emphasis on improving the chemical quality of water (e.g. pesticide pollution, emerging pollutants) and the quality of water in the environment, particularly with regard to freshwater biodiversity and recreational water use, such as for swimming. Many water-related diseases in developed countries have been eliminated through eradication programmes (particularly if vector-related), improvements in water supply and sanitation, and hygiene education (**Box 8.6**). The high quality of drinking water means that the main water-related health concern in many developed countries is exposure during recreation, when diseases such as leptospirosis and cryptosporidiosis may be contracted.

In developing countries, major economic investment in public water supply and sanitation has not been as common as in developed countries. Combined with more favourable climates for many water-related diseases, greater water resource pressures (e.g. lower levels of rainfall), and poorly implemented legislation regulating the standards of water quality, as well as a range of social and economic factors affecting access to clean water and good sanitation, the health impacts associated with water-related diseases are greater. While improvement to water access will help to improve health, measures should also be adopted that prevent the further spread of diseases such as malaria and yellow fever, which are responsible for thousands of deaths per year. Reducing illness associated with poor quality (or quantities) of water will help to improve the economic well-being of poor communities, allowing money that would otherwise be spent on treatment to be shifted to education, food, and improving household services.

Currently, the safety of your water is very much dependent on your geographical location. In industrialised and developed countries, high-quality public water supplies contribute to healthy societies; in many developing countries, public water supplies may not exist and a range of other factors (greater variety of pathogens, water scarcity, lack of political drivers, etc.) contribute to a greater health risk from water.

> **REFLECTIVE QUESTION**
>
> How should you wash your hands?

H SUMMARY

Both water quantity and water quality are important when considering the impact of water on human health and the health of the environment. Droughts and periods of water scarcity can lead to famine and death; too much water can also lead directly to death by drowning, but can have larger scale impacts through the spread of disease, food shortages, loss of shelter and livelihood disruption. Many organisms need water for some or all of their life cycle, and some cause disease in humans. Infectious diseases associated with water can be divided into five categories, based on their infection pathways:

- water-borne diseases – diseases spread by water, such as cholera and typhoid;
- water-carried diseases – diseases acquired by the accidental ingestion of, or exposure to, contaminated water, particularly during recreational activities (e.g. Legionnaire's disease);
- water-washed diseases – infections spread by lack of water (for personal hygiene), and insanitary conditions that may not directly involve water as a means of infection transfer (e.g. scabies and typhus);
- water-based diseases – infections transmitted by aquatic invertebrates (e.g. schistosomiasis);
- vector-based water-related diseases – infections, of which malaria is the most common, spread by insects that need water to complete their life cycle.

Chemicals in water can also affect health. Three causes of non-communicable diseases can be considered, including:

- environmentally available chemicals such as arsenic;
- anthropogenic chemicals – contaminants from agriculture and other 'traditional' human activities, such as pesticides and fertiliser runoff;
- anthropogenic chemicals – emerging contaminants such as endocrine disrupting compounds and disinfection by-products.

Treating water before use, through filtration and disinfection, can reduce infection rates. Separation of water and wastewater, with sanitation provision and facilities for good personal hygiene, also improves health by reducing the levels of many water-related communicable diseases. Water-based and vector-based disease levels do not respond as well to improvements in water quality; alternative strategies that prevent contact with the transmitting organisms are therefore required.

Developed and developing countries face different water-related health challenges. Developed countries are better able to adapt to physical water issues such as flooding and drought, and to biological threats, through well-developed treatment and distribution networks, meaning that the chemicals in water are of greatest concern. In developing countries, physical and biological issues prevail, although there are incidences of extensive chemical contamination of water supplies, such as arsenic in the Indian subcontinent.

FURTHER READING

Kolpin, D.W., Furlong, E.T., Meyer, M.T., Thurman, E.M., Zaugg. S.D., Barber, L.B. and Buxton, H.T. 2002. Pharmaceuticals, hormones, and other organic wastewater contaminants in US streams, 1999–2000: A national reconnaissance. *Environmental Science & Technology* 36: 1202–1211.

A comprehensive assessment of emerging contaminants in natural waterways.

Two books that provide further information on water-related diseases caused by pathogens are:

Mandal, B.K., Wilkins, E.G.L., Dunbar, E.M. and Mayon-White, R.T. 2005. *Lecture Notes on Infectious Diseases* (6th edition). Blackwell Publishing; London.

Mandell, G.L., Bennett, J.E. and Dolin, R. (eds). 2010. *Principles and Practice of Infectious Diseases (Volume 2) (7th edition)*. Elsevier Churchill Livingstone; Philadelphia, PA.

Smith D. (ed.). 1999. *Water-Supply and Public Health Engineering: Studies in the History of Civil Engineering (Volume 5)*. Ashgate Publishing; Hampshire.

Provides more detail on the historical association between water and health.

Fascinating research into infectious disease and water associations is presented in the following four texts:

Fewtrell, L. and Bartram, J. (eds). 2001. *Water Quality: Guidelines, Standards and – Assessment of Risk and Risk Management for Water-related Infectious Disease.* IWA Publishing; London.

Hunter, P.R., Waite, M. and Ronchi, E. (eds). 2002. *Drinking Water and Infectious Disease: Establishing the Links.* CRC Press; London.

Mara, D. and Horan, N.J. (eds). 2003. *Handbook of Water and Wastewater Microbiology.* Academic Press; New York.

Selendy, J.M.H. (ed.). 2011. *Water and Sanitation Related Diseases and the Environment: Challenges, Interventions and Preventive Measures.* Wiley-Blackwell; Chichester.

Classic papers

Carson, R. 1962. *Silent Spring.* Houghton Mifflin; New York.

The first report of concerns between widespread use of chemicals, particularly pesticides, and their potential impact upon the health of the environment.

Fraser, D.W. *et al.* 1977. Legionnaires disease – description of an epidemic of pneumonia. *New England Journal of Medicine* 297: 1189–1197.

A paper describing the early stages of identifying the aetiology of a newly identified water-related disease.

Murray, C.J.L. and Lopez, A.D. 1997. Global mortality, disability, and the contribution of risk factors: global burden of disease study. *Lancet* 349: 1436–1442.

A demonstration of the use of DALYs in disease burden studies by the authors credited with the development of DALYs.

PROJECT IDEAS

- Globally, different approaches are being adopted to reduce the impact of water-mediated diseases. The different approaches range from policy and regulation (e.g. an aim/legal requirement to ensure that a certain proportion of a population receives clean and safe drinking water), through engineered solutions (e.g. water treatment systems, construction of reservoirs, etc.) and eradication programmes to public education campaigns. All possess an inherent but often very different economic cost. Research the benefits and disadvantages of the different approaches adopted to help understand how successful different schemes can be in different financial situations.

- During disasters, diseases such as cholera and malaria may rise to epidemic proportions. Review existing guidelines, together with an evaluation of current understanding of disease transmission and control. What recommendations can you make following this review that could contribute significantly to improved disaster planning?

REFERENCES

Akanda, A.S. *et al.* 2011. Hydroclimatic influences on seasonal and spatial cholera transmission cycles: Implications for public health intervention in the Bengal Delta. *Water Resources Research* 47: W00H07, doi:10.1029/2010WR009914.

Alaani, S., Tafash, M., Busby, C., Hamdan, M. and Blaurock-Busch, E. 2011. Uranium and other contaminants in hair from the parents of children with congenital anomalies in Fallujah, Iraq. *Conflict and Health* 5: 15, doi:10.1186/1752-1505-5-15.

Austin, A.A., Fitzgerald, E.F., Pantea, C.I., Gensburg, L.J., Kim, N.K., Stark, A.D. and Hwang, S-A. 2011. Reproductive outcomes among former Love Canal residents, Niagara Falls, New York. *Environmental Research* 111: 693–701.

Ayliffe, G.A.J., Lowbury, E.J.C., Geddes, A.M. and Williams, J.D. 1992. *Control of Hospital Infection; A Practical Handbook (3rd edition)*. Chapman and Hall; London.

Bath, J.L., Eneh, P.N., Bakken, A.J., Knox, M.E., Schiedt, M.D. and Campbell, J.M. 2010. The impact of perception and knowledge on the treatment and prevention of intestinal worms in the Manikganj District of Bangladesh. *Yale Journal of Biology and Medicine* 83: 171–184.

Black, R.E., Deromana, G.L., Brown, K.H., Bravo, N., Bazalar, O.G. and Kanashiro, H.C. 1989. Incidence and etiology of infantile diarrhea and major routes of transmission in Huascar, Peru. *American Journal of Epidemiology* 129: 785–799.

Boorman, G.A. *et al.* 1999. Drinking water disinfection byproducts: Review and approach to toxicity evaluation. *Environmental Health Perspectives* 107: 207–217.

Brown, R.C., Lockwood, A.H. and Sonawane, B.R. 2005. Neurodegenerative diseases: An overview of environmental risk factors. *Environmental Health Perspectives* 113: 1250–1256.

Burnham, G. 1998. Onchocerciasis. *Lancet* 351: 1341–1346.

Carson, R. 1962. *Silent Spring*. Houghton Mifflin; New York.

Chorus, I. and Bartram, J. 1999. *Toxic Cyanobacteria in Water: A Guide to Their Public Health Consequences, Monitoring and Management*. E and FN Spon; London.

Costa, M. 1997. Toxicity and carcinogenicity of Cr(VI) in animal models and humans. *Critical Reviews in Toxicology* 27: 431–442.

Crisp, T.M. et al. 1998. Environmental endocrine disruption: An effects assessment and analysis. *Environmental Health Perspectives* 106: 11–56.

Crump, J.A., Luby, S.P. and Mintz, E.D. 2004. The global burden of typhoid fever. *Bulletin of the World Health Organization* 82: 346–353.

Elliott, P. et al. 2009. Geographic density of landfill sites and risk of congenital anomalies in England. *Occupational and Environmental Medicine* 66: 81–89.

Ellis, J., Oyston, P.C.F., Green, M. and Titball, R.W. 2002. Tularemia. *Clinical Microbiology Reviews* 15: 631–646.

Erickson, J.D., Mulinare, J., McClain, P.W., Fitch, T.G., James, L.M., McClearn, A.B. and Adams, M.J. 1984. Vietnam veterans risks for fathering babies with birth-defects. *JAMA – Journal of the American Medical Association* 252: 903–912.

Esrey, S.A., Potash, J.B., Roberts, L. and Shiff, C. 1991. Effects of improved water-supply and sanitation on ascariasis, diarrhea, dracunculiasis, hookworm infection, schistosomiasis, and trachoma. *Bulletin of the World Health Organization* 69: 609–621.

Fayer, R., Morgan, U. and Upton, S.J. 2000. Epidemiology of cryptosporidium: Transmission, detection and identification. *International Journal for Parasitology* 30: 1305–1322.

Fewtrell, L. and Bartram, J. (eds). 2001. *Water Quality: Guidelines, Standards and Health – Assessment of Risk and Risk Management for Water-related Infectious Disease*. IWA Publishing; London.

Fraser, D.W. et al. 1977. Legionnaires disease – description of an epidemic of pneumonia. *New England Journal of Medicine* 297: 1189–1197.

Gensburg, L.J., Pantea, C., Fitzgerald, E., Stark, A., Hwang, S-A. and Kim, N. 2009. Mortality among former Love Canal residents. *Environmental Health Perspectives* 117: 209–216.

Gilliom, R.J. et al. 2007. *Pesticides in the Nation's Streams and Ground Water, 1992–2001*. U.S. Geological Survey Scientific Investigations Report 2009-5189.

Gryseels, B., Polman, K., Clerinx, J. and Kestens, L. 2006. Human schistosomiasis. *Lancet* 368: 1106–1118.

Gubler, D.J. 1998. Dengue and dengue hemorrhagic fever. *Clinical Microbiology Reviews* 11: 480–496.

Gupta, S.K., Gupta, R.C., Seth, A.K., Gupta, A.B., Bassin, J.K. and Gupta, A. 2000. Methaemoglobinaemia in areas with high nitrate concentration in drinking water. *National Medical Journal of India* 13: 58–61.

Hall, T.F., Penfound, W.T. and Hess, A.D. 1946. Water level relationships of plants in the Tennessee Valley with particular reference to malaria control. *Journal. Tennessee Academy of Science* 21: 18–59.

Halliday, S. 2001. *The Great Stink of London: Sir Joseph Bazalgette and the Cleansing of the Victorian Metropolis*. The History Press Ltd; Stroud.

Hempel, S. 2007. *The Medical Detective: John Snow, Cholera and the Mystery of the Broad Street Pump*. Granta Books; London.

Hoekstra, A.Y. and Chapagain, A.K. 2008. *Globalization of Water: Sharing the Planet's Freshwater Resources*. Blackwell; Oxford.

Hunter, P.R., Waite, M. and Ronchi, E. (eds). 2002. *Drinking Water and Infectious Disease: Establishing the Links*. CRC Press; London.

IFRC (International Federation of Red Cross and Red Crescent Societies). 2009. *World Disasters Report 2009: Focus on Early Warning, Early Action*. IFRC; Geneva.

Kagbadouno, M.S., Camara, M., Bouyer, J., Courtin, F., Onikoyamou, M.F., Schofield, C.J. and Solano, P. 2011. Progress towards the eradication of tsetse from the Loos Islands, Guinea. *Parasites & Vectors* 4: 18.

Kavcar, P., Sofuoglu, A. and Sofuoglu, S.C. 2009. A health risk assessment for exposure to trace metals via drinking water ingestion pathway. *International Journal of Hygiene and Environmental Health* 212: 216–227.

Koelle, K., Rodo, X., Pascual, M., Yunus, M. and Mostafa, G. 2005. Refractory periods and climate forcing in cholera dynamics. *Nature* 436: 696–700.

Kolpin, D.W., Furlong, E.T., Meyer, M.T., Thurman, E.M., Zaugg, S.D., Barber, L.B. and Buxton, H.T. 2002. Pharmaceuticals, hormones, and other organic wastewater contaminants in US streams,

1999–2000: A national reconnaissance. *Environmental Science & Technology* 36: 1202–1211.

Kotloff, K.L. *et al.* 1999. Global burden of shigella infections: Implications for vaccine development and implementation of control strategies. *Bulletin of the World Health Organization* 77: 651–666.

Kuch, H.M. and Ballschmiter, K. 2001. Determination of endocrine-disrupting phenolic compounds and estrogens in surface and drinking water by HRGC-(NCI)-MS in the picogram per liter range. *Environmental Science & Technology* 35: 3201–3206.

Kuemmerer, K. 2009. The presence of pharmaceuticals in the environment due to human use – present knowledge and future challenges. *Journal of Environmental Management* 90: 2354–2366.

Levett, P.N. 2001. Leptospirosis. *Clinical Microbiology Reviews* 14: 296–326.

Mackintosh, C.A. and Hoffman, P.N. 1984. An extended model for transfer of microorganisms via the hands – differences between organisms and the effect of alcohol disinfection. *Journal of Hygiene* 92: 345–355.

Mandal, B.K., Wilkins, E.G.L., Dunbar, E.M. and Mayon-White, R.T. 2005. *Lecture Notes on Infectious Diseases* (6th edition). Blackwell Publishing; London.

Mandell, G.L., Bennett, J.E. and Dolin, R. (eds). 2010. *Principles and Practice of Infectious Diseases (Volume 2) (7th edition)*. Elsevier Churchill Livingstone; Philadelphia, PA.

Mara, D. and Horan, N.J. (eds). 2003. *Handbook of Water and Wastewater Microbiology*. Academic Press; New York.

Marr, J.S. and Cathey, J.T. 2010. New hypothesis for cause of an epidemic among Native Americans, New England, 1616–1619. *Emerging Infectious Diseases* 16: 281–286.

Moore, M.N. 2006. Do nanoparticles present ecotoxicological risks for the health of the aquatic environment? *Environment International* 32: 967–976.

Murray, C.J.L. and Lopez, A.D. 1997. Global mortality, disability, and the contribution of risk factors: Global burden of disease study. *Lancet* 349: 1436–1442.

Nájera, J.A., González-Silva, M. and Alonso, P.L. 2011. Some lessons for the future from the Global Malaria Eradication Programme (1955–1969). *PLoS Medicine* 8: e1000412.

Nowack, B. and Bucheli, T.D. 2007. Occurrence, behavior and effects of nanoparticles in the environment. *Environmental Pollution* 150: 5–22.

Pruss, A. 1998. Review of epidemiological studies on health effects from exposure to recreational water. *International Journal of Epidemiology* 27: 1–9.

Regli, S., Rose, J.B., Haas, C.N. and Gerba, C.P. 1991. Modeling the risk from giardia and viruses in drinking-water. *Journal American Water Works Association* 83: 76–84.

Resnikoff, S., Pascolini, D., Etya'ale, D., Kocur, I., Pararajasegaram, R., Pokharel, G.P. and Mariotti, S.P. 2004. Global data on visual impairment in the year 2002. *Bulletin of the World Health Organization* 82: 844–851.

Robertson, S.E., Hull, B.P., Tomori, O., Bele, O., Leduc, J.W. and Esteves, K. 1996. Yellow fever – a decade of reemergence. *JAMA – Journal of the American Medical Association* 276: 1157–1162.

Schubert, S.D., Suarez, M.J., Pegion, P.J., Koster, R.D. and Bacmeister, J.T. 2004. On the cause of the 1930s dust bowl. *Science* 303: 1855–1859.

Selendy, J.M.H. (ed.). 2011. *Water and Sanitation Related Diseases and the Environment: Challenges, Interventions and Preventive Measures*. Wiley-Blackwell; Chichester.

Smedley, P.L. and Kinniburgh, D.G. 2002. A review of the source, behaviour and distribution of arsenic in natural waters. *Applied Geochemistry* 17: 517–568.

Smith D. (ed.). 1999. *Water-Supply and Public Health Engineering: Studies in the History of Civil Engineering (Volume 5)*. Ashgate Publishing; Hampshire,

Smith, G.D. 2002. Commentary: Behind the Broad Street pump: Aetiology, epidemiology and prevention of cholera in mid-nineteenth century Britain. *International Journal of Epidemiology* 31: 920–932.

Universitat Politècnica de Catalunya. (2011). Japan earthquake caused a displacement of about two meters. *ScienceDaily*, 8 April. Retrieved 6 January 2013, from http://www.sciencedaily.com/releases/2011/04/110407121640.htm.

Vianna, N.J. and Polan, A.K. 1984. Incidence of low-birth-weight among Love Canal residents. *Science* 226: 1217–1219.

Vinten-Johansen, P., Brody, H., Paneth, N., Rachman, S. and Rip, M. 2003. *Cholera, Chloroform and the Science of Medicine: A Life of John Snow*. Oxford University Press; Oxford.

Ward, M.K., Feest, T.G., Ellis, H.A., Parkinson, I.S., Kerr, D.N.S., Herrington, J. and Goode, G.L. 1978. Osteomalacic dialysis osteodystrophy – evidence for a water-borne etiological agent, probably aluminum. *Lancet* 1: 841–845.

Watson, J.T., Gayer, M. and Connolly, M.A. 2007. Epidemics after natural disasters. *Emerging Infectious Diseases* 13: 1–5.

WHO. 2004. *The World Health Report 2004 – Changing History.* World Health Organization; Geneva.

WHO. 2008. *Guidelines for Drinking Water Quality. Volume 1: Recommendations (3rd edition).* World Health Organization; Geneva.

WHO. 2010a. Cholera vaccines: WHO position paper. *Weekly Epidemiological Record* No. 13, 85: 117–128 [online]. Available from: http://www.who.int/wer.

WHO. 2010b. *Malaria Fact Sheet No. 94 [online]* [accessed August 2011]. Available from: http://www.who.int/mediacentre/factsheets/fs094.

WHO. 2011a. *Global Alert and Response (GAR) – Cholera [online]* [accessed August 2011]. Available from: http://www.who.int/csr.

WHO. 2011b. *Cholera Fact Sheet No. 107 [online]* [accessed August 2011]. Available from: http://www.who.int/mediacentre/factsheets/fs107.

WHO/UNICEF. 2003. Basic needs and the right to health (chapter 5). *In:* UNESCO (ed.) *The First UN World Water Development Report: Water for People, Water for Life.* UNESCO Publishing and Berghahn Books; Paris, Oxford and New York.

CHAPTER NINE

Potable water and wastewater treatment

Nigel J. Horan

> **LEARNING OUTCOMES**
>
> After reading this chapter you should be able to:
>
> - show an awareness of the standards that potable waters must meet in order to be considered safe and wholesome
> - explain how potable water is produced to meet these standards, through a series of individual treatment stages
> - understand the desalination processes used to produce potable water from saline or brackish sources
> - describe major pollutants present in wastewater and their contribution to aquatic pollution
> - understand the importance of a discharge consent in achieving improvements to aquatic waters
> - appreciate our changing attitudes that view wastewater (and indeed all wastes) as a resource to be recovered
> - understand how wastewater treatment technologies are now driven by energy saving, greenhouse gas reduction and resource recovery measures, as well as meeting a discharge consent.

A INTRODUCTION

Previous chapters have identified the major source of **potable water**, namely surface waters (rivers and lakes), groundwater (including springs) or engineered storage reservoirs and dams. Ensuring an adequate supply of water that will serve the population and which will not run out during times of drought is a basic necessity. But in addition to adequacy, the water supplied must be clean, wholesome and palatable in order to protect the health of the consumer. Achieving this requires treatment, and the nature of the treatment is a function of the quality of the source water. Whereas groundwaters are *relatively* clean and need fewer treatment steps, surface waters are open to direct environmental input and so multiple treatment stages are generally needed to achieve the required water quality. In addition to the supply of safe drinking water there is the need to deal with wastewater. Wherever human populations settle, whether in villages, towns or cities, they are faced with the problem of disposing of their excreta in a safe manner. As populations have traditionally developed close to a source of potable water (typically a river or lake), this has usually been the first point of disposal. But as populations have increased this route has become increasingly polluted. The history of sanitary engineering is therefore the history of our attempts to remove this waste from its point of origin and dispose of it safely, with minimal impact on the environment. This chapter deals with the steps taken to treat water so that it is safe to drink. and then examines the principles of wastewater treatment and how we go about producing effluent that has a minimal impact on the aquatic environment.

B ENSURING A SAFE SUPPLY OF POTABLE WATER

Water treatment aims to remove three types of contaminant: microbial, inorganic and organic. Most importantly, water treatment should remove microbial contaminants, for instance, the **enteric pathogens** discussed in Chapter 8 such as the bacterium *Vibrio cholera*; the protozoa such as *Cryptosporidium parvum* and *Giardia lamblia*; and the viruses such as rotavirus and enterovirus. If present in water, onset of infection from these organisms is rapid, often as little as a few hours after consumption. By contrast, inorganic chemical contaminants such as arsenic, fluoride and lead may require many years of ingestion before symptoms develop. Toxic doses of these compounds are problematic, as they would render the water unpalatable. A third class of organic contaminants, such as pesticides and chlorinated hydrocarbons which may be present at remarkably low concentrations, are removed as a precaution, as they can induce symptoms in laboratory animals (Klaunig *et al.*, 1986), although these have not necessarily been observed in human populations. In all cases contaminants follow a **dose–response relationship,** which was recognised as early as the sixteenth century by the Swiss alchemist Paracelsus, who noted that the dose makes the poison. More recently the Royal Society of Chemistry (RSC, 2009) has rephrased this well-known expression and stated that:

> While there is no such thing as a safe chemical, it must be realized there is no chemical that cannot be used safely by limiting the dose or exposure. Poisons can be safely used and be of benefit to society when used appropriately.

The safe dose or exposure of a given chemical that may be ingested through drinking water is provided by the World Health Organization Guidelines for Drinking Water Quality (WHO, 2011). These guidelines provide a framework for drinking water safety, covering the management of emergencies and unforeseen events, as well as fact sheets on individual organisms, inorganic and organic chemicals. Water quality standards for most of the world's countries are based on the recommendations of the World Health Organization (**Table 9.1**). It is the role of the water process engineer to ensure that these standards are met at all times, and there is a wide range of treatment alternatives available to achieve this. The underlying principle behind water treatment is to provide multiple barriers, and not simply to rely on a single protection. The first barrier comes with the protection of the supply itself. Whether this is a simple spring or a large dam, preventing the ingress of faecal contamination, fertilisers and pesticides through river basin management can significantly improve the quality of the supply (see Chapter 4). Additional barriers are then provided by treatment and disinfection.

1 Flow train for water treatment

Depending on the quality of the raw water source, several unit processes are assembled to form a

Table 9.1 Water quality guidelines for safe drinking water for a number of parameters as recommended by the World Health Organization (see also **Table 8.2** in Chapter 8)

Parameter	Maximum concentration or value
Microbiological (number/100 mL)	
Escherichia coli	0
Enterococci	0
Chemical ($\mu g\ L^{-1}$)	
Antimony	20
Arsenic	10
Benzene	10
Boron	2400
Bromate	10
Cadmium	3
Chromium	50
Copper	2000
Fluoride	1500
Lead	10
Mercury	6
Pesticides	
Aldrin	0.03
Dieldrin	0.03
Heptachlor	0.03
Polycyclic aromatic hydrocarbons	0.10

flow train through which the water passes, becoming cleaner with each stage. A complete flow train will include one or more of the following processes: screening, clarification, filtration, activated carbon, ozone and chlorination (**Figure 9.1**). The major characteristics and applications of each of these processes are considered in more detail in the following sections.

1.1 Preliminary treatment processes

Prior to entering the water treatment works, surface water is subjected to a number of simple processes to remove larger physical contaminants. These processes are not required for groundwater, as they are mimicked by the action of the water percolating through the soil to the aquifer. The first treatment stage involves screening, which removes floating material such as leaves and twigs. This is accomplished by passing the water through a mesh or screen that retains particles larger than the mesh size. This material can then be removed by scraping or flushing the mesh. The screened water is stored for several days in a tank or reservoir which acts both to balance the flow and settle out larger suspended particles. This in turn will reduce the number of **pathogenic organisms** both by settlement and by die-off. For certain waters that are acidic and which contain high concentrations of iron and manganese in solution, an aeration stage is used to drive off dissolved carbon dioxide and replace it with oxygen. This process also oxidises soluble ferrous (Fe^{2+}) and manganese (Mn^{2+}) into their insoluble forms (Fe^{3+} and Mn^{3+}) that can now be removed easily by settlement.

1.2 Coagulation and flocculation

After storage the water will contain only finely suspended particles such as silt and mud of < 1 mm diameter, as well as colloidal material. The particles can be removed by clarification, which helps to reduce colour and also helps to remove much of the microbial contamination that is associated with the solid fraction. Clarification involves a process of **coagulation** and **flocculation** which relies on the fact that the solid particles carry a negative charge and so they can be coagulated by adding ions of an opposite (positive) charge. A number of inorganic and organic compounds are used as water coagulants. Ferric hydroxide, aluminum hydroxide (alum) and polyaluminium chloride are popular multivalent, inorganic coagulants. More recently, due to public worries as to the possible health impacts of aluminium in drinking water, ferric compounds are now proving more popular. Coagulation does not produce a strong material and so coagulant aids are employed to provide additional strength by a process of flocculation. High molecular

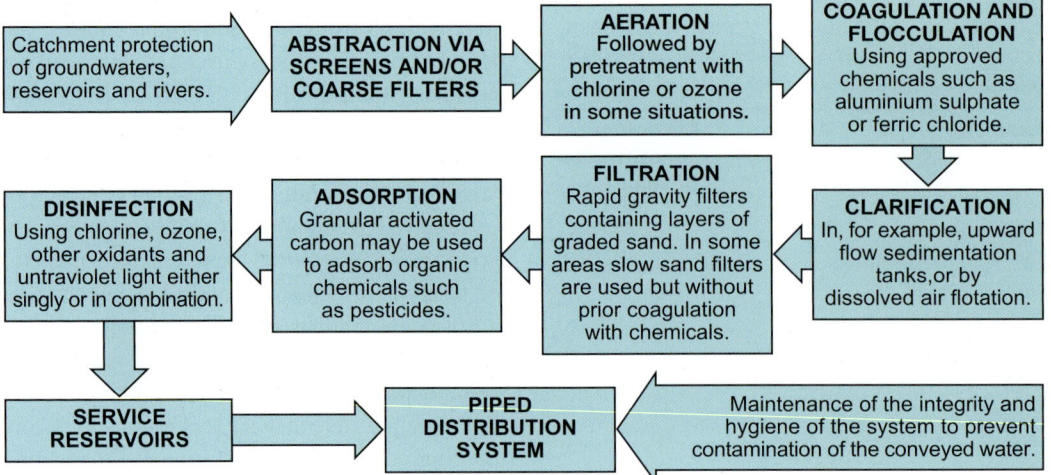

Figure 9.1 Typical process flow train for producing water of potable quality that meets the WHO guidelines summarised in **Table 9.1**.

weight synthetic polymers or polyelectrolytes such as polyquaternary amines are widely used in this role. They are long-chain organic molecules with charged groups along the chain length. They draw together the coagulated material to form a large and strong floc that can be readily removed.

1.3 Solids removal

The flocs formed during coagulation and flocculation must now be removed from the water and this can be achieved either by allowing them to sink in a settlement tank or by encouraging them to float in a flotation tank. The former can be rectangular, with the flow passing horizontally along the tank; circular, where the flow radiates from the centre; or hopper-bottomed with upward flow. The latter is a very popular process option and it comprises a large tank shaped like an inverted cone with vertical sidewalls and a conical base. This arrangement permits solids to settle to the bottom of the tank and the sloping base encourages them to move towards a central hopper at the apex of the cone.

The density of the flocculated particles is not high and the buoyancy of these particles can be increased by passing air through the water. If this air is introduced at high pressure, then small bubbles are produced as the pressure reduces to atmospheric within the tank. The gas bubbles attach to the flocculated particles, reducing their density further so that they float. The big advantage of dissolved air flotation is that the smaller particles that would take a long time to be removed by sedimentation can be removed more rapidly and more effectively. Once at the surface, the floated particles can be removed by skimming.

1.4 Filtration

The role of a filter is to sieve or strain suspended material from the water. The filter medium retains fine inorganic and organic particles while allowing the treated water to pass through. The filter medium may be comprised of sand or anthracite, and more recently excellent results have been obtained using clean, crushed, recycled glass (Horan and Lowe, 2007).

There are two types of filter: rapid gravity and slow sand. A rapid gravity filter (**Figure 9.2a**) is an intense and fully automated process that can treat flow velocities between 4 and 21 m hr^{-1} (see **Box 9.1** for an explanation of treatment flow rates). It is best suited for waters that have received

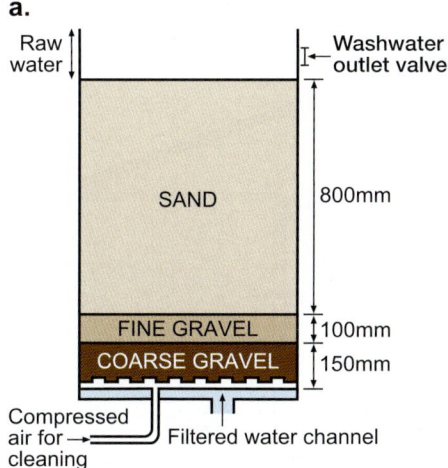

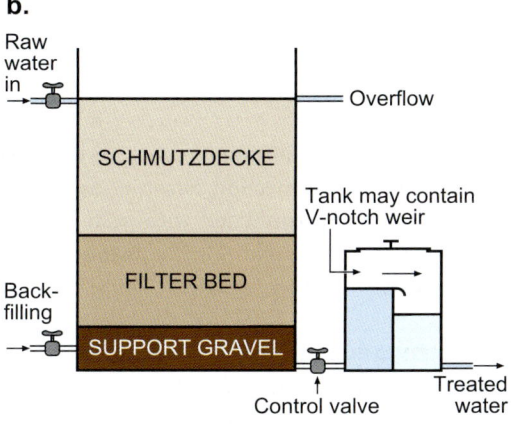

Figure 9.2 Schematic diagram of (a) rapid and (b) slow gravity filters that are able to take out smaller particles from water.

> **BOX 9.1 TECHNIQUES**
>
> ### Comparing process efficiencies
>
> The units (m hr^{-1}) are commonly used in both water and wastewater treatment to compare the flow rates that different unit processes can handle. They are obtained by dividing the flow rate (in m^3 hr^{-1}) by the area of treatment process (m^2) to give m^3 m^{-2} hr^{-1}, which is an areal loading rate. These units are exactly the same as m hr^{-1}, which is a flow velocity. The greater the areal loading rate or flow velocity, the greater the volume of water that a given process can treat. For instance, a primary settlement tank used in wastewater treatment is typically designed to ensure that the flow velocity does not exceed 2.5 m hr^{-1}. Thus, if a wastewater treatment plant has four primary settlement tanks, each of 30 m diameter, this provides a total area of 707 m^2. Thus the maximum flow that the plant can safely treat is 707 m^2 × 2.5 m hr^{-1}, or 1767.5 m^3 hr^{-1}.

pretreatment and coagulation. By contrast, a slow sand filter (**Figure 9.2b**) treats at rates between 0.1 to 0.4 m hr^{-1} and so it has a much larger footprint. It is applied to waters that have not received coagulation or, in some cases, pretreatment, provided that their turbidity is low (< 10 NTU; NTU are nephelometric turbidity units, which are a measure of the attenuation of light by the turbidity of the water).

The main components of a rapid gravity sand filter are: the filter media, the gravel support layers or down through the filter (upflow or downflow) and any suspended material in the water is retained on the filter both by **adsorption** and by filtration. As a result, the flow rate through the bed will be reduced as the bed starts to clog. When the **head loss** through the bed reaches a certain point the medium will require cleaning and this is achieved by a process of backwashing. Air is introduced at a high flow rate, counter current to the flow of water, and this thoroughly scours the bed, releasing trapped particles. The bed is then washed with water to remove these particles, with this backwash water being taken away for disposal. The efficiency of the filter is dependent on the effectiveness of the backwash cycle, and this relies on a well-designed underdrain. An underdrain has many functions. It supports a gravel layer and the filter medium. The underdrain collects the filtered water from the base of the filter and it also distributes air and water evenly across the base of the filter during backwashing (**Figure 9.3**).

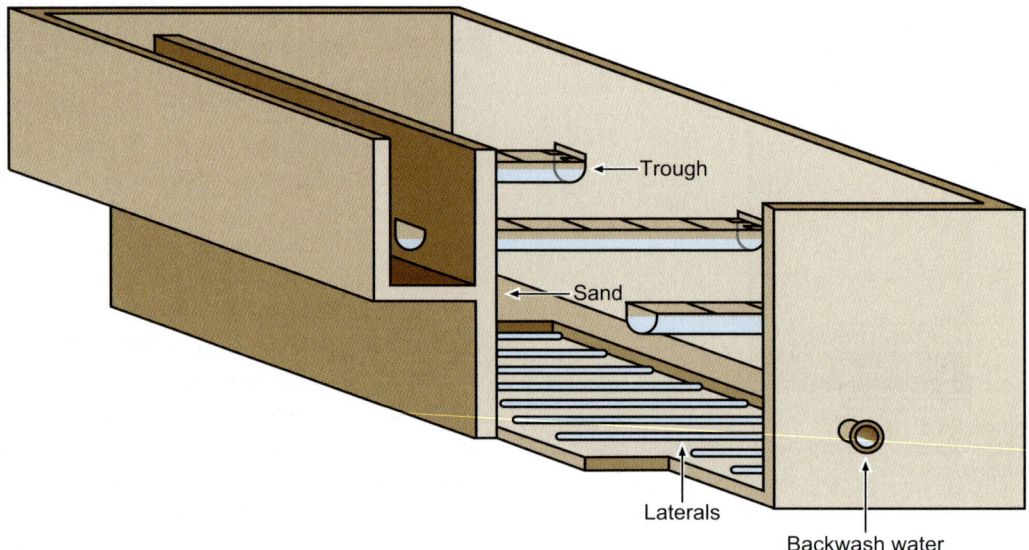

Figure 9.3 Underdrain design for a rapid gravity filter showing the many roles this has to perform.

By contrast, a slow sand filter is a shallow tank with a piped collection system at its base for taking away the treated water. These pipes are covered with a layer of gravel and then topped off with about a metre of sand. Water is fed onto the surface of the sand and then percolates under gravity through the bed. Thus the top layer of sand removes the larger particles from the flow. As this sand layer is exposed to the light, it permits the growth of algae that form a film on the surface. This film in turn attracts protozoa, bacteria and other microorganisms that form a gelatinous surface layer. Thus, slow sand filters have a biological action, as the microorganisms break down organic compounds held within the water and the algae provide oxygen, which aerates it. Over time the layer of **biofilm** that accumulates must be removed to permit the desired flow rates to be achieved. Unlike a rapid gravity filter with backwash, the top layer of sand on a slow sand filter is removed by manual scraping, followed by replacement with a new sand layer. It is because of the propensity of this biofilm to clog that slow sand filters are not used to treat coagulated water, as this can contain many fine, coagulated particles.

For many waters, filtration represents the final treatment before disinfection. However, for others which may be contaminated with pesticides or have taste problems, additional polishing stages are required.

1.5 Activated carbon

The activated carbon stage is a specialised process that relies on the adsorption abilities of activated carbon to remove organic material from water. Activated carbon has a very porous structure and its small pores will absorb a wide range of contaminants. These contaminants include pesticides and pesticide by-products, trihalomethanes and other disinfection by-products, algal toxins and chlorinated hydrocarbons. The activated carbon is usually used in a granular form, known as GAC and contained in filters that are operated in a similar manner to a rapid gravity filter. Activated carbon will eventually become saturated with absorbed products, at which stage it must be removed and replaced. The saturated material can be regenerated by heating to high temperatures $> 550°C$, which will destroy the absorbed organic contaminants.

1.6 Disinfection

Disinfection is the process that protects consumers from water-borne disease by eradicating any potential enteric pathogens in the treated water. It is not a sterilisation process and many bacteria will survive disinfection, but the disinfection choice and dose are intended to ensure that the more harmful enteric organisms are removed. A wide range of chemicals show disinfectant properties, but a key property for water treatment is that the disinfectant should not be harmful to humans, even at extremely low concentrations. Those most often used for treating potable water include chlorine (Cl_2), chlorine dioxide (ClO_2), hypochlorite (OCl^-) and ozone (O_3). In addition, **ultraviolet (UV) light** is often used, but this is a physical and not a chemical disinfectant. Disinfectants can be applied in two ways, either as a primary disinfectant to achieve the requisite degree of microbial kill or with the addition of a secondary disinfection stage that aims to maintain a residual concentration of disinfectant in the finished water in order to prevent either regrowth or ingress of pathogens in the distribution system. Physical disinfection such as UV cannot provide a residual effect that continues to disinfect the water down the supply chain. Chlorine is one of the most effective disinfectants that is easy to apply and which provides a residual concentration in the source water. However, chlorine can also react with organic compounds in the water to form disinfectant by-products, in particular the halomethanes. These compounds substitute one or more of the hydrogen molecules on a carbon atom with chlorine to produce mono-, di- and tri-halomethanes. There are public fears about the links between halomethanes and cancer (Pereira et al., 1982), although it is widely accepted that the risks to health from drinking unchlorinated waters are many orders of magnitude greater than the risks of cancer from water-borne halomethanes. Nevertheless, water companies make great efforts to reduce the risks of these compounds, especially where the source waters are rich in organics, such as those flowing from peatlands (Chow et al., 2003). Pre-treatment with GAC will help by reducing trace organic compounds in the water that might react with chlorine. In addition, other disinfectants might be employed, such as chloramines or ozone, which, although not as effective as chlorine, do not have by-products associated with them.

2 Membrane treatment

Membranes are being increasingly applied in the treatment of drinking waters because they can produce high-quality water in a single process with a relatively small footprint. A number of membrane types are used which are defined based on the membrane pore size. These are microfiltration (MF), ultrafiltration (UF), reverse osmosis (RO) and nanofiltration (NF) membranes. MF membranes have the largest pore size and will reject large particles as well as algae and protozoa. UF membranes can also reject bacteria, larger, soluble molecules such as proteins and many viruses. RO membranes are effectively non-porous and will exclude particles, bacteria, virus, soluble organics and even low molecular weight inorganic ions such as salt. Membranes are generally made of synthetic organic polymers and used as flat sheets or in tubular arrangements (**Figure 9.4**). Membranes offer a good process option for small water treatment plants that are faced with removing a wide variety of contaminants. Where the source water is rich in particles, then prior filtration is essential to prevent membrane clogging. In addition, the material retained by the membrane, known as the 'concentrate', can comprise as much as 15% of the treated water, or 'permeate', and requires careful disposal.

3 Desalination

Global demand for water is doubling every 20 years and growing faster than population. Due to changes in rainfall patterns, the movement of populations from rural to urban areas and increased demands from industry and agriculture,

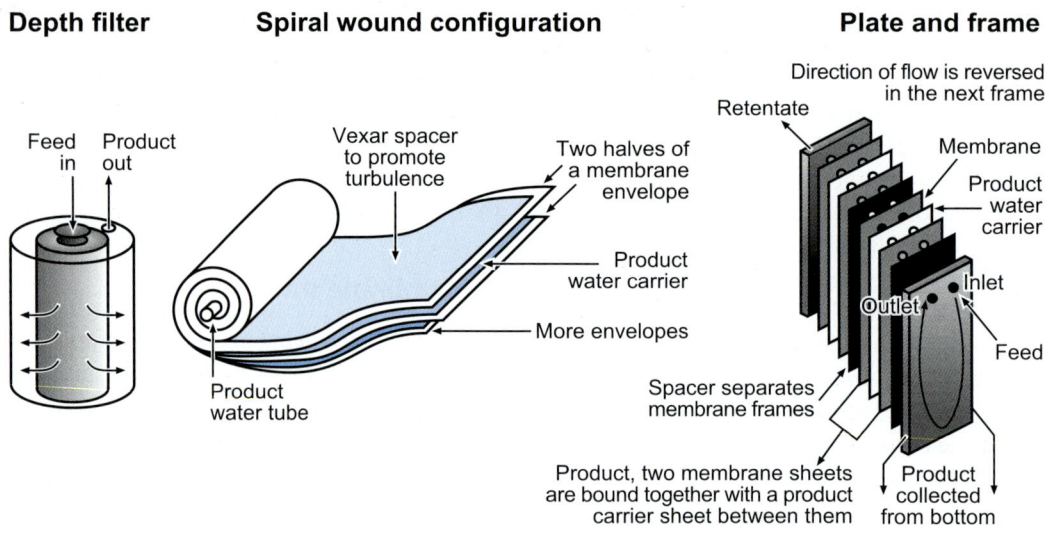

Figure 9.4 Arrangement of membranes used in a range of water treatment applications.

it has been predicted that by 2025, demand for clean water will exceed supply by 56%. As the oceans make up 97% of the world's supply of water, it is understandable that engineers look to this as a source to bridge the gap between water demand and water supply. But seawater typically contains around 35000 mg L^{-1} of salt, and brackish water between 2000 and 5000 mg L^{-1}, whereas the guidelines for drinking water quality require that salinity levels are below 250 mg L^{-1} (**Table 9.1**). The process of removing this salt from water is known as desalination and it requires large amounts of energy and specialised, expensive infrastructure. As a result, it is much more expensive (typically double the price) than water from more conventional sources. However, desalination already plays a major role in the provision of water for drinking and irrigation in areas such as the Middle East, which are energy rich but water poor; the world's largest desalination plant is currently found in the United Arab Emirates. But now many other nations, including the USA, China, Spain and Australia, are having to utilise this technology to deal with their own water scarcity problems. Indeed in 2010 the first desalination plant was opened in the UK, in East London, treating brackish water from the Thames Estuary, with the ability to produce 140,000 m^3 per day and serving 1 million people. However, the plant is intended to operate only during times of extended low rainfall and, to reduce its **carbon footprint**, it runs on recycled fats and oils recovered from London's restaurants.

Desalination can be achieved through distillation using heat, but reverse osmosis using membranes is now more popular and requires less energy. The RO membranes are used in a spiral wound arrangement that rolls up many layers of flat membrane sheets around a central pipe which delivers the saline water for treatment. The membranes are rated for their ability to reject compounds from contaminated water, and to achieve the guideline quality for chloride using a saline source water of 25000 mg Cl$^-$ L^{-1} requires a rejection rate of 99%. Polyamide thin-film composites membranes provide the necessary strength and durability and also offer the high rejection rates that are needed. Purification is achieved by applying a high pressure to the concentrated side of the membrane, which forces the purified water across the membrane to the dilute side. The impurities that remain on the concentrated side are taken away in the rejected water, and this can be as much as 50% of the total volume treated. The rejected water needs to be disposed of in such a way as to avoid environmental damage.

4 Wastewater as a potable water supply

One of the most environmentally benign sources of potable water is wastewater utilised through a process of direct reuse, thus completing the cycle and avoiding the need for water abstraction and wastewater discharge. Unplanned indirect reuse is practised daily throughout the world on rivers such as the Thames, the Yangtze, the Murray-Darling and Mississippi, where treated wastewater is discharged to watercourses which are then abstracted further downstream, either for irrigation or as water intake for a water treatment works. But direct reuse means that the treated wastewater is added directly to the normal drinking water distribution system. With the standards of treatment now available for wastewaters and potable waters, this concept is perfectly feasible, although it may not appeal to the majority of consumers. It has been practised with some success at a limited number of facilities worldwide, with perhaps the best documented being that of Windhoek, the capital city of Namibia (du Pisani, 2006). This city is the driest city below the Sahara, with no freshwater resources within a 500 km radius. Due to severe water shortages in the 1960s, the city introduced a water reclamation scheme in 1968 to supplement the potable water available to the city with domestic wastewater. The scheme has been very closely monitored and it has demonstrated that water of acceptable quality can be produced from domestic wastewater and that, if fully informed and consulted, consumers will accept this alternative in order to maintain their potable supplies. A more recent scheme has been put in place in Essex, the driest county in the UK and one with no potential for new water resources. The scheme involves the diversion of 40,000 m^3 of treated wastewater from the Langford treatment works, where it discharges to an estuary. This effluent receives additional treatment and UV disinfection before discharge to an adjacent river, four miles upstream of a water treatment works. Here it is abstracted for treatment and put into potable supply. When discharged to the river, the treated wastewater was of better quality than the river itself. However, by using this route, instead of transferring directly to a reservoir, it was possible to satisfy public misgivings about the scheme. This simple scheme augmented the available water resources by 8%, and at a fraction of the costs of desalination.

> **REFLECTIVE QUESTION**
>
> Can you describe the flow train to treat water so that it is potable?

C WASTEWATER TREATMENT

The ideal fate of excreted material is for it to be recycled to agricultural land, where it can complete the important nutrient cycles for nitrogen and phosphorus, as well as replenishing the soil with valuable organic material. Historically, this agricultural use was the traditional route for waste in the UK and Western Europe, with faecal material collected through chamber pots used indoors and pail closets or middens sited outdoors. The rapid growth of towns and cities during the Industrial Revolution of the eighteenth and nineteenth centuries meant that this agricultural route for disposal was overwhelmed and access to rural areas became limited. Thus, excreted material was either disposed of directly to watercourses or, in the poorer areas, directly onto the streets, where it would accumulate. As this material is a source of numerous pathogenic microorganisms and is responsible for many faeco-oral infections, it led to widespread epidemics, such as cholera, particularly among the poorer populations (see Chapter 8). It was in an attempt to improve the health of such people that the great Victorian reformers pioneered the widespread adoption of water-borne sewerage and the provision of a clean water supply. This measure is widely accepted as being the single largest contributor to improving the health and reducing the mortality of the population. Nevertheless, its uptake was slow, and even in the UK as recently as 1970 the Jeger Report, *Taken For Granted* (HMSO, 1970), a report of the Working Party on Sewage Disposal, established by the Ministry of Housing and Local Government, was critical of the 270,000 households in the UK that still lacked WCs but relied on earth and pail closets.

The water carriage system that was adopted serves the purpose of transporting excreta from its point of origin (the house, factory or office) to a point of ultimate disposal (the local watercourse). It has been in use for around 150 years and its major component is the sewerage network. The sewerage network comprises a system of drains which connect individual properties to a larger sewer running close to the house. In turn, the individual sewers connect to a much larger interceptor sewer that carries the material to a sewage treatment works. Of course this development is only made possible by the provision of mains water to a property, and it does come with a number of disadvantages. The sewerage network in many countries serves a dual function in that, as well as accepting sewage it can also be used to convey surface water drainage, and such systems are termed combined sewers. This is to differentiate them from separate systems, where a single piped network carries the sewage and a separate piped network is used to convey drainage. Both combined and separate sewerage systems are now under stress as the population continues to rise, because not only does this increase the amount of sewage to the system, but it also increases the amount of drainage as more and more land is covered with impermeable surfaces to provide housing, offices and industry. A final challenge comes from climate change, with more extreme storm events, which can overload the combined sewer network, leading to sewer flooding. The foul water can pour out onto streets, rivers or even properties when a sewer surcharges and floods.

Bringing together the sewage from large numbers of people to a single point source at the end of the main sewer generates a potentially huge pollution problem if it is not treated adequately prior to discharge to a watercourse. This section of the chapter will examine the principles of wastewater treatment and examine how an effluent can be produced that has minimal impact on the aquatic environment.

1 Components of a wastewater

Wastewater comprises two fractions, one soluble and one particulate, with the latter measured by the amount of material retained on a 7 μm filter paper. The particulate component is termed the 'total suspended solids' or usually just the 'total solids' (TS). Both the soluble and particulate fraction will comprise organic and inorganic material. If the material retained on the filter is heated to a temperature of 550°C or more it **volatilises**, as all the organic carbon is oxidised to carbon dioxide. The organic material lost in this way is known as the volatile suspended solids (VS) and measures the organic fraction of the waste. The inert material, or ash, that remains is a measure of the inorganic fraction. The inorganic fraction will contain cations, including metals such as sodium, magnesium and phosphorus, and ammonia nitrogen. It will also contain anions such as chloride and sulphate.

Wastewater discharged to a receiving watercourse may cause pollution in three ways:

1 The organic material in the wastewater acts as a food source for microorganisms in the receiving watercourse, which oxidise it to carbon dioxide and water (Equation 9.1). The oxygen for this reaction is derived from the watercourse itself and, as oxygen is sparingly soluble in water (at around 10 mg O_2 L^{-1} at 10°C), this is quickly exhausted. As a result, there is no oxygen remaining for other aquatic life, which either dies or moves elsewhere.

$$C_6H_{12}O_6 + 6O_2 \longrightarrow 6CO_2 + 6H_2O \qquad [9.1]$$

This effect is illustrated in **Figure 9.5**, where the dissolved oxygen in the river declines below the point at which organic material is discharged. If there is a single discharge, then this figure shows clearly that a river is able to recover from organic pollution once the oxygen demand is reduced. If discharge contains a large amount of organic material (for instance, a major industrial spillage or a broken sewer that discharges raw sewage), then the river can be polluted for many miles downstream. The oxygen sag curve illustrated in **Figure 9.5** can be used to ensure that the organic material in a sewage effluent is reduced to such an extent that it causes minimal impact on the oxygen concentration in a watercourse. It can also be used to ensure that treatment plant discharges are not sited too close together, thus ensuring that the river always has time to recover from an oxygen deficit.

The amount of oxygen needed to oxidise all the organic material in wastewater, as shown in Equation 9.1, is expressed in terms of its oxygen demand. This can be undertaken either biologically as the **biochemical oxygen demand** (**BOD**), which is measured using naturally occurring microorganisms (see section B1 in Chapter 4), or chemically as the **chemical oxygen demand** (**COD**), which uses strong chemical oxidising agents (see section B1 in Chapter 4). An individual excretes about 45 g BOD per day, and where a sewerage network is installed it has a per capita water usage that ranges from around 100 L in

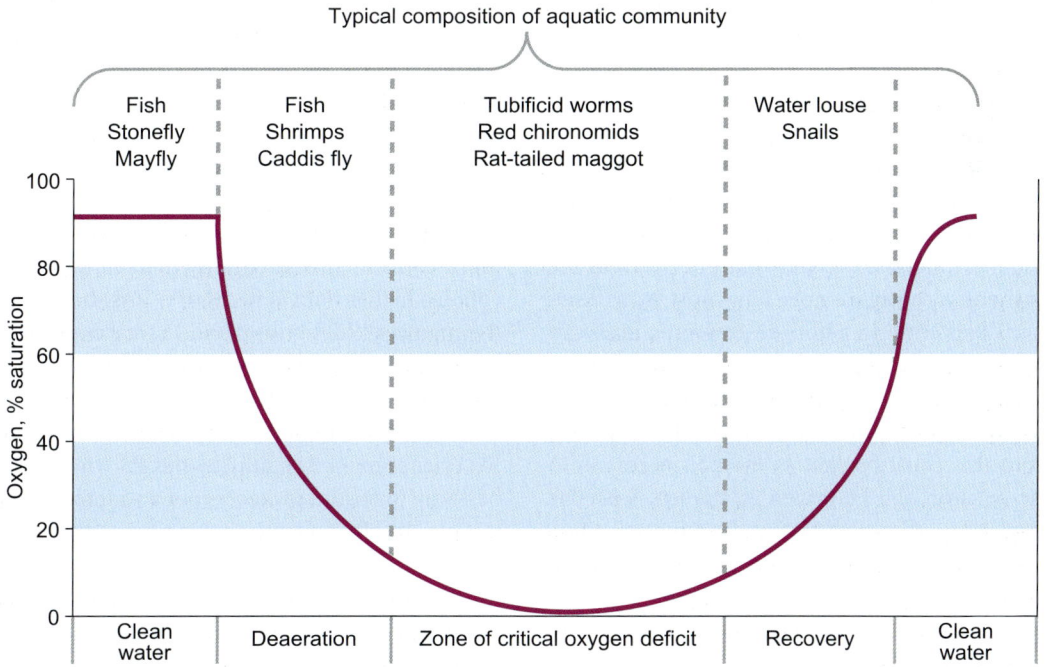

Figure 9.5 The oxygen sag curve that develops when wastewater is discharged to a watercourse with a concentration of organic material in excess of the available oxygen dissolved in the water.

areas of water shortage (such as the Middle East), 150 L in the UK, 320 L in Spain to as high as 575 L in the USA. Thus, the strength of sewage would be strong in the countries with sparse water usage (45 g BOD in 100 L water gives a concentration of 450 mg BOD L^{-1}), medium in the UK at 300 mg BOD L^{-1} and weak in the USA at 78 mg BOD L^{-1}. During wet periods, drainage in the sewer can often account for more than six times the per capita water usage and so the BOD concentration can be as low as 50 mgL^{-1} in the UK, although the amount (or BOD load) is unchanged. Usually if the BOD in the treated wastewater can be kept below 20 mgL^{-1} and if there is adequate dilution in the receiving watercourse (at least eight-fold), then it can be discharged with no detrimental impact to the oxygen concentration and, thus, aquatic life in the watercourse. Where rivers are much larger than this and provide much more dilution, then a higher BOD concentration in the treated effluent can be tolerated without harm. Thus, the performance expected from a treatment plant depends very much on the characteristics of the receiving watercourse.

2 The nutrients ammonia and phosphorus can encourage blooms of algae, in particular the blue-green algae or cyanophyta. This process is known as **eutrophication**, which is more common in hot countries, where sunlight is prevalent, and also more common in lakes than in rivers, as the nutrients are not removed by upstream flows (see Chapter 6). In addition, ammonia itself can be toxic to fish life, as the unionised form of ammonia (NH_3) interferes with oxygen transport across the gills of fish (Equation 9.2). It is apparent from this equation that as the pH increases, so the reaction will be driven to the left, with the formation of more unionised ammonia. Only at pH values greater than 8.5 does the unionised ammonia exceed 10% of the total ammonia fraction. If the concentration of ammonia-N in treated wastewater is kept below 1 to 4 mg L^{-1}, then such toxicity is prevented. In a similar way, if eutrophication is likely to be a problem, it can be averted by achieving a concentration of < 1 mg P L^{-1} in the treated wastewater.

$$NH_3 + H_2O \longrightarrow NH_4^+ + OH^- \qquad [9.2]$$

3 Suspended solids contain organic material and thus contribute to a BOD demand, and 10 mg TS can contribute up to 6 mg BOD. Thus, if a treatment plant is to achieve an effluent quality of < 20 mg L^{-1} it must also ensure that the solids are removed, and generally < 30 mg L^{-1} can be achieved with ease through simple settlement. Pollution from TS is not clearly defined and it has a number of effects. It will make the watercourse more turbid and thus prevent sunlight penetrating; it will also settle on the **benthic layer** and can make it difficult for invertebrate species to feed and reproduce (see Chapter 6).

In order to protect the receiving water quality, wastewater must be treated to reduce these contaminants to concentrations that will pose no pollution threat. The specific values are very much a function of the climate, the watercourse and the size of the effluent discharge. It is the role of the regulator (such as the US Environmental Protection Agency or the UK Environment Agency) to set standards that are appropriate, and these standards are known as consents to discharge. As wastewater treatment is an expensive process that requires energy and emits greenhouse gases, there is no environmental benefit in setting a consent higher than is necessary. It is the role of the engineer to design and build a treatment plant that will meet these consents. If the consents are not met, the watercourse is not protected, and so money spent on the facility has been wasted. Experience in many countries has shown that the best way to ensure protection of a watercourse is to have an effective monitoring procedure with random sampling, and powers to prosecute if the consent is not met. It is sometimes said that people do what you inspect, not what you expect. It is necessary, therefore, to be very clear in the legislation as to exactly how consent will be measured. It is not adequate simply to state that the treatment plant must achieve an effluent of 20 mg BOD L^{-1}. It is far more useful to state that (for instance): 'the treatment plant must achieve a 95 percentile compliance with an effluent standard of 20 mg BOD L^{-1} based on a random spot sample analysed in a quality assured laboratory'.

In addition to the above three parameters, there are other contaminants that are of importance only in specific situations. For instance, excreted enteric pathogens do not impact on the water quality in terms of the aquatic organisms, but if the water is used for recreational purposes such as bathing or canoeing, then they pose a risk to human health. In addition, as the quality of watercourses in developed countries improves and fish return to them, we are now recognising impacts from other pollutants that were not previously apparent. For instance, women who take the contraceptive pill or receive hormone replacement therapy excrete oestrogens in their urine. Oestrogen will pass through a wastewater treatment process with very little removal. When present in the receiving watercourse, it can cause many problems in fish which breed below the discharge point, such as birth defects and **hermaphrodite** characteristics. All of these problems are dealt with as they arise, new effluent standards are promulgated and additional treatment processes are added to the flow train.

2 Types of wastewater treatment system

There are many different process options for treating wastewater, but a feature of all of them is that as the effluent consent becomes tighter and the land available to construct a facility reduces, then more highly engineered systems are needed. Thus, simple treatment options such as pond systems and reed beds are relatively low in capital and operating costs but require an extremely large land area. They can also provide reasonable treatment, and where the consent is based on a composite sample with compliance based on an average, they can meet a standard of 20 mg BOD L^{-1} and 30 mg TS L^{-1}. However, they are not able to meet tighter standards where effluent quality is based on achieving a 95 percentile compliance from a spot sample, and neither can they meet consents for ammonia or phosphorus. For example, an effluent that achieves an average BOD of 20 mg L^{-1} would imply that on many occasions the effluent quality was in excess of 20 mg L^{-1}. In order to meet a 95 percentile effluent compliance, the average must be half this 95 percentile value (**Figure 9.6**). So to achieve a 95 percentile compliance with a BOD of 20 mg L^{-1} means that an average of 10 mg L^{-1} must be achieved, and this cannot be met by simple treatment systems.

Larger, more complex options, such as trickling filters, activated sludge or membrane systems, can routinely achieve effluent quality of 10 mg L^{-1} BOD, 10 mg L^{-1} TS and < 1 mg L^{-1} ammonia with a 95 percentile compliance (although this is not always necessary for environmental protection). They also have a very small site footprint. However, their capital costs and operating costs are higher than those for simple systems. They also have a large energy demand, and the current challenge for engineers is to reduce this demand through process efficiency and energy recovery from the wastewater (**Table 9.2**).

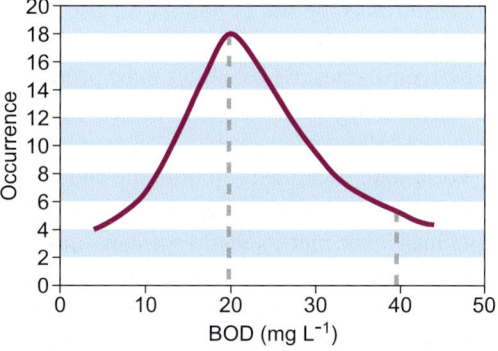

Figure 9.6 Annual distribution of biochemical oxygen demand (BOD) in the effluent from a wastewater treatment plant, illustrating the influence of compliance at 50 and 95 percentiles.

Table 9.2 The performance of a range of wastewater treatment systems, showing improving performance as capital and operating cost increases

Treatment process	Indicative effluent quality at a 95 percentile compliance				
	BOD (mg L^{-1})	TS (mg L^{-1})	Ammonia-N (mg L^{-1})	Phosphorus (mg L^{-1})	Faecal coliform (log removal)*
Waste stabilisation ponds	50	50	12	12	>6
Reed beds	40	40	12	12	>4
Trickling filters	20	30	<4	12	<3
Activated sludge	15	15	<1	<1	<3
Membrane bioreactors	<4	<4	<1	<1	>6

Note: * A 1-log removal means that 90% of faecal coliforms are removed, 2-log removal is 99%, etc.

For those areas where wastewater infrastructure is not yet in place, there is also a great debate as to whether to build several smaller treatment facilities close to the point of production or a single large facility treating all the wastes from a river basin (Verstraete et al., 2009). The arguments in favour of the smaller, decentralised treatment are that they do not require such a large sewerage network and this reduces the pumping (and thus energy) costs. It currently takes about 0.5 to 1 kWh m^{-3} for water to reach the home, and a similar amount to collect it from the home at a wastewater treatment works. The arguments in favour of centralised treatment are that it reduces the number of discharge points to the environment (and thus the risk of pollution); also economies of scale mean cheaper per capita treatment. The large volumes of wastewater to be processed permit the introduction of novel technologies for energy and resource recovery.

Wastewater treatment up until the end of the twentieth century focused simply on producing an effluent that met its discharge consent. This was very effective but has left us with processes that have a large energy demand and are wasteful of resources. As a result, a new approach has emerged in the twenty-first century, one that does not see wastewater as a problem but, rather, as a resource. This resource can provide energy (e.g. see **Box 9.2**) and it can permit the recovery of useful materials and the essential nutrients nitrogen and phosphorus. But such an approach is only possible at large works where the value of the recovered material will cover the capital costs of the processes required to achieve this. For this

> **BOX 9.2 THE FUTURE OF WATER**
>
> ### Energy from warm sewers
>
> The wastewater in the sewer has a huge embedded energy. Water entering the house has an average temperature of 10°C, but in the sewer it has an average temperature of 10 to 15°C in winter and 20 to 25°C in summer. This temperature rise is due to warming from hot taps, heating systems and warm body temperatures. To achieve this temperature rise has required an energy input of around 47 kWh m^{-3}, which represents a staggering 30% of the thermal energy input to a building. Sewer heat recovery is now being practised in a number of locations to utilise this energy (Meggers and Leibundgut, 2011), and heat exchangers in the sewer can provide space or hot water heating to local residential and commercial buildings.

reason, the remainder of this chapter will focus on large, centralised treatment systems only, and consider how we now treat wastewater as a resource and what a Resource Recovery Centre of the future may look like.

3 The stages in the treatment process

Wastewater treatment is solely a solids removal process. Wastewater passes through a number of treatment stages, each of which removes smaller solid particles, until ultimately only colloidal and soluble solids remain. At this point a biological treatment stage is needed in which microorganisms are utilised that are able to oxidise this soluble and colloidal material, using it both as a source of energy and to provide cellular carbon and nitrogen. The new microorganisms that grow can then in turn be removed through a solids removal process. The initial stages of treatment are all physical solids removal processes and are designed exclusively based on the daily flow rate that they receive. They are termed preliminary and primary treatment. The biological stage is known as secondary treatment and it is designed based on the daily load of organic material (as BOD) and ammonia that must be treated. These three stages are able to produce an effluent with a BOD < 20 mg L^{-1}, a TS < 30 mg L^{-1} and an ammonia < 1 mg L^{-1}, all with a 95 percentile compliance. If additional treatment is required, for instance, to reduce the BOD concentration for discharge to a very sensitive watercourse, or to remove enteric pathogens for discharge to recreational waters, this is termed tertiary treatment.

3.1 Physical treatment processes

The flow arriving at a wastewater treatment plant comprises: (i) a domestic contribution, (ii) flow from commerce and industry, and (iii) infiltration of water into the sewerage network. These three components make up what is known as the dry weather flow (DWF) and represent the flow in the absence of rainfall, traditionally expressed with units of either m^3 per day or L s^{-1}. Rainfall will add significantly to this flow in the form of drainage and will serve to dilute the wastewater. To ensure that the sewerage network does not flood during times of heavy and extended rainfall, it contains a number of overflows whenever the sewer passes close to a watercourse. These are meant to operate intermittently during extreme storm events and

a.

b.
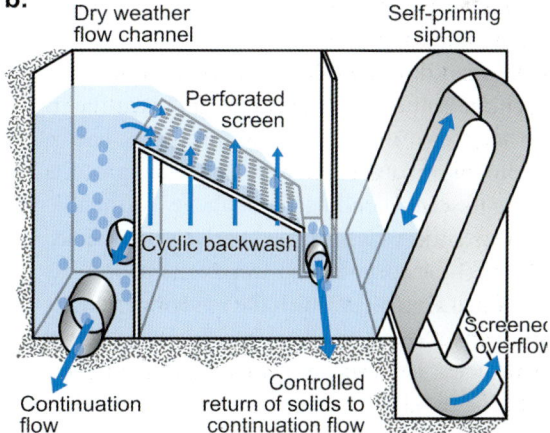

Figure 9.7 (a) A combined sewer overflow serves to protect the sewer from flooding, although it does discharge untreated sewage to the environment during storm events. (b) Innovation processes are now available to remove coarse material from combined sewer overflows to prevent aesthetic pollution.

Figure 9.8 Screens are the first stage of treatment, and remove solid material which is larger than the holes in the screen, in this case 6 mm.

they discharge untreated, but very dilute, sewage to watercourses (**Figure 9.7**). The overflow is known as a combined sewer overflow (CSO) (see Chapter 4, section F2.1). At the sewage works itself all the incoming flow passes by a weir that is set to receive any flows above 6 × DWF. This is known as a 'Formula A' weir, and again protects the works from flooding. Flows above 'Formula A' pass over this weir untreated to the watercourse.

The first stage in the treatment process is termed a 'screen', in which the wastewater passes through a moving drum or plate that is perforated with holes, typically of 6 mm diameter (**Figure 9.8**). Thus, all solid material > 6 mm, known as 'screenings', is removed from the flow on this screen. It is then compressed to remove water, and traditionally is sent to landfill for ultimate disposal. In the UK more than 1000 tonnes of screenings are produced each day, and their carbon footprint is high at 4.6 tonnes CO_{2Eq} per tonne of screenings. Thus, water companies are now looking at the potential for reducing this volume by, for instance, composting or anaerobic digestion, both of which also generate valuable end-products.

The screened wastewater is subject to grit removal, usually through settlement in tanks or long channels. If the velocity of the wastewater is reduced to around 0.3 m s^{-1}, then the heavy grit particles will settle out and can be removed by scraping the base of the tank or channel. The resultant grit can be recovered and reused. The grit can be used around site for repair work to the roads, or it can be further processed and used as a building material or as valuable horticultural grit.

The screened and degritted wastewater now passes on to the primary treatment stage to receive additional or full treatment. However, not all the sewage will receive full treatment and the capacity of the primary and secondary stage is restricted to 3 × DWF, what is known as flow to full treatment (FFT). For the majority of the time the flow to the works is less than 3 DWF, but on those occasions when it exceeds this figure, flows > 3 DWF will pass to large storm tanks for storage. Typically the tanks can provide storage for about two hours, but if the storm event is longer than this any additional flow to the storm tanks overflows directly to watercourses. Once the storm has ended, the storm tank contents are pumped back to the inlet, where they receive full treatment. It may appear from the above that a lot of wastewater goes directly to watercourses without any form of treatment, and this is quite true as it involves flows from CSOs, 'Formula A' weirs and storm tank discharges. However, this happens only when the flow is greater than 6 DWF (and often much greater). Therefore the sewage is diluted at least five-fold with rainwater. In addition, if the storm is of a magnitude to produce this volume of drainage, then the river also must have a very high flow rate. This provides a lot of dilution for the wastewater, and also ensures that it is carried away rapidly by the river. Nevertheless, this source of pollution remains a concern that will inevitably receive more attention in the future.

Primary treatment is a very simple process whereby the wastewater is held in a large tank under quiescent (still) conditions for a minimum of two hours (**Figure 9.9**). Under these conditions the majority of the solids in the wastewater will

Figure 9.9 Primary settlement tanks proved a relatively cheap and simple option for removing up to half the pollution in sewage as well as recovering a primary sludge that is a valuable source of embedded energy.

Figure 9.10 Attached growth processes maintain a viable population of microorganisms by the provision of an inert support such as plastic or mineral media. Shown here are rotating biological contactors.

settle to the bottom of the tank to form a primary or raw sludge. Screened, degritted sewage entering the primary tank will typically contain around 140 mg L^{-1} solids, and between 40 and 60% of this is removed. The solids that settle in the tanks are also rich in organic material, and so primary tanks can achieve removals of up to 30% of the incoming BOD. Thus, the settled sewage leaving the primary tanks might have a TS of around 70 mg L^{-1} and a BOD of 110 mg L^{-1}. The raw sludge at the base of the tank is removed at a TS of about 15,000 mg L^{-1}, showing how effective this process is at thickening the sludge.

Primary sludge is an extremely valuable resource. Over 80% of the solids are volatile, and so it is rich in energy and nitrogen and it passes on to the sludge handling process to recover these resources. The settled sewage passes forward to receive biological treatment.

3.2 Biological treatment processes

There is a wide range of process options that provide biological treatment, but they all have the same aim. This is to maintain a large population of microorganisms in contact with the settled sewage and to provide adequate oxygen to meet the BOD demand of this wastewater within the time it is retained in the basin. An attached growth process uses an inert support material to which the microorganisms attach and build up to quite a substantial thickness, known as a biofilm. The inert supports range from natural materials such as crushed granite or basalt, to blast furnace slag or synthetic materials such as plastic. Attached growth processes include trickling filters, reed beds and rotating biological contactors (**Figure 9.10**). In all these processes oxygen is supplied naturally from the atmosphere and the process efficiency depends on the rate of oxygen transfer through the biofilm. Introduction of mechanical aeration can improve process efficiency and thus shrink the process footprint. Aerated attached growth processes include biological aerated flooded filters (BAFF) and submerged aerated filters (SAF).

In a suspended growth process the microorganisms are suspended within the settled sewage and air is introduced either by surface aeration using motor-driven impellers, or from the base of the tank through diffuser systems. The suspension of microorganisms is known as the mixed liquor suspended solids (MLSS) and is usually maintained at a concentration between 3000 and 4000 mg L^{-1}. In order to maintain the concentration of organism at this level it is necessary to recycle a fraction of MLSS from the base of the final settlement tank back to the aeration basin. This is known as the return sludge. A system that utilises suspended growth with return sludge is known as the activated sludge process, and it has many process variants (**Figure 9.11**). Thus activated sludge uses energy for two processes, introducing air into the mixed liquor and pumping the return sludge from the final settlement tank. The challenge with wastewater treatment is to deliver the high-quality effluent associated with activated sludge, but at a reduced energy input, and this need is driving much process innovation (see **Box 9.3**).

In both attached and suspended growth processes it is necessary to remove a fraction of the biomass each day. This promotes an actively growing population of microorganisms and reduces the overall demand for oxygen. In attached growth systems the microorganisms slough off naturally from the inert support as the biofilm becomes too large to support the amount of BOD and oxygen available and it is sheared by the velocity of wastewater passing the support media. This is known as humus sludge. In suspended growth systems a fraction of the MLSS is pumped (or 'wasted') from the system each day. This is known as the waste or surplus activated sludge.

Figure 9.11 The activated sludge process with air introduced into the tank. It is one of the most popular treatment processes worldwide, due to the quality of the final effluent that it can produce.

There is a relationship between the amount of humus or waste activated sludge and the amount of BOD removed in the process. This relationship is termed the yield. Typically between 0.7 and 1.0 kg of sludge is produced for each kg of BOD removed.

3.3 Final or secondary settlement

The effluent leaving an attached or suspended growth process has a high TS concentration, which represents new microorganisms that have grown on the BOD in the settled sewage (as expressed by the yield). It is necessary to remove these solids before the treated wastewater is discharged to the watercourse. Fortunately, these microorganisms settle very well. The large particles of microorganisms have a strong negative charge of up to $-70mV$ and are able to interact with positively charged material such as cations to form large agglomerates or flocs. They can have settlement velocities as high as 3 m hr^{-1}, and these solids are removed in settlement tanks that are very similar in appearance and behaviour to primary tanks. A final settlement tank (FST) is able to produce an effluent with a TS of 10 mg L^{-1} or less, which represents a solids removal efficiency of 99.75% for an MLSS of 4000 mg L^{-1}. As with a primary tank, these solids settle to the bottom of the tank to form a sludge with a solids concentration of around 1.2% or 12,000 mg L^{-1}. For a humus tank serving attached growth processes, all of this sludge is removed from the tank, or wasted. If the tanks serve the activated sludge process, then a fraction of the solids are returned to the aeration basin as return sludge, and the rest are wasted.

This secondary sludge, like primary sludge, is also a good source of energy and comprises more than 75% volatile solids. However, it is more difficult to recover this energy because the microorganisms that make up the waste sludge have strong cell walls that resist destruction. The secondary sludge now passes forward for further treatment. The final effluent leaving the tank is generally discharged directly to the watercourse, but in certain situations it may pass forward for additional or tertiary treatment, such as disinfection for pathogen removal, or sand filtration to achieve a better removal of TS.

BOX 9.3 CASE STUDIES

Energy-neutral wastewater treatment

Esholt wastewater treatment works treats the wastewater from Bradford in West Yorkshire, UK, and has a population equivalent of 700,000 with a flow of up to 300 L s^{-1}. It was upgraded in 2008 to meet tighter discharge consents for BOD, TS and ammonia, and energy saving was central to the plant design philosophy. The influent to the works has a large available head with a fall of 10 m between the inlet works and the primary tank. Energy is recovered along this steep gradient by passing the flow through two 90 kW screw generators or hydro turbines, installed in series. A flow of 267 L s^{-1} passes sequentially through the two screws over the total fall of around 10 metres and this generates 180 kW, around 5% of the site's energy needs. Better design and control of aeration and oxygen recovery by denitrification in the activated sludge process has resulted in an average energy efficiency of 4.5 kg O$_2$ per kWh, compared to older installations which routinely achieve around 2.7 kg O$_2$ per kWh. Esholt now generates 7.3 GWh per year of renewable electricity, about 30% of the site's needs. The addition of a new thermal hydrolysis process prior to digestion of the 26,000 tonnes of sludge produced means that the site is on target to be energy self-sufficient by 2015, the first such site in the UK.

3.4 Membrane treatment

The final settlement stage of activated sludge is recognised as the weak link in the process, as an FST is expensive to construct and can frequently experience problems with the settlement of the sludge, which leads to a large loss of solids in the final effluent. It is also expensive to return sludge from the tank to the aeration basin. An alternative for removing solids from the aeration stage is to use a membrane with a small pore size, typically around 0.45 μm, that allows the treated effluent to pass but which restricts the passage of particles larger than the pore size. Inclusion of a membrane means that it is not necessary to recirculate solids back for aeration, as the membrane itself separates the solids retention time (as they are held in the basin) from the liquid retention time (as the liquid passes through the membrane and is discharged). It also means that effluent quality is extremely good with TS < 1 mg L^{-1} and BOD frequently < 4 mg L^{-1}. The small pore size also completely removes bacterial pathogens, which are generally > 1 μm in size, as well as many enteric viruses. As a result the treated wastewater can be directly reused. For instance, King County in the US state of Washington operates a large membrane plant serving 1.5 million people, including the city of Seattle. It produces 26,500 m^3 per day of high-quality water for non-potable reuse that is available for use by golf courses, farms and vineyards. This water reuse reduces withdrawals from the river, thus making more flow available for salmon.

3.5 Sludge handling

Primary and secondary sludge requires further treatment before it can be sent to its final destination. As it is produced with a solids concentration of just 1 or 2% this means that up to 99% of the sludge is water. Thus, the first stage is to remove this water by a process of thickening. Thickening is achieved by the addition of a large molecular weight organic compound that contains a lot of highly charged molecules, known as polyelectrolyte. These can be cationic or anionic, depending on their charge. They are used in conjunction with an inorganic cation such as iron or aluminium to provide additional charged material. Together these compounds cause the sludge to coagulate and then flocculate into large clumps that help to release the water from the sludge. The water is then removed either by **centrifuge** (spinning) or by running the sludge in a thin layer onto a belt with small holes that allows the water to pass through (**Figure 9.12**). In this way a sludge of up to 6% DS can be achieved. The water that is recovered is recycled back to the head of the works and is known as sludge return liquor.

4 Energy and agricultural use of sludge

Historically, the process of sludge treatment and handling has been one of the most expensive aspects of the treatment process, accounting for up to 60% of the total costs of treatment. Now, because of the rising cost of energy, this sludge is one of the most valuable aspects of the treatment process. Indeed, sludge handling now generates surplus energy and this extra energy can be used in upstream processes, thus reducing the energy demand of the whole process.

The key to energy recovery is the process of anaerobic digestion. An anaerobic digester is a very simple reactor in engineering terms but one that houses very complex microbiology and biochemistry. The digestion process is undertaken in the absence of oxygen (hence the term 'anaerobic'). Microbial energy-yielding reactions always involve oxidative processes in which electrons are removed from an energy source and passed to an electron acceptor, which becomes reduced. Oxygen is the traditional electron acceptor, accepting electrons to become reduced to water. In the absence of oxygen this causes a problem for the microorganisms, which are forced to use carbon in place of oxygen as an electron acceptor. Carbon can be reduced to many forms, often of commercial importance, such as ethanol, butanol or propanol. In the anaerobic digestion

Figure 9.12 A belt thickener allows the water contained in primary and waste activated sludge to drain away to produce a thickened sludge with a TS of up to 6%. The sludge requires conditioning first, to allow the particles to agglomerate and release the trapped water.

of sludge the desired end-product is the most reduced form of carbon, methane, often termed 'biogas'. Methane is energy rich, with 1 m^3 methane releasing 9.97 kWh energy on burning.

In order to generate the required end-product a four-stage biochemical transformation must be accomplished. Initially the large polymeric molecules of protein, carbohydrate and lipids are broken down to their component monomers of amino acids, simple sugars, glycerol and fatty acid by a process known as hydrolysis, which involves the insertion of a water molecule. Primary sludge undergoes rapid hydrolysis because the major components of a primary sludge have already undergone a preliminary digestion in the gut. By contrast, hydrolysis of secondary sludge is slow, as the microbial cell wall is very resistant to microbial attack. Thus the anaerobic digestion process is slow and the sludge must be retained for between 15 and 20 days at 35°C to fully degrade the sludge. As a result it is now common at larger wastewater treatment works to include an additional process to speed up hydrolysis. For instance, thermal hydrolysis subjects the sludge to a temperature of 135°C and a pressure of 6 bar for up to 30 minutes. The pressure is then rapidly released, causing rupture of the microbial cell wall and the release of the cell contents. This means that much more energy can be recovered from the sludge in subsequent digestion. But such expensive and complex options as thermal hydrolysis are viable only for large works generating large quantities of sludge. The simple monomers can now be broken down further by oxidation to a number of volatile fatty acids (VFAs), such as acetic, propionic and butyric acid, by a process known as acidogenesis. These VFAs are then converted to acetic acid through acetogenesis and the acetic acid is cleaved to produce carbon dioxide and methane by methanogenesis.

Additional methane is also produced by hydrogenotrophic methanogens that are able to reduce carbon dioxide to methane with hydrogen. Each dry tonne of sludge can generate between 80 to 150 m^3 methane through anaerobic digestion, which has an energy value of up to 1.50 MWh, although typical yields in practice are around 90 m^3. This biogas contains other contaminants, such as carbon dioxide, hydrogen sulphide, ammonia and siloxanes, and will require cleaning up before it is used to generate energy. However, there are many uses for the biogas. It can be injected directly into the gas grid, or compressed for use as a vehicle fuel or sold as bottled gas. Its more widespread usage at the moment involves burning in a combined heat and power (CHP) engine. Rather like the engine of a car, a CHP engine, known as the prime mover, burns biogas, which turns an engine. This in turn drives a generator that produces electricity. This electricity can then be used on site to provide aeration and other activities that require electricity. The exhaust gases from the CHP, together with the engine cooling water, are then passed through heat exchangers to recover the heat, which is used to heat the digester itself, to provide heat for thermal hydrolysis and for space heating (**Figure 9.13**). Large sludge treatment centres fitted with thermal hydrolysis are now operated with a surplus of energy that can be exported to grid.

The digestion process destroys up to 60% of the VS content of the sludge, converting it to biogas. This still leaves the inert material and the remaining VS that is not destroyed, together with the new organisms that were produced during the digestion process. This material is known as 'biosolids' and it is produced at between 3 and 4% DS (see **Box 9.4**). A feature of the digestion process is that it renders the solids

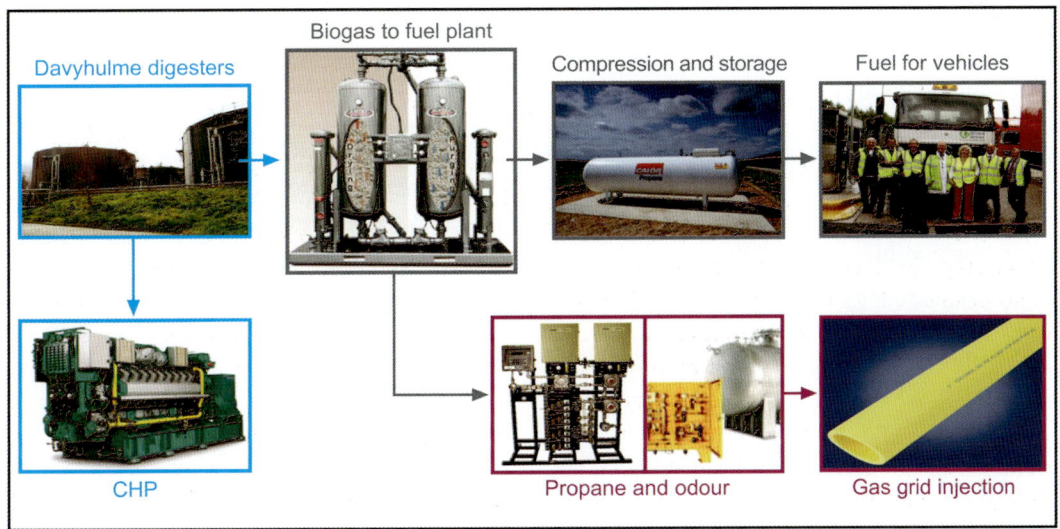

Figure 9.13 A flow scheme for the anaerobic digestion of sludge, demonstrating the range of options for energy recovery.

very easy to dewater and, using a similar system of belt presses or centrifuges that were applied during thickening, the biosolids can be dewatered to 30% DS or more to produce a sludge cake. The ideal route for this material is to recycle it to agricultural land, where it provides a valuable soil conditioner and source of nutrients and trace elements. During the digestion process, the breakdown of large polymers and bacterial cell walls also releases a lot of ammonia and phosphorus. These soluble materials are recovered during dewatering of the biosolids, and the dewatering liquor can contain ammonia as high as 1400 mg L^{-1} and phosphorus concentrations as high as 100 mg L^{-1}. Both of these nutrients can be precipitated in the form of magnesium ammonium phosphate or struvite, which is a valuable slow-release phosphate fertiliser. A number of commercial processes achieve this by dosing the dewatering liquor with either magnesium chloride or magnesium hydroxide and mixing this with air introduced in a rolling motion to form pellets of struvite that settle rapidly and can be recovered and dried. Thus, the end-products recovered from sludge handling are biosolids recycled to land, struvite sold as fertiliser and energy in the form of biogas.

> **REFLECTIVE QUESTION**
>
> What are the processes of wastewater treatment?

D SUMMARY

The world's freshwater supplies are a finite resource but are rarely regarded and treated as such. Technology is available to produce palatable and safe drinking water from a wide range of source supplies, but this comes at a large cost and with its own pollution problems, particularly

---

BOX 9.4 TECHNIQUES

Recovering energy from waste by anaerobic digestion

Energy can be recovered from any biodegradable source of organic material through anaerobic digestion, and the amount of energy can be estimated from the organic content and the yield. For instance, a typical mixture of primary and waste activated sludge has a TS content of 6% and a VS content of 74%. When digested at 35°C with a retention time of 15 days, a methane yield of about 0.4 Nm^3 kg^{-1} is obtained (where N indicates that the volume of methane has been normalised to STP). An efficient digester will achieve a 65% VS destruction. Thus, the composition of 1 m^3 of sludge is:

Dry solids = 60 kg, VS = 44.4 kg, ash = 15.6 kg, water = 940 L

As a result of digestion 28.86 kg VS will be destroyed and so 11.54 m^3 of methane is produced. With an energy content of 9.97 kWh m^{-3} methane, then digestion of 1 m^3 of sludge will generate 115 kWh energy.

The composition of the biosolids leaving the digester will be:

Ash = 15.6 kg, VS = 15.54 kg, dry solids = 31.14 kg, water = 940L.

Thus the dry solids content is now 3.2% with a VS of 49.9%.

in relation to climate change and energy use. The flow train for water treatment includes screening and settlement, coagulation and flocculation, solids removal and filtration, and disinfection. Activated carbon may be used to remove organic contaminants, while desalination is used in some countries to treat seawater where water resources are scarce, but desalination is very energy intensive.

Water and wastewater management now seeks to safeguard and manage the world's water resources in a more sustainable manner while providing for increasing populations and economies. Water is a resource that can be used and then reused. In certain parts of the world treated wastewater is now recycled directly to the water treatment works for potable reuse. Introducing wastewater into water use means that less is released back into the environment, and wastewater treatment becomes important not only as a pollution prevention measure but also as an important tool in water resource conservation.

Many of the world's rivers, lakes, estuaries and seas are polluted as a result of the discharge of untreated human and industrial effluent. Reducing pollution from wastewater is not technically challenging, using physical and biological treatment processes, but it is expensive and, if not operated correctly, with proper monitoring, the technology will not achieve its aim. Wastewater is now viewed as a resource that can be recovered, and treatment plants can now be built that are energy neutral and with much reduced operating costs. Over a period of less than a decade, wastewater treatment has gone from being an expensive, energy-intensive process with a number of waste outputs, to being an almost energy-neutral treatment process with negligible waste outputs and almost complete recycling of all input materials. Although the capital costs remain high, the money is well spent, as it increases both the amenity and the commercial value of the aquatic environment, providing enjoyment for both present and future generations. Indeed, the additional cost to the consumer to fund the development of wastewater treatment facilities has been small and demonstrates how it might be feasible to move towards a zero-waste economy operating largely on renewable energy, with a negligible carbon footprint.

FURTHER READING

Agler, M.T., Wrenn, B.A., Zinder, S.H. and Angenent, L.T. 2011. Waste to bioproduct conversion with undefined mixed cultures: The carboxylate platform. *Trends in Biotechnology* 29: 70–78.

The microbiology of anaerobic digestion permits a wide range of end-products to be produced. This research paper outlines clearly both the range of products that are available and how the process can be operated to produce them.

Binnie, C. and Kimber, M. 2009. *Basic water treatment*. Thomas Telford; London.

A good practical guide to the range of unit operations that comprise a modern water treatment works, with clear explanations of their role and mode of operation.

Eikelboom, D.H. 2000. *Process control of activated sludge plants by microscopic investigation*. IWA Publishing; London.

The author was the first to appreciate that there are many different organisms responsible for bulking of activated sludge and that they could be identified by their morphology. This is a practical guide to the identification of these bacteria.

Henze, M., van Loosdrecht, M.C.M., Ekama, G. and Brdjanovic, D. 2008. *Biological wastewater treatment: Principles, modelling and design*. IWA Publishing; London.

Provides a detailed guide to the approach taken in Europe and North America for the design of wastewater treatment processes.

Horan, N.J. 1990. *Biological wastewater treatment systems: Theory and operation.* John Wiley and Sons; Chichester.

A concise guide to understanding the biological principles which guide the design and operation of wastewater treatment facilities.

Judd, S. 2006. *The MBR Book: Principles and applications of membrane bioreactors in water and wastewater treatment.* Elsevier; Oxford.

A comprehensive description of the principles of MBRs, with a good outline of the commercial processes that are now widely available.

Mara, D.D. 2011. *Good practice in water and environmental management: Natural wastewater treatment.* CIWEM; London

Natural wastewater treatment systems find many applications in small communities and this handbook reviews the design principles for such treatment processes.

Mara, D.D. and Horan, N.J. (eds). 2003. *The handbook of water and wastewater microbiology.* Academic Press; London.

A selection of articles from international experts introducing all aspects of water and wastewater microbiology.

Metcalf & Eddy Inc. 2003. *Wastewater engineering: Treatment and re-use (4th edition).* McGraw Hill; New York.

The classic textbook, comprehensive in its scope and coverage of the design of processes used for the treatment of wastewater and sludge.

MWH. 2005. *Water treatment: Principles and design (2nd edition).* John Wiley and Sons; Chichester.

This is becoming the classic textbook for water treatment, with a broad and comprehensive coverage of the treatment options.

Classic papers

Anderson, J. 2003. The environmental benefits of water recycling and reuse. *Water Science and Technology: Water Supply* 3: 1–10.

One of the first contributions that outlined the possibilities for utilising treated effluents through recycling and reuse.

Ardern, E. and Lockett, W.T. 1914. Experiments on the oxidation of sewage without the aid of filters. *Journal of the Society of the Chemical Industry* 33: 523.

This paper, presented in Manchester to the Society of Chemical Industry, is acknowledged as the invention of the activated sludge process, which is now the most successful wastewater treatment process option.

Barnard, J.L. 1976. A review of biological phosphorus removal in the activated sludge process. *Water SA* 2: 136–144.

The discovery by Barnard of the mechanism for biological P-removal has led to huge improvements in the quality of treated effluents and helped to reduce the impact of eutrophication wherever it is practised.

Marais, G.V.R. and Ekama, G.A. 1976. The activated sludge process Part 1 – steady-state behaviour. *Water SA* 2: 164–200.

This work was the start of a unified model of the activated sludge process, which led to computer-based design and optimisation of treatment plants.

PROJECT IDEAS

- Design a poster that explains the steps of water treatment for potable supply and another that outlines wastewater treatment steps.

- Find out how much sludge material from wastewater treatment is reused in your local area and perform a study to outline the potential for use of a larger proportion of the sludge. What markets are there for sludge products?

- Design an educational video that outlines how wastewater treatment could be an almost energy-neutral process.

REFERENCES

Agler, M.T., Wrenn, B.A., Zinder, S.H. and Angenent, L.T. 2011. Waste to bioproduct conversion with undefined mixed cultures: The carboxylate platform. *Trends in Biotechnology* 29: 70–78.

Anderson, J. 2003. The environmental benefits of water recycling and reuse. *Water Science and Technology: Water Supply* 3: 1–10.

Ardern, E. and Lockett, W.T. 1914. Experiments on the oxidation of sewage without the aid of filters. *Journal of the Society of the Chemical Industry* 33: 523–539.

Barnard, J.L. 1976. A review of biological phosphorus removal in the activated sludge process. *Water SA* 2: 136–144.

Binnie, C. and Kimber, M. 2009. *Basic water treatment*. Thomas Telford; London.

Chow, A.T., Tanji, K.K. and Gao, S. 2003. Production of dissolved organic carbon (DOC) and trihalomethane (THM) precursor from peat soils. *Water Research* 37: 4475–4485.

du Pisani, P.L. 2006. Direct reclamation of potable water at Windhoek's Goreangab reclamation plant. *Desalination* 188: 79–88.

Eikelboom, D.H. 2000. *Process control of activated sludge plants by microscopic investigation*. IWA Publishing; London.

Henze, M., van Loosdrecht, M.C.M., Ekama, G. and Brdjanovic, D. 2008. *Biological wastewater treatment: Principles, modelling and design*. IWA Publishing; London.

HMSO. 1970. *Working party on sewage disposal (The Jeger Report). Taken for granted*. HMSO; London.

Horan, N.J. 1990. *Biological wastewater treatment systems: Theory and operation*. John Wiley and Sons; Chichester.

Horan, N.J. and Lowe, M. 2007. Full-scale trials of recycled glass as tertiary filter medium for wastewater treatment. *Water Research* 41: 253–259.

Judd, S. 2006. *The MBR Book: Principles and applications of membrane bioreactors in water and wastewater treatment*. Elsevier; Oxford.

Klaunig, J.E., Ruch, R.J. and Pereira, M.A. 1986. Carcinogenicity of chlorinated methane and ethane compounds administered in drinking water to mice. *Environmental Health Perspectives* 69: 89–95.

Mara, D.D. 2011. *Good practice in water and environmental management: Natural wastewater treatment*. CIWEM; London.

Mara, D.D. and Horan, N.J. (eds). 2003. *The handbook of water and wastewater microbiology*. Academic Press; London.

Marais, G.V.R. and Ekama, G.A. 1976. The activated sludge process Part 1 – steady-state behaviour. *Water SA* 2: 164–200.

Meggers, F. and Leibundgut, H. 2011. The potential of wastewater heat and exergy: Decentralized high-temperature recovery with a heat pump. *Energy and Buildings* 43: 879–886.

Metcalf & Eddy Inc. 2003. *Wastewater engineering: treatment and re-use (4th edition)*. McGraw Hill; New York.

MWH. 2005. *Water treatment: Principles and design (2nd edition)*. John Wiley and Sons; Chichester.

Pereira, M.A., Lin, L.H., Lippitt, J.M. and Herren, S.L. 1982. Trihalomethanes as initiators and promotors of carcinogenesis. *Environmental Health Perspectives* 46: 151–156.

RSC. 2009. *What is a poison?* Note produced by a Working Party of the Environment, Health and Safety Committee [EHSC] of the Royal Society of Chemistry [online]. Available from: http://www.rsc.org/images/PoisonV2_tcm18-37982.pdf.

Verstraete, W., Van de Caveye, P. and Diamantis, V. 2009. Maximum use of resources present in domestic 'used water'. *Bioresoure Technology* 100: 5537–5545.

WHO. 2011. *Guidelines for drinking-water quality (4th edition)*. World Health Organization; Geneva.

CHAPTER TEN

Water economics

Sonja S. Teelucksingh, Nesha C. Beharry-Borg and Dabo Guan

> **LEARNING OUTCOMES**
>
> After reading this chapter you should be able to:
>
> - understand the role of water in an economic system
> - explain the total economic value of water resources
> - understand the different types of economic valuation techniques that can be applied to water resources
> - describe the methodology of the Millennium Ecosystem Assessment and how this methodology can be applied to different ecosystems.

A INTRODUCTION

The fluctuations in supply, coupled with a myriad of competing demands, make the study of water resources perfectly suited to the discipline of environmental economics. This chapter uses concepts from environmental economics but with a focus on water as the environmental component of concern. Water is a core environmental component and is crucial for economic development, and it is therefore essential to understand water economics if we are to develop an integrated approach to understanding both water and development.

Since its much-cited use in the Brundtland Report (United Nations, 1987), plenty of attention has been paid to the concept of sustainable development. The notion of development itself can be seen through the perspective of human needs rather than solely in terms of economic growth. This enables us to pay more attention to the way growth is achieved, how it is distributed and how livelihoods are affected in the development process. Sustainability has frequently been defined as 'the maintenance of social well-being' for both current and future generations. That definition is an elusive phrase that can be interpreted in a number of different ways. However, despite contradictory interpretations, a common international consensus is that sustainable development is something for which economies should strive. This is made clear by the list of eight

Millennium Development Goals derived by the United Nations in 2000 and signed up to by 189 nations. One of the goals is to 'ensure environmental sustainability'. The UN Conference on Sustainable Development in 2012 assessed global progress towards these development goals, and to address emerging challenges. One of the challenges identified for urgent action was that of water scarcity (see Chapters 1 and 7). Water availability can represent a significant constraint on the sustainable development of an economy. This, coupled with increasing population, places water resources under growing demands and leads to the need for improved water management as a critical element in the achievement of sustainable development. Water resource management affects every aspect of an economy. At the same time, the patterns of water demand relate directly to economic structures and sectors. Water stress and water insecurity have particular implications for developing countries, in the context of those dominated by rural subsistence-based communities heavily dependent on agriculture and characterised by a lack of water infrastructure.

There has recently been a major global research effort designed to both provide a snapshot of current linkages between biodiversity and human well-being and to suggest ways for its future conservation (TEEB, 2010). This global study, known as 'The Economics of Ecosystems and Biodiversity' (TEEB), came up with an interesting approach to the concept of the value of an environmental resource, called the three-tiered approach (**Figure 10.1**). The methodology suggests that values need to be recognised, but while understanding that different people in different situations can apply different values to the same resources. Once recognised, the value needs to be demonstrated in economic terms by the use of the appropriate economic valuation tools. Finally, once demonstrated, these ecosystem values then need to be embedded into the decision-making processes by the management of the governance frameworks and institutional structures which oversee the resource, and via the use of economic

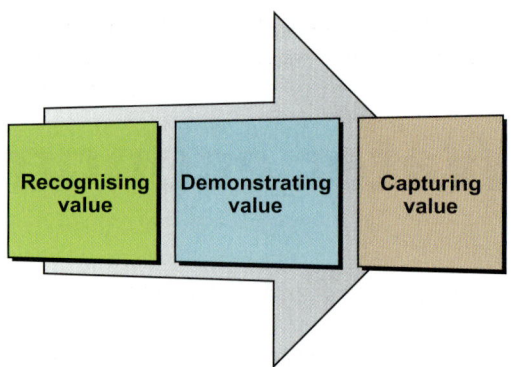

Figure 10.1 The three-tiered approach to economic value.

instruments such as payments for protecting the environment so as to enhance water quality.

This chapter follows these principles to provide an overview of the main issues in the specialised subdiscipline that has become known as the economics of water. Firstly, we *recognise value* by discussing the role of water within an economic system, in terms of both consumer and producer demands. Coupled with demand must come a supply to meet this demand. How natural systems supply water, and how these systems can be threatened by human activity, is then discussed. The intersection of demand and supply inevitably raises the question of the pricing of water. Given the role of water in an economic system, we then ask the question – what is the *value* of water? Different philosophical conceptions exist around the notion of value. Environmental economists speak of the 'total economic value' of an environmental good, which deconstructs into a series of sub-classifications that combine to form this total value. The Millennium Ecosystem Assessment (2005) offered an alternative approach, classifying environmental goods into what has become known as 'ecosystem goods and services', with a focus on the flow of benefits from ecosystems, via these goods and services, to human well-being. In the language of water resources, we then focus upon a case study of wetlands and the ecosystem goods and services that they provide, which include the provision of freshwater, water regulation and water purification. Whichever typology is adopted, there exists a range of economic valuation techniques that can be applied to water resources to *demonstrate value*. Once values are estimated, the question then becomes how to integrate these values into decision-making processes, or how to *capture value*, where the complex issue of water management is addressed. Complexity is to be found primarily in a governance sense, where water resources transcend political boundaries, therefore demanding a collective approach to the overall management of what is the world's most precious resource.

B WATER WITHIN THE ECONOMY

1 Supply and demand

Water by its very nature is defined as a **common property resource**. It is widely accepted that those resources upon which poor rural households from developing countries depend for their daily livelihoods are open access or common property, with a major problem facing developing countries being the degradation of these common resources. It is inevitable that a high dependence on any open access or common property resource, together with a lack of, or improperly designed/enforced, **property rights**, can lead to conflicts over resource use and ownership, leading to an increasing need for comprehensive management strategies (see Chapter 11). Water is mobile across political entities (e.g. the River Danube flows through ten European countries). Water is also characterised by uneven spatial distribution across the globe (see Chapters 1 and 7).

Like many environmental resources, freshwater is characterised by a limited supply, together with an array of competing wants by a range of **stakeholders**. When these stakeholder demands inevitably conflict, the issue becomes more complicated. Pollution that crosses political boundaries is known as **transboundary pollution**. This can occur, for example, with marine pollution in small island states that are geographically close to one another. Transboundary pollution is a problem that results from the lack of clearly defined property rights in areas where environmental resources are shared or impacted upon. This brings into the analysis the question of governance (see Chapter 11 for a more detailed treatment). Constitutional settings in many developing countries are characteristically weak. This has direct implications for environmental resource use and management. Institutional and government failures are one of the reasons identified for environmental destruction, through environmentally adverse policies or the inability to resolve competing objectives (for example, where expansion of industrial activity in the pursuit of economic development can involve practices that pollute groundwater). This is further compounded by the different spatial scales being considered. It is inevitable that trade-offs in use, impacts upon supply and factors affecting demand can occur at local, regional and international stakeholder levels.

The supply of water originates from two main sources: surface water from rivers and lakes and groundwater (**Figure 10.2**). Each of these sources has characteristics of both renewable and non-renewable resources (Dalhuisen et al., 2000). Surface water and groundwater is to some extent renewable, through seasonal rainfall, as well as through inputs back into the surface water system of treated wastewater, via water management and purification systems (see Chapter 9). Surface water is easily accessible but can be highly vulnerable to agricultural, industrial and human wastes, is seasonally variable, and its supply is susceptible to periods of drought. As described in Chapter 5, groundwater in some places is considered to be non-renewable. Groundwater needs to be extracted, and at an optimal rate, but could be available at times of drought. The question in any geographical setting then becomes the extent

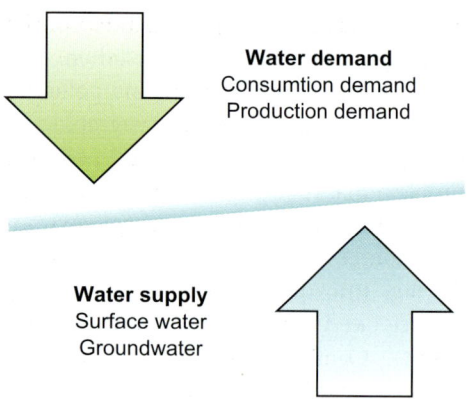

Figure 10.2 The balance between water demand and water supply.

to which surface and groundwater can be simultaneously and optimally used (**conjunctive use**) to meet economic demand.

The main distinction in the economic demand for water is the types of activities for which the water is demanded. We can distinguish between productive purposes (where water is used as a resource in the production of goods and services) and consumptive purposes (where water is used by households for uses such as drinking and sanitation). Though varying across countries, a significant component of water demand is the demand of households. Households are not, however, homogeneous units, and the extent of residential water demand is impacted upon by a host of characteristics such as size of household, level of income and type of accommodation (see Chapter 7). On the productive side, water demand varies according to type of industry, with industrial demands for water in manufacturing processes and agricultural demands for irrigation and the watering of crops and animals dominating the demand.

The differences in use of water for consumption and for production raise the possibility of using water of different qualities for different purposes. In this way water could be classed as a mixed resource rather than as a uniform one. Water with a quality suitable for drinking (**potable water**) may not be demanded for many residential and industrial purposes. For example, there is no need to flush a toilet or wash a car with potable water. However, in many developed countries water for toilet flushing is supplied at the same quality as water for all other household purposes. The largest part of water demand is for activities that do not require potable water (Dalhuisen et al., 2000). However, water is usually supplied as a uniform good, in particular in developing countries where no distinction is made as to the type of water provided for these different uses. Little is known about consumer demand for different qualities of water. It is feasible to assume that, for different aspects of water quality, there exists a range of supply and demand characteristics that must be examined in their own right (Dalhuisen et al., 2000).

2 Water industry

The water industry forms the bridge between the supply and demand for water, through its activities of collecting, treating, storing, transporting and distributing water. By definition this is an industry that demands a significant level of investment and that is also associated with high costs. Because of this, water supply in many developing countries is characterised by a monopoly and is under some sort of central government control. However, the water supply chain can itself be disaggregated and various elements of it can be privatised. For example, the collection of water, the treatment of water and the transport and distribution of water can all be undertaken by different organisations that come from the private sector. An interesting issue then becomes the extent to which private versus public organisations exist in the water industry, and which mixture is in the best interests of water security.

Given the role of the water industry in harnessing water supply and meeting water demands, the question now becomes the optimum price at which this is achieved. The water industry traditionally passes on the costs of physical supply (e.g. collecting, treating, pumping and piping the water) to homes and businesses. However, the water itself as a publicly provided good is free, with no royalties implemented to extract the resource. Cheap water therefore reflects the lower capital costs of the water industry rather than an abundance of the resource. This means that the economic value of water does not enter the transaction which takes place on the market, and therefore is not captured in the price. Consumers of water are therefore unaware of the actual cost of the resource to society at large, or of its relative scarcity. Economists therefore apply a range of methodological tools to the valuation of water resources. The history of non-market valuation in the United States is in fact closely linked with water projects (Hanemann, 2006).

3 Water, climate and the economy

Water resources affect, and have always affected, every aspect of economic life (e.g. see Chapter 1). There exists a delicate balance between water demand and water supply at all levels, from the household itself, the community, the country and on the global scale. Forecasts of economic growth and population tell us that, by 2030, global demand for water will outstrip its supply by 40%, and that for one-third of the global population (mostly in developing countries) this deficit will be more than 50% (Water Resources Group, 2009). There is therefore an urgent need to address the challenges of water security, for both current and future livelihoods.

As a part of natural variability, precipitation, and thus the availability of water resources throughout time and space, varies within bounds determined by given climate conditions. In this context, climate change has a direct impact on water supply (see Chapter 2). Many of the anticipated impacts of climate change upon human livelihoods will operate through water. Changes in the global hydrological cycle as a result of changing temperatures can lead to changes in water quantity and quality through accelerated glacier melt, changes in precipitation, temperature and evaporative demand, runoff and recharge patterns and rates, and saltwater intrusion into coastal aquifers (Bates *et al.*, 2008). There is also the increased risk of extreme meteorological events such as floods, droughts and tropical cyclones (see Chapter 2). These impact upon human livelihoods, as humans depend on the provisioning, regulating, cultural and supporting **ecosystem services** provided by freshwater ecosystems. Such impacts may be felt more strongly in some places than in others. Water resources in Small Island Developing States have been identified as highly vulnerable to climate change, as described in **Box 10.1**.

The predictions of the Fourth Intergovernmental Panel on Climate Change indicated that water resources will be one of the sectors most affected by changes in climate (Bates *et al.*, 2008; see Chapter 2). Water scarcity is estimated to increase as climate change effects are felt. The growing world population, unsustainable practices and inefficient allocation of water threaten to induce and/or intensify water scarcity, with disastrous consequences for the environment and societies. Health issues are also relevant, where lack of access to potable water and proper sanitation, in particular, can cause the persistence of associated ailments (see Chapter 8). It therefore becomes necessary for society to adapt to the range of water-related hazards that accompany climate change.

The negative environmental effects have the potential for disrupting human social and economic activity everywhere on Earth, but are particularly threatening to coastal communities (e.g. **Box 10.1**). Healthy ecosystems are at the front line of defence against climate change, through both (1) protection from natural hazards and (2) the ability to reduce economic vulnerabilities through their contribution to human well-being and sustainable livelihoods. There has therefore been a shift from a focus on natural hazards in particular to a focus on the series of factors that can

> **BOX 10.1 CONTEMPORARY CHALLENGES**
>
> ### Areas at risk: Small Island Developing States
>
> Small Island Developing States are found in the Caribbean, the Pacific, and close to the African continent. Recognised by the Millennium Development Goals, the Millennium Ecosystem Assessment and the IPCC reports as a unique category of analysis, Small Island Developing States face a particular set of economic, environmental and developmental challenges. Their key underlying characteristic is that of the vulnerability of local livelihoods to environmental and economic changes. Small populations are coupled with high population densities, concentrated in coastal zones that comprise much of the land area. An inevitably high ratio of coastal to total land area means that island ecosystems are characterised by highly coupled terrestrial and marine ecosystems (McElroy *et al.*, 1990). There is a heavy reliance on natural resource exploitation, with tourism (and, increasingly, eco-tourism) dominating island economies (Mimura *et al.*, 2007). They also exhibit a high degree of economic vulnerability to the world economy, due to their dependence on international trade. These characteristics make Small Island Developing States especially vulnerable to the effects of climate change, sea level rise and extreme events (Nurse *et al.*, 2007). Due to the heavy reliance on natural resource exploitation for economic livelihoods at both micro- and macro-levels and the high prevalence of coastal zone areas, environmental shifts such as ecosystem changes, natural disasters and climate change impacts can have extreme economic and welfare effects in Small Island Developing States (van Beukering *et al.*, 2007; Bates *et al.*, 2008). In particular, there is an increasing focus on the adaptation of coastal communities to climate impacts (Galloway and Scally, 2006; Nicholls *et al.*, 2010).
>
> Small Island Developing States are most at risk from sea level rises, increased average global temperatures affecting the normal seasonal cycles, increased incidences and intensity of storms and hurricanes, along with the consequent increased risks of storm surges, and increased incidences of droughts and floods. These impacts can have a direct effect in the area of water resources and water services where, under most climate scenarios, water resources in Small Island Developing States are forecast to be highly compromised (Arnell, 2004; Mimura *et al.*, 2007; Bates *et al.*, 2008; Sallof and Muller, 2009). The IPCC reports explicitly identify Small Island Developing States as a 'hotspot' area where climate change effects are present or imminent and where urgent action is required in the water sector (IPCC, 2007; Overmars and Gottleib, 2009).
>
> The characteristics of Small Island Developing States constrain effective management of the water sector; parallel to this is the pattern of demand for water resources in Small Island Developing States. Water demand can be intensified due to the large economic reliance of many Small Island Developing States on water-intensive industries such as the tourism sector, and to a lesser extent the agricultural sector, as their main developmental option. At the community level, many coastal livelihoods can be closely linked to these industries. IPCC Scenario Analyses confirm that for both of these sectors climate change impacts will be negative (Nurse *et al.*, 2007). Interestingly, it has also been suggested that an increasing demand for water may be much more significant than forecast, due to climate change impacts (Overmars and Gottleib, 2009).

change the coping capacities (resilience) of social systems, and the ways in which these capacities can be affected by policy decisions at community, national, regional and international levels.

Water resource management affects every aspect of the economy. At the same time, the patterns of water demand relate directly to economic structures and sectors. Increasing pressure of demand on water resources, coupled with climate-induced changes in supply, will have different effects, depending on geographical, economic and social criteria. While climate change can directly affect the supply of freshwater, it is non-climatic drivers that most impact upon the demand. Effective water management and principles for adaptation with a goal to achieving water security therefore require an analysis of both the supply and demand patterns of a country and the climatic, economic and other non-climatic drivers that determine them.

> **REFLECTIVE QUESTION**
>
> When you pay for your water what are you actually paying for?

C WATER PRICING

In *The Wealth of Nations*, published in 1776, Adam Smith pointed out a well-known paradox regarding the usefulness of water and its price: 'Nothing is more useful than water, but it will purchase scarce anything; scarce anything can be had in exchange for it.' Water, like many other goods we might want, is scarce. Under drought conditions it becomes a bit scarcer than normal. For any other good, we would expect supply reductions to correspond to increasing prices. When frosts damage citrus crops, orange juice prices rise. When hurricanes knock out oil refineries, gasoline prices shoot upwards. However, water is one of the exceptional goods where price is not an indicator of its scarcity.

1 Water as an economic good

Water is vital for life and is crucial for economic development. Increasing urbanisation, population and industrial development have increased demand for water, resulting in a considerable decrease in annual renewable water resource per capita. As the price of water is kept very low, the consumer takes this low price to be the real value of water and so increases consumption. Water has become scarce, but its use is rampant. It has become important to sort out an efficient pricing structure for water.

In the Dublin conference of 1992, for the first time in a United Nations setting, water was declared as an economic good. However, water had been treated as an economic good for many centuries before this. In Europe and the early period of the United States, private water supply companies thrived in a wide variety of settings. However, the nineteenth-century 'sanitary revolution' saw the demand for public ownership and management of most of these companies in the name of public welfare. Heavy emphasis on the nature of water as a public good led to the development of heavily subsidised public systems. However, by the 1980s the World Bank and other institutions started stressing the need for privatisation in supplying water. For the World Bank, privatisation increases efficiency and brings about proper allocation (see Chapter 11 for alternative views).

Pricing water is one of the ways to increase efficiency, equity and sustainability in the water sector. For a proper price, government intervention is also required so that equity and other important issues are properly addressed. Agenda 21 of the United Nations looks for sustainable development as a way to reverse environmental degradation as well as poverty. The main concern is to eradicate poverty from the global map by giving the poor more access to the resources they need to live sustainably. According to the definition of Agenda 21, sustainability includes social development, economic development and environmental protection. It is here that water pricing, by encouraging judicious use of resources, can contribute to environmental protection and can promote economic efficiency and social equity.

2 Pricing water: from benefit to cost

In a competitive market, price is determined by the demand and supply of goods and services. Water resources invoke an economic demand as they are used by people in the production of goods and services, such as agricultural output, human health, recreation and quality of life. Supplying water to production or residential usage is usually associated with costs for transmission, treatment and maintenance.

Providing or protecting water resources involves active employment of capital, labour and other scarce resources. Utilisation of these resources to provide water supplies means that they are not available to be used for other purposes. The economic concept of the 'value' of water is thus couched in terms of society's willingness to make trade-offs between competing uses of limited resources, and in terms of aggregating over individuals' willingness to make these trade-offs. The economist's task of estimating the benefits or loss of benefits resulting from a policy intervention is easiest when the benefits and costs are revealed explicitly through prices in established markets. When it comes to measuring environmental and some other impacts, however, valuing benefits is more difficult, and requires indirect methods (see section D).

On the other hand, the economist's notion of cost, or more precisely, of opportunity cost, is linked with – but distinct from – everyday usage of the word. Opportunity cost is an indication of what must be sacrificed in order to obtain something. In the water resources context, it is a measure of the value of whatever must be sacrificed to make those resources available. It has been observed over and over again in diverse markets for goods and services of various kinds that the incremental costs of providing an additional unit increase as the total quantity supplied increases. In the language of economics, there are increasing (or upward-sloping) **marginal costs**. The costs of a cubic metre of water flowing out of a kitchen tap include the costs of transmission, treatment and distribution; some portion of the capital cost of reservoirs and treatment systems, both those in existence today and those future facilities necessitated by current patterns of use; and the opportunity cost in both use and non-use value of that cubic metre of water in other potential functions. This is the long-run marginal cost (LRMC) of supplying water.

In a competitive market – which, as we have explained above, is not the context for most water resources – the quantity of a good or service provided and its price are jointly determined by the forces of supply and demand, which are closely linked with costs and benefits, as described here. In fact, the downward-sloping marginal benefit curve is the demand curve, and the upward-sloping marginal cost curve is the supply curve. Where these intersect, where demand and supply balance one another, markets achieve equilibrium, determining the quantity provided and the price in the process. And that particular combination of price and quantity maximises the difference between benefits and costs; that is, it maximises what economists call net benefits (the sum of consumer surplus and producer surplus). This is the definition of economic efficiency, and the efficient quantity and the efficient price of any good or service.

3 Types of water tariffs

Since water is not traded in markets, we would not expect prices to adjust automatically to reflect periods of scarcity as they do for other goods and services. Instead, most water pricing is regulated by public institutions – city councils, public utility commissions, water boards and other entities.

A set of procedural rules used to determine the condition of services and charges among various classes is known as a water tariff. Tariff structure varies according to the state as the institution involved in its provision also differs. The water charges can be in the form of non-volumetric water tax, non-volumetric flat rate tariff, uniform metered tariff and metered block tariffs or can be a combination of the above charges. The most popular tariffs in OECD countries are two-part tariffs and the increasing block tariff. Countries such as Australia, Austria, Denmark, Finland and the United Kingdom that have successful water pricing schemes use a two-part tariff structure with a variable part and a fixed part. The fixed part varies according to some characteristic of the user and the variable element according to the average cost pricing. Sometimes the variable part is replaced by an 'increasing block tariff'. This increases the tariff structure. Advantages of the two-part tariff structure are that it brings in a stable revenue. The fixed element encourages conservation, for it charges the consumer according to their consumption.

In the increasing block tariff there are two or more blocks, for which prices are charged differently, and with each successive block the price increases. The utility must decide on the number of blocks, volume of water use associated with each block, and prices to be charged for each block, when designing an increasing block tariff structure (Boland and Whittington, 1998). The increasing block tariff is a progressive tariff. There is a lifeline block for the poor at below cost rate, and beyond this minimum volume the charge increases. The lifeline block works as a subsidy for the poor and promotes welfare. Thus the increasing block tariff encourages equity. The system enables poorer households to get access to low-rate water since they possess fewer water-consuming appliances (Whittington, 1997), and also allows for rich-to-poor and industrial-to-household subsidies (Bolland and Whittington, 1998).

> **REFLECTIVE QUESTION**
>
> What factors ought to be involved when setting a price for water?

D WATER AS AN ECOSYSTEM SERVICE

1 Total Economic Value

Many have incorrectly equated economic value with market price. Were this true, it would imply that only marketed goods and services had value. Given the essentialness of water for all life, the notion of the *value* of water (as opposed to its price) has occupied great minds for centuries. Adam Smith, in *The Wealth of Nations*, observed that the concept of value itself was subject to a myriad of sometimes paradoxical interpretations, where 'the things which have the greatest value in use have frequently little or no value in exchange; and, on the contrary, those which have the greatest value in exchange have frequently little or no value in use'. Smith used the example of diamonds and water as the illustration for this concept, where diamonds could command a great many other goods in exchange but had little use value, whereas water, of which 'nothing is more useful', could not be used in exchange. Benjamin Franklin attributed value to scarcity, with the famous proverbial saying 'when the well's dry, we know the worth of water'.

Economists, in considering the value of nature and its components, have attempted to demystify the concept of value by defining what has become known as **Total Economic Value**, and disaggregating this into various categories and subcategories. The Total Economic Value of environmental goods consists of *use value* and *non-use value* (**Figure 10.3**).

The use value is a value related to the present or future use by individuals. It can be subdivided into direct use values and indirect use values. Direct use values are derived from the actual use of a resource (such as water consumption); indirect use values refer to the benefits derived from ecosystem functions (such as river basin protection). A separate category is made up of *option values*, which are values attributed by individuals who have knowledge that a resource will be

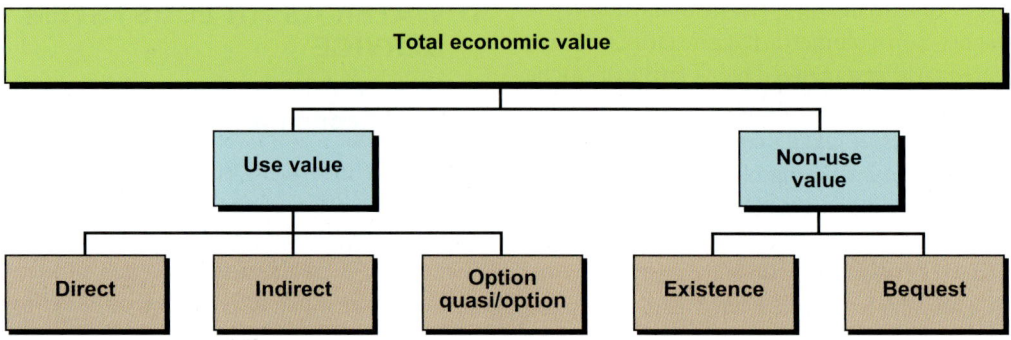

Figure 10.3 A typology of economic value.

available for future use. Thus it can be considered as an assurance that a resource will be able to supply benefits in the future. For example, the continued availability of a certain amount of forest acreage to sequester carbon and mitigate the impacts of climate change would provide future benefits to society. The quasi-option use value shown in **Figure 10.3** (sometimes also classified as a non-use value) represents the value derived from the preservation of the future potential use of the resource, given some expectation of an increase in knowledge. For example, the continued availability of certain species may have value to humanity, but this value may not be obvious now. However, with increased scientific knowledge, these species may come to play a significant role in human welfare in the future.

The non-use values (**Figure 10.3**) are associated with the benefits derived simply from the knowledge that a natural resource – such as a river basin – is maintained. By definition, such a value is not associated with the use of the resource or the tangible benefits deriving from its use. It can be subdivided into two parts that overlap, according to its definition. First, there are *existence values*, which are not connected to the real or potential use of the good, but reflect a value that is inherent in the fact that it will continue to exist independently from any possible present or future use by individuals. Secondly, *bequest values* are associated with the benefits to individuals derived from the awareness that future generations may benefit from the use of the resource. These can also be altruistic values when the resource in question should, in principle, be available to other individuals in the current generation.

2 Ecosystem services and goods

The Millennium Ecosystem Assessment (MEA), in its survey of the world's natural resources, provided an innovative methodological framework for analysing environmental resources. The emphasis was placed strongly on the concept of human well-being, the contributions made by ecosystems through their supply of ecosystem goods and services, and the drivers of change of ecosystems that would then impact upon human well-being (**Figure 10.4**). The MEA approach provided a framework that addressed the many mechanisms through which nature contributes to human well-being at different levels of spatial scale, and the factors (both direct and indirect) that can impact upon these mechanisms. The MEA methodology has resulted in a growing body of literature on ecosystem services.

The MEA identified four categories of ecosystem services, namely provisioning, regulating, cultural and supporting services (Millennium Ecosystem Assessment, 2005). These contribute in different ways to the constituents of human well-being. Chapter 6 (section D) lists some eco-

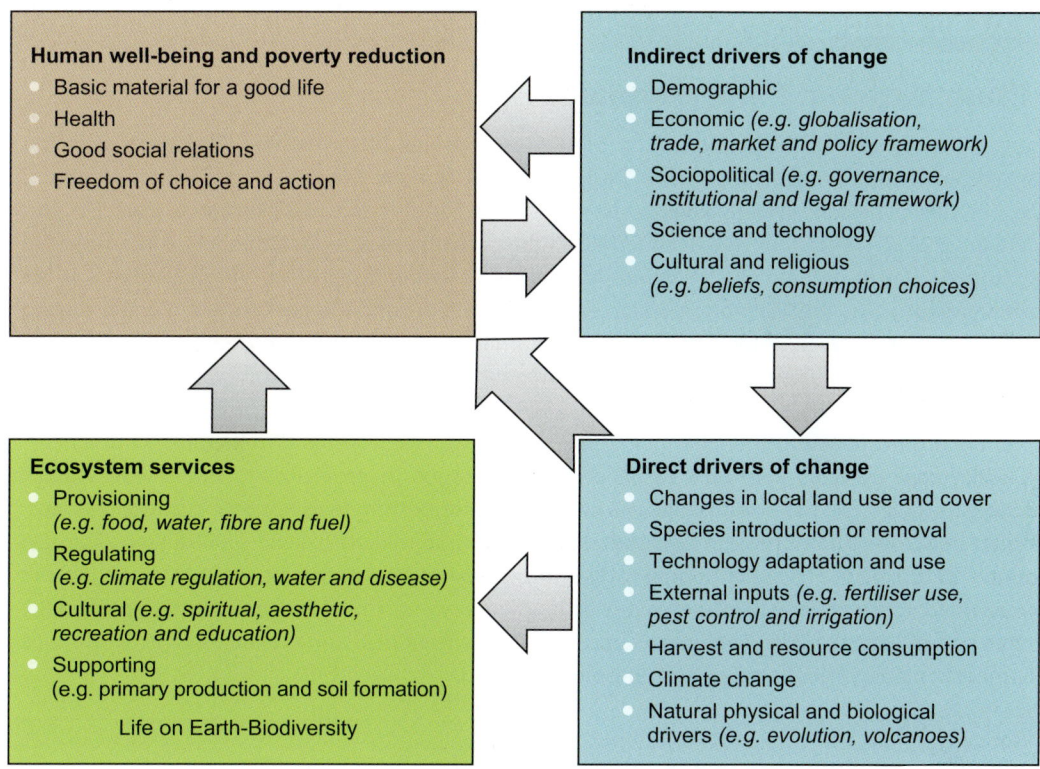

Figure 10.4 Linkages between ecosystem services, human well-being and drivers of change.
Source: After Millennium Ecosystem Assessment, 2005.

system services that aquatic ecosystems provide. The notion of human well-being and livelihood is a complex one. The MEA presents a comprehensive overview of the factors that both comprise and affect the state of human well-being. Furthermore, the MEA methodology hypothesises that drivers of change impact upon ecosystems and the goods and services they provide, thus affecting human well-being. If we are able to quantitatively or qualitatively define these linkages, the impacts upon human well-being of decisions that affect ecosystems at any spatial scale can be mapped. In addition, we can theorise a number of feedback loops: the supply of ecosystem services itself can impact upon the drivers of change; the state of human well-being can affect ecosystem services via increased or decreased demand placed on these services; and human well-being is directly linked to the drivers of change.

As an example of the ecosystem service framework we can examine wetland ecosystems. Wetland ecosystems are areas such as swamps or peatlands. Wetland ecosystems are at the source of many water resource services. A major supply of renewable freshwater comes from inland wetland ecosystems. Many wetlands are responsible for water purification, treating and detoxifying a host of different waste products through their natural processes.

Against this background, the MEA approach can be used to categorise the wide range of ecosystem service benefits of wetland ecosystems to local, regional and international communities (Millennium Ecosystem Assessment, 2005).

BOX 10.2 CASE STUDIES

Tram Chim National Park, Vietnam

There are many examples of the monetisation of services provided by wetlands. One such example is that carried out for the Tram Chim National Park (Figure 10.5) in the Mekong River Delta in Vietnam, where the benefits generated by improvements in wetland conditions were estimated (Nan Do and Bennett, 2010). In 2012 Tram Chim was declared as a Ramsar wetland site (meaning it has international protected status), due to its biodiversity value. It consists of swamps and low-lying reed beds in an area frequently flooded by the Mekong. A 53 km dyke was built around the park to retain water during the dry season in 1985, which helped to restore the damage done during the Vietnam War. In 1996 the height of the dyke was raised further to prevent fires which have put the ecological system at risk. The site was designated as a National Park in 1999. To improve wetland diversity, the Park's management board proposed further changes to the dyke system and wetland management practices. Using an ecosystem services approach to account for changes in provisioning (e.g. clean water), regulating (e.g. reduced fire risk), cultural (e.g. tourism) and supporting (e.g. nutrient cycling) services, the estimated benefits of this programme ranged from US$0.15 million to US$ 0.96 million (Nan Do and Bennett, 2010). These results suggest that implementation of the improved management programme would improve social welfare as well as biodiversity (e.g. improved crane bird populations). Through this sort of exercise it is possible to support management decision making, as the economics of environmental management strategies can be established.

Figure 10.5 Part of Tram Chim National Park.
Source: Photo courtesy of Shutterstock.

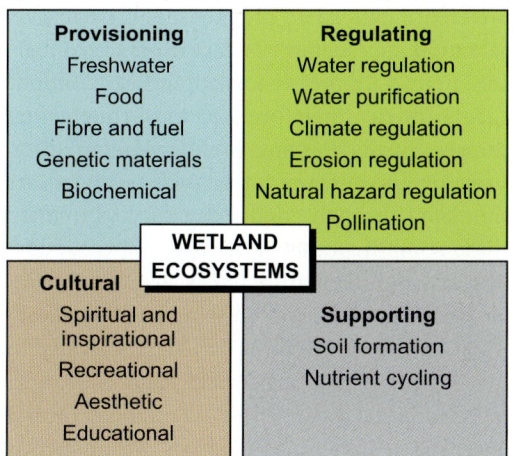

Figure 10.6 The goods and services provided by wetland ecosystems.

By identifying such services, economic tools can then enable us to trace the benefit flows from wetland ecosystems to human beneficiaries (see e.g. **Box 10.2**). **Figure 10.6** itemises some of the main services that wetland ecosystems can provide. As **Figures 10.6** and **10.7** show, water resources flowing from wetland ecosystems are considered as both provisioning and regulating ecosystem services. The concept of the total economic value of an environmental resource has given way to the concept of goods and services flowing from ecosystems upon which human well-being depends (**Figure 10.7**). However, these two concepts are related. Water resources are considered mainly in the provisioning and regulating categories of the MEA. Freshwater for human consumption represents a provisioning service of wetland ecosystems, which also coincides with direct use values and option values. Water regulation and purification are considered regulating services of wetland ecosystems, and these coincide with indirect use values and option values. Whichever methodological approach is adopted, the components that make up the value of the resource need to be identified before economic valuation can be undertaken.

> **REFLECTIVE QUESTION**
>
> What is the difference between use value and non-use value?

Group	Service	Direct use	Indirect use	Option	Non-use
Provisioning	Freshwater Food Fibre and fuel Biochemical	✓	NA	✓	NA
Regulation	Water regulation Water purification Climate regulation Erosion regulation Natural hazard regulation Pollination	NA	✓	✓	NA
Cultural	Spiritual and inspirational Recreational Aesthetic Educational	✓	NA	✓	✓
Habitat/supporting	Soil formation Nutrient cycling	Valued through other ES categories			

Figure 10.7 Total Economic Value versus ecosystem services of wetland ecosystems.

EVALUATION TECHNIQUES FOR WATER RESOURCES

There exists a methodological toolbox for the valuation of non-marketed environmental goods and services such as wetland ecosystems and water resources. Some of the techniques are more capable of revealing the values of some types of ecosystem goods and services than others. Furthermore, it is undeniable that, no matter the technique, some of these values themselves, in the context of human welfare, are by definition notoriously difficult to reveal. Whatever the outcome of the valuation exercise, it is now frequently the case that the resulting estimates are considered as baseline estimates only. Furthermore, the **precautionary principle** will dictate that, in the absence of rigorous economic values, and despite pressing immediate economic needs, policy should lean in the direction of conservation rather than exploitation.

There are three main approaches to economic valuation: direct market valuation, revealed preference and stated preference. Direct market valuation approaches use data from actual markets. In revealed preference approaches, economic agents 'reveal' their preferences through their choices. Finally, in stated preference approaches, a market and demand for ecosystem services are simulated through conducting surveys that include some hypothetical changes in ecosystem services originating from policy changes. A number of studies have used stated preference methods when market information has been unavailable. In converse situations, where such data exist, revealed preference methods are applied. Moreover, stated preference methods are frequently used to elicit non-use or cultural values, as these values do not have recorded behaviour, unlike use values, which can be better measured by revealed preference methods. These approaches are described below. **Figure 10.8** suggests a classification of economic values related to freshwater resources and the various market and non-market approaches used to capture the magnitude of the respective benefits.

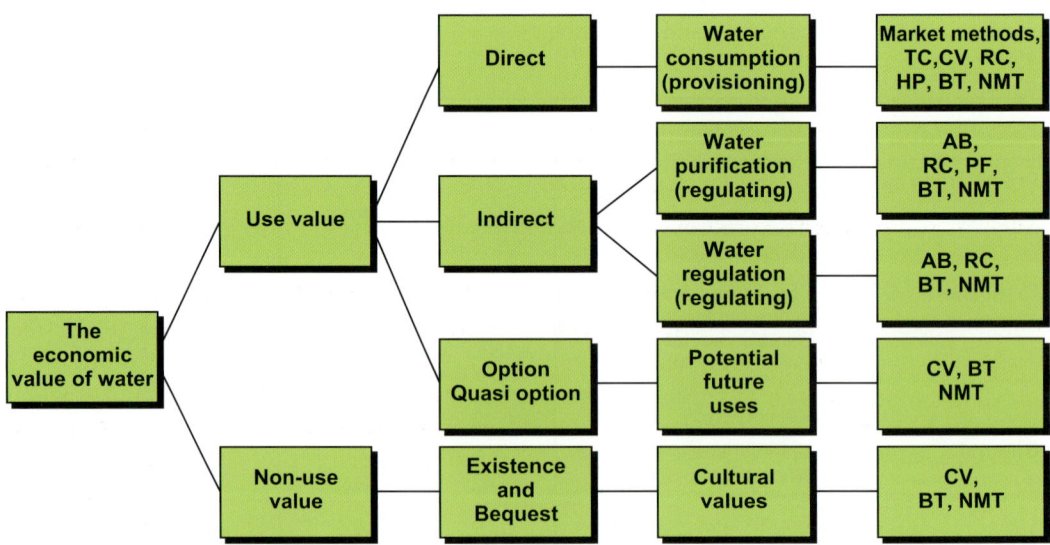

Figure 10.8 Economic valuation techniques for water resources. The initials in the boxes on the right-hand side are described in the main text.

1 Direct market valuation

A good part of the value of ecosystems is represented by commercial and financial gains and losses through the market that can be measured by market-based methods of economic valuation. Some of the market valuation approaches include: averting behaviour, replacement/restoration, the production factor method and dose–response. The averting behaviour (AB) (or 'preventive expenditure') technique measures the expenditure incurred in order to avert damage to the natural environment, human infrastructure or human health. AB should be seen as a minimum estimate of impact costs, since it does not measure the consumer surplus (i.e. the additional amount above actual expenditure that consumers would be willing to pay in order to protect a particular good or service). The replacement/restoration cost (RC) technique can be used to measure the costs incurred in restoring or replacing productive assets or restoring the natural environment or human health as a result of the impacts of environmental degradation. As with preventive expenditure, restoration costing is a relatively simple technique to use, and has the added advantage over preventive expenditure of being an objective valuation of an impact – i.e. the impact has occurred, or at least is known. Use of the replacement costs method relies on replacement or restoration measures being available and the costs of those measures being known. As such, the method is unlikely to be appropriate for costing the impacts on irreplaceable assets. Another shortcoming of the technique is that actual replacement or restoration costs do not necessarily bear any relationship to the willingness of individuals to pay to replace or restore something.

The production factor method estimates the economic value of an environmental commodity through an 'impact-pathway' approach, in which a change in the environmental attribute is linked to impacts on 'endpoints' that are relevant for human well-being. For example, the benefits of tree planting, through reduced erosion, are measured first by the link between soil cover and erosion rates and then by the link between erosion rates and agricultural productivity. Such methods can be very useful for valuing many services provided by ecosystems, including forestry (timber and non-timber), agriculture (value of diversity in crops and use of genetic material) and marine systems (losses from overfishing, species invasion).

2 Stated preference valuation

Market prices and costs can provide estimates of the increase in the value of commercial activities. However, some environmental goods and services do not affect markets, or market data are not available to value them. The development of non-market valuation has been closely linked with water projects and the development of cost-benefit analysis. Hanemann (2006) noted that the positive and negative environmental impacts of water projects have been frequently assessed using non-market valuation since the 1980s. In such cases methods have been developed to derive consumers' preferences. These are revealed preference and stated preference (SP) valuation methods.

Contingent valuation (CV) is currently the most-used stated preference technique for the valuation of environmental goods, where individuals state their *willingness to pay* or *willingness to accept* compensation for a good or service. This valuation is often done during survey work with individuals or groups. Only stated preference methods like CV can elicit the monetary valuation of the non-use values, which typically leave no 'behavioural market trace'. Furthermore, CV allows environmental changes to be valued even if they have not yet occurred. It allows the specification of hypothetical policy scenarios or states of nature that lie outside the current or past institutional arrangements or levels of provision. Finally, CV allows one to enrich the information base by submitting the process of value formation to public discussion. Against this is the criticism that the values are hypothetical (payments are not actually made or cash paid out), and so it might be that the method is subject to many biases. **Box 10.3** provides an important example of contingent valuation for water management in the USA.

The conjoint choice, choice experiment or attribute-based method (these are all the same) is also a commonly used stated preference method. This method elicits information on values by asking individuals to choose between alternatives. This can be done using conjoint ranking, where individuals rank alternatives in order of preference, and conjoint rating, which indicates their strength of preference on a **cardinal scale**. **Box 10.4** provides an example of a choice experiment used to measure farmers' minimum willingness to accept requirements under a possible local payment for an ecosystem services programme.

According to Bateman *et al.* (2002), choosing which of the two stated preference approaches to use depends on: the kind of value needed (i.e. total or relative), information availability (CV has a larger literature), welfare and/or welfare-consistent estimates, cognitive processing and sampling means (number of responses per individual). Freeman (2003) was 'cautiously optimistic' about the stated preference method and reported that others are attracted to stated preference methods because of the 'relatively easy and inexpensive way to get usable values for environmental resources'. In a similar vein, Whittington (2002) concluded that stated preference is vital to a developing country's policy application, but it is far from being a high-quality option at a low cost.

3 Revealed preference valuation

The revealed preference methods include hedonic pricing (HP), the travel cost (TC) method and benefit transfer (BT). HP estimates the economic value of an environmental commodity, such as clean water, clean air or an attractive view, by studying the relation between such attributes and house prices (Palmquist, 1991). Hedonic price estimation has been applied to elicit environmental/ecosystem values associated with recreation, landscape values and genetic and species diversity. Hedonic techniques are

---

BOX 10.3 CASE STUDIES

Economic effects of regulating flow releases from Glen Canyon Dam, USA

The Glen Canyon Dam is upstream from the Grand Canyon National Park and the Glen Canyon National Recreation Area. The hydropower operations in the dam were thought to be impacting upon fisheries and rafting, particularly as water was stored for times of peak energy demand rather than for good baseflow in the river network. The US Bureau of Reclamation decided to use contingent valuation to establish how much the recreation was worth and therefore to understand whether providing more baseflows would be worth more than the water stored for the energy required at peak times. Visitor surveys were conducted in the 1980s and results suggested that a change in practice might be worth $2 million per year (Bishop *et al.*, 1989). However, the study also led to a more nuanced outcome; rather than the debate being about recreation versus power generation, it was realised that flows could be optimised so as to maximise economic value across all uses and that society was also concerned with other impacts, such as native vegetation and birds, erosion and endangered fish. A willingness-to-pay survey was conducted for households across the USA which showed strong support for promoting a more natural flow regime for the river network, including large spring flows (Loomis, 1997).

BOX 10.4 CASE STUDIES

Water-quality supply in the UK uplands

A project was undertaken to understand whether farmers are willing to accept privately funded compensation payments for protecting water-quality supplies in the UK uplands. This study by Beharry-Borg *et al.* (2012) examined the feasibility of using payments financed by a water company to pay farmers to adjust agricultural land management practices to protect water quality. The choice experiment method was used to measure farmers' minimum willingness to accept requirements under a possible local payment for an ecosystem services programme to adjust agricultural land management practices in Nidderdale and the Washburn Valley (Yorkshire, UK). Choice data for the study were drawn from three geographical locations in the Nidderdale catchment in the north of England. These were Fearby, Middlesmoor, Pateley Bridge and the Washburn area.

Each land management attribute was chosen based on water-quality monitoring data and hydrological modelling results for Nidderdale, together with feedback on an early version of the questionnaire from a pilot group of farmers. The results of monitoring of water-quality indicators and hydrological modelling showed that changes in the input of nitrogen and ditch blocking were the land management practices that were expected to have the greatest impact on water quality in the Nidderdale area, affecting nitrate levels and discoloration (i.e. reducing the colour) of drinking water, respectively. The ditches are found in organic upland soils and peat bogs. The ditches result in a deepening of the water table and cause the organic matter to decompose, releasing brown dissolved organic carbon into water courses (**Figure 10.9**). Each land management attribute and its levels was chosen to satisfy scientific understanding of the impact of such measures on water quality, and also to represent a feasible range of land management changes that could be considered by farmers.

Some mathematical models were used to examine the data derived from the choice experiment with farmers. Multinomial logit and latent class models were used to analyse 97 farmers' responses to establish their preferences, quantify willingness to accept

Figure 10.9 Ditch through a peatland in Nidderdale, UK. These ditches have increased water discolouration downstream. Many of these ditches are now being blocked for multiple benefits.

continued

compensation (WTA) values and explain the presence of unobserved heterogeneity. Results from the multinomial logit model showed that, on the whole, farmers had significant aversions to reducing applications of artificial fertiliser and farmyard manure and, in general, the aversion to reducing farmyard manure application was stronger than the aversion to reducing applications of artificial fertiliser. Somewhat surprisingly, on average across all the farmers surveyed, there appeared to be no particular preference for or aversion towards blocking drains in the upland landscape, and no particular preference for or aversion towards the length of management agreement offered. This overall lack of preference for or aversion to drain blocking may arise because decision rights over drain blocking are typically held by large landowners, rather than by tenant farmers. The compensation payment analysis showed that, provided that adequate levels of compensation payments were offered, farmers could be encouraged to adopt land management practices that have the potential to enhance water quality. In other words, farmers could be encouraged to reduce applications of artificial fertiliser and farmyard manure and to block drains that had previously resulted in deterioration of water quality.

The findings suggested that there is likely to be considerable variation across farmers in their strength of preference for, or strength of aversion to, different elements of land management programmes which aim to protect drinking water quality. In addition, the results suggest that farmers can be encouraged to voluntarily adopt more environmentally sound land management practices and thereby enhance delivery of ecosystem services such as drinking water provision or water-quality regulation, through the use of incentive payments. However, these initial results provide only a broad indication of the levels of incentive payments necessary to achieve participation. The analysis tested could be used in a cost-effectiveness framework to address the specific costs of water treatment, or the predicted cost of implementing payment programmes to farmers to change their land management practices, to achieve specified requirements for water colour and nitrate levels at a sub-catchment level. Within such a framework private sector water companies would be able to use these estimates to consider a number of policy scenarios and estimate the costs of each, so as to identify their preferred policy option.

particularly employed in valuing visual amenity, quality of soil assets and exposure to air pollution. The TC method estimates the economic value of recreational sites by looking at the generalised TCs of visiting these sites; the valuation is then based on deriving a demand curve for the site in question, through the use of various economic and statistical models. Where the individual makes a choice involving more than one site, the discrete choice models have used the random utility theory framework to value not only visits to different sites but also the attributes of sites, such as water quality. The TC technique has been widely applied, especially in North America.

The method of BT consists of exporting previous benefit estimates (either from stated preference or revealed preference) from one site to another, at one point in time, with regard to the researcher's area of interest. In BT estimates there are three possible forms of transfers: transfer of an average of willingness to accept compensation estimates from one primary study, transfer of the willingness to accept compensation function, and transfer of willingness to accept compensation estimates by aggregating other estimates employing meta-analyses (Bateman et al., 2002). BT involves the use of economic values from one specific area with a known resource and policy conditions, and applying these to another site in similar circumstances. Generally, the first site is known as the 'study site' and the second as the 'policy site'. Sites differ in characteristics, and one has to be cautious when applying these from one site to another. On the one hand, this method reduces the cost of starting a completely new valuation study, whereas on the other hand, the compilation of a comprehensive database often proves costly.

4 Other techniques

There also exists a host of non-monetary techniques that have attempted to assess the importance of ecosystem services; a comprehensive discussion of these is provided by Christie *et al.* (2008). Such methods include consultative and participatory approaches such as focus groups, citizens' juries, health-based approaches, **Q-methodologies** and **Delphi surveys**. While these methods do not yield monetary indicators, as do the economic techniques outlined above, they can provide useful insights into how water resources are perceived and utilised. The possibility also exists that these non-economic methods can serve to complement the economic methods which can then, with added non-economic insights, yield more accurate, rigorous and robust monetary estimates.

> **REFLECTIVE QUESTION**
>
> What are the main valuation techniques for water?

F ECONOMIC MANAGEMENT OF WATER FOR THE PRESENT AND FUTURE

The essential nature of water to all life requires a coherent and consistent management of water resources at a variety of stakeholder and spatial scales, with trade-offs among competing wants and uses. This brings up the need for the creation, strengthening or maintenance of the relevant institutions and instruments in order to manage these different interests. Underlying all of this is the need for the recognition, demonstration and capture of values of water resources. The public-good and transboundary nature of water resources brings the critical need for water governance. In recognition of this growing need, there has been a contemporary shift towards diverse governance mechanisms in the water sector, with a focus on multiple stakeholders at different levels of spatial and political scale. In Western Europe in particular, regional governance is playing an increasing role in water governance and management (Lieberherr, 2011).

There exist several ways in which the government can intervene in the water sector (Chapter 11 discusses this issue in detail). By the provision of supply, public institutions hold the responsibility for water infrastructure and water pricing as well as the management of the upstream or downstream private sector institutions that may have developed along the water chain. Governments can apply taxes, grant subsidies and provide information (Dalhuisen *et al.*, 2000). They can also regulate or intervene in the demand chain by, for example, encouraging water conservation among residential users or imposing restrictions on industrial or agricultural activities that may impact upon water supply and water quality. An increasingly popular scheme for the management of ecosystem services across competing wants, using economic incentives, is that of payments for ecosystem services (e.g. **Box 10.3**). Much current work is dealing with establishing the design and potential operation of such payment systems from theoretical and empirical perspectives, considering the efficiency of alignment between payments and outcome, and estimating public willingness to pay for the ecosystem services so secured.

It is important to note explicitly that this type of economic management framework for water is by definition multi-disciplinary.

In particular, the disciplines of humanities, social sciences and natural sciences all play overlapping roles. Analyses of natural systems, traditionally linked to ecology and the natural sciences, can now be complemented by a coordinated multi-disciplinary approach including humanities and social science disciplines such as economics, sociology, anthropology and governance. Via such a holistic approach, it is possible to better undertake a coherent analysis of the impacts of changes in water resources on human welfare, within the context of sustainable livelihoods, and the institutional and policy settings within which such changes occur. Ultimately, research towards these goals will inform policy on a progression towards poverty reduction, sustainable development and the achievement of the Millennium Development Goals (UN Water, 2010).

> **REFLECTIVE QUESTION**
>
> Why is the economic management of water a complex interdisciplinary process?

G SUMMARY

Water availability can represent a significant constraint on the sustainable development of an economy. Water by its very nature is defined as a common property resource. Resource allocation issues, coupled with increasing levels of economic growth and populations that place water resources under increasing demands, lead to the need for improved water management as a critical element in the achievement of sustainable development. Value and price hardly equate in the context of environmental goods and services. To this end, economists have produced a methodology of Total Economic Value in which the use and non-use components of environmental goods and services can be disaggregated. The Millennium Ecosystem Assessment took this methodology one step further by categorising ecosystem goods and services into provisioning, regulating, cultural and supporting. No matter what the methodology used, there exist a range of economic valuation tools that can be used to monetise ecosystem goods and services. Valuation techniques can include market-based and non-market-based methods. In fact, the history of non-market valuation is linked to water sector evaluations. In terms of water resources, we can for each category of value or for each category of ecosystem services apply particular techniques to produce economic values. Non-monetary techniques are also becoming increasingly popular, in an effort to complement the range of available economic techniques. Once values are established, the work then turns to how these values can be embedded into the decision-making processes. This brings forth the crucial concept of 'water governance', where a contemporary shift has taken place towards diverse governance mechanisms in the water sector, with a focus on multiple actors at different levels of spatial and political scale. An increasingly popular scheme for the management of ecosystem services across competing wants, using economic incentives, is that of payments for ecosystem services. Ultimately, what is urgently required is a coherent analysis of the impacts of changes in water resources on human welfare, within the context of sustainable livelihoods and the institutional and policy settings within which such changes occur, with movements towards water security placing the world economy firmly on a path to poverty reduction, sustainable development and the achievement of the Millennium Development Goals.

FURTHER READING

Bates, B.C., Kundzewicz, Z.W., Wu, S. and Palutikof, J.P. (eds) 2008. *Climate change and water.* Technical Paper of the Intergovernmental Panel on Climate Change, IPCC Secretariat; Geneva.

This report of the IPCC examines in some detail the linkages between climate change and water resources, including detailed sectoral analyses and an overview of global and regional impacts. It suggests adaptation and mitigation strategies to climate change through water management strategies.

Hanemann, W.H. 2006. The economic conception of water. In Rogers, P.P., Llamas, M.R., and Martinez-Cortina, L. (eds) *Water crisis: myth or reality?* Taylor & Francis; London.

This paper presents the economists' view of water and how water can be managed from the economists' perspective. By appealing to various economic thoughts and concepts, the article provides a history of economic thinking from the perspective of water issues.

Millennium Ecosystem Assessment. 2005. *Ecosystems and human well-being: wetlands and water synthesis.* World Resources Institute; Washington, D.C.

This synthesis report of the Millennium Ecosystem Assessment for the Ramsar Convention uses the ecosystem services methodology and provides key findings on the range of ecosystem services provided by global wetlands (including water provision, regulation and purification). The report includes current status and trends, direct and indirect drivers of change, future scenarios, and impacts on human well-being.

Classic papers

Barbier, E.B., Acreman, M. and Knowler, D. 1997. *Economic valuation of wetlands: a guide for policy makers and planners.* Ramsar Convention Bureau; Switzerland.

This handbook was one of the seminal contributions to the economic valuation literature and wetlands analysis. Though this came before the Millennium Ecosystem Assessment and its methodologies on ecosystem services and human well-being, this handbook provides a clear discussion of the concept of Total Economic Value, the outputs, values and benefits associated with wetlands, and corresponding valuation techniques.

Dalhuisen, J.M., Groot, H.L.F. de and Nijkamp, P. 2000. The economics of water: a survey of issues. *International Journal of Development Planning Literature* 15: 3–20.

This paper surveys the central issues of the economics of water, including demand and supply characteristics, institutional capacities, pricing and rate structures of water, and provides a case for government intervention due to market failures.

PROJECT IDEAS

- Design a project to work out the economic value of water-quality improvements in a local lake. What factors would you need to consider and what data are required?

- Investigate whether the price you or your organisation (e.g. university) currently pay for your water is fair.

- Test the different techniques for valuing water in your local area and see whether they provide very different results.

REFERENCES

Barbier, E.B., Acreman, M. and Knowler, D. 1997. *Economic valuation of wetlands: a guide for policy makers and planners.* Ramsar Convention Bureau; Switzerland.

Bateman, I. *et al.* 2002. *A manual: economic valuation with stated preference techniques.* Edward Elgar Publishing; Massachusetts.

Bates, B.C., Kundzewicz, Z.W., Wu, S. and Palutikof, J.P. (eds). 2008. *Climate change and water.* Technical Paper of the Intergovernmental Panel on Climate Change, IPCC Secretariat; Geneva.

Beharry-Borg, N., Smart, J.C.R., Termansen, M. and Hubacek, K. 2012. Evaluating farmers' likely participation in a payment program for water quality protection in the UK uplands. *Regional Environmental Change*, doi: 10.1007/s10113-012-0282-9.

Bishop, R., Brown, C., Welsh M. and Boyle, K. 1989. Grand Canyon and Glen Canyon Dam operations: an economic evaluation. In Boyle, K. and Heekin, T. (eds) W-133, Benefits and costs in natural resources planning, Interim Report #2. Department of Agricultural and Resource Economics, University of Maine, Orono.

Christie, M. *et al.* 2008. *An evaluation of economic and non-economic techniques for assessing the importance of biodiversity to people in developing countries [online].* DEFRA; London. Available from: http://sciencesearch.defra.gov.uk/Document.aspx?Document=WC0709_7562_FRP.pdf.

Dalhuisen, J.M., Groot, H.L.F. de and Nijkamp, P. 2000. The economics of water: a survey of issues. *International Journal of Development Planning Literature* 15: 3–20.

Freeman, A.M. 2003. *The measurement of environmental and resource values: theory and methods (2nd edition).* Resources For the Future; Washington, DC.

Hanemann, W.H. 2006. The economic conception of water. In Rogers, P.P., Llamas, M.R. and Martinez-Cortina, L. (eds) *Water crisis: myth or reality?* Taylor & Francis; London.

Lieberherr, E. 2011. Regionalization and water governance: a case study of a Swiss wastewater utility. *Procedia Social and Behavioral Sciences* 14: 73–89.

Loomis, J.B. 1997. Use of non-market valuation studies in water resource management assessments. *Water Resources Update* 109: 5–9.

Millennium Ecosystem Assessment. 2005. *Ecosystems and human well-being: wetlands and water synthesis.* World Resources Institute; Washington, DC.

Nan Do, T. and Bennett, J. 2010. Using choice experiments to estimate wetland values in Viet Nam: implementation and practical issues. In Bennett, J. and Birol, E. (eds) *Choice experiments in developing countries, implementation, challenges and policy implementation.* Edward Elgar; Cheltenham.

Smith, A. 1776. *The wealth of nations: an inquiry into the nature and causes of the wealth of nations.* W. Strahan and T. Cadell; London.

TEEB. 2010. *Mainstreaming the economics of nature: a synthesis of the approach, conclusions and recommendations of TEEB.* United Nations Environment Programme.

United Nations. 1987. *Our common future.* Oxford University Press; Oxford.

CHAPTER ELEVEN

Water conflict, law and governance

Kitriphar Tongper and Anamika Barua

LEARNING OUTCOMES

After reading this chapter you should be able to:

■ distinguish between various types of water conflicts and their causes
■ understand what water rights are and how they are related to water access
■ outline major international laws and institutions dealing with transboundary water conflict and to what extent they have been successful in resolving conflicts.

A INTRODUCTION

Like many other resources that sustain and support life, water fit for human consumption and use has become a scarce resource even though it is supposedly abundant in supply. As long as water was abundantly available for use among people, there were no problems related to sharing or **property rights**. However, this resource, which is one of the most important in sustaining life, has become increasingly scarce and the issue of sharing water has become a very important one. Towards this end, various schemes and plans have been designed to ensure that everyone gets their fair share of water. These plans, however, are not always perfect and they often do not address a number of issues. The lack of proper sharing mechanisms has often been the cause of conflict situations related to water sharing and access. Since water resources, like rivers, go beyond national, political boundaries, the conflicts related to water sharing are often **transboundary,** involving two or more nations. As it is also a vital resource, when there is scarcity, conflict situations arise even within a country's boundaries.

This chapter deals with the types and causes of water conflicts, distinguishes between various types of water rights and how these rights determine access to water. Through various case studies, the chapter discusses the role of international treaties, laws and institutions in resolving water conflicts both within and between nations.

B WATER CONFLICTS

Water conflict is a term describing a conflict between countries, states or groups over an access to water resources. As pressure on freshwater supplies rises, due to population growth, economic development and pollution, access to water, and its allocation and use, are becoming increasingly critical concerns that can drive conflict and which may have profound consequences on societal stability (Haftendorn, 2000). While water has not been a major cause of violent conflicts historically, water-related tensions can emerge between and within states. They occur on three interdependent levels:

- the local level: e.g. between societal groups over access to a water point; or between the state and people affected by the construction of a dam;
- the national level: e.g. between different interest groups (farmers, industry, tourism, environmentalists) in relation to national policies affecting water management, for example, over the reallocation of water between economic sectors;
- the international level: e.g. between upstream and downstream states over the use of shared rivers.

Water-related tensions can occur when water is scarce, but even when the resource is not severely limited, its allocation and use can still be hotly contested. The coexistence of a variety of uses and users – such as agriculture, industry, different clans or ethnic groups, and rural and urban users – increases the likelihood of conflicting interests over water. These can drive or exacerbate existing threats of conflict on intra- and inter-state levels. Many regions around the world deal with shortages of water. However, some areas deal more with conflicts over inadequate water supplies and disputes over shared water supplies. In regions where countries compete for access to water, the relations between countries are likely to be unstable. Over 200 bodies of water are shared by two or more countries or areas. In the absence of strong institutions and agreements, changes within a basin can lead to transboundary tensions. Strife over water is plaguing states, including those within the Middle East, Eastern Europe and South East Asia (Richardson, 2010). Notably, these are regions where the amount of water available per person within several countries is 'stressed' ($< 1000m^3$ per person per year) or 'scarce' ($< 500 m^3$ per person per year (see Chapter 1 and **Figure 1.7**)). Although conflicts arise mainly because of the use of common water resources, a distinction can be made between conflicts arising through **consumptive water use,** pollution and distribution. **Table 11.1** describes the causes of conflicts.

1 Conflict through use

A conflict through consumptive use could be found in a situation where one state using the river clashes with another state, citing environmental concern or loss of water due to the other state's activities. **Box 11.1** provides an example.

Table 11.1 Causes of conflicts

Conflict type	Conflict through consumptive use	Conflict through pollution	Relative distribution conflict	Absolute distribution conflict
Conflict cause	Water use	Water quality	Water distribution	Water distribution and availability
Examples	Cauvery water dispute, India (See **Box 11.1**)	River Rhine	Ganga basin	Jordan basin

Source: After Haftendorn, 2000.

> **BOX 11.1 CASE STUDIES**
>
> ### Conflict through consumptive use: Cauvery water dispute, India
>
> Cauvery is the name of the 802 km-long river which originates from Talacauvery in Kodagu district in the southern Indian state of Karnataka and flows mainly through Karnataka and another state, Tamil Nadu. Its basin also covers areas in the other southern state of Kerala, and some areas of the union territory of Pondicherry. Both Karnataka and Tamil Nadu are heavily dependent on the river to meet all their water needs, especially agricultural needs. The use of the river waters was regulated by agreements dating back to 1892 and 1924, when India was a colony of the British Empire and the two states of Karnataka and Tamil Nadu were known as the Princely state of Mysore and the Madras Presidency, respectively. The River Cauvery is dependent on rains, and during periods of heavy monsoon rains there is excessive flow of water, even leading to floods. But when the monsoon fails there is severe shortage of water, leading to a drought situation: the water needs of farmers are not met and tension and conflict arises in the two states.
>
> According to the 1892 and 1924 agreements the approximate river water allotments were 75% to Tamil Nadu and Pondicherry, 23% to Karnataka, and the rest to Kerala. The state of Karnataka has always indicated that it feels it did not get its due share of water in comparison with Tamil Nadu, and it demanded a renegotiation of the settlement. Tamil Nadu, on the other hand, pleaded that it has already developed 12,000 km^2 of land and depends on the existing pattern of water usage.
>
> On 2 June 1990 a tribunal known as the Cauvery Water Disputes Tribunal (CWDT) was constituted by the Government of India to adjudicate the water dispute regarding the inter-state River Cauvery and its river valley. The Tribunal gave an interim award on 25 June 1991 of 5.8 billion m^3, which Karnataka had to ensure reached Tamil Nadu in a water year. The award also stipulated the weekly and monthly flows to be ensured by Karnataka for each month of the water year. The tribunal further directed Karnataka not to increase its irrigated land area. Violence followed the interim order and Karnataka witnessed the worst-ever anti-Tamil riots. In 1992, 1993 and 1994 the rain was sufficient to pacify the dispute between the two states. But in 1995 the monsoons failed badly in Karnataka and the state found itself unable to fulfil the interim order (Rahman, 2006). The resulting problem was solved only when the Supreme Court recommended the intervention of the Prime Minister to provide a political solution.
>
> The CWDT submitted its reports and decision under Section 5 (2) of the Inter-State River Water Disputes Act, 1956 on 5 February 2007, allocating 11.9 billion m^3 of water annually to Tamil Nadu, 7.6 billion m^3 to Karnataka, 0.8 billion m^3 to Kerala and 0.2 billion m^3 to the Union Territory of Pondicherry (MoWR, 2007). The tribunal also reserved 0.3 billion m^3 for environmental protection and 0.1 billion m^3 for inevitable escape into the sea. However, all four parties to the dispute were not satisfied with the verdict and they have decided to file review petitions seeking a possible renegotiation.

Most frequently, activities such as the construction of a dam or the channelling of the river flow leads to international conflict.

2 Conflict through pollution

The possibility of conflict through pollution arises when industry, urbanisation or mechanised agriculture in the upstream area leads to harmful consequences for water quality for the lower-lying states. An example of the conflicts on the Rhine from the 1970s to 1990s is provided in **Box 11.2**.

3 Relative distribution conflict

A relative conflict of distribution presents itself where a disparity over the use of water exists between upper- and lower-lying states. There are many examples of this type of conflict around the world, often on large rivers (e.g. see Chapter 3). One of the world's largest rivers is the Ganges and this is used as an example of a relative distribution conflict in **Box 11.3**.

BOX 11.2 CASE STUDIES

Conflict through pollution: the Rhine

Although the Rhine is a relatively small river, it is one of Europe's best-known and most important rivers. Its length is 1,320 km, of which 880 km is navigable. From its source in Switzerland, the Rhine flows via France, Germany and the Netherlands into the North Sea. The catchment area of 170,000 km^2 also covers parts of Italy, Austria, Liechtenstein, Luxembourg and Belgium. Hence, the Rhine is a river whose drainage area falls in a rich, highly industrialised area of Europe. Furthermore, the interests of the upper- and lower-lying areas around the river are particularly distinct, as a result of their different uses of the Rhine. The primary cause of pollution of the Rhine is from the chemical industry in the upper-lying states of Switzerland and Germany, as well as from French potassium mines in Alsace and the German coal works in the Ruhr and Lippe (Frijters and Leentvaar, 2002). Many of the costs of this pollution are borne by the lower-lying states such as the Netherlands, whose primary use of the Rhine is for drinking water and agriculture. Rising toxic sediment in the Rhine forced costs upon the city of Rotterdam (in the Netherlands), which had to remove the toxic mud from Rotterdam harbour to special waste depots.

The conflicting interests of the upper- and lower-lying parties along the Rhine prevented a quick solution. In the late 1960s and early 1970s, the Netherlands, being one of the most negatively affected lower-lying states, called for continued measures to combat the increasing pollution. In 1976 the environment ministers of the Rhine states signed two agreements for the protection of the Rhine against chemical pollution and the reduction of salt loads. Under the Chemical Agreement a financing proposal was put forward with regard to the clean-up costs, whereby France and Germany assumed 30% each, the Netherlands 34% and Switzerland 6% of the costs. In exchange, the Netherlands agreed not to sue Germany and Switzerland. However, in response to local resistance to the agreed measures, France declined to ratify the Chemical Agreement in 1979. It was only after the 1983 elections and a change in the French government that the process of ratification was once again taken up. An additional catalyst, a series of chemical accidents on the Rhine in 1986, increased public awareness (Haftendorn, 2000), and in 1987 the Chemical Agreement finally came into force. In 1991 it was replaced by a second agreement calling for an even greater reduction in salt levels.

BOX 11.3 CASE STUDIES

Relative distribution conflict: the Ganges basin

The Ganges basin conflict is primarily a conflict of water sharing between India and Bangladesh. It can also be characterised as a conflict of interests between upstream and downstream river use. India, which manages the upper part of the drainage basin, developed plans for water diversions for its own irrigation, navigability, and water-supply interests. Bangladesh, however, had interests in protecting the historic flow of the river for its own downstream uses. The potential clash between upstream development and downstream historic use set the stage for very long-drawn-out attempts at conflict management (Priscoli and Wolf, 2009).

The attempts at conflict management started from the time when Bangladesh was a part of Pakistan and was known as East Pakistan. In 1960 an expert-level meeting was held between India and Pakistan, followed by three more meetings in 1962. While the meetings were going on, India informed Pakistan that it had started the construction of the Farakka Barrage (Figure 11.1), designed to divert 1100 $m^3 s^{-1}$ from the Ganges basin into the Bhagirathi-Hooghly River during the dry season to flush out the accumulating silt, which would improve navigability for Calcutta city port and keep the city's water supply low in salts. The meetings continued with not much headway being made and the Farakka Barrage was completed in 1970 (but not yet in operation). Bangladesh came into being in 1971, after gaining independence from Pakistan. The governments of India and Bangladesh had agreed to establish the Indo-Bangladesh Joint Rivers Commission, but the question of the Ganges was excluded and was supposed to be handled only at the level of the Prime Ministers of the two countries.

The Prime Ministers of India and Bangladesh met in New Delhi in 1974 and made a declaration that during the period of minimum flow in the Ganges, there might not be enough water for both an Indian diversion and Bangladeshi needs. They also expressed their determination to work for a mutually

Figure 11.1 The Farakka Barrage.
Source: Photo by Ritesh Maity, Calcutta, India.

continued

acceptable allocation of the water available during the periods of minimum flow before the opening of the Farakka Barrage. The problems, however, persisted and there were a series of five commission meetings between the two countries from June 1974 to January 1975, with one ministerial-level meeting in April 1975. In 1975 the two sides agreed to a limited trial operation of the barrage. However, without renewing or negotiating a new agreement with Bangladesh, India continued to divert the Ganges waters at Farakka after the trial run. This had serious consequences for Bangladesh and there were setbacks to agriculture, fisheries, navigation and industry. There were more meetings, but the conflict was unresolved and in January 1976 Bangladesh lodged a formal protest against India with the General Assembly of the United Nations. The United Nations encouraged the parties to meet urgently for negotiations and on 18 April 1977 an understanding was reached on fundamental issues, followed by the signing of the Ganges Water Agreement on 5 November 1977. This agreement broadly covered the sharing of the waters of the Ganges at Farakka and the finding of a long-term solution for augmentation of the dry season flows of the Ganges (Priscoli and Wolf, 2009).

The agreement was to cover a period of five years and could be extended by mutual agreement, but no solution had been reached by the end of the five-year period. Following this, both sides agreed not to extend the 1977 agreement, but to initiate fresh attempts to arrive at a solution within 18 months. This was followed by an Indo-Bangladesh Memorandum of Understanding, signed on 22 November 1985, on the sharing of the Ganges dry season flow through 1988 and establishing a Joint Committee of Experts to help resolve development issues. On 12 December 1996 a new treaty was signed by the Prime Ministers of the two states and this agreement was based generally on the 1985 accord, which delineates a flow regime under varying conditions. The agreement also established a 30-year water-sharing arrangement and recognised the rights of Bangladesh as a downstream river user.

4 Absolute distribution conflict

An absolute conflict of distribution exists when there simply is not enough water to meet all the legitimate needs of the contesting states sharing a water resource or resources. **Box 11.4** provides a case study.

5 Future conflicts

As water becomes increasingly scarce, conflicts are expected to arise. Firstly, there may be local conflicts arising from competition among users over access to a water point. Secondly, there will be sectoral conflicts where industry, agriculture and citizens are in competition for larger volumes of the resource on a much wider level. Thirdly, there may be more international conflicts between countries over the use of shared rivers and water resources. The distribution of water resources throughout the world is increasingly becoming a political issue, and a lack of understanding among countries over what constitutes an equitable and fair share of available water resources and supply is a major problem for the future.

REFLECTIVE QUESTION

Are there any water conflict examples that are local to you – if so, what types of conflict are they?

> **BOX 11.4 CASE STUDIES**
>
> ### Absolute distribution conflict: the Jordan basin
>
> The River Jordan supplies Israel and Jordan with the vast majority of their water. As the population of Israel grew, the demand for water and reliance on the Jordan River grew. In the early 1950s Israel created a system called the National Water Carrier to transport water from the River Jordan, which was not well received by the Arab Nations. In 1955, Syrian artillery units opened fire on the Israeli construction team. In an attempt to settle the water dispute, US President Eisenhower appointed Eric Johnston as mediator. Negotiations between Arab states and Israel on regional water-sharing agreements continued for more than two years, with no success beyond a cease-fire.
>
> In response to the completion of the Israeli National Water Carrier, in 1964 the Arab League attempted to divert two tributaries of the Jordan – the Banias and the Hasbani – to the north. Israel in turn prevented this by military action. Meanwhile, with international financial aid, Jordan built a side canal on the eastern Jordan shores, the Abdullah Canal, for irrigation of cultivated land in northwestern Jordan and to satisfy the water needs of Amman. In order to guarantee a continuous water flow, two dams were proposed for the Yarmuk River, the most important tributary of the Jordan and which forms the border with Syria. One of these was completed in 1967 and subsequently destroyed by Israeli in the Six-Day War. The Six-Day War of 1967 changed the entire water scenario in the region (Haftendorn, 2000). Israel now controlled the Jordan waters, along with unlimited access to the underwater reservoirs of west Jordan.
>
> With the 1994 Israeli–Jordanian Peace Agreement, the division of the transboundary waters was contractually defined for the first time. According to the agreement, Jordan received the right to use the largest part of the Yarmuk, up to 45 million m^3 yr^{-1}, and Israel the largest part of the River Jordan discharge, up to 40 million m^3 yr^{-1}. To this end, the two parties declared their intention to jointly construct a dam on the Yarmuk River on the Israeli–Jordanian–Syrian border, in spite of Syrian objections. In addition, a joint water commission was established. However, to date there has been no agreement with Syria.

C RIGHTS TO ACCESS WATER

Having seen that conflicts arise over water, it is only natural to assume that such conflicts sometimes occur over claims and counter-claims over the right to use available water. This brings into focus the issue of rights over water in different situations and at different times. The term 'water right' is not easy to define, due to the fact that there is no universally agreed definition. The term is used to mean different things in different contexts and different jurisdictions. In its simplest conception, a water right is frequently understood to be a legal right to abstract and use a quantity of water from a natural source such as a river, stream or aquifer.

1 History of water rights

A whole body of legal arrangements has grown up over many generations that is specifically related to the use of and right to use water. The history of water rights can be traced back to Roman law, one of the elements of which was to confer a privileged position on the owners of land adjacent to watercourses. This, in turn, had a major influence on conceptions of water rights in the influential European legal tradition, prior to the introduction of modern water rights regimes. Indeed some of these influences can still be observed. For example, Roman law denied the possibility of private ownership of running water. The Institutes of Justinian, published in AD 533–534, held that running water was a part of the 'negative community' of things that could not be owned, along with air, the seas and wildlife. At the same time, it was recognised that things in the negative community could be used and that the 'usufruct' or right to use the advantage of the resource needed to be regulated so as to provide order and prevent over-exploitation.

Roman law also distinguished the more important, perennial streams and rivers from the less important seasonal water bodies. The former were considered to be common or public, while the latter were private. The right to use a public stream or river was open to all those who had access to them. Roman law, however, recognised the right of the government to prohibit the use of any public water and required an authorisation for taking water from navigable rivers.

2 Water rights in modern society

Societies in the past have devised their own ways of dealing with the issue of allocating water resources for use among their members, and these formed a system of water rights applicable within particular societies. Remnants of these systems can still be found in customary practices which still wield some influence in the allocation of water in developing countries. Today, different countries have devised different arrangements related to water rights, and over the years these systems have become increasingly complex. Issues such as population growth creating scarcity, pollution and so on have contributed to the complexity of legal systems and arrangements in different countries, and modern governments are increasingly challenged to make the allocation and apportioning of water usage fair and equitable. It has also been observed that European conceptions of water and water law have strongly influenced the development of formal water laws around the world, through the two principal European legal traditions: the civil law and the common law (Hodgson, 2006).

The civil law tradition, which is sometimes referred to as the Romano-Germanic family, is found in most European countries (including the former socialist countries of Eastern and Central Europe), nearly all countries in Latin America, large parts of Africa, Indonesia and Japan, as well as the countries of the former Soviet Union (Hodgson, 2006). The common law tradition emerged from the law of England. Countries in which the common law tradition applies include Australia, Canada, India, New Zealand, Pakistan, Singapore and the United States of America, and the remaining African countries that are not in the civil law tradition, as well as other Commonwealth countries and a number of countries in the Middle East.

3 Types of water rights

There are fundamental differences between the nature and source of water rights in different countries. In some cases water rights are linked to landownership; this means that those who buy land are assured of having rights over the water resources available there. But there are systems where the rights to water and other resources are separate from the right of land ownership. Generally, there are two kinds of rights related to water: *riparian water rights* and *prior appropriation water rights*. Riparian rights refer to the system of allocating water among those who own the land at or near the source of the water. The origins of the system of riparian rights can be traced back to the English common law. The prior appropriation water rights, on the other hand, refer to the system of allocating water rights from a source of water on a first-use basis. Thus, prior appropriation rights are not linked to ownership of the land over which the water sources are present. Instead, the first person to use the water from a water source for a beneficial use will have the right to continue to use that quantity of water for that purpose. Other users may use the remaining water for their own beneficial purpose, provided that they do not impinge on the rights of previous users. Both types of water rights exist in the USA: riparian rights in the eastern states and prior appropriation rights in the western states.

Riparian rights are property rights that have come to be regarded as inherent in a riparian parcel of land. A simple definition of a riparian parcel of land is the land which borders a natural body of water. Ownership of a riparian parcel of land means that a person owns the land, the beach, the bottom land covered by water, the right to fish, hunt, swim and boat on the surface of the entire body of water, in common with all other riparian property owners. Riparian rights also include the right to make reasonable use of the water. But it is also implied that if the available water is not enough to meet the needs of all users, allotments are made in proportion to frontage on the water source (this is known as *correlative rights*). Riparian rights cannot be sold or transferred other than with the adjoining land.

The doctrine of prior appropriation was developed in the nineteenth century to serve the practical demands of water users in the western United States. It originated in the customs of miners on federal public lands who accorded the best rights to those who first used water, just as they had accorded mining rights to those who first located ore deposits. The doctrine of prior appropriation was later extended to include farmers and other users, even on private lands. The flexibility of the common law tradition is such that this new, more suitable water rights doctrine was accepted as the law in a number of states in the United States (Hodgson, 2006). All of the states in which the doctrine of prior appropriation applies have statutory administrative procedures to provide an orderly method for appropriating water and regulating established water rights (Getches, 1997). The doctrine of prior appropriation clearly indicated that water rights and land rights were different. Water rights under this doctrine are acquired on the basis of beneficial use rather than landownership, and those rights continue as long as the beneficial use is maintained. The right of individuals to use water under the prior appropriation system is based on the application of a quantity of water to a beneficial use.

Groundwater is managed in a similar way to surface water both by appropriation rights and by the principle of riparian rights (typically using correlative rights), in different places (e.g. different states of the USA) to restrict abstraction. However, good groundwater data are required on flow directions, recharge rates and so on (see Chapter 5) in order to support the legal framework and resolve or avoid disputes. Under the doctrine of appropriation, expansion of groundwater pumping operations would be illegal, as it could adversely affect either others who use the groundwater or, as groundwater often supplies river baseflow, other surface water users.

In some places water rights are not clearly defined. India, for example, does not have any specific law defining ownership and rights over water sources. The rights are derived from several pieces of legislation and customary practices and beliefs. In India water is a state-based subject rather than a national one and states have the exclusive power to regulate water supplies, irrigation and canals, drainage and embankments, water storage, hydropower and fisheries. There are nevertheless restrictions with regard to the use of inter-state rivers. The central government is entitled to legislate on certain issues. These include shipping and navigation on national waterways, as well as powers to regulate the use of tidal and territorial waters. The Indian constitution also provides that the Centre can legislate with regard to adjudication of inter-state water disputes.

4 Water allocation

Following the discussion of water rights, it will be understood that conditions leading to the increasing scarcity of water bring into focus another important issue, that of water allocation, or the manner in which water is distributed to those who have a right to it. Water allocation has been defined by Burchi and Andrea (2003) as 'the function of assigning water from a given source to a given user or number of users for abstracting it and applying it to a given use'. They also pointed out that within a system, where the state is responsible for a country's water resources, the decision of who should abstract water and for what use rests with a public authority.

The way in which water is allocated is no longer as simple as it was in early human societies. The term 'water use' today does not only refer to abstraction of water for basic human needs, but extends to irrigation for agricultural purposes, industrial uses, hydroelectricity and so on. Modern civilisation has seen levels of development that have never been experienced before. Global population is close to 7 billion. To compound the issue of scarcity there is the additional problem of pollution, which has also become a global menace (see Chapters 4 and 5). Pollution has (to a very large extent) had an impact on the availability of clean sources of water suitable for human use.

As the resource comes under greater pressure, better institutions are needed to allocate and reallocate water between uses and users. Both government agencies and user groups dealing with water have tended to be sectorally defined, focusing exclusively on irrigation, or domestic water supply, or environment. Even within the irrigation sector, allocation between irrigation systems has often been unclear. As long as there is surplus water in the river basin, this is not problematic, as different users do not interfere with each other. But as overall water use increases, there is interference, not only between surface abstractions along a river, but also between surface and groundwater use (Bruns and Meinzen-Dick, 2000). The allocation and use of water resources today falls under the control of governments. Governments in most countries today have a specific department monitoring water resources.

However, the manner in which water resources are allocated by governments differs from one country to another and there is no uniform method applicable to all countries. This creates challenges for social justice. According to Meinzen-Dick and Mendoza (1996) and Meinzen-Dick and Rosegrant (1997), institutional arrangements for water allocation, and particularly for reallocation, can be grouped into three broad categories. Firstly, in *user-based allocation*, water users join together to coordinate their actions, managing water resources as a form of common property. Secondly, in *agency allocation*, water is treated as public property, with government agencies assuming authority for directing who does and does not receive water in accordance with policies and procedures. Thirdly, in *market allocation*, which corresponds with private property held by individuals or organisations, water may be allocated and reallocated through private transactions, with users trading water through short- or long-term agreements, and reallocating rights in response to prices. These three forms of water allocation institutions may be combined in various ways at different locations and levels of water management.

5 Water and social justice

Water interacts with people in several ways – as a vital resource for life, agriculture, sanitation, health and overall well-being. In spite of the fact that water plays a significant role in people's lives, figures suggests that at least 1.1 billion people across the world are living without access to safe, clean drinking water and over 2 billion without access to adequate sanitation (WHO, 2004). Over 2 million people die each year from water-related diseases, and there is a much bigger annual loss of meaningful productive healthy years – perhaps 80 million work years are lost each year (McDonald *et al.*, 2011). Hence, management of water resources has critical implications for people's lives and livelihoods, for overall economic development and for social prosperity (Rasul and Choudhury, 2010). Social theories of justice, equity and fairness underscore the need to ensure social justice in water resource management (Syme *et al.*, 1999; Tisdell, 2003). Social justice and water refers to fairness of access to water resources and the equality of burden from poor water quality and water hazards.

Competing water needs have triggered conflicts between disparate water users, such as commercial and subsistence, rich and the poor. The rural poor, who depend heavily on public water bodies such as rivers, streams, khals (natural channels of water) and beels (permanent backwater lakes in the floodplain) for sustenance and well-being, have often been marginalised and even deprived of access to public water bodies in the process of water resource development, such as the construction of dams, irrigation canals and flood control structures (Phansalkar, 2007). Competing needs have also been a bone of contention between different sectors and different regions, such as domestic and agriculture, agriculture and industry, agriculture and fisheries, upstream and downstream, rural and urban areas, and fisheries and flood control.

The movement towards privatisation of water, which is gaining momentum worldwide, may make water access much more difficult and unaffordable for poor people in developing countries if it is not governed properly. On the other hand, it may work very well if a fair pricing structure is in place (see section C3 in Chapter 10). Proponents say that such a system is the only way to distribute water to the world's thirsty. However, experience in some locations shows that selling water on the open market does not address the needs of poor, thirsty people. On the contrary, privatised water is sometimes delivered only to those who can pay for it, such as wealthy cities and individuals and water-intensive industries, like agriculture and high-tech. Therefore, privatisation has to be associated with social justice so that we can avoid an outcome which suggests that those who can afford to pay for water are assured of a supply of clean and safe water, while those who cannot afford to do so have to manage with whatever is available. Such a situation would obviously go against the principle of social justice, which stands for the equitable treatment of all citizens.

On 28 July 2010 the United Nations General Assembly voted overwhelmingly to adopt a resolution recognising the human right to drinking water and sanitation; 122 countries voted in favour of the resolution, none opposed and 41 abstained. The General Assembly also voted to call on member states to provide financial resources and technology to help realise this right in poorer countries. This was lauded as a historic landmark in the fight for water justice, and while the resolution itself was not binding, it demonstrated the intent of the General Assembly. It is expected that this resolution will serve as a basis for a more binding declaration or convention in the near future, which will ensure that the right to water becomes a human right, for which each and every government is responsible.

Recent EU legislation known as the Water Framework Directive places emphasis on environmental protection and makes it illegal for land managers or water users to deteriorate the ecological status of water bodies. Efforts must be made to improve the status of water bodies towards 'good ecological status', and groundwater use must be sustainable. This means that a river basin approach is required to water management, and water use across the entire basin needs to be considered so that rivers, lakes and estuaries are protected for the sake of the environment. Thus, water allocation for the environmental good must now be considered, monitored and acted upon within EU member states and there must be clearly delineated responsibilities and regulatory agencies within each state (see **Box 6.3** in Chapter 6 for further details).

> **REFLECTIVE QUESTION**
>
> What types of water rights and water allocations are there?

D WATER GOVERNANCE

'Water governance refers to the range of political, social, economic and administrative systems that are in place to develop and manage water resources, and the delivery of water services at different levels' (Rogers and Hall, 2003). As stated by the United Nations (2006), water governance has four dimensions:

- a social dimension concerned with 'equitable use'
- an economic dimension concerned with 'efficient use'
- an environmental dimension concerned with 'sustainable use'
- a political dimension concerned with 'equal democratic opportunities'.

Each of these dimensions is 'anchored in governance systems across three levels: government, civil society and the private sector'. To realise 'effective governance', the UN report *Water: a shared responsibility* (2006) proposes a checklist that includes the following:

- participation
- transparency
- equity
- effectiveness and efficiency
- rule of law
- accountability
- coherency
- responsiveness
- integration
- ethical considerations.

The absence of some or all of these practices has resulted in 'bad' or 'poor' governance, a simple definition of which is the inability and/or unwillingness to alter patterns of resource allocation, use and management, despite clear evidence of resource degradation, uneconomic behaviour and abiding poverty and social inequality (UN, 2006).

An effective water governance framework that would be universally applicable is very difficult to conceptualise, since each and every country has its own unique set of issues and problems related to water allocation and distribution. However, the generally accepted principle is that there should be a healthy balance between, policy, law and institutions, which are considered to be the main components of effective water governance. Whatever its policies towards water management, a country needs to develop each of these areas – policy, law, institutions, regulations, contracts and compliance – in order to have effective water governance. The establishment of water governance capacity may follow different patterns in different countries. Not every country pursues the same sequence in terms of adopting policies on water, enacting the laws to realise the policies and establishing the institutions to implement the law (Isa and Stein, 2009).

The use of the word 'policy' here refers to a government's plan to address the issue of water sharing and use within its boundaries. This plan is usually reflected in legislation and executive orders. Law is an essential component of effective water governance because when a legal system is created it formalises the processes through law. It is also very important that institutions should be set up to carry out the laws that are created. Institutions may exist at the local, national and even international levels. When a balance is struck between policy, law and institutions, the establishment of an effective water governance capacity is a very strong possibility.

The case of South Africa is relevant to understanding how a process of policy and legal reform contributed to the progressive development of a country's water governance capacity. In 1994, the Minister of Water Affairs and Forestry appointed a Policy and Strategy Team that reviewed all South African water laws and published a document entitled 'You and Your Water Rights'. This document sought to assist the public in making meaningful contributions to policy development, and set out the main principles and provisions of the existing legal structure and also contextualised these against their origin and historical development (Isa and Stein, 2009). As a result of the call for a public response to the review, simple and concise statements were published for comment which would constitute a framework for the development of a new, detailed policy and a new national law. The 1997 White Paper which followed confirmed that water is an indivisible resource and a national asset, and abolished the system of riparian rights through which water ownership was tied to landownership along rivers. It recognised water to meet basic human needs and maintain ecosystems as a right. It also recognised the authority of the country to prioritise water uses to meet the requirements of neighbouring countries and promoted an integrated system for managing water quality, quantity and supply. These principles became law in subsequent years and a National Water Act included the following: the national strategy, a classification of water and reserves, a mechanism for allocating and regulating water uses through a licensing system, a system of prices, catchment and management agencies, and water users' associations.

> **REFLECTIVE QUESTION**
>
> Do you feel that water governance in your area has all the elements of the UN checklist?

E INTERNATIONAL INSTITUTIONS AND LAWS TO RESOLVE WATER CONFLICT

'International water law' (also known as international watercourse law or international law of water resources) is a term used to identify those legal rules that regulate the use of water resources shared by two or more countries. The primary role of international water law is to determine a state's entitlement to the benefits of the watercourse (substantive rules) and to establish certain requirements for states' behaviour while developing the resource (procedural rules) (Vinogradov *et al.*, 2003). What distinguishes international law from domestic law is that the former is both created and enforced by states (at the international level) primarily in order to regulate inter-state relations in various areas of human activity, while the latter involves matters within a state's borders (Cosgrove, 2003). International law offers a series of means to resolve international disputes, both diplomatic (negotiations, consultation, good offices, mediation, fact-finding, inquiry, conciliation and the use of joint bodies and institutions) and legal (arbitration and adjudication). Generally, water conflicts are settled through negotiations, with an agreement as the final outcome (Cosgrove, 2003). Few water laws are drafted to deal with all future water conflicts, and there are some water regulations which are drafted specifically to deal with a particular conflict. Two examples of general water regulations are the 1992 Helsinki Convention (**Box 11.5**) which seeks to protect the Baltic Sea and the 1997 United Nations Convention on the Law of the Non-Navigational Uses of International Water Courses (see **Box 11.5**). The aim of both of these conventions is to establish general principles for the use of transboundary water resources.

1 International treaties

Most transboundary water resources are subject to a treaty regime. Around 3,500 international agreements govern the use of most of the world's shared waters. Some of the agreements are watercourse-specific (e.g. the Columbia River Treaty, see **Box 11.5**), some are boundary agreements (e.g. the 1909 Canada–United States Boundary Waters Treaty, see **Box 11.5**), and some are umbrella agreements regulating all regional waters. For example, the Convention on the Protection and Use of Transboundary and International Lakes, also known as the Water

> **BOX 11.5 CASE STUDIES**
>
> ## Examples of international water agreements
>
> *1. Helsinki Convention on the Protection of the Marine Environment of the Baltic Sea Area (1992 Helsinki Convention)*
>
> The 1992 Helsinki Convention came into force on 17 January 2000. The governing body of the Convention is the Helsinki Commission – Baltic Marine Environment Protection Commission – also known as HELCOM. The present contracting parties to HELCOM are all nine Baltic Sea riparian states: Denmark, Estonia, Finland, Germany, Latvia, Lithuania, Poland, Russia and Sweden, plus the EU (BFN, 2008). While the earlier 1974 Helsinki Convention was concerned primarily with issues of technical pollution control and the pollution of the Baltic Sea, the 1992 Convention, which still operates, is concerned with the entire marine environment of the Baltic Sea area. Its purpose is to prevent and eliminate pollution in order to promote the ecological restoration of the Baltic Sea area and the preservation of its ecological balance. Its geographical scope covers not only the entire Baltic Sea, including the sea floor and coastal zones, but also its drainage area.
>
> *2. United Nations Convention on the Law of the Non-Navigational Uses of International Water Courses*
>
> The 1997 UN Water Convention, for which 103 member states voted in favour, is the only treaty governing shared freshwater resources that is of universal applicability. It was the first global water law. It is a framework convention, in the sense that it provides a framework of principles and rules that may be applied and adjusted to suit the characteristics of particular international watercourses. The Convention applies to uses of international watercourses and of their waters for purposes other than navigation and to measures of protection, preservation and management related to the uses of those watercourses and their waters. A key principle is that all states bordering an 'international watercourse' can utilise the resource in an 'equitable and reasonable manner' in order to achieve 'optimal and sustainable utilisation'. However, the state is obligated to undertake all necessary measures to ensure that such utilisation does not lead to any of the other riparian states suffering 'significant harm'. States also need to tolerate some disadvantages caused by the legitimate use of the water resource by the other riparian states, as long as this damage is not 'considerable'. The convention represents an important contribution to the strengthening of the rule of law in this increasingly critical field of international relations and to the protection and preservation of international watercourses. In an era of increasing water scarcity, it is to be hoped that the Convention's influence will continue to grow (McCaffre, 2008).
>
> *3. The Columbia River Treaty*
>
> The 1964 Columbia River Treaty is an international agreement between Canada and the United States for the cooperative development and operation of water resources in the Columbia River basin. The treaty was signed to mandate the construction and operation of three water-storage dams in British Columbia (Canadian storage) for the purpose of providing flood control and optimum hydropower generation in the Columbia River basin in Canada and/or the United States. The treaty also authorised the construction of Libby Dam on the Kootenai River in Montana for flood control and other purposes

continued

in the United States. The Libby Dam creates power and flood control benefits downstream in Canada and the United States, and these benefits have no payment requirements. All four dams were built. The treaty has no end date, but has a minimum length of 60 years. The treaty provides that either Canada or the United States can unilaterally terminate the treaty by providing a minimum of 10 years' advance, written notice. However, if both the Canadian and United States' federal governments agree, the treaty can be renegotiated or terminated at any time. Unless terminated or renegotiated, most of the treaty's current provisions will continue indefinitely (Vinogradov et al., 2003).

4. The 1909 Canada–United States Boundary Waters Treaty

The Boundary Waters Treaty of 1909 was a treaty between the United States and Great Britain relating to the use of boundary waters and to settle all questions pending between the United States and the Dominion of Canada involving the rights, obligations or interests of both countries along their common frontier (IJC, 2012). The Treaty created the independent International Joint Commission (IJC) to prevent and resolve boundary waters disputes between Canada and the United States. The IJC makes decisions on applications for projects such as dams in boundary waters and regulates the operations of many of those projects. The Treaty covers all the 'boundary waters which are defined as the waters from main shore to main shore of the lakes and rivers and connecting waterways, or the portions thereof, along with the international boundary between the United States and the Dominion of Canada passes, including all bays, arms, and inlets thereof, but not including tributary waters which in their natural channels would flow into such lakes, rivers, and waterways, or waters flowing from such lakes, rivers, and waterways, or the waters of rivers flowing across the boundary' (preliminary article – text of the Treaty). The Treaty was supposed to remain in force for five years from the day of ratification and until terminated by 12 months' written notice given by either party. The Treaty still stands, more than 100 years after agreement.

Convention, is an international environmental agreement to improve national attempts and measures for the protection and management of transboundary surface waters and groundwaters. Parties are obliged to cooperate, create joint bodies and operate with mutual assistance and exchange of information. All these treaties act as instruments for dispute resolution to maintain cordial relations between neighbouring states.

2 The role of international institutions in conflict management

Water is a resource that transcends national boundaries and its short supply leads to conflict situations. The different case studies described above have highlighted that water disputes leading to conflict have become part of modern civilisation. The disputes are usually resolved by the parties to the dispute using negotiations and agreements, without any third party involvement. However, there are many instances where parties to a dispute enter into negotiations with third party involvement. There is no international tribunal or institution to govern the relations of states in the utilisation and protection of transboundary freshwater resources. Therefore, international law is the only instrument available that provides a legal framework while at the same time making it possible to determine respective legal rights and obligations, along with the mechanisms for ensuring compliance and resolving disputes between states.

However, international treaty regimes should be flexible enough to reflect the constantly changing natural status of water resources, as well as the growing human impact and demands on them. Ideally, the regimes should contain built-in flexibility that would allow them to adapt to changing conditions such as fluctuations in precipitation, droughts, floods and other emergency situations (Vinogradov *et al.*, 2003).

> **REFLECTIVE QUESTION**
>
> What are the key international mechanisms for dealing with water disputes?

F SUMMARY

Water for human consumption and use is becoming increasingly scarce on a global scale and 90% of the projected increase in population by 2050 is most likely to be in developing countries, where even at present levels the majority of people do not have safe and sustainable access to drinking water and sanitation. The available water supply, which is not able to cater for everyone at present, will be stretched beyond the capacity to cope in the very near future. Competing needs between the industrial and agricultural sectors, especially in developing countries will lead to conflicts over this most basic of resources, with little possibility of arriving at solutions acceptable to all parties involved.

Problems related to the use and sharing of water resources will remain, since there is no universally acceptable formula for the sharing and use of water. Issues like water rights, water allocation and water access are being treated in different ways, in different locations, depending on the available legal framework and traditional and customary practices. Developed countries have a way of managing water resources which can be characterised as being streamlined and uniform. However, in developing countries there are different ways of dealing with the issue, and often traditional and customary beliefs and practices play a very important role in managing and sharing water resources, and this may mean that the models adopted are not very equitable and fair.

The need for international institutions and laws to resolve water conflicts and work towards a more equitable and fair distribution of water resources is an urgent one. Although it is clear that there is no single uniform formula that can provide a solution for all countries and regions, the presence of laws and institutions at the international level will undoubtedly go a long way towards working out a solution to the future and present water-sharing problem.

FURTHER READING

Bruns, B.R. and Meinzen-Dick, R.S. (eds). 2000. *Negotiating water rights.* Intermediate Technology Publications Ltd; London.

Getches, D.H. 1997. *Water law in a nutshell.* West Publishing; Minnesota.

Priscoli, D. and Wolf, A.T. 2009. *Managing and transforming water conflicts.* Cambridge University Press; New York.

These three books provide a good overview on types of water rights, water laws and water conflicts.

Classic papers

Falkenmark, M. 1986. Fresh water – time for a modified approach. *Ambio* 15: 192–200.

Hodgson, S. 2006. *Modern water rights: theory and practice.* Food and Agriculture Organisation of the United Nations; Rome.

These two papers have been highly influential in understanding water rights and water laws.

Wolf, A. (ed.). 2001. *Conflict prevention and resolution in water systems.* Edward Elgar; Cheltenham.

A collection of classic papers on water conflict at various scales.

PROJECT IDEAS

- Perform a study into the viability of water rights trading in your own country. Who are the players on the market and what might the drivers be? What are the regulatory or social barriers? What might the impacts be on water provision? Are there sufficient data and what extra data might be required?

- Provide a report on a case study of a current water conflict, outlining the nature of the conflict, the organisations involved and possible mechanisms for resolving the conflict, listing their pros and cons.

REFERENCES

BFN. 2008. *International Agreements and Programs.* BFN Federal Agency for Nature Conservation [online] [accessed on 22 February 2012]. Available from: http://www.bfn.de/0310_helsinki+M52087573ab0.html.

Bruns, B. and Meinzen-Dick, R.S. 2000. Introduction. *In:* B. Bruns and R.S. Meinzen-Dick (eds) *Negotiating water rights.* Intermediate Technology Publications Ltd; London, 39–40.

Bruns, B.R. and Meinzen-Dick, R.S. (eds). 2000. *Negotiating water rights.* Intermediate Technology Publications Ltd; London.

Burchi, S. and Andrea, A.D. 2003. *Preparing national regulations for water resources management principles and practice.* Food and Agriculture Organisation; Rome.

Cosgrove, W. 2003. *Water security and peace – a synthesis of studies prepared under the PCCP- water for peace process.* UNESCO; France.

Falkenmark, M. 1986. Fresh water – time for a modified approach. *Ambio* 15: 192–200.

Frijters, I. and Leentvaar, J. 2002. *Rhine case study.* UNESCO; The Netherlands.

Getches, D.H. 1997. *Water law in a nutshell.* West Publishing; Minnesota.

Haftendorn, H. 2000. Water and international conflict. *Third World Quarterly* 21: 51–68.

Hodgson, S. 2006. *Modern water rights: theory and practice.* Food and Agriculture Organisation of the United Nations; Rome.

IJC. 2012. *International Joint Commission, Boundary Water Treaty* [online] [accessed on 2 March 2012]. Available from: http://bwt.ijc.org/index.php?page=Treaty-Text&hl=eng.

Isa, A. and Stein, R. 2009. *Rule: reforming water.* International Union for Conservation of Nature; Switzerland.

McCaffre, S. 2008. *United Nations* [online] [accessed on 21 February 2012]. Available from: http://untreaty.un.org/cod/avl/ha/clnuiw/clnuiw.html.

McDonald, A., Clarke, M., Boden, P. and Kay, D. 2011. Social justice and water. *In:* R. Hester and R.M. Harrison, *Sustainable water.* Royal Society of Chemistry; London, 93–113.

Meinzen-Dick, R. and Mendoza, M. 1996. Alternative water allocation mechanisms; Indian and international experiences. *Economics and Political Weekly* 31: 25–30.

Meinzen-Dick, R. and Rosegrant, M.W. 1997. Alternative allocation mechanisms for intersectoral water management. *In:* J. Richter, P. Wolff, H. Franzen and F. Heim (eds) *Strategies for intersectoral water management in developing countries; challenges and consequences for agriculture.* Deutsche Stiftung für internationale Entwicklung; Germany, 256–273.

MoWR (Ministry of Water Resources, Government of India). 2007. *The Report of the Cauvery Water Disputes Tribunals.* Ministry of Water Resources, Government of India; New Delhi.

Phansalkar, S.J. 2007. Water equity and development. *International Journal of Rural Management* 3: 1–25.

Priscoli, D. and Wolf, A.T. 2009. *Managing and transforming water conflicts.* Cambridge University Press; New York.

Rahman, M. 2006. The Ganga water conflict. *Asterisko* 1/2: 195–208.

Rasul, G. and Choudhury, J.U. 2010. *Equity and social justice in water resource management in Bangladesh.* International Institute of Environment and Development; London.

Richardson, M. 2010. The coming water crisis in Asia [online]. Institute of Southeast Asian Studies; Singapore. Available from: http://www.iseas.edu.sg/viewpoint/mr18oct10.pdf.

Rogers, P. and Hall, A.W. 2003. *TEC background paper: Effective water governance.* Global Water Partnership; Sweden (TEC background paper; no. 7). 37 pp.

Syme, G.J., Nancarrow, B.E. and McCreddin, J.A. 1999. Defining the component of fairness in the allocation of water to environmental and human use. *Journal of Environmental Management* 57: 51–71.

Tisdell, J.G. 2003. Equity and social justice in water doctrines. *Social Justice Research* 16: 401–416.

United Nations. 2006. *Water: a shared responsibility.* UNESCO and Berghahn Books; Paris and New York.

Vinogradov, S., Wouters, P. and Jones, P. 2003. *Transforming potential conflict into cooperation potential: the role of international water law.* UNESCO; Paris.

Wolf, A. (ed.). 2001. *Conflict prevention and resolution in water systems.* Edward Elgar; Cheltenham.

WHO (World Health Organization). 2004. *Water, sanitation and hygiene links to health, facts and figures [online].* Available from:http://www.who.int/water_sanitation_health/factsfigures2005.pdf.

CHAPTER TWELVE

The future of water: water footprints and virtual water

Martin R. Tillotson, Megan Beresford, Dabo Guan and Joseph Holden

---

LEARNING OUTCOMES

After reading this chapter you should be able to:

- describe virtual water
- explain the water footprint concept and provide examples at different scales
- show an awareness of some global and regional issues associated with water footprints.

A INTRODUCTION

This concluding chapter introduces some emerging topics which may shape how water is evaluated on a global and regional scale in the future. In particular it focuses on the water footprint as a tool for understanding water use and flows of water uses around the world. It concludes by summarising the need for integrated water resource approaches, bringing forward themes highlighted by the other chapters in this book.

B THE BASIC WATER FOOTPRINT CONCEPT

Traditional measures of water use simply rely on measuring the amount of water withdrawal from rivers, lakes and groundwater (blue water) within a country or region. However, these measures ignore indirect water uses and the water used to deal with pollution from producing products or services (grey water) or within soils and vegetation (green water). The concept of a water footprint was first introduced by Hoekstra (2003) when he described it in terms of a nation's use of global water resources, which is the sum of the domestic water plus the net virtual water import. The net virtual import refers to the amount of water used in the production of a service or good which is imported into a country (Hoekstra, 2003). Virtual water (a term first coined by Allan, 1996), also known as embodied or embedded water, refers to the water that has been used in producing a product. The term is now accepted in the field of water management (Hoekstra and Chapagain, 2008), but can be confusing because the water use is real rather than virtual.

The idea of calculating the water footprint of a country was based on the analogy of an **ecological footprint**, which calculates the amount of land needed for the production of goods and services to sustain a country but also the land needed to assimilate the waste produced by the population. For the water footprint the same concepts apply, including water use in treating waste. The concept of the water footprint is aimed at improving water management on a global level (Hoekstra, 2007). The water footprint can be viewed as an indicator of total freshwater use (Hoekstra et al., 2011) and is governed ultimately by human consumption. This differentiates it from other measures which associate water demand with production within a country rather than consumption by a country. Research on the water footprint aims to show links between global trade and water management and between human consumption and water use (Hoekstra, 2007). However, the water footprint can also be calculated for smaller levels than country levels – for example, cities, companies, products or even individuals.

The water footprint can be split into three categories – blue, green and grey (Hoekstra et al., 2011). The blue water footprint involves 'clean' water consumption from freshwater sources such as surface and groundwater which does not return back to catchment from which it was taken in a clean form. The green water footprint includes consumption of water from rainfall that is stored on or in the soil or plants. Green water evaporates or is lost through plant transpiration and can be made productive for crop use. The grey water footprint represents the water used in dealing with pollution created throughout a supply chain and is defined by the volume of freshwater that it takes to assimilate the load of pollutants to return the water source back to its original quality (Hoekstra et al., 2011). For a particular good, product or service the total water footprint is therefore a summation of the combined green, blue and grey water footprints at a particular point in space and time. It is important to note that in water footprinting the definition of grey water is different from the definition used in traditional environmental management, where it refers to water used for things such as washing clothes and bathing but where the water can then be further used for other purposes such as for watering gardens. Therefore when you read about grey water in the literature or on websites you need to check exactly what it is referring to.

Figure 12.1 shows a schematic of some of the direct and indirect water uses for a home that contribute to a water footprint in a typical northern European household. The interesting point here is that the metrics of traditional water use are based entirely on blue water, which accounts for only about 3% of consumption. Of the 97% or so of virtual water consumed by a typical household, over three-quarters is derived from agricultural products (Chapagain and Orr, 2008). The water footprint of a range of products and goods has been calculated (Mekonnen and Hoekstra, 2011, 2012); **Table 12.1** presents a selection of these and it is interesting to note that:

1 The majority of water consumption is related to the green water footprint, i.e. primary production.
2 More intensively manufactured products, or those involving more complex supply chains, have proportionately greater blue and/or grey water consumption.
3 The combined water footprints are presented as global averages. This is because environmental factors, farming and manufacturing practices vary significantly.

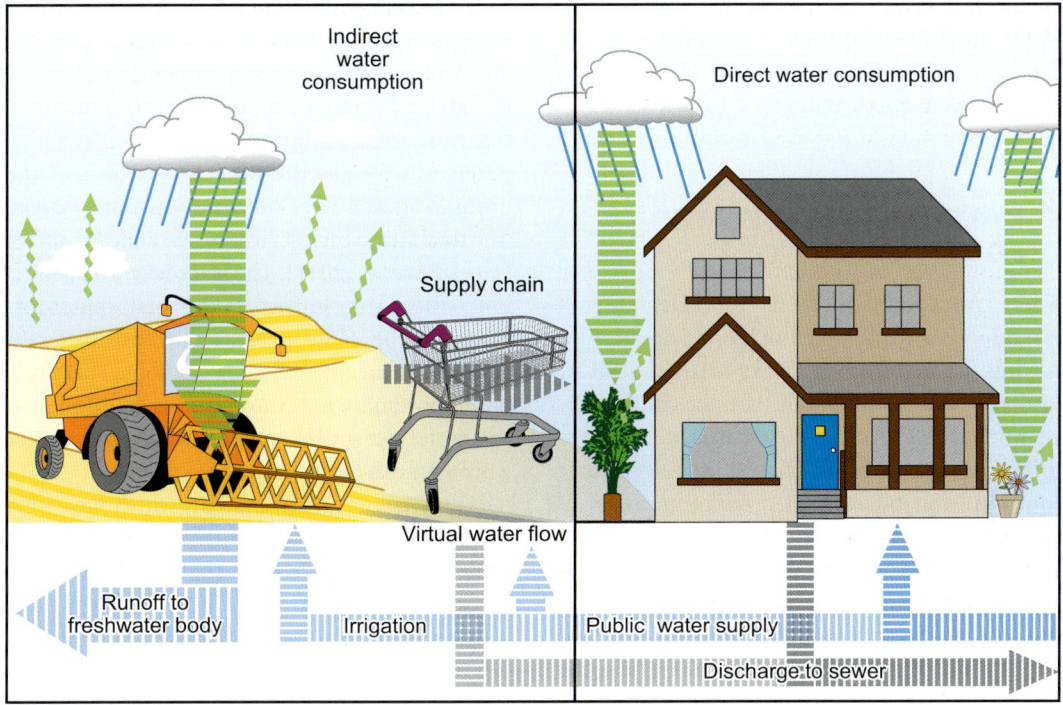

Figure 12.1 Examples of indirect and direct water use for a home.

The water footprint gives a volumetric measure of freshwater use and water pollution that consumers and producers can relate to (Hoekstra et al., 2011). However, it should be noted that the water footprint does not provide an environmental impact assessment, as the severity of impacts of water use depends on the vulnerability of the catchment. The use of the water footprint concept for policy making must therefore include this caveat (Wichelns, 2010). There is no distinction in the water footprint between sustainable and unsustainable water use, as a certain volume of water used in one catchment may not have a detrimental effect, but elsewhere the case may be different (Hoekstra, 2007).

Table 12.1 Water footprint composition data for a selection of products

Product	Water footprint litres	Unit per	Green %	Blue %	Grey %
Lettuce	237	kg	77	7	16
Milk	1020	L	85	8	7
Eggs	3300	kg	79	7	13
Cotton (fabric)	10000	kg	54	33	13
Chocolate	17196	kg	98	1	1
Coffee (roasted)	18900	kg	96	1	3

> **REFLECTIVE QUESTION**
>
> What is a water footprint?

C CALCULATING A WATER FOOTPRINT

There are several methods for calculating a water footprint, and for a full list and description readers are referred to Hoekstra *et al.* (2011). However, the main methods are the chain summation approach, the stepwise accumulative approach and the input–output model. A water footprint can be calculated for a number of different products and processes, for a group of consumers, a geographical area, a nation, a river basin and for a business. The water footprint of a product is measured in the volume of water used per unit of product (see **Box 12.1**), whereas the water footprint of a consumer, nation or business is expressed as the volume of water used per year (Hoekstra, 2008).

Blue water data can be sourced from historical data such as water meter readings. Models are sometimes used to calculate the green water footprint based on data on soil moisture, plant type and growth processes, climate conditions and so on (e.g. Fader *et al.*, 2011). These models predict green water use based on our best understanding of the physical processes involved. Grey water data can be calculated from measuring the concentrations of pollutants in the flow that are disposed into natural systems, detected by measuring local ambient water quality standards (Gerbens-Leenes and Hoekstra, 2008). All of these data on blue, green and grey water use can be very difficult to collect and water footprint assessments in many cases are in their early stages of development. In cases where data on grey and green water are not available, blue water must be used as the primary source of information, as this can be acquired from water bills, meter readings or regional water provision statistics (but see Chapter 7).

The chain summation approach is simple and is used when there is just one output product and when the process steps in the production system can be fully attributed to the production of that product. It simply involves working out how much water is used in the production of the product at each stage of the production process. The totals from each stage are then added together (Hoekstra *et al.*, 2011). The place where a product is produced may impact on the total water footprint, as the same process may use different amounts of water in different locations. Therefore, it is important to note the locations of production and water use so that the final water footprint of a product can be mapped.

There are not many systems which have only one final output product or where the inputs can simply be ascribed to the product. The stepwise accumulative approach can be used when there is more than one product input to create the final output product. It takes the sum of the water footprints of the input products used in the last stage of processing to calculate the output product's water footprint. It may also be used when one input product creates a number of output products, and here the water footprint of the input product is distributed to the output products accordingly (Hoekstra *et al.*, 2011). This approach is more realistic but is complicated, and it is more difficult to obtain water footprint data for the intermediate processing stages without carrying out water footprint investigations for each stage itself.

The environmental input–output analysis is another method to calculate a water footprint from a top-down accounting perspective. It is usually applied to calculate a water footprint for a country or a city. Sometimes it can be used to calculate the water footprint for a company in combination with the chain summation approach. The environmental input–output approach provides a complete description of the national and/or international supply chain. The method is based on final consumption and ensures that water used in production is assigned

BOX 12.1 TECHNIQUES

Calculating the water footprint of a soft drink

The following example is based on the scientific literature (Ercin *et al.*, 2011) to illustrate the detailed accounting process involved in calculating the total water footprint of a product. In this case it is for a hypothetical sugar-containing carbonated soft drink. The calculation is based on green, blue and grey water consumption and includes the following elements:

1. At the factory in the *direct* production of the packaged product. This includes:
   - blue water incorporated into the product
   - blue water consumed in the production process
   - grey water produced as a result of the production process.

2. At the factory in the *indirect* support of production of the packaged product. This is deemed an overhead:
   - blue water used by factory employees, for example, for drinking, eating, washing
   - grey water produced as a result of the above activities.

3. In the supply chain *directly* concerned with the production and supply of the ingredients and packaging which make up the final product. The direct supply chain inputs into the factory are:
   - sugar and other ingredients (caffeine, phosphoric acid, lemon oil, orange oil, vanilla extract, carbon dioxide)
   - packaging materials (PET bottle, bottle cap, labelling/adhesive, other packaging).

4. In the supply chain *indirectly* associated with production, for example, energy, transport, construction materials.

Table 12.2 presents the high-level outcomes and summation of the manufacturing and supply chain analysis described above.

Table 12.2 Water footprint (litres) of a hypothetical 0.5 litre PET bottle containing a sugar carbonated soft drink (data from Ercin *et al.*, 2011)

		Green	Blue	Grey	Total
Direct production		0.00	0.50	0.00	0.50
Indirect production		0.00	0.00	0.00	0.00
Direct supply chain					
Sugar	Min	0.00	7.00	2.40	26.00
	Max	117.90	123.50	12.00	167.00
Other Ingredients		133.51	0.30	0.00	133.81
Packaging		1.034	0.256	6.0694	7.3594
Indirect supply chain		0.0012	0.005	0.7184	0.7246
Total	Min	134.55	8.06	9.19	168.39
	Max	252.45	124.56	18.79	309.39

A number of observations can be made:

- Direct blue water consumption equates to the volume of water incorporated into the soft drink, which in this case is 0.5 litres.
- In practice it was very difficult to derive accurate water footprint data for many of the elements involved. Values, based on other literature, or assumptions were made, particularly where the contribution of a particular element is small. In this case an assumption has been made that net indirect consumption of water is zero.
- The type of sugar (i.e. beet, cane or high fructose maize syrup and its country of origin) used in the soft drink is the most significant variable in the overall water footprint calculation. A much more detailed water footprint analysis underpins this data and is summarised in **Figure 12.2**. This type of variability is extremely common in primary production, and mean values are often presented.
- While the agriculturally derived ingredients make up the majority of the water footprint, they constitute only a tiny fraction of the mass of the final product.
- The overwhelming majority of the water footprint (99.7–99.8%) is associated with the supply chain, rather than with on-site production. However, common practice in water efficiency assessment is to focus purely on site consumption.

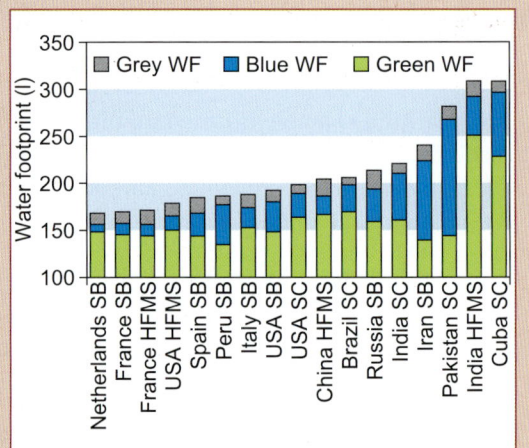

Figure 12.2 Green, blue and grey water footprint contributions for sugar beet (SB), sugar cane (SC) and high fructose maize syrup (HFMS) sourced from different geographical locations. Volumes have been normalised for sugar content of a 0.5 litre soft drink.

Source: Reproduced with kind permission from Springer Science+Business Media: *Water Resources Management*, Corporate water footprint accounting and impact assessment: The case of the water footprint of a sugar-containing carbonated beverage, v25, 2011, 721–741, Ercin, A.E., Aldaya, M.M. and Hoekstra, A.Y. Figure 3.

to the end-product consumed (and its location). For example, the water required for the production of work clothes provided by a car manufacturer is assigned to the final consumption of the car. This can, for instance, lead to considerable bias in the allocation of virtual water to different countries. A multi-regional environmental input–output model provides a methodological framework to comprehensively deal with international trade interlinkages across countries and sectors. Such models provide a description of the entire global supply chain and are able to trace water across any supply chain layer.

Environmental input–output analysis has a long history in water accounting studies. An early study undertaken by Hartman (1965) examined features of input–output models in terms of their usefulness for analysing regional water consumption and allocation. More recent studies of water-related input–output models include those in **Table 12.3**.

The water footprint of a consumer is the total volume of freshwater consumed or polluted in the production and use of goods and services. For a consumer it is necessary to calculate their direct water footprint which is made up of the water

Table 12.3 Recent water accounting studies using environmental input–output models

Reference	Study
Lenzen and Foran (2001)	Water consumption accounts provided a detailed analysis on Australia's water requirements by private household, government, export and import.
Hubacek and Sun (2005)	Compared water supply and demand for all major watersheds in China using hydro-ecological regions to match watersheds with administrative boundaries.
Velázquez (2006)	To identify the key water-consuming economic sectors in Andalusia and distinguish direct and indirect water use.
Guan and Hubacek (2008)	Advanced the work of Hubacek and Sun (2005) by taking the pollution absorption capacity into account, using North China as a case study.
Hubacek *et al.* (2009)	Examined environmental implications of urbanisation and lifestyle change in China by calculating ecological and water footprints, which was one of the first studies on water footprints using the input–output model.
Lenzen and Peters (2010)	Analysed the direct and indirect water requirements of two Australian cities in their domestic hinterland in a highly spatially disaggregated model.
Yu *et al.* (2010)	Assessed regional and global water footprint for the UK by applying a uni-directional multi-regional input–output framework.

use in the home. Interestingly, Crawford (2010) found that over 50 years the total operational water requirement for consumers living in a house (direct water use) equalled that of the water used in the original construction of the house. To calculate the total water footprint of a consumer the direct water use needs to be added to the indirect water footprint derived from the water used in producing the products (e.g., food, coffee, clothes, energy supply, furniture) that are used by the consumer (Godfrey and Chalmers, 2012). The water footprint of each product or service privately consumed is added to the proportion of shared water footprint of public goods and services (e.g. street lighting, local parks).

The water footprint of a nation includes the internal water footprint, which is the total water in goods and services used by the population, minus the water used by the country in producing goods that are exported. The external water footprint of a nation must also be measured to determine the overall water footprint. The external water footprint of a nation is equal to the volume of water resources used in producing goods and services in other countries that are imported into the nation and used by that nation.

> **REFLECTIVE QUESTION**
>
> How would you calculate the water footprint of a packet of sugar? How would you calculate your own water footprint?

D THE GLOBAL WATER FOOTPRINT PICTURE

The water footprints of nations are shown in **Figure 12.3** (average per person for domestic use and consumer goods) and **Figure 12.4** (total national water footprint). American citizens consume twice the global average of water, while Chinese citizens consume around half the global average per person. Trade between countries means that products consumed in many rich countries in Europe, the USA and Japan, for example, impact on water resources in developing countries. Many countries in Africa are net exporters of water and yet suffer from water shortages. **Figure 12.5** shows how much of the water footprint in different countries is dependent on water imports. Much of Western Europe

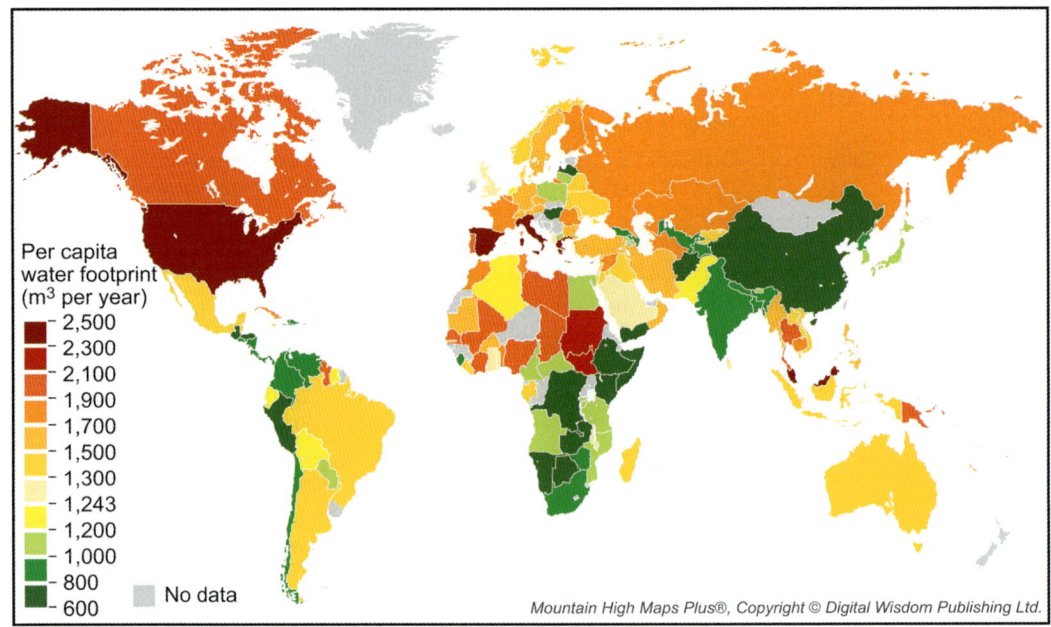

Figure 12.3 The water footprint of nations averaged per person for drinking, washing, food consumption and consumption of other consumer goods.
Source: From Jones (2010) based on data from the Water Footprint Network.

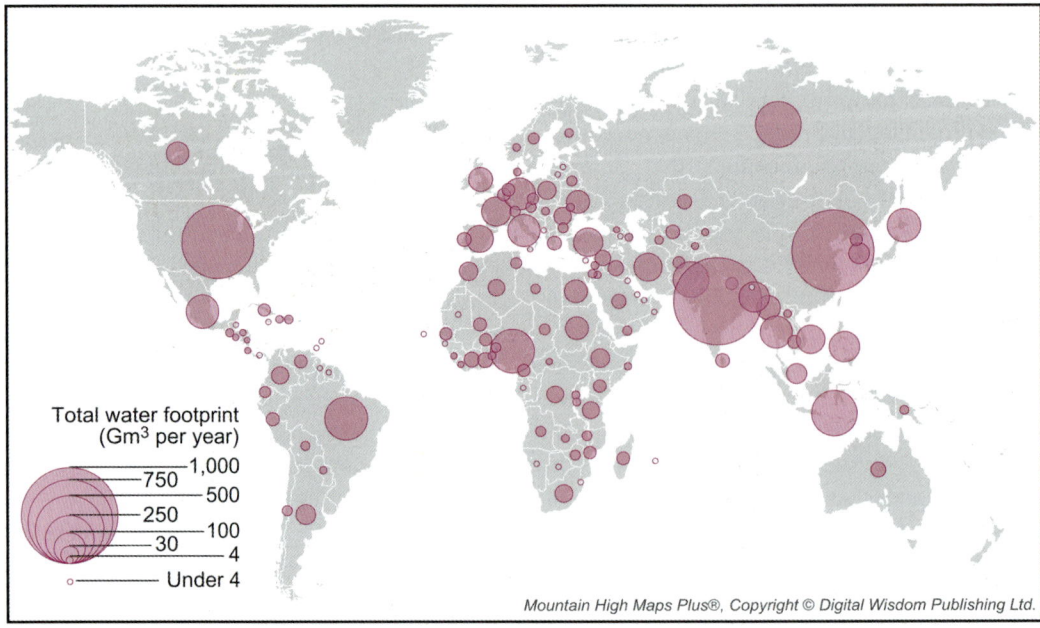

Figure 12.4 The total water footprint of nations.
Source: From Jones (2010) based on data from the Water Footprint Network.

THE FUTURE OF WATER 341

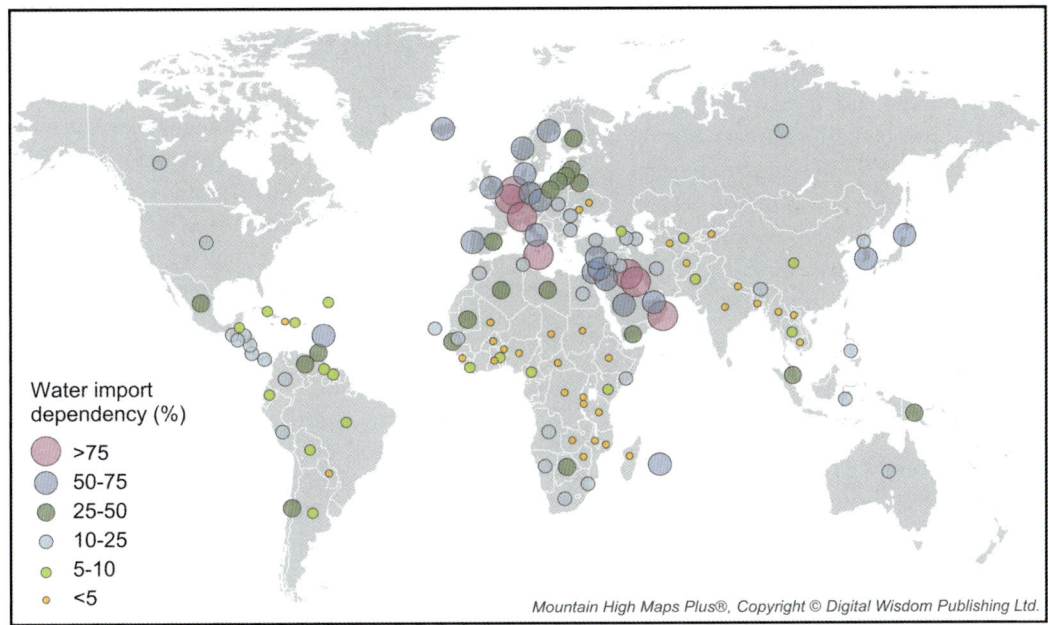

Figure 12.5 The proportion of national water footprints dependent on imports.
Source: From Jones (2010) based on data from the Water Footprint Network.

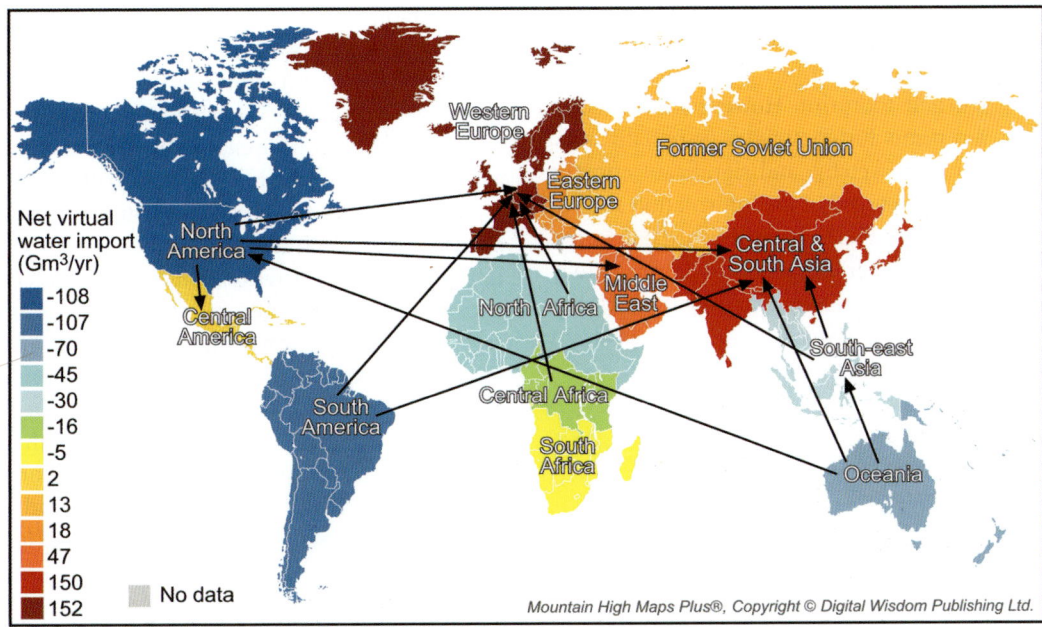

Figure 12.6 Virtual water balances for major global regions for the water flows associated with agricultural trade. The arrows show some of the largest net movements of virtual water.
Source: From Jones (2010) after Hoekstra and Chapagain (2008).

and the Middle East are highly dependent on virtual water imports (Jones, 2010). However, through such trade a nation can save domestic water resources and it may be that in some instances this is better for global water resources. If the virtual water flow occurs from regions which use water productively (e.g. little water required to grow a particular crop) to regions which are not very productive with water use (e.g. requiring lots of water to grow the same crop), then that will benefit the global water footprint. It is estimated that 263–352 billion m³ of water is saved per year through such trade (Fader *et al.*, 2011; Hoekstra, 2010). Regional net virtual water balances are shown in **Figure 12.6** related to agricultural trade (which represents 80% of virtual water flows), with some of the largest flows shown by the arrows. Thus, global policy makers could look at virtual water flows and the water footprint (to include green and grey water use) to see how trade can be used to effectively reduce global water resource use and also support those areas that are water scarce.

In 2010 the UN told big business to prepare for legislation on water footprints, and many multinational corporations have committed to reducing their footprint. However, the data on water footprints for companies (especially large ones) is rather sketchy, and significant effort is required to pull together the information so as to enable full footprints to be established. However, evaluating the water footprint of a company may well benefit that company, allowing it to become more efficient and save costs, as well as establish best practice for the global good.

> **REFLECTIVE QUESTION**
>
> How does trade affect net virtual water balances across the world?

E VIRTUAL WATER FLOW AND ITS APPLICATION TO CHINA

In order to illustrate how informative an assessment of virtual water flows might be we turn to a case study which looks at the geographical regions of China. China's economy has been growing at the rate of 9.8% per year during the 25 years since 1987, with highly active interregional and international trade. The most direct and significant outcome of China's rapid economic growth has been the significant improvement in the quality of life for Chinese people. China's population has experienced a transition from 'poverty' to 'adequate food and clothing'. Today, growing parts of the population are getting closer to 'well-to-do' lifestyles. These segments of society are not only satisfied with enough food and clothes, but also seek a higher quality of life through such things as high-nutrient food/processed foods, comfortable living, health care and other quality services. All these developments have impacted upon natural resource availability, especially water. China is trying to feed and satisfy the needs of 1.3 billion inhabitants, which amounts to 22% of the total world population, with only 7% of the world's arable land and 6% of the world's freshwater resources (Fischer *et al.*, 1998). The vast majority of the population lives on the best agricultural land, which lies on the plains in the northern and eastern parts of the country. Developing large-scale industries in these regions is infringing upon the environment. Water is already considered a scarce natural resource in those regions of China. North China (see **Figure 12.7** for locations of Chinese regions) has a population of 311 million people but a water availability of as little as 271 m³ per person per year, one-eighth of the national level and one-25th of the world average. Note, as described in Chapters 1 and 7, that areas with water availability per capita of less than 1000 m³ are thought to be water scarce. Significant amounts of 'virtual water' are traded between regions or exported to other countries, which may impact further on water scarcity, particularly for North China.

THE FUTURE OF WATER 343

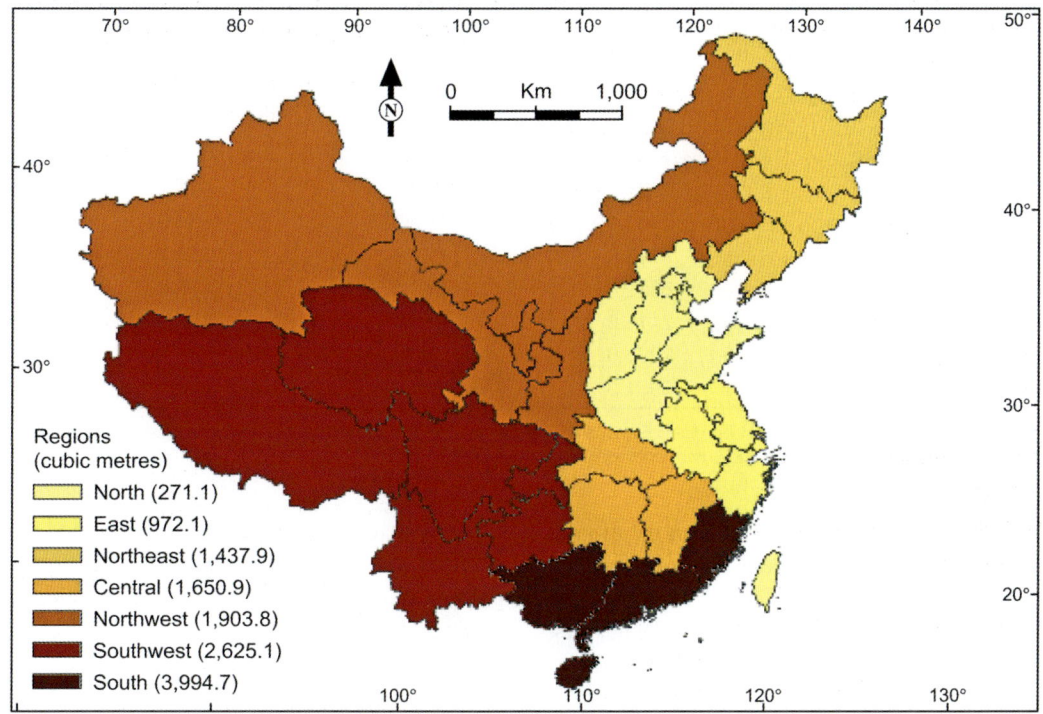

Figure 12.7 Map showing China separated into seven regions, shaded by their freshwater availability per person per year in m³.

Table 12.4 lists and compares the per capita water availability for each of the economic regions of China. As North China is not endowed with abundant water resources it should, theoretically, import more water-intensive goods, such as agricultural products, and export less water-intensive goods in order to maintain the trade balance. However, the authors of this chapter have found that annually 14,429 million m³ of virtual water are exported out of North China (through production of exports). On the other hand, the import of virtual water is only 2302 million m³, which reduces the net flow to other regions to 8443 million m³. Indeed, this means that North China uses up almost 10% of its total water resource for exports to other regions, mainly through the trade of water-intensive commodities such as agricultural crops, processed food and chemical products. By contrast, South China (e.g. Guangdong province) is endowed with rich water resources but imports only 1557 million m³ of virtual water, 50% through the trade of water-intensive products (e.g. agricultural crops). On the other hand, Guangdong exports non-intensive water commodities, such as instruments and social services, to maintain its financial trade balance (Guan and Hubacek, 2007).

We can also calculate grey water consumption in the production of exports from both North and South China. The pollutants generated from producing exported goods will stay in the region, resulting in negative impacts on the local environment. In other words, a region virtually accepts the discharge of pollutants from other regions by exporting goods. North China imports as well as exports commodities from other regions, and while doing so generates enormous amounts of grey water. Imports into North China

Table 12.4 Availability of water resources in China

Region	Total fresh water resource (10^8 m^3)	Population in year 2000 (in millions)	Per capita water (in m^3 per year)
North	843.5	311.1	271.1
Northeast	1529	106.3	1437.9
East	1926.2	198.1	972.1
Central	2761.2	167.3	1650.9
South	5190.8	129.9	3994.7
Southwest	6389.8	243.4	2625.1
Northwest	2115.6	111.1	1903.8
China average	2271.0		
World average	6981.0		

Source: Reprinted from *Ecological Economics*, v61, Guan, D. and Hubacek, K., Assessment of regional trade and virtual water flows in China, 159–170, copyright 2007, with permission from Elsevier.

lead to the generation of 808 million m³ of grey water in other regions where the commodities are produced, while North China's exports result in 577 million m³ of grey water. Hence the net grey water generated by North China is 231 million m³. By comparison, South China, which is a water-abundant region, generates 162 million m³ (net) of grey water. In this case, South China virtually accepts 162 million m³ of grey water to satisfy other regions' consumption (Guan and Hubacek, 2007).

The consumption of grey water has led to many rivers in North China no longer supporting other uses. In order to remediate these rivers the government has been investing more than 40 billion Yuan annually to clean up and improve the quality of those rivers. In terms of economic benefits, it is interesting to note that one-third of the total profits earned through trade in North China have been spent in reducing grey water consumption. Water-constrained regions like North China need to reduce the production of water-intensive products, and increase the production and export of light-industry goods in order to maintain the trade balance (Guan and Hubacek, 2007).

The above analysis of China using the North and South as distinctive regions with water scarcity and water abundance shows how assessment of virtual water could be incorporated into decision making relating to trade in order to meet present and future consumption and production levels.

> **REFLECTIVE QUESTION**
>
> Why is current industrial activity in North China unsustainable from a water perspective?

F MOVING BEYOND THE STANDARD WATER FOOTPRINT

It is possible to envisage other types of water footprint in addition to the main one described above which have excellent potential to support decision making. 'Footprint' is a measure of impact, which can also be applied to risk and disaster management. Flooding in one location can impact upon the whole economy. Neglecting these knock-on costs (i.e. the true footprint of the flood) means that we might be ignoring the economic benefits and beneficiaries of flood risk management interventions. In 2007, for example, floods cost the UK economy about £3.2 billion directly, but the wider effect might actually add another 50 to 250% to that. The flood footprint is a measure of the exclusive total socio-economic impact that is directly and indirectly caused by a flood event to the flooding region and wider economic systems and social networks.

G THE FUTURE OF WATER RESOURCES: AN INTEGRATED APPROACH

While water is fundamental to life it has also shaped the development of civilisations (Chapter 1). It will continue to do so. Ever since the beginning of civilisation there have been conflicts over water rights (Chapters 1 and Chapter 11), and access to water provides opportunities for economic advantage. Indeed de Villiers (1999) argued that if you do not have enough freshwater you need to do one of the following: import it or make more of it (e.g. desalinisation), reduce demand and use less (typically by water conservation or pricing, or perhaps even population control), or steal water from others! While some of the ways in which we currently manage our water resources make economic sense, others do not (Chapters 7 and 10). Indeed, evaluations of the basic economics of water do not necessarily get us around social justice issues (Chapter 11), but there are economic instruments that can help – through having tiered tariffs, for example (Chapter 7). The world is more globalised and interconnected than ever before, through trade and communications. These processes impact dramatically upon water resource use and allocation. As we have seen in this chapter there are new ways of looking at water resources and water economics, such as water footprinting and virtual water flows and also the concept of ecosystem services (Chapter 10), which may help to redefine the way we make decisions on, govern, or at least accept water resource allocation and use on global to local scales, which means that we can avoid the last of de Villiers' (1999) strategies (the stealing of water). These new mapping and accounting approaches allow us to see the impacts of water use in one location for societies in other locations and for the global good. Equally, they allow us to see where inequities lie and where water use needs to change (e.g. the North China example discussed above). Crucially, policy makers need to be influenced in this regard so that their local situation can be seen in light of the global situation. Of course water management is only one part of the decision-making process, along with other economic, practical and social factors, but, as water is so fundamental, it needs to be central to decision making. Whatever the situation, it is quite clear that excessive overuse of water by the few (including direct and indirect uses) needs to be curbed where it adversely affects the many.

Many of the chapters in this book deal directly with issues that are associated with sustainability. Here we refer to sustainability in the sense of meeting the needs of the present without compromising the ability of future generations to meet their own needs. Abstracting groundwater that is ancient, and non-renewable in the short term, is not sustainable (Chapter 5). This is why most assessments of available freshwater resources for countries or per person per year are based on 'renewable' freshwater resources. Managing and forecasting water demand is clearly an important approach to sustainability. This requires scientists and engineers who understand water cycle processes to work with climate change modellers and with those who work on population dynamics, migration, land use, cropping systems and so on in order that the science base is best prepared for the challenges ahead and so that the policy community can be best informed. Chapter 7 outlined several strategies for demand management, while Chapter 1 noted that a focus on innovation in the agricultural sector was crucial for global water resources because that is where most **consumptive water use** takes place. Food waste and overconsumption of water will have to be reduced. The world is interconnected and trade means that even if you live in an area that has abundant water, you might be contributing to water scarcity for other people in different parts of the world through your use of products from those water-scarce areas.

The misuse and pollution of water (e.g. Chapters 4–6) adds unnecessary pressure to the water resource system, makes water resources less sustainable, worsens risks to human health (Chapter 8), damages the economy due to reduced provision of ecosystem services (Chapter 9), causes conflict (Chapter 10) and reduces resilience to climate change (Chapter 2). It also impacts upon the world's vast oceans. In Chapter 1 it was noted that only a tiny proportion of the water on Earth is available as an accessible freshwater resource for us to use. However, our pollution of this resource has entered the oceans and causes damage to marine life. For example, in Chapter 4 the nutrient enrichment of the Gulf of Mexico caused in large part by agricultural pollution across the US Midwest was noted. The United Nations estimates that 10% of the plastic we produce ends up in the oceans, with around 18,000 pieces per square kilometre. Such plastic can be harmful to sea creatures, not only due to entanglement but also when it is eaten. Many tiny pieces of plastic that have broken off larger pieces as a result of wave action and abrasion can be mistaken for plankton (in some parts of ocean centres there may be six times as many tonnes of plastic than plankton) and eaten, causing the bioaccumulation (see Chapter 6) of toxins in sea creatures. The ocean currents move the floating debris of plastic bags, bottles, cartons, caps and so on which often later collect in concentrated areas in the centre of the surface ocean circulation systems. Some of these remote ocean locations can have more than 335,000 pieces of plastic per square kilometre, demonstrating the far-reaching impacts of humans upon water across the world.

Contemporary climate change is impacting upon the water cycle and will have major consequences for many parts of the world in terms of water availability and water hazards (Chapter 2), which in turn will impact upon economic performance. Climate change impacts need to be understood in the context of current and future projected trade patterns and virtual water flows so as to establish whether patterns are sustainable in the long term. Both climate and land management change (through population growth, development, agricultural expansion, etc.) affect how water moves through river basins and impacts upon river flow (Chapter 3) and surface water quality (Chapter 4), which has knock-on impacts for aquatic ecosystems (Chapter 6) and associated costs in terms of economics, as well as costs for human health (Chapter 8) and wider ecosystem services (Chapter 11). These processes also impact upon groundwater, while the expanding need for water is also causing water quantity and quality problems for groundwater abstraction (Chapter 5).

One of the greatest water challenges is that of safe and clean sanitation systems for large parts of the world where currently the systems are totally inadequate. Such systems are crucial for health and well-being, and while it is possible to move towards carbon-neutral systems that recover resources from wastewater in the developed world (Chapter 9) there is still an enormous way to go before large proportions of the world's population have adequate sanitation infrastructures (Chapter 8). There is no way that the UN Millennium Development Goals of halving the number of people without access to safe water by 2015 will be met. But we should not give in – we should be spurred into even greater action.

In many senses the figures on water resources which have been highlighted in this book are frightening. There truly is a global water crisis. It cannot be otherwise when so many people in the world have no access to safe water or adequate sanitation and while the population is set to increase by another 50% within the next four decades. However, it is no good sitting back and letting disaster happen. We have to take action now, be innovative with technology, fund searches for innovation and look for new approaches (the Bill and Melinda Gates Foundation research programme on reinventing the toilet is a good example; see http://www.gatesfoundation.org/watersanitationhygiene/Pages/home.aspx for updates). Plenty of examples have been mentioned in the other chapters of this book of recent innovation and discoveries which show that new ideas are emerging throughout the wide-ranging subject of water resources (this chapter probably provides the first-ever publication of the idea of the flood footprint, for example) and novel ideas should be appropriately supported, and this requires political will. Therefore, we have to ensure that environmental scientists, social and political scientists, businesses, policy makers and local communities work together to tackle the world's water resource issues. An integrated approach is essential because our water problems are multifaceted and interrelated; they are also vital to the survival of every person on the planet. We hope, therefore, that you have found it useful to read all 12 chapters in this book, as they cover physical, environmental, technological, social, political and economic aspects of water resources. The purpose of the book has been to bring together these subject areas related to water in one volume as a learning and reference aid. We hope you agree that it has been highly beneficial to do so and now feel willing, able and inspired to learn more about the interdisciplinary subject of water.

H SUMMARY

Virtual water is the water used in the production of a product or service. By consuming that product or service your indirect water use will increase by the amount of virtual water associated with that product. Therefore, your personal water footprint will increase beyond direct water use (e.g. drinking water, washing), to indirect uses. For example, if you bought a book there would be water use associated with making the paper and ink, ranging from water abstracted for manufacture (blue water), water used to grow the trees from which the paper was derived (green water) and pollution of water used in the printing process (grey water). Water footprints can be calculated for individuals, groups, businesses, regions and countries. Examining virtual water flows between these groups, and in particular mapping totals and net flows, can be highly informative in terms of optimising the efficiency of water use and examining where there are water-scarce regions which are highly vulnerable because they are net exporters of water. These sorts of tools can also be excellent for businesses and individuals to help them establish how we can be more efficient with our total water consumption (direct and indirect) and how trade can be used to reduce total global water consumption. This is a necessity as we move into a future with increasing population and growing concern over climate change impacts, both for ourselves as humans and for the wider, global ecosystem too.

FURTHER READING

Allan, J.A. 2011. *Virtual water: tackling the threat to our planet's most precious resource.* I. B. Tauris; London. An important book by the person who came up with the term 'virtual water'.

Fader, M., Gerten, D., Thammer, M., Heinke, J., Lotze-Campen, H., Lucht, W. and Cramer, W. 2011. Internal and external green-blue agricultural water footprints of nations, and related water and land savings through trade. *Hydrology and Earth System Sciences* 15: 1641–1660.

An interesting modelling study to establish agricultural water flows through trade.

The three papers below are useful case studies of regional virtual water flows to follow up the Chinese example provided in this chapter:

Feng, K., Hubacek, K., Minx, J., Siu, Y.L., Chapagain, A., Yu, Y., Guan, D. and Barrett, J. 2011. Spatially explicit analysis of water footprints in the UK. *Water* 3: 47–63.

Feng, K., Siu, Y.K., Guan, D.A. and Hubacek, K. 2012. Assessing regional virtual water flows and water footprints in the Yellow River basin, China: a consumption based approach. *Applied Geography* 32: 691–701.

Guan, D. and Hubacek, K. 2007. Assessment of regional trade and virtual water flows in China. *Ecological Economics* 61: 159–170.

Two books by the group that originally developed the water footprint concepts and that provide excellent guidance on the water footprint topic and ideas around virtual water flows:

Hoekstra, A.Y. and Chapagain, A.K. 2008. *Globalization of water: sharing the planet's freshwater resources.* Blackwell Publishing; Oxford.

Hoekstra, A.Y., Chapagain, A.K., Aldaya, M.M. and Mekonnen, M.M. 2011. *The water footprint assessment manual – setting the global standard.* Earthscan; London.

Classic papers

Allan, J.A. 1996. The political economy of water: reasons for optimism but long-term caution. *In*: Allan, J.A. (ed.), *Water, peace and the Middle East: negotiating resources in the Jordan basin.* Tauris Academic Studies; London.

One of the first articles to use the term 'virtual water'.

PROJECT IDEAS

- Quantify the amount of water used for producing exports in a country or region and trade to other places for final consumption. Discuss the findings with reference to international trade theories from the perspectives of water resources availability, economic structure changes and technological developments.

- Calculate the water footprint of different dietary lifestyles for individuals (different groups of population or across different countries). Discuss the findings within the context of current national and international sustainable development policies.

- Determine the possible impacts of energy-related policies in your country or region for water efficiency.

REFERENCES

Allan, J.A. 1996. The political economy of water: reasons for optimism but long-term caution. *In*: Allan, J.A. (ed.), *Water, peace and the Middle East: negotiating resources in the Jordan basin.* Tauris Academic Studies; London.

Allan, J.A. 2011. *Virtual water: tackling the threat to our planet's most precious resource.* I. B. Tauris; London.

Chapagain, A.K. and Orr, S. 2008. *UK water footprint: the impact of the UK's food and fibre consumption on global water resources*, Volume 1. WWF-UK; Godalming, UK.

Crawford, R.H. 2010. Life cycle water analysis of an Australian residential building and its occupants. *Proceedings of the Seventh Australian Conference on Life Cycle Assessment: Revealing the secrets of a green market, Melbourne, 9–10 March 2011.*

de Villiers, M. 1999. *Water wars. Is the world's water running out?* Phoenix Press; London.

Ercin, A.E., Aldaya, M.M. and Hoekstra, A.Y. 2011. Corporate water footprint accounting and impact assessment: the case of the water footprint of a sugar-containing carbonated beverage. *Water Resources Management* 25: 721–741.

Fader, M., Gerten, D., Thammer, M., Heinke, J., Lotze-Campen, H., Lucht, W. and Cramer, W. 2011. Internal and external green-blue agricultural water footprints of nations, and related water and land savings through trade. *Hydrology and Earth System Sciences* 15: 1641–1660.

Feng, K., Hubacek, K., Minx, J., Siu, Y. L., Chapagain, A., Yu, Y., Guan, D. and Barrett, J. 2011. Spatially explicit analysis of water footprints in the UK. *Water* 3: 47–63.

Feng, K., Siu, Y.K., Guan, D.A. and Hubacek, K. 2012. Assessing regional virtual water flows and water footprints in the Yellow River basin, China: a consumption based approach. *Applied Geography* 32: 691–701.

Fischer, G., Chen, Y. and Laixiang, S. 1998. *Interim report: the balance of cultivated land in China during 1988–1995*. International Institute for Applied Systems Analysis; Austria.

Gerbens-Leenes, P.W. and Hoekstra, A.Y. 2008. *Business water footprint accounting: a tool to assess how production of goods and services impacts on freshwater resources worldwide [online]* [accessed 23 January 2012]. Available from: http://www.waterfootprint.org/Reports/Report27-BusinessWaterFootprint.pdf.

Godfrey, J.M. and Chalmers, K. (eds). 2012. *Water accounting: international approaches to policy and decision-making*. Edward Elgar Publishing; Cheltenham.

Guan, D. and Hubacek, K. 2007. Assessment of regional trade and virtual water flows in China. *Ecological Economics* 61: 159–170.

Guan, D. and Hubacek, K. 2008. A new and integrated hydro-economic accounting and analytical framework for water resources: a case study for North China. *Journal of Environmental Management* 88: 1300–1313.

Hartman, L.M. 1965. The input–output model and regional water management. *Journal of Farm Economics* 47: 1583–1591.

Hoekstra, A.Y. 2003. *Virtual Water Trade [online]*. The Netherlands; IHE Delft [accessed 23 January 2012]. Available from: http://www.waterfootprint.org/Reports/Report12.pdf.

Hoekstra, A.Y. 2007. Human appropriation of natural capital: comparing ecological footprint and water footprint analysis. *Ecological Economics* 68: 1963–1974.

Hoekstra, A.Y. 2008. *Water neutral: reducing and offsetting the impacts of water footprints [online]* [accessed 23 January 2012]. Available from: http://www.waterfootprint.org/Reports/Report28-WaterNeutral.pdf.

Hoekstra, A.Y. 2010. The globalization of water. *In:* Jones, J.A.A., *Water sustainability: a global perspective*. Hodder Education; London.

Hoekstra, A.Y. and Chapagain, A.K. 2008. *Globalization of water: sharing the planet's freshwater resources*. Blackwell Publishing; Oxford.

Hoekstra, A.Y., Chapagain, A.K., Aldaya, M.M. and Mekonnen, M.M. 2011. *The water footprint assessment manual – setting the global standard*. Earthscan; London.

Hubacek, K. and Sun, L. 2005. Economic and societal changes in China and its effects on water use. *Journal of Industrial Ecology* 9: 187–200.

Hubacek, K., Guan, D., Barrett, J. and Wiedmann, T. 2009. Environmental implications of urbanization and lifestyle change in China: ecological and water footprints. *Journal of Cleaner Production* 17: 1241–1248.

Jones, J.A.A. 2010. *Water sustainability: a global perspective*. Hodder Education; London.

Lenzen, M. and Foran, B. 2001. An input–output analysis of Australian water usage. *Water Policy* 3: 321–340.

Lenzen, M. and Peters, G.M. 2010. How city dwellers affect their resource hinterland. *Journal of Industrial Ecology* 14: 73–90.

Mekonnen, M.M. and Hoekstra, A.Y. 2011. The green, blue and grey water footprint of crops and derived crop products. *Hydrology and Earth System Sciences* 15: 1577–1600.

Mekonnen, M.M. and Hoekstra, A.Y. 2012. *Water footprint [online]* [accessed 16 August 2012]. Available from: http://www.waterfootprint.org/?page=files/productgallery.

Mekonnen, M.M. and Hoekstra, A.Y. 2012. A global assessment of the water footprint of farm animal products. *Ecosystems* 15: 401–415.

Velázquez, E. 2006. An input–output model of water consumption: analysing intersectoral water relationships in Andalusia. *Ecological Economics* 56: 226–240.

Wichelns, D. 2010. Virtual water and water footprints offer limited insight regarding important policy questions. *Water Resources Development* 26: 639–651.

Yu, Y., Hubacek, K., Feng, K. and Guan, D. 2010. Assessing regional and global water footprints for the UK. *Ecological Economics* 69: 1140–1147.

Glossary

Absolute water scarcity According to the United Nations' use of the term, this occurs when the renewable freshwater available is less than 500 m^3 per person per year.

Acid neutralising capacity The ability of a substance to change an acidic substance into a neutral one, as well as buffering an acidic solution.

Adiabatic cooling This is the process whereby a rising parcel of air cools, due to the expansion of the parcel.

Adit A passage into a mine which is nearly horizontal and may be used for access or drainage.

Adsorption–desorption Adsorption is the uptake or attracting of a substance to the surface of another material, while desorption is the release of a substance from the surface of another material. This may be through the surface or from the surface itself.

Advect To move a body of mass horizontally.

Aerosols Microscopic particles contained within the atmosphere that interfere with the Earth's incoming energy from the Sun and the outgoing energy being emitted from the Earth's surface. Aerosols can result in either cooling or warming of the Earth's climate.

Albedo The proportion of incoming energy from the Sun that is reflected by a surface. Snow has a high albedo, whereas tarmac has a low albedo.

Allochthonous Something which originates from a different location to the one in which it is found.

Alluvial fan A depositional landform which develops when a fast-moving stream is given the opportunity to slow its speed and become more spread out in its arrangement. They often occur as a river exits a constrained geomorphology such as a canyon and enters a flatter, wider expanse of land.

Apertures Gaps in rocks (or other substances) through which light can pass.

Aquifer Rock that is porous enough to absorb and retain water and permeable enough to allow groundwater to pass through it freely.

Artesian well A well drilled into an aquifer which is trapped between two layers of impermeable rock. The aquifer receives its water from a higher altitude than the one at which it is situated and this creates pressure which forces the water to rise upwards in the well. Due to this natural pressure there is no need for a pump in an artesian well with the water rising above the level of the water table. Dependent upon the pressure which exists, the water level can reach the same level as the ground surface.

Asymptote A straight line which approaches a curve but never quite meets it. The distance between the curve and the straight line approaches zero as they tend to infinity.

Autochthonous Something which originates from the location in which it is found.

Avulsion With reference to fluvial geomorphology, this is the action whereby a river channel rapidly abandons its current location in favour of a new channel, often as a result of becoming choked with sediment.

Baseflow The stable portion of a river's discharge that is fed by groundwater.

Bases Substances that have a pH higher than 7 and release hydroxide ions (OH^-).

Bedload transport The movement of material in a river which has partial contact with the river bed. Movement may be through the processes of rolling, sliding and saltating.

Benthic layer (aka Benthic zone) Ecological zone found at the deepest point of a body of water such as the sea floor or a lake bottom.

Benthos Benthic organisms (see *Benthic*).

Bioaccumulate The buildup of toxic substances, such as heavy metals and pesticides, in the tissue of living things.

Bioavailable (Bioavailability) The potential for adsorption and use of a nutrient by an organism.

Biochemical oxygen demand (BOD) A measure of the amount of oxygen required in a given set of conditions for aerobic organisms in water to break down the organic matter present.

Biofilm A buildup of organisms, including bacteria, algae and fungi, in a slimy, thin layer on solid surfaces in aqueous environments.

Biofuel Fuel which is the product of recently decaying living matter, providing a renewable source of fuel.

Boundary layer The section of the atmosphere just above the Earth's surface. As a result of this there is a sharing of heat and moisture from the Earth's surface into the atmosphere.

Buffer zone A section of land which is permanently vegetated in an effort to improve water, soil and air quality, often through the retention of pollutants.

Buffering capacity This is the ability of a substance to resist pH changes and in the context of water is related to the ability of the water to maintain a steady pH.

C3 plants Plants which directly fix carbon dioxide. The carbon is fixed into a compound containing three carbon atoms. Eighty-five per cent of plants are C_3, including most temperate zone plants and broadleaf plants. C_3 plants are disadvantaged in hot, dry conditions.

Capillary action The process whereby a liquid can flow in a narrow space without the need for the influence of other forces or in opposition to other forces, such as gravity.

Capillary water The water which is retained in soil pore spaces and in a continuous layer around soil particles. The water is held so strongly that gravity cannot drain it. The tension at which the capillary water is held in the soil may be greater than 30 times the atmospheric pressure.

Carbon footprint The amount of carbon that an individual or an organisation uses over a given time period. It is often expressed in terms of the equivalent amount of carbon dioxide.

Cardinal scale A method of measurement using numbers that denote a quantity such as one, two, three.

Catchment Hydrological unit which may be outlined by topography and which is the natural drainage area that is drained by a stream or river.

Centrifuge A piece of apparatus which spins samples around, applying centrifugal forces to them to separate fluids.

Chemical oxygen demand (COD) A test used in environmental sciences to indirectly determine the amount of organic compounds in the water. This method is often used to determine the amount of organic pollution which is found in surface waters.

Cloud seeding The process whereby ice crystals or silver iodide crystals are spread in the upper layers of clouds to act as cloud condensation nuclei and stimulate the creation of precipitation.

Coagulation The process in water treatment to remove impurities whereby chemicals are used to aid the clumping together of suspended solids so that they can be filtered out.

Common property resource A natural resource which is owned by the general public. The use of the resource is often regulated by a country's government and no individual has exclusive rights to its use. Common property resources include those from public waters, land and air.

Complexes Molecules which are made up of two or more component molecules and are joined with a weak bond in place of a covalent bond.

Condensation The change in state from a gaseous vapour to a liquid. It is the process of becoming more dense. Pertaining to water, it occurs when humid air comes into contact with a cold surface.

Conglomerate In geological terms this is the term used to describe a rock which contains large individual clasts in a matrix of finer-grained particles which have been cemented together in formation.

Conjunctive use Pertaining to the joint use of surface and groundwater supplies to service a need and sharing of water storage between groundwater and surface water zones.

Connate water In geological and sedimentological terms this is water which has been trapped in the pores of rock during the formation process.

Consumers In the ecological food chain these organisms gain their energy by eating other organisms.

Consumptive water use The withdrawal and use of water where the water is not returned to the system that provided the source of the water.

Convective The molecular motion responsible for the transfer of energy, such as heat, through a fluid.

Coriolis effect The rotation of the Earth on its axis causes a moving object or fluid to be apparently deflected. Deflection occurs to the right in the northern hemisphere and to the left in the southern hemisphere and is stronger at higher latitudes.

Covalent bond A chemical bond formed between two non-metal atoms. It involves the sharing of a pair of electrons between the atoms.

Critical load A threshold in environmental pollution. When it is exceeded a system will begin to fail, due to an excessive level of pollution being in the system, which affects its functioning.

Cyclonic The atmospheric characteristics found in a low-pressure weather system.

Delphi survey A method used for decision making whereby a series of questionnaires are completed on a particular subject by experts in that field to result in an informed outcome.

Delta The depositional landform produced at the mouth of a river where it meets another water body such as the ocean or a lake. It is usually triangular in shape.

Desalination The process whereby salt and dissolved solids are removed from water, particularly seawater. This process is often undertaken to allow otherwise unsuitable water sources to be used for domestic purposes.

Desertification The occurrence whereby semi-arid land which was once habitable and productive becomes desert. Prolonged drought, possibly due to a changing climate and also to the overuse of the land by human activity, results in the top layers of soil becoming dry and therefore more easily eroded, especially by the wind.

Detergent A water-soluble chemical which is used for the purposes of cleaning through the action of combining with particles of dirt and making them more soluble.

Detritivores Organisms which gain their energy by feeding on dead and decaying organic matter, including plant litter.

Dew-point temperature The temperature at which a cooling air parcel becomes saturated with water vapour, and at which water vapour condenses to form liquid water.

Diagenesis The physical, chemical and biological changes which sediment undergoes after its initial deposition and when it is in the process of becoming sedimentary rock.

Diffuse pollution Pollution which comes from sources which are difficult to precisely locate. The pollution is likely to have come from a number of locations over a wide area. An example of diffuse pollution is nitrate leaching from agricultural land.

Displacement flow The process whereby water infiltrating at the top of a slope rapidly pushes out water in the soil at the bottom of the slope. This has an influence over storm hydrographs.

Domestication The influence of humans over the evolution of a plant or animal species.

Dose–response relationship An expression of the amount of toxin that can enter an environmental system and the extent of the response of the system to that amount of toxin.

Drainage basin The area of land drained by a single river system.

Drainage density The measure of the total length of all the rivers and streams in a drainage basin divided by the total drainage basin area.

Ecological footprint A measure of the demands that humanity places on the Earth's natural resources and ecosystems. It is a standardised measure which determines the area of land and water required to sustain a human population of a given size.

Ecosystem services Services provided by ecosystems that support human life in some way (e.g. food, medicine and clean water). Evaluating these services helps to focus public attention on environmental issues that could result in the loss of these services.

Effective rainfall The rainfall that is not stored on the land surface nor infiltrates into the soil. It produces the overland flow which generates quick-responding storm flow in the river.

Endemic In ecological terms this refers to species of plants or animals which belong to a specific location due to their evolution in biogeographically isolated areas, such as the Galapagos Islands.

Endorheic basin An internal drainage basin which is a closed hydrological system, retaining water rather than allowing it to flow to other water bodies such as other rivers or the ocean. They are most likely to be found in desert locations, although they can form in other climatic zones, and can be seasonal or permanent.

Ensemble Pertaining to climatic modelling this refers to a group of models which represent a climatic system being considered all at once. Each model has a slight difference in its representation of the system so as to determine how accurate the model is in representing certain parameters.

Enteric pathogens Bacteria which occur in the intestines of humans and animals and cause disease.

Epidemiology The study of health and disease patterns and causes and application of this study to control health and disease problems.

Epilimnion Within a water body this is the layer at the surface which is warmer and less dense than the layer which is trapped below it and is cooler and more dense.

Equivalent hydraulic fracture A hypothetical fracture with smooth walls with the equivalent size of a more complex real fracture, which tends to have rough walls and is therefore difficult to measure and assess.

Eutrophication The enrichment of plant nutrients in freshwater and marine water bodies; this results in the accelerated bloom of plants in the water, leading to a detrimental effect on other organisms in the water.

Evapotranspiration The process whereby water from the Earth's surface is lost to the atmosphere as water vapour through evaporation and transpiration by plants.

Exorheic basin The dominant type of drainage basin in the world, these systems are open and drain to other rivers and, ultimately, the sea.

Field capacity The term used to describe the maximum amount of water which can be held as capillary water in soil once all gravitational water has drained away.

Flocculation The following stage after coagulation; this is the process whereby a liquid is slowly mixed so that smaller particles agglomerate to form larger, suspended particles which are visible.

Folded mountains Mountain ranges which are formed through the coming together of two continental tectonic plates. As they both have the same density this results in the rock layers being pushed inwards and upwards.

Food web A representation of the connectivity of food chains and trophic levels in an ecological community.

Genotype The entire genetic make-up of an individual organism.

Germ theory of disease A theory developed in the 1800s stating that the presence and actions of microorganisms in the body are the cause of many diseases. The research of Louis Pasteur and Robert Koch provided scientific evidence for the theory.

Glacial mass balance The difference over time between the total accumulation and ablation of a glacier. The mass balance can be either positive or negative, depending upon whether accumulation exceeds ablation or vice versa.

Greenhouse gases These gases are present in the Earth's atmosphere and can absorb and emit infrared radiation and therefore aid in the regulation of the Earth's temperature. Water vapour, carbon dioxide, methane, nitrous oxide and ozone are some of the most abundant greenhouse gases present in the atmosphere.

Gross primary productivity This term refers to the total amount of energy used by plants during photosynthesis to fix biomass in an ecosystem within a specific time period.

Head loss This is the measure of the reduction in the total pressure of a fluid as it moves through a pipe. It is the result of friction between adjacent particles in the fluid moving past each other, the friction between the fluid and the pipe which it is moving through, and also the turbulence which occurs as a result of the redirection of fluids. The head loss is proportional to the length of the pipe which the fluid is flowing through.

Headroom With respect to water availability this term refers to the difference which exists between supply and demand. Headroom can be classified as Target headroom or Available headroom, each representing the supply–demand balance of water utility.

Helicoidal flow The flow of water in a river meander which is corkscrew in pattern. As a result of this the water level on the outside of the meander bend becomes super-elevated, and the return flow of the water is along the channel bed.

Hermaphrodite An organism which has both female and male reproductive organs.

Hydraulic conductivity A measure of the transmission of water through soil at a specific hydraulic gradient. It is denoted by the letter K.

Hydraulic gradient This term refers to the change in hydraulic pressure per unit distance in a specific direction.

Hydraulic head A term representing the energy of water at a given point in an aquifer. It is often measured via the level at which water stands in an unconfined or confined well. It is the sum of three component parts: the pressure head, the elevation head and the velocity head at a specific point.

Hydrological pathways The generic term used to describe the ways in which water can flow through a catchment. These pathways can be categorised into three main groups: above the ground, including evaporation, interception, stemflow and throughfall; overland flow; and below the ground, including infiltration, macropore flow and subsurface matrix flow.

Hygroscopic aerosols A specific type of aerosol which is water soluble and has the ability to attract water vapour from the air.

Hygroscopic water Water which exists as a film around soil particles. In this form it is not available to plants, as a result of the strong attraction between the water molecules and the soil particles.

Hyper-arid This refers to an area of the world which has very little vegetation, and annual rainfall which is rarely greater than 100 mm. There may be long periods of a number of years where there is no rainfall at all.

Hypolimnion Within a water body this represents a lower layer which is cooler and denser than the water found above it. There is little mixing between this layer and the one above it, due to its characteristics.

Hyporheic zone A zone found below a river channel where there is a mixing of ground water and channel water in the substrate.

Hypoxia The condition whereby the dissolved oxygen available is below the required level to support most life forms. It is the result of the oxygen being used in the decomposition of organic matter exceeding oxygen replenishment through photosynthesis and from the atmosphere. The concentration of oxygen in hypoxic waters is usually less than 2–3 ppm.

Igneous rock Rock which is formed through the solidifying of magma once it is has escaped a volcano.

Infiltration (Infiltrate) The process whereby water is absorbed into the soil and then transported downwards through the soil profile.

Infiltration capacity The upper rate at which water can flow into the soil from the surface. This rate can change through time, depending on how wet the soil is and the surface conditions.

Infiltration-excess overland flow Where the rate of rainfall or irrigation water supply to the soil surface exceeds the infiltration capacity, leading excess water to flow across the surface. This is also known as Hortonian overland flow.

Infiltration rate The time it takes for a unit of water added to a surface to enter a unit of soil.

Infrared radiation The energy that is released by all solids, liquids and gases as heat.

Interception The process whereby precipitation is captured by plants as it falls and, instead of reaching the ground, is lost back to the atmosphere through evapotranspiration.

Interflow The movement of water which has infiltrated the surface layers of the soil and then moves laterally in the unsaturated zone of the soil above the water table.

Intertropical Convergence Zone (ITCZ) The region where the trade winds from the northern and southern hemispheres converge. Conditions are favourable for cloudiness and heavy rainfall.

Ion A charged atom or molecule. The charge results from an imbalance in the number of electrons, which are negatively charged, as compared to the number of protons, which are positively charged. If there is one more proton than electrons, then the atom or molecule is said to be positively charged and it will be attracted to an atom or molecule that is negatively charged.

Irrigation The process whereby water is artificially applied to the land, often used to aid crop production in arid climates.

Jet stream Narrow, high-speed winds caused by sharp temperature gradients, located within Rossby waves in the upper atmosphere. They can be thousands of kilometres long and hundreds of kilometres wide.

Karst A distinctive landscape created by the dissolving of rocks. Such landscapes often include sinkholes, and underground caverns.

Latent heat of vaporisation The amount of energy which is required to change a specific mass of a liquid into a gas without a change in temperature.

Lentic A word used to describe a feature or organism associated with slow-moving or static water bodies such as lakes.

Levees Raised embankments which are constructed to protect land from rising flood waters.

Lithification The process whereby newly deposited unconsolidated sediments are transformed into cohesive rock.

Littoral zone The zone in a lake closest to the shore and where light can reach the bottom, allowing for a diverse range of plants and algae to exist.

Long profile A graphical representation of the change in altitude of a river channel along its longitudinal course from head to mouth.

Lotic Word used to describe a feature or organism associated with fast-moving water bodies such as rivers.

Lysimeter An instrument used to measure the water which flows through or is lost from the soil and which can also be used to measure the chemical composition of the water as it flows through.

Macropore flow The transfer of water through the soil between large pores greater than 0.1 mm in diameter.

Macropores Openings within the soil which are greater than 0.1 mm in diameter and can result in rapid flow. They can be produced as a result of structural failings in the soil or biological activity such as the burrowing of animals, earthworms and plant roots.

Marginal costs The increase or decrease in the total cost of production, should one extra unit of a product be produced.

Matrix flow The transfer of water through the soil between microscopic pores smaller than 0.1 mm in diameter.

Meiofauna Benthic organisms which range in size from approximately 0.1 mm to 1 mm.

Mesohabitats These units combine to make up a reach of a river and have similar characteristics, such as deep, slow-flowing pools.

Metabolism The various chemical transformations that take place to maintain the cells of living organisms.

Metalloids Elements whose properties are considered to be a mix between metal and non-metal and which as a result are difficult to classify. Boron is considered a metalloid.

Metamorphic See *Metamorphic rocks*.

Metamorphic rocks These rocks are produced under the Earth's surface, where heat and pressure result in structural and mineralogical alterations.

Miasmatic theory of disease A historic theory of disease which suggests that diseases were the result of miasma, a form of 'bad air'. The spread of cholera and the Black Death were attributed to this theory.

Nanomaterials This term is used to broadly refer to materials which have a structural component which is less than 100 nm. The properties of these materials differ from others in the sense that they have an increased relative surface area and quantum effects which can result in changes in other properties.

Natural greenhouse effect Atmospheric greenhouse gases, such as carbon dioxide and water vapour, absorb 90% of the long-wave radiation emitted from Earth, resulting in an average global temperature approximately 35°C warmer than would be experienced without the natural levels of greenhouse gases present.

Nektonic (Nekton) The term used to refer to organisms which move around in the water column, rather than being restricted to one region within it.

Net primary production This is the difference between the biomass which is fixed by plants during photosynthesis (*gross primary production*) minus the energy which is used by plants during respiration.

Neuston The term used to refer to organisms which exist on or just below the surface of a water body.

Non-point source pollution See *Diffuse pollution*.

North Atlantic Oscillation index A climate system centred over the North Atlantic Ocean basin. In normal circumstances there is a north–south pressure gradient, with low pressure over Iceland and high pressure over the Azores. This pressure difference drives the mean surface winds and winter weather systems. It has been observed that changes in pressure over Iceland and the Azores occur out of phase with each other, especially in the wintertime, leading to the Oscillation. The changes in the pressure can have an impact on marine and terrestrial ecosystems. The North Atlantic Oscillation can be either strongly positive or strongly negative.

Nosocomial Originating from or occurring within a hospital. The term is usually used with reference to an infection.

Orographic The term used when considering things in relation to mountains or their influence.

Osmoregulation The process whereby a balance is kept by cells and simple organisms between the fluid and the electrolytes they contain and their surrounding conditions.

Overland flow The process whereby water runs over the surface of the land. Overland flow is often classified into two types: *Infiltration-*

excess overland flow and *Saturation-excess overland flow*.

Oxbow lake A body of water produced when a meander becomes cut off from the rest of the river channel as a result of a change in the river's course.

Palaeolimnological Pertaining to ancient lakes through a study of their sediments and fossils.

Parasites Organisms which exist by living on or inside another organism, taking nourishment from the host organism but not giving anything back to it.

Parent rock The original rock from which another material may be formed. It is usually referred to in relation to soil formation and the parent rock which was weathered to produce the initial soil particles.

Partial contributing area concept With respect to the process of *Infiltration-excess overland flow* this is the notion that this will occur only in localised areas, rather than over a whole catchment.

Partial pressure The pressure that would be exerted by a single gas if that gas were the only one present in a specific volume.

Patches Also known as microhabitats, this is another spatial unit in relation to river systems. These habitats may be a metre square in size.

Pathogenic organisms The term used to describe organisms which can cause disease.

Pelagic photic zone This is the zone at the top of a body of water where the abundance of light allows for *Photosynthesis* to take place.

Percolates The ability of a liquid such as water to move through small holes and pores in a substance such as soil.

Permafrost The term used to describe soil and bedrock materials which are frozen below 0°C for a minimum of two years. Permafrost is predominantly found in high latitudes.

Permeability The ability of a substance to transmit fluids through it. The permeability of soil can be influenced by its composition; for example, a sandy soil is more permeable than a clayey soil.

Photolysis The action of the breakdown and separation of molecules by light.

Photosynthesis The process by which organisms such as plants and algae (autotrophs) create carbohydrates and release oxygen using light energy, carbon dioxide and water.

Phreatic zone The zone within an aquifer which is filled with water.

Piezometric Pertaining to the measurement of water pressure inside an aquifer.

Planktonic The name given to organisms which are suspended in the water column and move by the action of drifting and floating.

Plants These are multicellular organisms which can produce their own energy via the chemical process of *Photosynthesis*.

Plutonic The term used to refer to igneous rocks which have solidified at a great depth below the Earth's surface.

Point-source pollution Contamination of a water body which has originated from a specific location such as a pipe from an industrial plant.

Pore spaces The gaps which exist between soil particles.

Pore water pressure The pressure of water which is held in soil pores or rock aquifers under saturated conditions. Pore water pressure is measured relative to atmospheric pressure.

Potable water Water of a high enough quality that it can be consumed by humans without risk.

Precautionary principle A decision-making approach which believes that lack of scientific evidence for warnings about future threats of serious damage should not be used as an excuse to avoid action in order to prevent damage from happening; action should be taken as early as possible.

Precipitation The condensation of water vapour to form water droplets in the atmosphere, which are then deposited on the Earth's surface in a liquid (e.g. rain) or solid form (e.g. snow, hail).

Probabilistic technique This method of analysis makes use of stochastic properties and considers those things which could happen by chance. A common probabilistic technique used in modelling is that of running Monte Carlo simulations.

Producers In ecological terms this refers to organisms which are able to produce their own energy through processes such as *Photosynthesis*.

Profundal zone A deep zone of water found in oceans and lakes. Due to the depth, light penetration is limited.

Property rights Rights pertaining to the authority that determines how a resource can be used.

Q-methodology A research method which takes into consideration the individual's viewpoint.

Rating equation The equation for a curve of best fit derived from the plotting of known river discharge and water level values. From this it is possible to determine the river discharge value at a particular point along a river if the water level is at a particular height.

Recharge The process whereby water moves downwards from the surface to groundwater aquifers to replenish the water stored there.

Recovery efficiency This is the ratio between usable water and the injection of water through aquifer storage and recovery (ASR). It indicates the level of mixing which occurs between stored water and native water in an aquifer. Efficiency may begin low but improve with subsequent flushes of water.

Recurrence interval The period of time in which a flood of a particular magnitude is likely to return. For example, a 1-in-100-year flood event would happen, on average, once every 100 years.

Riffles The accumulation of coarse sediment which forms bar deposits across a river, which tend to be spaced apart by between five and seven times the channel width.

Riming The process of cloud growth whereby supercooled water droplets are deposited on the ice crystal surfaces in the clouds. This method leads to the rapid growth of clouds.

Riparian Pertaining to the banks of a river or stream.

River basin See *Catchment*.

River discharge The volume of water flowing through a given point on a river at a given time.

River reach A length of a river which shares similar physical characteristics.

River regime This term refers to the annual variation in river discharge.

Rossby waves Large upper-atmosphere undulations which disturb the belt of prevailing westerly winds associated with the Ferrel cell. They contain jet streams.

Runoff See *Overland flow*.

Salinity The concentration of salt dissolved in water.

Saltating A method of sediment transport whereby sediment grains are bounced along a bed surface.

Saturation-excess overland flow Where all of the pore spaces within the soil become filled with water, and therefore saturated, forcing the excess water to flow across the surface.

Saturation vapour pressure In a closed container evaporation will occur until there are the same number of molecules returning to the liquid as there are escaping from it. This is the point where the vapour is said to be saturated. The vapour pressure differs according to temperature.

Sedimentary rocks These rocks are formed through a buildup of sediments, whether that be through transport and deposition in water, precipitates from a solution or secretions from organisms. These sediments build up over time and are subject to forces to become consolidated and solidified.

Sensible heat This refers to the heat which can increase the temperature of a substance, but its phase remains unaltered.

Severe water scarcity This occurs when the availability of renewable freshwater is less than 500 m^3 per person per year within a region or country.

Shear displacement The relative movement of one layer compared to an immediately adjacent layer along a plane. Such a displacement would usually result in a weakening of a substance.

Shear stress A stress acting upon a particle in the same direction as the surface it is resting upon. In rivers the shear stress is the velocity of flowing water. When a sediment particle can be lifted from the river bed then the critical flow velocity (critical shear stress) is reached.

Soil water tension This is the force which would be required to remove from the soil the water which is held in a film around the soil particles.

Solutes Substances which are dissolved within another substance to form a solution.

Solute load The amount of dissolved material which is carried in stream or river water at a given point in space for a set period of time. It is usually measured in units of mass (e.g. kilograms).

Sorption The process whereby one substance is taken up by another substance which is in a different phase state.

Specific capacity The term used to measure the performance of a well. It is derived by dividing the yield of the well by the amount of drawdown.

Specific discharge A measure of the amount of groundwater discharge across a cross-sectional area of a porous medium. It is also known as the Darcy velocity.

Specific heat The amount of heat which is required to increase the temperature of a given unit mass of a substance by 1°C.

Stage Referring to the height of the water level in a river channel at a given point in time.

Stakeholders Referring to those who have an interest in a resource which is available. It may be an individual or a group which have a claim on a resource.

Stomata Openings in a plant through the surface of the leaves. Through these openings oxygen can be taken in and water and carbon dioxide can be expelled.

Stratification The development of layers in a body of water, due to different densities in the water.

Stream order The dimensionless numbering method used to identify the size of a perennial stream and its position in the stream network.

Sublimation The process by which a solid changes directly into a gas.

Suspended sediment Particles larger than 0.45 µm which are suspended in water.

Sustainable urban drainage systems A technique used within urban environments to mitigate against the issues of flooding and surface water quality which can occur through the traditional approach of routing runoff through pipe networks directly to a watercourse, as well as that of water harvesting. Methods used include the creation of permeable surfaces, underground storage areas and wetlands.

Symbiosis The biological process whereby two organisms can exist in the same geographic location without causing harm to the survival and productivity of the other.

Thermocline A sharp temperature gradient which exists in a body of water (ocean or lake) where there is a dramatic difference between the water temperature above and the water temperature below.

Throughflow The movement of water draining through the soil in a downslope direction.

Total dissolved solids The inorganic and organic particles which are dissolved in water. A measure of them can be used as an indicator of water quality.

Total Economic Value An approach in environmental economics which does not restrict the value placed on a natural resource to its monetary worth. Instead the Total Economic Value considers the sum of all the values that a resource can have, those which come from its direct and indirect use, and those which come from the mere existence of the resource.

Total hydraulic potential See *Hydraulic head*.

Toxicity Pertaining to how toxic a substance is.

Transboundary Describes something which moves across boundaries or borders; it may be used to refer to freshwater resources or pollution.

Transboundary pollution Predominantly relating to air pollution (but also relates to water pollution), this term is used to describe pollution which originates in one jurisdiction but can be transported across a boundary or border between countries.

Transmissivity The volume of horizontal water flow through an aquifer.

Transpiration The evaporation of water which is released into the atmosphere through the pores of plant leaves.

Tropopause The boundary in the Earth's atmosphere which separates the *Troposphere* from the Stratosphere.

Troposphere The lower layer of the atmosphere that extends between 6 and 15 km in altitude above the Earth's surface.

Tsunami An energetic sea wave triggered by an earthquake, landslide or meteor impact in the ocean, which can become very large once it reaches shallow water and cause devastation in coastal zones.

Turbidity A measure of the clarity of water. Fine particles suspended within the water reduce the penetration of light into the water body.

Ultraviolet (UV) light On the spectrum of light this form of light has a shorter wavelength than natural light but a longer wavelength than natural light. It is the type of light that the Earth receives from the sun.

Vadose zone The zone of water which is below the Earth's surface but remains above the water table and is therefore aerated.

Variable source area concept The theory put forward that the areas in a catchment where runoff is generated will vary in space (location and size) and time. The areas where overland flow occurs are influenced by a number of factors, including rainfall, temperature, topography and vegetation.

Vesicle A cavity filled with fluid or air.

Volatilisation (also volatilise) The process whereby a substance is evaporated and becomes a vapour.

Water scarcity According to the United Nations' use of the term, this occurs when the renewable freshwater available is less than 1,000m^3 per person per year. However, other definitions suggest that water scarcity occurs below 500 m^3 per person per year. The UN defines the latter as absolute water scarcity.

Water softening The process of reducing the levels of metal ions in the water, predominantly calcium and magnesium, which otherwise mean that it is hard.

Water stress According to the United Nations' use of the term, this occurs when the renewable freshwater available is less than 1,700m^3 per person per year.

Water table The upper limit of the saturated zone of the soil or rock.

Watershed See *River Catchment*.

Weathering The process whereby the physical and chemical erosion of rock causes it to break down into smaller particles.

Index

Absolute distribution conflict 320, 321
Abstraction, groundwater 123, 124, 127, 136, 137, 138, 139, 141, 142, 143, 145, 151, 153, 156, 346
Acid mine drainage 79, 80, 83, 105, 106, 107, 149
Acid rain 79, 80, 94, 109, 110, 112
Acidification 109, 110, 111, 113, 180, 188
Acidity, juvenile 107
Acidity, vestigial 106
Activated sludge process 283, 284
Adiabatic cooling 21
Advect 23
Aerosol, hygroscopic 22
Aerosols 19, 30, 33, 35, 40, 41, 44, 87, 217
Afforestation 40, 183
Aggregation 23, 24, 205
Albedo 24, 33, 34, 35, 41
Algae 81, 166, 167, 168, 169, 175–178, 182, 183, 186, 187, 224, 270, 271, 276
Aquaculture 165, 190, 191, 192, 194
Aquifer 55, 79, 123, 124, 127–129, 130, 131, 132–157, 244, 267, 297, 321
Aquifer Storage and Recovery (ASR) 143, 145
Aquifer, artesian 130
Aquifer, confined 130, 131, 133, 141, 142, 148
Aquifer, fractured 131, 135, 136, 155
Aquifer, perched 130
Aquifer, unconsolidated 131, 141
Arsenic 143, 145, 149, 150, 151, 239–242, 259, 266, 267

Attribute-based method see conjoint choice method
Averting Behaviour Technique (AB) 307

Baseflow 52, 58, 62, 63, 75, 82, 124, 126, 127, 134, 308, 323
Basin, endorheic 54, 58
Basin, exorheic 58
Bedload transport 70, 71
Benefit Transfer Method 308
Benthic zone 165
Benthos 95, 166, 187
Bergeron-Findeisen Process 23, 24
Biochemical Oxygen Demand (BOD) 84, 154, 275, 277, 100, 276, 278, 279, 282, 285
Biota 109, 110, 111, 161, 165, 170, 180
Buffer zones 34, 97

Capillary action 4, 5, 6, 52
Capillary water 52
Carbohydrates 6
Carbon, activated 267, 270, 289
Catchment see river basin
Catchment, groundwater 127
Cauvery Water Disputes Tribunal (CWDT) 317
Chain-summation approach 336
Channel, river 40, 49, 50, 52, 57, 67, 68, 69, 70–72, 74–76, 180, 181, 182, 204
Channels, braided 69
Channels, Straight 69

Chemical Agreement 318
Chemical Oxygen Demand (COD) 84, 154, 185, 191, 275, 277
Chemistry, surface water 80, 81, 84, 87, 88, 89, 91, 92, 115
Chemosynthesis 6
Chlorination 94, 228, 229, 233, 267
Choice Experiment Method see conjoint choice method
Cholera 154, 224, 225, 226, 227, 232, 249, 258, 260, 266, 274
Civil law tradition 322
Clarification 267
Clausius- Clapeyron relation 22, 41
Climate change 3, 4, 14, 15, 17, 24, 25, 26, 30, 36, 37, 40–44, 49, 50, 66, 75, 81, 102, 112, 114, 115, 143, 152 , 165, 188, 204, 215, 217, 218, 238, 244, 250, 274, 289, 297, 298, 299, 302, 346, 347
Cloud seeding 215
Clouds, frozen see clouds, glaciated
Clouds, glaciated 23
Clouds, mixed 23
Clouds, warm 23
Coagulation 188, 247, 267, 268, 269, 289
Coarse Particulate Organic Matter (CPOM) 169
Collision-coalescence 23, 24
Combined Sewer Overflow (CSO) 101, 114, 279, 280
Common law tradition 322, 323
Condensation 20, 21, 22, 23, 33, 41, 44
Conductivity, hydraulic 133, 134, 135, 142
Conjoint Choice Method 308, 309
Conservation 9, 94, 217, 289, 299, 301, 306, 311
Consumers, organisms 166, 168, 169, 173, 174, 189
Consumers, water 96, 165, 205, 217, 273, 297, 307, 336, 339
Consumption, water 12, 13, 17, 123, 204, 205, 207, 208, 244, 246, 266, 299, 301, 305, 315, 330, 334, 336–340, 343–345
Consumptive purposes 296
Contaminant, chemical 184, 239, 246, 249, 259
Contaminant, emerging 101, 247
Contaminant, groundwater 123, 149, 151, 155
Contaminant, inorganic 154, 266
Contaminant, microbial 266, 267
Contaminant, organic 150, 152, 248, 266, 270, 289
Contamination 99, 107, 108, 136, 138, 143, 150–152, 154, 184, 224, 232, 233, 234, 244–247, 249, 259, 266, 267
Conventions, water 110, 325, 327, 328, 329,
Coriolis effect 21

Covalent bond 2. 16
Cyanobacteria 168, 186, 244, 245

Darcy's Law 124, 133, 135, 138, 156
Deforestation 40, 63, 35, 72, 183, 237
Degradation products, biological 153
Dehydration 6, 232
Demand management 204, 213, 217, 218
Dengue fever 231, 237, 249
Deposition 71, 72, 75, 91, 92, 93, 109, 110–113, 131
Desalination 215, 265, 271–273, 289
Desertification 30,35,40
Detritivory 168, 169
Dew-point temperature 21
Diffuse source 93, 94, 101, 150
Digestion, anaerobic 280, 285, 286, 287, 288
Disability Adjusted Life Year (DALY) 255, 256
Discharge consent 265, 278, 284
Discharge, groundwater 20, 124, 126, 133
Disdrometers 55
Disease 14, 94, 96, 154, 190, 191, 192, 209, 223, 224, 226–239, 245, 246, 248–250, 255–259, 324
Disease, water-based 226, 235, 258
Disease, water-borne 8, 12, 226, 232, 233, 239, 249, 258, 271
Disease, water-carried 226, 234, 258
Disease, water-vector 235
Disease, water-washed 226, 234, 258
Disinfection 214, 224, 234, 241, 248, 259, 266, 270, 271, 273
Displacement flow 57
Distribution 115, 123, 156, 215, 224, 257, 259, 273, 295, 296, 300, 316, 318–321, 326, 330
Domestic Consumption Monitor 205, 207, 208
Drainage 63
Drainage density 68
Drainage network 57, 64, 68, 69, 70, 75
Drainage network, dendritic 69
Drainage network, parallel 69, 70
Drinking Water Directive 241
Drought 6, 14, 15, 24, 25, 27, 29, 30–35, 42, 44, 56, 92, 132, 194, 204, 209, 215–218, 223, 249, 250, 253, 258, 259,266, 295, 297–299, 317, 330
Dry Weather Flow (DWF) 101, 279
Dust Bowl 30, 31, 32, 33, 249

Economics, environmental 293
Ecosystem, river 161, 162, 181, 182, 189
Ecosystem, wetland 303, 305, 306
Effective rainfall 61, 62
Efficient use 326

Effluent 84, 94, 97, 103, 185, 192, 240, 246, 247, 248, 266, 273–279, 283–285, 289
El Niño 26, 29, 30, 35, 92, 204
El Niño Southern Oscillation (ENSO) 26, 92, 204
Elevational potential 52
Endocrine disrupting compounds (EDC's) 247, 248, 249
Energy balance 37, 44
Energy budget, global 21, 35
Energy transfers 44
Environmental warming 188
Epilimnion 87, 187
Equitable use 326
Erosion 1, 54, 68, 70, 71–73, 94, 95, 131, 140, 185, 307 308
Eutrophication 94, 96, 98, 165, 177, 185, 187, 192, 245, 276
Evaporation 3, 4, 12–14, 20, 21, 25, 26, 33, 35, 41, 44, 55, 56, 58, 59, 79, 87, 88, 125, 127, 217
Evapotranspiration 20, 22, 24, 25, 33, 38–40, 42, 43, 56, 156
Expansion, agricultural 40, 65, 364

Feedback 24, 25, 33, 35, 38, 39, 41–45, 303
Field capacity 126
Filtration 81, 102, 145, 156, 157, 233, 248, 259, 268–271, 284, 289
First-order acidity balance (FAB) 111, 112
Flocculation 108, 247, 267, 268, 285
Flood management 10
Flood protection 75
Flood solution 67
Flooding 9, 29, 42, 63, 66, 67, 71, 74, 75, 80, 98, 104, 105, 114, 223, 249, 251, 252, 253, 255, 259, 274, 279, 280, 344, 345
Flooding, coastal 66, 75
Flooding, fluvial 66
Flooding, groundwater 66, 75
Flooding, pluvial 66, 75
Floods 8, 24, 25, 26, 27, 29, 30, 42, 49, 50, 66, 67, 105, 182, 249, 274, 297, 298, 317, 330, 344
Floods Directive 67
Flow train 266, 267, 268, 273, 277, 289
Flowpath 63, 89
Fluid potential 132
Fluoride 240, 241, 242, 266, 267
Flux 19, 20–24, 37, 41, 85, 86, 99, 114, 149, 190
Flux, sensible heat 39
Food web 165, 166, 168, 170–175, 178, 182, 189, 194, 196
Food web, aquatic 166, 170, 171, 173, 174, 175, 178

Food web, connectance 173, 174
Footprint, flood 345, 347
Footprint, water 333–349
Force, adhesive 4, 5
Force, cohesive 4, 5
Fractions 207, 274

Gauge 53, 55, 60, 61
Gauging, dilution 60
Germ theory of disease 224, 226
Glacial mass balance 37
Goods and services, ecosystem 97, 188, 294, 296, 300, 301, 302, 303, 305, 306, 307, 312, 334, 338, 339
Governance 294, 295, 311, 312, 315, 326, 327
Gradient, hydraulic 54, 123, 133–135
Greenhouse effect, natural 21
Greenhouse gas 3, 19, 21, 26, 35, 37, 41, 44, 265, 276
Gross primary productivity 25, 176, 177, 183
Groundwater Directive 130
Groundwater Flow models 136, 137
Groundwater quality 123, 145
Groundwater resources 124, 127, 136, 154, 204
Groundwater, potable 123, 145, 156
Growth process, attached 282, 283, 284
Growth process, suspended 282, 283, 284

Headroom 216, 218
Heat transfer processes 44
Hedonic pricing method 308
Helicoidal flow 70
Hurricane Katrina 251, 252
Hydraulic head see total hydraulic potential
Hydrocarbons 100, 154, 241, 243, 266, 267, 270
Hydrogen bond 2, 3, 16
Hydrograph 49, 57, 61, 62, 63, 67, 75, 126, 127
Hydrological pathways 49, 50, 80, 81, 115
Hydrological processes 74, 84, 89, 115
Hypolimnion 87, 187
Hypoxia 246

Iceberg sourcing 215
Impeller meter 60
Infections, water-related 266, 227–229, 233, 234–236, 257, 258, 274
Infiltration 50, 57, 58,63, 75, 103, 104, 124, 125–127, 138, 139, 143, 279
Infiltration capacity 50, 51, 57, 58, 63, 75, 125
Infiltration excess overland flow 50, 51, 57, 58, 63, 77, 125
Infiltration rate 50, 61, 62, 77
Infrared radiation 21

Input-output models 336, 338, 339
Integrated catchment management 127
Interflow see throughflow
Intertropical Convergence Zone 34
Invasions, biological 189
Irrigation 8, 11, 13, 14, 33, 40, 43, 73, 107, 124, 128, 138, 140, 152, 156, 204, 209, 217, 235, 236, 244, 250, 272, 296, 319, 321, 324, 325
Irrigation systems 14, 235, 324

Jet stream 26, 27, 29

La Niña 26, 29, 32, 35
Land management 30, 35, 40, 62–68, 75, 91, 94, 99, 112, 113, 239, 245 250, 304, 309, 310, 346
Land-atmosphere coupling 25, 35
Landscape engineering 8
Latent heat 3, 21, 39, 41, 44
Law 113, 315, 321–323, 326–330
Law, international water 327
Legionnaires' disease 234, 258
Legislation 10, 79, 80, 81, 109, 113, 114, 193, 224, 226, 241, 246, 258, 276, 323, 325, 326, 342
Littoral zone 164, 168
Long Term Yield (LTY) 214
Long-run marginal cost (LRMC) 300
Lotic 161
Lysimeters 56

Macrophytes 107, 166, 168, 178, 183, 186, 187, 193
Macropore flow 52, 54, 63
Macropores 52, 54, 58, 63, 81
Malaria 190, 226, 230, 235, 236, 237, 238, 249, 250, 255, 258, 260
Marginal cost 300
Matric potential 52, 53
Matrix flow 52, 75, 145
Matrix throughflow 57
Meander 10, 69, 70, 71, 72, 74, 75
Membrane treatment 271, 285
Membranes 227, 233, 271–273, 277, 278, 285
Miasmatic theory of disease 224, 225
Microbiological load 249
Millennium Development Goals 298, 312
Millennium Ecosystem Assessment (MEA) 180, 293, 294, 302, 303, 312
Mine remediation treatment 108
Mineral dissolution reactions 146, 147, 148, 156

Nanomaterials 248
National Water Act 327

Nekton 166
Net benefits 300
Neuston 166
Nitrate 82, 86, 90, 91, 94, 95, 96, 99, 113, 117, 145, 150, 151, 177, 185, 186, 188, 242, 245, 309, 310
Non Aqueous Phase Liquids (NAPL's) 152, 153
Nutrient cycling 178, 273
Nutrient spiralling 178, 179

Opportunity cost 300
Overland flow 50, 51, 57, 58, 63, 66, 75, 89, 95, 96, 125, 127
Oxbow lake 71, 164
Oxygen sag curve 275

Parasitism 170, 174
Partial contributing area concept 51
Particulate 81, 82, 87, 96, 169, 170, 179, 274
Pelagic zone 164
Permafrost 43, 93
Pesticides 93, 94, 96, 97, 100, 101, 154, 241, 243, 245–247, 258, 259, 266, 267, 270
Phi-index 62
Photosynthesis 6, 7, 16, 37, 38, 81, 89, 166, 175, 176, 186, 187
Phreatic zone 124, 246, 130
Phytoplankton 166, 168, 186, 187, 189
Piezometers 52, 53
Piston see displacement flow
Plague 230, 235
Plankton 81, 166, 168, 178, 186, 187, 189, 346
Pollution 33, 54, 79, 80, 81, 82, 84, 93, 94, 95, 97, 98, 99, 100, 101, 104, 105, 110, 111, 112, 114, 115, 123, 180, 193, 204, 223, 246, 258, 265, 274, 275, 276, 278, 281, 288, 289, 316, 318, 322, 324, 328, 33, 334, 335, 339, 346, 347
Pollution, air 33, 91, 110, 111, 310
Pollution, atmospheric 91, 109
Pollution, diffuse 93, 97, 102, 114
Pollution, groundwater 124, 149, 154, 157
pollution, mine drainage 106, 107, 108
Pollution, nitrate 99
Pollution, non-point source 93, 94, 98
Pollution, point source 184
Pollution, transboundary 295
Pollution, urban 9, 98, 101
Pollution, urban diffuse 100, 101
Pools, landform 71, 72
Pore water pressure 52
Preventative expenditure technique see averting behaviour technique (AB)

Primary treatment 279, 281
Privatisation 299, 325
Probabilistic techniques 26
Producers, aquatic 166, 167, 169, 173, 177, 179, 189
Producers, primary 165, 166–168, 176
Production factor method 307
Production, primary 165, 168, 175, 176, 177, 178, 180, 185, 186, 188, 194, 334, 338
Production, secondary 177, 178, 180, 182, 191, 194
Productive purposes 296
Profundal zone 165
Proxy data 30, 31

Radon 241, 243
Rapid gravity filter 268, 270
Rating equation 59, 60
Recharge, artificial 124, 143–145, 157
Recharge, groundwater 126, 127, 180
Recharge, rainfall 124, 125, 156
Recovery efficiency 145
Redox reaction 83
Reforestation 40
Rehabilitation, river 9, 72, 75, 76
Religion 8
Replacement/restoration cost technique 307
Reservoir 10, 20, 21, 37, 79, 80, 85, 87, 94, 115, 124, 138, 156, 181, 182, 183, 191, 214, 215, 245, 260, 266, 273, 300, 321
Respiration, aerobic 89
Respiration, community 178, 194
Restoration 9, 50, 74, 77, 104, 307, 328
Riffles 71, 72
Riming 23
Riparian zone 166, 175
River basin 14, 49–78, 37, 44, 45, 80, 81, 85, 86, 88, 89, 91, 92, 102, 105, 113, 114, 115, 127, 134, 162, 173, 181, 183, 193, 196, 266, 278, 301, 302, 309, 324, 325, 327, 328, 334, 335, 336, 346, 339
River channel modifications 72
River Continuum Concept 162
River discharge 20, 30, 33, 43, 49, 50, 55–66, 75, 89, 214
River flow models 62, 63
River flow regulation 181
River regimes 57, 58, 63, 75, 181
Rock, metamorphic 110, 130, 131
Rock, sedimentary 88, 89, 130, 131, 132, 135, 146, 147, 148
Rossby waves 27
Runoff 56–58, 63, 79–115, 124, 126, 143, 145, 151, 154, 180, 181, 185, 217, 232, 239, 245, 247, 259, 297
Runoff efficiency 56
Runoff, river 22, 24, 33, 38–41, 43, 44, 181

Salinity 14, 22, 145, 244, 272
Sanitation 12, 223, 224, 226, 227–233, 235, 254, 255–259, 296, 297, 324, 325, 330, 346, 347
Saturation 21, 23, 58, 61, 62, 113, 125, 175
Saturation vapour pressure 21, 22, 23
Saturation-excess overland flow 51, 57, 58, 63, 66, 75, 81, 125
Schistosomiasis 190, 226, 230, 235, 236, 258
Screening 267, 280
Secondary settlement 284
Sediment Oxygen Demand (SED) 84
Sediment transport 49, 70, 71, 181
Sediment transport mechanism 70
Services, ecosystem goods and see Goods and services, ecosystem
Sewage network 226, 256, 274
Sewers, combined 101, 114, 274, 279, 280
Single mass curve analysis 214
Slow sand filter 224, 269, 270
Sludge 108, 224, 245, 281–289
Sludge, activated 105, 224, 277, 278, 283–286, 288
Small Island Developing States 297, 298
Soil and Water Assessment Tool (SWAT) 65
Soil suction 52, 53
Soil water potential 52
Soil water tension 52
Soil zone 126, 146, 148
Solute concentration 80, 81, 85, 93, 115
Solute load 85, 86
Solutes 80–90, 95, 115, 116
Specific heat 3
Stage-discharge equation see rating equation
Steady-state water chemistry model (SSWC) 111
Stepwise accumulative approach 336
Stewardship, agricultural 97–99
Stomata 21, 37, 38, 39,
Stormflow 61, 62, 75
Stratification 85, 87
Stratification, thermal 187, 189
Stream order 68, 179
Stream water 12, 84, 85, 88, 89, 91, 92, 110, 127
Sublimation 23
Submergence potential 52
Surface tension 4, 5
Suspended sediment 60, 81, 82, 166, 191
Suspended transport 70

Sustainability 151, 216, 293, 294, 299, 345
Sustainable Urban Drainage Systems (SUDS) 64, 101–105
Sustainable use 15, 156, 193, 326

Teleconnections 26, 29, 30
Tensiometers 53
The Critical Load Approach 110, 111, 112
The Economics of Ecosystems and Biodiversity (TEEB) 294
Thermal expansion 3
Thermal stratification 2, 87, 187, 189,
Thermocline 87
Throughflow 51, 57, 58, 75, 89, 126, 127
Top-down demand estimation 208, 209
Total dissolved solids (TDS) 82–85, 88, 146–149
Total economic value 293, 294, 301, 305, 312
Total hydraulic potential 52, 54, 123, 132, 133, 134, 136, 156
Transpiration 14, 20, 21, 33, 37, 38, 39, 43, 55, 58, 63, 125, 126, 127
Travel cost method (TC) 308, 310
Treatment processes, biological 282, 289
Treatment processes, physical 279, 289
Trellised drainage system 69
Tropopause 27
Troposphere 26, 27, 33, 41
Tsunami 66, 249, 251, 252, 253
Turbidity 82, 106, 175, 183, 185, 186, 187, 269
Typhoid fever 154, 226, 227, 232, 233, 249, 258

Unit hydrograph model 61–63
Urban diffuse pollutants 100, 101

Vadose zone 124
Valuation, contingent (CV) 307, 308
Valuation, direct market 306, 307
Valuation, revealed preference 306, 308
Valuation, stated preference 306
Value, bequest 302
Value, existence 302
Value, non-use 300–302, 305, 307
Value, option 301, 302, 305
Value, use 301, 305, 306
Vapour pressure 21, 22, 23, 39, 41
Vapour pressure deficit (VPD) 39
Variability, climatic 19, 24, 26, 34, 39, 45, 91
Variability, multi-decadal 30
Variable source area concept 51, 62
Vulnerability, groundwater 123, 124, 154, 155, 157

Wastewater treatment 98, 145, 154, 224, 245, 247, 248, 265–291
Water allocation 323–326
Water allocation, agency 324
Water allocation, market 324
Water allocation, user-based 324
Water balance 20, 55, 56, 88, 123, 127, 209, 214, 215, 216, 341, 342
Water balance evaluation 127
Water budget 49, 55, 56, 63
Water crisis 12, 13, 16, 346
Water cycle 3, 19–48, 56, 61, 114, 181, 345, 346
Water cycle, global 19, 20, 37, 40, 44, 61, 114
Water cycle, hydraulogical 20, 33, 41, 89, 123, 124, 156, 297
Water demand 12, 13, 40, 43, 44, 203–22 272, 294, 296–299, 334, 345
Water Framework Directive (WFD) 10, 9 05, 113, 130, 193, 204, 257, 325
Water rights 8, 315, 321–326, 330, 331, 345
Water rights, correlative 323
Water rights, prior appropriation 322, 323
Water rights, riparian 322, 323, 327
Water table 51, 54, 57, 63, 81, 105, 106, 107, 13 133, 134, 137, 138, 139, 140, 141, 142, 152, 156, 1, 244, 309
Water treatment 98, 104, 115, 145, 150, 154, 188, 4, 228, 233, 240, 245–248, 265–291, 310
Water use, in-stream 12, 98, 204
Water use, off-stream 12, 204
Water, availability 1, 10, 11, 13, 14, 39, 41, 43, 88, 107 188, 210, 213, 249, 294, 312, 316, 324, 342, 343, 346
Water, blue 333, 334, 336, 337, 338, 347
Water, conflict 12, 40, 315, 316, 320, 327, 330
Water, consumptive use 12, 316, 317, 345
Water, economic valuation 293, 294, 295, 297, 305–308, 311, 312
Water, economics 293–314, 345
Water, efficiency 13, 14, 16, 40, 56, 77, 203, 213, 215, 216, 219, 338, 348
Water, embedded see water, virtual
Water, embodied see water, virtual
Water, exchange see flux
Water, global use 13, 14
Water, green 333, 334, 336, 347
Water, grey 104, 217, 333, 334, 336, 337, 338, 342, 343, 347
Water, hard 146, 147
Water, hygroscopic 50
Water, industry 208, 215, 216, 296, 297

INDEX

Water, management 8, 114, 123, 127, 154, 203, 219, 245, 289, 294, 295, 299, 308, 312, 316, 324, 325, 326, 334, 345
Water, molecules 1–5, 52, 286
Water, potable 37, 203, 204, 205, 212, 215, 265–291, 296, 297
Water, pricing 294, 299–301, 308, 311, 325, 345
Water, resource development 214, 299, 325
Water, resource planning 61, 213, 215, 218
Water, risk 254–256
Water, rituals 1, 8
Water, scarcity 11, 13, 40, 210, 211, 213, 258, 272, 294, 297, 299, 301, 315, 322, 323, 324, 328, 342, 344, 345
Water, shortage 13, 209, 210, 273, 276, 316, 317, 339
Water, soft 146, 148
Water, storage 6, 37, 43, 56, 58, 64, 127, 135, 143, 180, 181, 238, 250, 323, 328
Water, stress 11, 12, 13, 210, 211, 213, 217, 250, 294
Water, stress 11, 12, 13, 39, 210, 211, 213, 217, 250, 294
Water, surface 4, 29, 75, 79–122, 127, 138, 140, 143, 145, 154, 156, 158, 193, 194, 203, 214, 232, 233, 235, 238, 240, 244, 245, 265, 267, 274, 295, 323, 329, 346
Water, tariff 205, 206, 300, 301, 345
Water, treaties 320, 327–330
Water, virtual 333–349
Watercourse 8, 9, 63, 94, 97, 104, 108, 156, 183, 273–284, 321, 327, 328
Water-harvesting techniques 14, 104, 214, 217
Watersheds see river basin
Well efficiency 13, 14, 16, 40, 56, 77, 142, 203, 213–216, 219, 338, 348
Well loss 141, 142
Wells, groundwater abstraction 139, 140–143, 153, 154, 157
World Health Organisation (WHO) 151, 240–243, 245, 255, 256, 257, 266, 267, 268

New eBook Library Collection

eFocus on the Environment

30 day free trials available!

Growing concerns about climate change, pollution and environmental degradation have put the environment at the top of the agenda in the early decades of the 21st Century. This collection includes books from 18 disciplines, offering the broadest possible understanding of environmental issues. Its coverage ranges from philosophical works exploring environmental values to works offering practical suggestions to immediate policy questions.

Key features:

- Global coverage with extensive treatment of both the developed and developing world

- Particular attention to the concept of sustainability, at both a micro and a macro level

- Includes several key reference works, such as: *Routledge Handbook of Climate Change and Society, Fifty Key Thinkers on the Environment, The Complete Guide to Climate Change,* and *The Environment Dictionary*

- Looks at the impact of tourism and other major industries.

eFocus on the Environment is available as a subscription package with 10 new eBooks added per year.

Recommend this package to your librarian today!

Order now for guaranteed capped price increase.

www.ebooksubscriptions.com

For a complete list of titles, visit:
www.ebooksubscriptions.com/eFocusEnvironment

For more information, pricing enquiries or to order a free trial, please contact your local online sales team:

UK and Rest of the World
Tel: +44 (0) 20 7017 6062
Email: online.sales@tandf.co.uk

United States, Canada and South America
Tel: 1-888-318-2367
Email: e-reference@taylorandfrancis.com

Taylor & Francis eBooks
Taylor & Francis Group

Routledge Paperbacks Direct

Bringing you the cream of our hardback publishing at paperback prices

This exciting new initiative makes the best of our hardback publishing available in paperback format for authors and individual customers.

Routledge Paperbacks Direct is an ever-evolving programme with new titles being added regularly.

To take a look at the titles available, visit our website.

www.routledgepaperbacksdirect.com

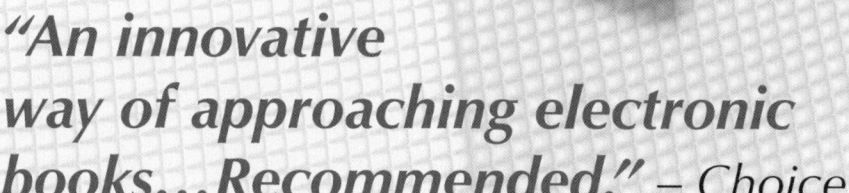

ROUTLEDGE INTERNATIONAL HANDBOOKS

Routledge International Handbooks is an outstanding, award-winning series that provides cutting-edge overviews of classic research, current research and future trends in Social Science, Humanities and STM.

Each *Handbook*:

- is introduced and contextualised by leading figures in the field
- features specially commissioned original essays
- draws upon an international team of expert contributors
- provides a comprehensive overview of a sub-discipline.

Routledge International Handbooks aim to address new developments in the sphere, while at the same time providing an authoritative guide to theory and method, the key sub-disciplines and the primary debates of today.

If you would like more information on our on-going *Handbooks* publishing programme, please contact us.

Tel: +44 (0)20 701 76566
Email: reference@routledge.com

www.routledge.com/reference